LONDON MATHEMATICAL SOCIETY LECTURE NOTE SERIES

Managing Editor: Professor N.J. Hitchin, Mathematics Institute,
University of Oxford, 24–29 St Giles, Oxford OX1 3TG, United Kingdom

The titles below are available from booksellers, or, in case of difficulty, from Cambridge University Press.

London Mathematical Society Lecture Note Series. 264

New Trends in Algebraic Geometry

EuroConference on Algebraic Geometry
Warwick, July 1996

Edited by

Klaus Hulek (chief editor)
Universität Hannover

Fabrizio Catanese
Georg-August-Universität, Göttingen

Chris Peters
Université de Grenoble

Miles Reid
University of Warwick

CAMBRIDGE
UNIVERSITY PRESS

CAMBRIDGE UNIVERSITY PRESS
Cambridge, New York, Melbourne, Madrid, Cape Town, Singapore, São Paulo

Cambridge University Press
The Edinburgh Building, Cambridge CB2 2RU, UK

Published in the United States of America by Cambridge University Press, New York

www.cambridge.org
Information on this title: www.cambridge.org/9780521646598

First published 1999

A catalogue record for this publication is available from the British Library

ISBN-13 978-0-521-64659-8 paperback
ISBN-10 0-521-64659-6 paperback

Transferred to digital printing 2005

Contents

Foreword

The volume contains a selection of seventeen survey and research articles from the July 1996 Warwick European algebraic geometry conference. These papers give a lively picture of current research trends in algebraic geometry, and between them cover many of the outstanding hot topics in the modern subject. Several of the papers are expository accounts of substantial new areas of advance in mathematics, carefully written to be accessible to the general reader. The book will be of interest to a wide range of students and nonexperts in different areas of mathematics, geometry and physics, and is required reading for all specialists in algebraic geometry.

The European algebraic geometry conference was one of the climactic events of the 1995–96 EPSRC Warwick algebraic geometry symposium, and turned out to be one of the major algebraic geometry events of the 1990s. The scientific committee consisted of A. Beauville (Paris), F. Catanese (Pisa), K. Hulek (Hannover) and C. Peters (Grenoble) representing AGE (Algebraic Geometry in Europe, an EU HCM–TMR network) and N.J. Hitchin (Oxford), J.D.S. Jones and M. Reid (Warwick) representing Warwick and British mathematics. The conference attracted 178 participants from 22 countries and featured 33 lectures from a star-studded cast of speakers, including most of the authors represented in this volume.

The expository papers Five of the articles are expository in intention: among these a beautiful short exposition by Paranjape of the new and very simple approach to the resolution of singularities; a detailed essay by Ito and Nakamura on the ubiquitous ADE classification, centred around simple surface singularities; a discussion by Morrison of the new special Lagrangian approach giving geometric foundations to mirror symmetry; and two deep and informative survey articles by Behrend and Siebert on Gromov–Witten invariants, treated from the contrasting viewpoints of algebraic and symplectic geometry.

Some main overall topics Many of the papers in this volume group around a small number of main research topics. Gromov–Witten invariants was one of the main new breakthroughs in geometry in the 1990s; they can be developed from several different starting points in symplectic or algebraic geometry. The survey of Siebert covers the analytic background to the symplectic point of view, and outlines the proof that the two approaches define the same invariants. Behrend's paper explains the approach in algebraic geometry to the invariants via moduli stacks and the virtual fundamental class, which essentially amounts to a very sophisticated way of doing intersection theory calculations. The papers by Paoletti and Wilson give parallel applications of Gromov–Witten invariants to higher dimensional varieties: Wilson's paper determines the Gromov–Witten invariants that arise from extremal rays of the Mori cone of Calabi–Yau 3-folds, whereas Paoletti proves that Mori extremal rays have nonzero associated Gromov–Witten invariants in many higher dimensional cases. The upshot is that extremal rays arising in algebraic geometry are in fact in many cases invariant in the wider symplectic and topological setting.

Another area of recent spectacular progress in geometry and theoretical physics is Calabi–Yau 3-folds and mirror symmetry. This was another major theme of the EuroConference that is well represented in this volume. The paper by Voisin, which is an extraordinary computational tour-de-force, proves the generic Torelli theorem for the most classical of all Calabi–Yau 3-folds, the quintic hypersurface in $\mathbb{P}^4$. The survey by Morrison explains, among other things, the Strominger–Yau–Zaslow special Lagrangian interpretation of mirror symmetry. Beauville's paper gives the first known construction of a Calabi–Yau 3-fold having the quaternion group of order 8 as its fundamental group. The paper by Batyrev proves that the Betti numbers of a Calabi–Yau 3-fold are birationally invariant, using the methods of p-adic integration and the Weil conjectures; the idea of the paper is quite startling at first sight (and not much less so at second sight), but it is an early precursor of Kontsevich's idea of motivic integration, as worked out in papers of Denef and Loeser. Several other papers in this volume (those of Ito and Nakamura, Mukai, Shioda and Wilson) are implicitly or explicitly related to Calabi–Yau 3-folds in one way or another.

Other topics   The remaining papers, while not necessarily strictly related in subject matter, include some remarkable achievements that illustrate the breadth and depth of current research in algebraic geometry. Shioda extends his well-known results on the Mordell–Weil lattices of elliptic surfaces to higher genus fibrations, in a paper that will undoubtedly have substantial repercussions in areas as diverse as number theory, classification of surfaces, lattice theory and singularity theory. Faber continues his study of tautological classes on the moduli space of curves and Abelian varieties, and gives an algorithmic treatment of their intersection numbers, that parallels in many respects the Schubert calculus; he obtains the best currently known partial results determining the class of the Schottky locus. Gizatullin initiates a fascinating study of representations of the Cremona group of the plane by birational transformations of spaces of plane curves. Eyssidieux gives a study, in terms of Gromov's Kähler hyperbolicity, of universal inequalities holding between the Chern classes of vector bundles over Hermitian symmetric spaces of noncompact type admitting a variation of Hodge structures. Küchle and Steffens' paper contains new twists on the idea of Seshadri constants, a notion of local ampleness arising in recent attempts on the Fujita conjecture; they use in particular an ingenious scaling trick to provide improved criteria for the very ampleness of adjoint line bundles.

Manetti's paper continues his long-term study of surfaces of general type constructed as iterated double covers of $\mathbb{P}^2$. He obtains many constructions of families of surfaces, and proves that these give complete connected components of their moduli spaces, provided that certain naturally occuring degenerations of the double covers are included. This idea is used here to establish a bigger-than-polynomial lower bound on the growth of the number of connected components of moduli spaces. In more recent work, he has extended these ideas in a spectacular way to exhibit the first examples of algebraic surfaces that are proved to be diffeomorphic but not deformation equivalent.

The Fourier–Mukai transform is now firmly established as one of the most important new devices in algebraic geometry. The idea, roughly speaking, is that a sufficiently good moduli family of vector bundles (say) on a variety X induces a correspondence between X and the moduli space $\widehat{X}$. In favourable cases, this correspondence gives an equivalence of categories between coherent sheaves on X and on $\widehat{X}$ (more precisely, between their derived categories). The model for this theory is provided by the case originally treated by Mukai, when X is an Abelian variety and $\widehat{X}$ its dual; Mukai named the transform by analogy with the classical Fourier transform, which takes functions on a real vector space to functions on its dual. It is believed that, in addition to its many fruitful applications in algebraic geometry proper, this correspondence and its generalisations to other categories of geometry will eventually provide the language for mathematical interpretations of the various "dualities" invented by the physicists, for example, between special Lagrangian geometry on a Calabi–Yau 3-fold and coherent algebraic geometry on its mirror partner (which, as described in Morrison's article, is conjecturally a fine moduli space for special Lagrangian tori). Mukai's magic paper in this volume presents a Fourier–Mukai transform for K3 surfaces, in terms of moduli of semi-rigid sheaves; under some minor numerical assumptions, he establishes the existence of a dual K3 surface, the fact that the Fourier–Mukai transform is an equivalence of derived categories, and the biduality result in appropriate cases.

The paper of Ito and Nakamura is the longest in the volume; it combines a detailed and wide-ranging expository essay on the ADE classification with an algebraic treatment of the McKay correspondence for the Kleinian quotient singularities $\mathbb{C}^2/G$ in terms of the G-orbit Hilbert scheme. The contents of their expository section will probably come as a surprise to algebraic geometers, since alongside traditional aspects of simple singularities and their ADE homologues in algebraic groups and representation of quivers, they lay particular emphasis on partition functions in conformal field theory with modular invariance under $SL(2, \mathbb{Z})$ and on II_1 factors in von Neumann algebras. Their study of the G-Hilbert scheme makes explicit for the first time many aspects of the McKay correspondence relating the exceptional locus of the Kleinian quotient singularities $\mathbb{C}^2/G$ with the irreducible representations of G; for example, the way in which the points of the minimal resolution can be viewed as defined by polynomial equations in the character spaces of the corresponding irreducible representations, or the significance in algebraic terms of tensoring with the given representation of G. Ito and Nakamura and their coworkers are currently involved in generalising many aspects of the G-orbit Hilbert scheme approach to the resolution of Gorenstein quotient singularities and the McKay correspondence to finite subgroups of $SL(3, \mathbb{C})$, and this paper serves as a model for what one hopes to achieve.

Thanks to all our sponsors   The principal financial support for the Euro-Conference was a grant of ECU40,000 from EU TMR (Transfer and Mobility of Researchers), contract number ERBFMMACT 950029; we are very grateful for this support, without which the conference could not have taken place. The main funding for the 1995–96 Warwick algebraic geometry symposium was provided by British EPSRC (Engineering and physical sciences research council). Naturally enough, the symposium was one of the principal activities of the Warwick group of AGE (European Union HCM project Algebraic Geometry in Europe, Contract number ERBCHRXCT 940557), and financial support from Warwick AGE and the other groups of AGE was a crucial element in the success of the symposium and the EuroConference. We also benefitted from two visiting fellowships for Nakamura and Klyachko from the Royal Society (the UK Academy of Science). Many other participants were covered by their own research grants.

    The University of Warwick, and the Warwick Mathematics Institute also provided substantial financial backing. All aspects of the conference were enhanced by the expert logistic and organisational help provided by the Warwick Math Research Centre's incomparable staff, Elaine Greaves Coelho, Peta McAllister and Hazel Graley.

Klaus Hulek and Miles Reid, November 1998

Birational Calabi–Yau n-folds
have equal Betti numbers

Victor V. Batyrev

Abstract

Let X and Y be two smooth projective n-dimensional algebraic varieties X and Y over $\mathbb{C}$ with trivial canonical line bundles. We use methods of p-adic analysis on algebraic varieties over local number fields to prove that if X and Y are birational, they have the same Betti numbers.

1 Introduction

The purpose of this note is to show how to use the elementary theory of p-adic integrals on algebraic varieties to prove cohomological properties of birational algebraic varieties over $\mathbb{C}$. We prove the following theorem, which was used by Beauville in his recent explanation of a Yau–Zaslow formula for the number of rational curves on a K3 surface [1] (see also [3, 12]):

Theorem 1.1 *Let X and Y be smooth n-dimensional irreducible projective algebraic varieties over $\mathbb{C}$. Assume that the canonical line bundles Ω_X^n and Ω_Y^n are trivial and that X and Y are birational. Then X and Y have the same Betti numbers, that is,*

$$H^i(X, \mathbb{C}) \cong H^i(Y, \mathbb{C}) \quad \text{for all } i \geq 0.$$

Note that Theorem 1.1 is obvious for $n = 1$, and for $n = 2$, it follows from the uniqueness of minimal models of surfaces with $\kappa \geq 0$, that is, from the property that any birational map between two such minimal models extends to an isomorphism [5]. Although n-folds with $\kappa \geq 0$ no longer have a unique minimal model for $n \geq 3$, Theorem 1.1 can be proved for $n = 3$ using a result of Kawamata ([6], §6): he showed that any two birational minimal models of 3-folds can be connected by a sequence of flops (see also [7]), and simple topological arguments show that if two projective 3-folds with at worst $\mathbb{Q}$-factorial terminal singularities are birational via a flop, then their singular Betti numbers are equal. Since one still knows very little about flops in dimension $n \geq 4$, it seems unlikely that a consideration of flops could help to prove Theorem 1.1 in dimension $n \geq 4$. Moreover, Theorem 1.1 is false in general for projective algebraic varieties with at worst $\mathbb{Q}$-factorial Gorenstein terminal singularities of dimension $n \geq 4$. For this reason, the condition in Theorem 1.1 that X and Y are smooth becomes very important in the case $n \geq 4$. We remark that in the case of holomorphic symplectic manifolds some stronger result is obtained in [4].

I would like to thank Professors A. Beauville, B. Fantechi, L. Göttsche, K. Hulek, Y. Kawamata, M. Kontsevich, S. Mori, M. Reid and D. van Straten for their interest, fruitful discussions and stimulating e-mails.

2   Gauge forms and p-adic measures

Let F be a local number field, that is, a finite extension of the p-adic field $\mathbb{Q}_p$ for some prime $p \in \mathbb{Z}$. Let $R \subset F$ be the maximal compact subring, $\mathfrak{q} \subset R$ the maximal ideal, $F_{\mathfrak{q}} = R/\mathfrak{q}$ the residue field with $|F_{\mathfrak{q}}| = q = p^r$. We write

$$N_{F/\mathbb{Q}_p} \colon F \to \mathbb{Q}_p$$

for the standard norm, and $\| \cdot \| \colon F \to \mathbb{R}_{\geq 0}$ for the multiplicative p-adic norm

$$a \mapsto \|a\| = p^{-\operatorname{Ord}(N_{F/\mathbb{Q}_p}(a))}.$$

Here Ord is the p-adic valuation.

Definition 2.1 Let $\mathcal{X}$ be an arbitrary flat reduced algebraic S-scheme, where $S = \operatorname{Spec} R$. We denote by $\mathcal{X}(R)$ the set of S-morphisms $S \to \mathcal{X}$ (or sections of $\mathcal{X} \to S$). We call $\mathcal{X}(R)$ the set of R-*integral points* in $\mathcal{X}$. The set of sections of the morphism $\mathcal{X} \times_S \operatorname{Spec} F \to \operatorname{Spec} F$ is denoted by $\mathcal{X}(F)$ and called the set of F-*rational points* in $\mathcal{X}$.

Remark 2.2   (i) If $\mathcal{X}$ is an affine S-scheme, then one can identify $\mathcal{X}(R)$ with the subset

$$\{ x \in \mathcal{X}(F) \mid f(x) \in R \text{ for all } f \in \Gamma(\mathcal{X}, \mathcal{O}_{\mathcal{X}}) \} \subset \mathcal{X}(F).$$

(ii) If $\mathcal{X}$ is a projective (or proper) S-scheme, then $\mathcal{X}(R) = \mathcal{X}(F)$.

Now let X be a smooth n-dimensional algebraic variety over F. We assume that X admits an extension $\mathcal{X}$ to a regular S-scheme. Denote by Ω_X^n the canonical line bundle of X and by $\Omega_{\mathcal{X}/S}^n$ the relative dualizing sheaf on $\mathcal{X}$. Recall the following definition introduced by Weil [11]:

Definition 2.3 A global section $w \in \Gamma(\mathcal{X}, \Omega^n_{\mathcal{X}/S})$ is called a *gauge form* if it has no zeros in $\mathcal{X}$. By definition, a gauge form w defines an isomorphism $\mathcal{O}_{\mathcal{X}} \cong \Omega^n_{\mathcal{X}/S}$, sending 1 to w. Clearly, a gauge form exists if and only if the line bundle $\Omega^n_{\mathcal{X}/S}$ is trivial.

Weil observed that a gauge form w determines a canonical p-adic measure $\mathrm{d}\mu_w$ on the locally compact p-adic topological space $\mathcal{X}(F)$ of F-rational points in $\mathcal{X}$. The p-adic measure $\mathrm{d}\mu_w$ is defined as follows:

Let $x \in \mathcal{X}(F)$ be an F-point, $t_1, \ldots, t_n$ local p-adic analytic parameters at x. Then $t_1, \ldots, t_n$ define a p-adic homeomorphism $\theta \colon U \to \mathbf{A}^n(F)$ of an open subset $\mathcal{U} \subset \mathcal{X}(F)$ containing x with an open subset $\theta(\mathcal{U}) \subset \mathbf{A}^n(F)$. We stress that the subsets $\mathcal{U} \subset \mathcal{X}(F)$ and $\theta(\mathcal{U}) \subset \mathbf{A}^n(F)$ are considered to be open in the p-adic topology, not in the Zariski topology. We write

$$w = \theta^* \left(g \mathrm{d}t_1 \wedge \cdots \wedge \mathrm{d}t_n \right),$$

where $g = g(t)$ is a p-adic analytic function on $\theta(\mathcal{U})$ having no zeros. Then the p-adic measure $\mathrm{d}\mu_w$ on $\mathcal{U}$ is defined to be the pullback with respect to θ of the p-adic measure $\|g(t)\| \mathbf{dt}$ on $\theta(\mathcal{U})$, where $\mathbf{dt}$ is the standard p-adic Haar measure on $\mathbf{A}^n(F)$ normalized by the condition

$$\int_{\mathbf{A}^n(R)} \mathbf{dt} = 1.$$

It is a standard exercise using the Jacobian to check that two p-adic measures $\mathrm{d}\mu'_w, \mathrm{d}\mu''_w$ constructed as above on any two open subsets $\mathcal{U}', \mathcal{U}'' \subset \mathcal{X}(F)$ coincide on the intersection $\mathcal{U}' \cap \mathcal{U}''$.

Definition 2.4 The measure $\mathrm{d}\mu_w$ on $\mathcal{X}(F)$ constructed above is called the *Weil p-adic measure* associated with the gauge form w.

Theorem 2.5 ([11], Theorem 2.2.5) *Let $\mathcal{X}$ be a regular S-scheme, w a gauge form on $\mathcal{X}$, and $\mathrm{d}\mu_w$ the corresponding Weil p-adic measure on $\mathcal{X}(F)$. Then*

$$\int_{\mathcal{X}(R)} \mathrm{d}\mu_w = \frac{|\mathcal{X}(F_q)|}{q^n},$$

where $\mathcal{X}(F_q)$ is the set of closed points of $\mathcal{X}$ over the finite residue field F_q.

Proof Let

$$\varphi \colon \mathcal{X}(R) \to \mathcal{X}(F_q) \quad \text{given by} \quad x \mapsto \overline{x} \in \mathcal{X}(F_q)$$

be the natural surjective mapping. The proof is based on the idea that if $\overline{x} \in \mathcal{X}(F_q)$ is a closed F_q-point of $\mathcal{X}$ and $g_1, \ldots, g_n$ are generators of the maximal ideal of $\overline{x}$ in $\mathcal{O}_{\mathcal{X},\overline{x}}$ modulo the ideal $\mathfrak{q}$, then the elements $g_1, \ldots, g_n$ define a p-adic analytic homeomorphism

$$\gamma \colon \varphi^{-1}(\overline{x}) \to \mathbb{A}^n(\mathfrak{q}),$$

where $\varphi^{-1}(\overline{x})$ is the fiber of φ over $\overline{x}$ and $\mathbb{A}^n(\mathfrak{q})$ is the set of all R-integral points of $\mathbb{A}^n$ whose coordinates belong to the ideal $\mathfrak{q} \subset R$. Moreover, the p-adic norm of the Jacobian of γ is identically equal to 1 on the whole fiber $\varphi^{-1}(\overline{x})$. In order to see the latter we remark that the elements define an étale morphism $g \colon V \to \mathbb{A}^n$ of some Zariski open neighbourhood V of $\overline{x} \in \mathcal{X}$. Since $\varphi^{-1}(\overline{x}) \subset V(R)$ and $g^*(\mathrm{d}t_1, \wedge \cdots \wedge \mathrm{d}t_n) = h\omega$, where h is invertible in V, we obtain that h has p-adic norm 1 on $\varphi^{-1}(\overline{x})$. So, using the p-adic analytic homeomorphism γ, we obtain

$$\int_{\varphi^{-1}(\overline{x})} \mathrm{d}\mu_\omega = \int_{\mathbb{A}^n(\mathfrak{q})} \mathrm{d}t = \frac{1}{q^n}$$

for each $\overline{x} \in \mathcal{X}(F_\mathfrak{q})$. $\square$

Now we consider a slightly more general situation. We assume only that $\mathcal{X}$ is a regular scheme over S, but do not assume the existence of a gauge form on $\mathcal{X}$ (that is, of an isomorphism $\mathcal{O}_\mathcal{X} \cong \Omega^n_{\mathcal{X}/S}$). Nevertheless under these weaker assumptions we can define a unique natural p-adic measure $\mathrm{d}\mu$ at least on the compact $\mathcal{X}(R) \subset \mathcal{X}(F)$ – although possibly not on the whole p-adic topological space $\mathcal{X}(F)$!

Let $\mathcal{U}_1, \ldots, \mathcal{U}_k$ be a finite covering of $\mathcal{X}$ by Zariski open S-subschemes such that the restriction of $\Omega^n_{\mathcal{X}/S}$ on each $\mathcal{U}_i$ is isomorphic to $\mathcal{O}_{\mathcal{U}_i}$. Then each $\mathcal{U}_i$ admits a gauge form ω_i and we define a p-adic measure $\mathrm{d}\mu_i$ on each compact $\mathcal{U}_i(R)$ as the restriction of the Weil p-adic measure $\mathrm{d}\mu_{\omega_i}$ associated with ω_i on $\mathcal{U}_i(F)$. We note that the gauge forms ω_i are defined uniquely up to elements $s_i \in \Gamma(\mathcal{U}_i, \mathcal{O}^*_\mathcal{X})$. On the other hand, the p-adic norm $\|s_i(x)\|$ equals 1 for any element $s_i \in \Gamma(\mathcal{U}_i, \mathcal{O}^*_\mathcal{X})$ and any R-rational point $x \in \mathcal{U}_i(R)$. Therefore, the p-adic measure on $\mathcal{U}_i(R)$ that we constructed does not depend on the choice of a gauge form ω_i. Moreover, the p-adic measures $\mathrm{d}\mu_i$ on $\mathcal{U}_i(R)$ glue together to a p-adic measure $\mathrm{d}\mu$ on the whole compact $\mathcal{X}(R)$, since one has

$$\mathcal{U}_i(R) \cap \mathcal{U}_j(R) = (\mathcal{U}_i \cap \mathcal{U}_j)(R) \quad \text{for } i, j = 1, \ldots, k$$

and

$$\mathcal{U}_1(R) \cup \cdots \cup \mathcal{U}_k(R) = (\mathcal{U}_1 \cup \cdots \cup \mathcal{U}_k)(R) = \mathcal{X}(R).$$

Definition 2.6 The p-adic measure constructed above defined on the set $\mathcal{X}(R)$ of R-integral points of an S-scheme $\mathcal{X}$ is called the *canonical p-adic measure*.

For the canonical p-adic measure $d\mu$, we obtain the same property as for the Weil p-adic measure $d\mu_\omega$:

Theorem 2.7

$$\int_{\mathcal{X}(R)} d\mu = \frac{|\mathcal{X}(F_q)|}{q^n}.$$

Proof Using a covering of $\mathcal{X}$ by some Zariski open subsets $\mathcal{U}_1, \ldots, \mathcal{U}_k$, we obtain

$$\int_{\mathcal{X}(R)} d\mu = \sum_{i_1} \int_{\mathcal{U}_{i_1}(R)} d\mu - \sum_{i_1 < i_2} \int_{(\mathcal{U}_{i_1} \cap \mathcal{U}_{i_2})(R)} d\mu + \cdots + (-1)^{k-1} \int_{(\mathcal{U}_1 \cap \cdots \cap \mathcal{U}_k)(R)} d\mu$$

and

$$|\mathcal{X}(F_q)| = \sum_{i_1} |\mathcal{U}_{i_1}(F_q)| - \sum_{i_1 < i_2} |(\mathcal{U}_{i_1} \cap \mathcal{U}_{i_2})(F_q)|$$
$$+ \cdots + (-1)^{k-1} |(\mathcal{U}_1 \cap \cdots \cap \mathcal{U}_k)(F_q)|.$$

It remains to apply Theorem 2.5 to every intersection $\mathcal{U}_{i_1} \cap \cdots \cap \mathcal{U}_{i_s}$. $\square$

Theorem 2.8 *Let $\mathcal{X}$ be a regular integral S-scheme and $\mathcal{Z} \subset \mathcal{X}$ a closed reduced subscheme of codimension ≥ 1. Then the subset $\mathcal{Z}(R) \subset \mathcal{X}(R)$ has zero measure with respect to the canonical p-adic measure $d\mu$ on $\mathcal{X}(R)$.*

Proof Using a covering of $\mathcal{X}$ by Zariski open affine subsets $\mathcal{U}_1, \ldots, \mathcal{U}_k$, we can always reduce to the case when $\mathcal{X}$ is an affine regular integral S-scheme and $\mathcal{Z} \subset \mathcal{X}$ an irreducible principal divisor defined by an equation $f = 0$, where f is a prime element of $A = \Gamma(\mathcal{X}, \mathcal{O}_{\mathcal{X}})$.

Consider the special case $\mathcal{X} = \mathbb{A}_S^n = \mathrm{Spec}\, R[X_1, \ldots, X_n]$ and $\mathcal{Z} = \mathbb{A}_S^{n-1} = \mathrm{Spec}\, R[X_2, \ldots, X_n]$, that is, $f = X_1$. For every positive integer m, we denote by $\mathcal{Z}_m(R)$ the subset in $\mathbb{A}^n(R)$ consisting of all points $x = (x_1, \ldots, x_n) \in R^n$ such that $x_1 \in q^m$. One computes the p-adic integral in the straightforward way:

$$\int_{\mathcal{Z}_m(R)} d\mathbf{x} = \int_{\mathbb{A}^1(q^m)} dx_1 \prod_{i=2}^n \left(\int_{\mathbb{A}^1(R)} dx_i \right) = \frac{1}{q^m}.$$

On the other hand, we have

$$\mathcal{Z}(R) = \bigcap_{m=1}^\infty \mathcal{Z}_m(R).$$

Hence

$$\int_{Z(R)} \mathbf{dx} = \lim_{m \to \infty} \int_{Z_m(R)} \mathbf{dx} = 0,$$

and in this case the statement is proved. Using the Noether normalization theorem reduces the more general case to the above special one. $\square$

3 The Betti numbers

Proposition 3.1 *Let X and Y be birational smooth projective n-dimensional algebraic varieties over $\mathbb{C}$ having trivial canonical line bundles. Then there exist Zariski open dense subsets $U \subset X$ and $V \subset Y$ such that U is isomorphic to V and $\mathrm{codim}_X(X \setminus U), \mathrm{codim}_Y(Y \setminus V) \geq 2$.*

Proof Consider a birational rational map $\varphi \colon X \dashrightarrow Y$. Since X is smooth and Y is projective, φ is regular at the general point of any prime divisor of X, so that there exists a maximal Zariski open dense subset $U \subset X$ with $\mathrm{codim}_X(X \setminus U) \geq 2$ such that φ extends to a regular morphism $\varphi_0 \colon U \to Y$. Since $\varphi^* \omega_Y$ is proportional to ω_X, the morphism φ_0 is étale, that is, φ_0 is an open embedding of U into the maximal open subset $V \subset Y$ where φ^{-1} is defined. Similarly φ^{-1} induces an open embedding of V into U, so we conclude that φ_0 is an isomorphism of U onto V. $\square$

Proof of Theorem 1.1 Let X and Y be smooth projective birational varieties of dimension n over $\mathbb{C}$ with trivial canonical bundles. By Proposition 3.1, there exist Zariski open dense subsets $U \subset X$ and $V \subset Y$ with $\mathrm{codim}_X(X \setminus U) \geq 2$ and $\mathrm{codim}_Y(Y \setminus V) \geq 2$ and an isomorphism $\varphi \colon U \to V$.

By standard arguments, one can choose a finitely generated $\mathbb{Z}$-subalgebra $\mathcal{R} \subset \mathbb{C}$ such that the projective varieties X and Y and the Zariski open subsets $U \subset X$ and $V \subset Y$ are obtained by base change $* \times_S \mathrm{Spec}\, \mathbb{C}$ from regular projective schemes $\mathcal{X}$ and $\mathcal{Y}$ over $\mathcal{S} := \mathrm{Spec}\, \mathcal{R}$ together with Zariski open subschemes $\mathcal{U} \subset \mathcal{X}$ and $\mathcal{V} \subset \mathcal{Y}$ over $\mathcal{S}$. Moreover, one can choose $\mathcal{R}$ in such a way that both relative canonical line bundles $\Omega^n_{\mathcal{X}/\mathcal{S}}$ and $\Omega^n_{\mathcal{Y}/\mathcal{S}}$ are trivial, both codimensions $\mathrm{codim}_{\mathcal{X}}(\mathcal{X} \setminus \mathcal{U})$ and $\mathrm{codim}_{\mathcal{Y}}(\mathcal{Y} \setminus \mathcal{V})$ are ≥ 2, and the isomorphism $\varphi \colon U \to V$ is obtained by base change from an isomorphism $\Phi \colon \mathcal{U} \to \mathcal{V}$ over $\mathcal{S}$.

For almost all prime numbers $p \in \mathbb{N}$, there exists a regular R-integral point $\pi \in \mathcal{S} \times_{\mathrm{Spec}\, \mathbb{Z}} \mathrm{Spec}\, \mathbb{Z}_p$, where R is the maximal compact subring in a local p-adic field F; let $\mathfrak{q}$ be the maximal ideal of R. By an appropriate choice of $\pi \in S \times_{\mathrm{Spec}\,\mathbb{Z}} \mathrm{Spec}\,\mathbb{Z}_p$, we can ensure that both $\mathcal{X}$ and $\mathcal{Y}$ have good reduction modulo $\mathfrak{q}$. Moreover, we can assume that the maximal ideal $I(\overline{\pi})$ of the unique closed point in

$$S := \mathrm{Spec}\,R \overset{\pi}{\hookrightarrow} S \times_{\mathrm{Spec}\,\mathbb{Z}} \mathrm{Spec}\,\mathbb{Z}_p$$

is obtained by base change from some maximal ideal $J(\overline{\pi}) \subset \mathcal{R}$ lying over the prime ideal $(p) \subset \mathbb{Z}$.

Let $\omega_{\mathcal{X}}$ and $\omega_{\mathcal{Y}}$ be gauge forms on $\mathcal{X}$ and $\mathcal{Y}$ respectively and $\omega_{\mathcal{U}}$ and $\omega_{\mathcal{V}}$ their restriction to $\mathcal{U}$ (respectively $\mathcal{V}$). Since Φ^* is an isomorphism over S, $\Phi^*\omega_{\mathcal{V}}$ is another gauge form on $\mathcal{U}$. Hence there exists a nowhere vanishing regular function $h \in \Gamma(\mathcal{U}, \mathcal{O}_{\mathcal{X}}^*)$ such that

$$\Phi^*\omega_{\mathcal{V}} = h\omega_{\mathcal{U}}.$$

The property $\mathrm{codim}_{\mathcal{X}}(\mathcal{X} \setminus \mathcal{U}) \geq 2$ implies that h is an element of $\Gamma(\mathcal{X}, \mathcal{O}_{\mathcal{X}}^*) = \mathcal{R}^*$. Hence, one has $\|h(x)\| = 1$ for all $x \in \mathcal{X}(F)$, that is, the Weil p-adic measures on $\mathcal{U}(F)$ associated with $\Phi^*\omega_{\mathcal{V}}$ and $\omega_{\mathcal{U}}$ are the same. The latter implies the following equality of the p-adic integrals

$$\int_{\mathcal{U}(F)} d\mu_{\mathcal{X}} = \int_{\mathcal{V}(F)} d\mu_{\mathcal{Y}}.$$

By Theorem 2.8 and Remark 2.2, (ii), we obtain

$$\int_{\mathcal{U}(F)} d\mu_{\mathcal{X}} = \int_{\mathcal{X}(F)} d\mu_{\mathcal{X}} = \int_{\mathcal{X}(R)} d\mu_{\mathcal{X}}$$

and

$$\int_{\mathcal{V}(\mathcal{F})} d\mu_{\mathcal{Y}} = \int_{\mathcal{Y}(\mathcal{F})} d\mu_{\mathcal{Y}} = \int_{\mathcal{Y}(R)} d\mu_{\mathcal{Y}}.$$

Now, applying the formula in Theorem 2.7, we arrive at the equality

$$\frac{|\mathcal{X}(F_q)|}{q^n} = \frac{|\mathcal{Y}(F_q)|}{q^n}.$$

This shows that the numbers of F_q-rational points in $\mathcal{X}$ and $\mathcal{Y}$ modulo the ideal $J(\overline{\pi}) \subset \mathcal{R}$ are the same. We now repeat the same argument, replacing R by its cyclotomic extension $\mathcal{R}^{(r)} \subset \mathbb{C}$ obtained by adjoining all complex $(q^r - 1)$th roots of unity; we deduce that the projective schemes $\mathcal{X}$ and $\mathcal{Y}$ have the same number of rational points over $F_q^{(r)}$, where $F_q^{(r)}$ is the extension of the finite field F_q of degree r. We deduce in particular that the Weil zeta functions

$$Z(\mathcal{X}, p, t) = \exp\left(\sum_{r=1}^{\infty} |\mathcal{X}(F_q^{(r)})| \frac{t^r}{r}\right)$$

and

$$Z(\mathcal{Y}, p, t) = \exp\left(\sum_{r=1}^{\infty} |\mathcal{Y}(F_q^{(r)})| \frac{t^r}{r}\right)$$

are the same. Using the Weil conjectures proved by Deligne [9] and the comparison theorem between the étale and singular cohomology, we obtain

$$Z(\mathcal{X}, p, t) = \frac{P_1(t)P_3(t)\cdots P_{2n-1}(t)}{P_0(t)P_2(t)\cdots P_{2n}(t)} \tag{1}$$

and

$$Z(\mathcal{Y}, p, t) = \frac{Q_1(t)Q_3(t)\cdots Q_{2n-1}(t)}{Q_0(t)Q_2(t)\cdots Q_{2n}(t)},$$

where $P_i(t)$ and $Q_i(t)$ are polynomials with integer coefficients having the properties

$$\deg P_i(t) = \dim H^i(X, \mathbb{C}), \quad \deg Q_i(t) = \dim H^i(Y, \mathbb{C}) \quad \text{for all } i \geq 0. \tag{2}$$

Since the standard Archimedean absolute value of each root of polynomials $P_i(t)$ and $Q_i(t)$ must be $q^{-i/2}$ and $P_i(0) = Q_i(0) = 1$ for all $i \geq 0$, the equality $Z(\mathcal{X}, p, t) = Z(\mathcal{Y}, p, t)$ implies $P_i(t) = Q_i(t)$ for all $i \geq 0$. Therefore, we have $\dim H^i(X, \mathbb{C}) = \dim H^i(Y, \mathbb{C})$ for all $i \geq 0$. $\square$

4   Further results

Definition 4.1 Let $\varphi\colon X \dashrightarrow Y$ be a birational map between smooth algebraic varieties X and Y. We say that φ *does not change the canonical class*, if for some Hironaka resolution $\alpha\colon Z \to X$ of the indeterminacies of φ the composite $\alpha \circ \varphi$ extends to a morphism $\beta\colon Z \to Y$ such that $\beta^*\Omega_Y^n \cong \alpha^*\Omega_X^n$.

The statement of Theorem 1.1 can be generalized to the case of birational smooth projective algebraic varieties which do not necessarily have trivial canonical classes as follows:

Theorem 4.2 *Let X and Y be irreducible birational smooth n-dimensional projective algebraic varieties over $\mathbb{C}$. Assume that the exists a birational rational map $\varphi\colon X \dashrightarrow Y$ which does not change the canonical class. Then X and Y have the same Betti numbers.*

Proof We repeat the same arguments as in the proof of Theorem 1.1 with the only difference that instead of the Weil p-adic measures associated with gauge forms we consider the canonical p-adic measures (see Definition 2.6). Using the birational morphisms $\alpha\colon \mathcal{Z} \to \mathcal{X}$ and $\beta\colon \mathcal{Z} \to \mathcal{Y}$ having the property

$$\beta^* \Omega^n_{\mathcal{Y}/S} \cong \alpha^* \Omega^n_{\mathcal{X}/S},$$

we conclude that for some prime $p \in \mathbb{N}$, the integrals of the canonical p-adic measures $\mu_{\mathcal{X}}$ and $\mu_{\mathcal{Y}}$ over $\mathcal{X}(R)$ and $\mathcal{Y}(R)$ are equal, since there exists a dense Zariski open subset $\mathcal{U} \subset \mathcal{Z}$ on which we have $\alpha^* \mu_{\mathcal{X}} = \beta^* \mu_{\mathcal{Y}}$. By Theorem 2.7, the zeta functions of $\mathcal{X}$ and $\mathcal{Y}$ must be the same. $\quad\square$

Another immediate application of our method is related to the McKay correspondence [10].

Theorem 4.3 *Let $G \subset \mathrm{SL}(n, \mathbb{C})$ be a finite subgroup. Assume that there exist two different resolutions of singularities on $W := \mathbb{C}^n/G$:*

$$f\colon X \to W, \quad g\colon Y \to W$$

such that both canonical line bundles Ω^n_X and Ω^n_Y are trivial. Then the Euler numbers of X and Y are the same.

Proof We extend the varieties X and Y to regular schemes over a scheme S of finite type over $\mathrm{Spec}\,\mathbb{Z}$. Moreover, one can choose S in such a way that the birational morphisms f and g extend to birational S-morphisms

$$F\colon \mathcal{X} \to \mathcal{W}, \quad G\colon \mathcal{Y} \to \mathcal{W},$$

where $\mathcal{W}$ is a scheme over S extending W. Using the same arguments as in the proof of Theorem 1.1, one obtains that there exists a prime $p \in \mathbb{N}$ such that $Z(\mathcal{X}, p, t) = Z(\mathcal{Y}, p, t)$. On the other hand, in view of (2), the Euler number is determined by the Weil zeta function (1) as the degree of the numerator minus the degree of the denominator. Hence $e(X) = e(Y)$. $\quad\square$

With a little bit more work one can prove an even more precise statement:

Theorem 4.4 *Let $G \subset \mathrm{SL}(n, \mathbb{C})$ be a finite subgroup and $W := \mathbb{C}^n/G$. Assume that there exists a resolution*

$$f\colon X \to W$$

with trivial canonical line bundle Ω^n_X. Then the Euler number of X equals the number of conjugacy classes in G.

Remark 4.5 As we saw in the proof of Theorem 3.1, the Weil zeta functions of $Z(\mathcal{X}, p, t)$ and $Z(\mathcal{Y}, p, t)$ are equal for almost all primes $p \in \operatorname{Spec} \mathbb{Z}$. This fact being expressed in terms of the associated L-functions indicates that the isomorphism $H^i(X, \mathbb{C}) \cong H^i(Y, \mathbb{C})$ for all $i \geq 0$ which we have established must have some deeper motivic nature. Recently Kontsevich suggested an idea of a motivic integration [8], developed by Denef and Loeser [2]. In particular, this technique allows to prove that not only the Betti numbers, but also the Hodge numbers of X and Y in 1.1 must be the same.

References

[1] A. BEAUVILLE, *Counting rational curves on K3 surfaces*, preprint, alg-geom/9701019.

[2] J. DENEF AND F. LOESER, *Germs of arcs on singular algebraic varieties and motivic integration*, Preprint 1996, math.AG/9803039, 27 pp., to appear in Invent. Math.

[3] B. FANTECHI, L. GÖTTSCHE, D. VAN STRATEN, *Euler number of the compactified Jacobian and multiplicity of rational curves*, alg-geom/9708012.

[4] D. HUYBRECHTS, *Compact hyper-Kähler manifolds: Basic Results*, alg-geom/9705025.

[5] Y. KAWAMATA, K. MATSUDA AND K. MATSUKI, *Introduction to the Minimal Model Program*, Adv. Studies in Pure Math. **10** (1987), 283-360.

[6] Y. KAWAMATA, *Crepant blowing ups of three dimensional canonical singularities, and applications to degenerations of surfaces*, Ann. of Math. (2) **127** (1988), 93-163.

[7] J. KOLLÁR, *Flops*, Nagoya Math. J. 113 (1989), 15–36.

[8] M. KONTSEVICH, Lecture at Orsay (December 7, 1995).

[9] P. DELIGNE, *La conjecture de Weil. I*, Inst. Hautes Etudes Sci. Publ. Math., **43** (1974), 273-307.

[10] M. REID, *The McKay correspondence and the physicists' Euler number*, Lect. notes given at Univ. of Utah and MSRI (1992).

[11] A. WEIL, *Adèles and algebraic groups*, Progr. Math. **23**, Birkhäuser, Boston 1982.

[12] S.-T. YAU, E. ZASLOW, *BPS states, string duality, and nodal curves on K3*, Nucl. Phys. **B471** (1996), 503-512.

Victor V. Batyrev,
Mathematisches Institut, Universität Tübingen,
Auf der Morgenstelle 10, 72076 Tübingen, Germany
e-mail: batyrev@bastau.mathematik.uni-tuebingen.de

A Calabi–Yau threefold with non-Abelian fundamental group

Arnaud Beauville[*]

Abstract

This note, written in 1994, answers a question of Dolgachev by constructing a Calabi–Yau threefold whose fundamental group is the quaternion group H_8. The construction is reminiscent of Reid's unpublished construction of a surface with $p_g = 0$, $K^2 = 2$ and $\pi_1 = H_8$; I explain below the link between the two problems.

1   The example

Let $H_8 = \{\pm 1, \pm i, \pm j, \pm k\}$ be the quaternion group of order 8, and V its regular representation. We denote by $\widehat{H}_8$ the group of characters $\chi \colon H_8 \to \mathbb{C}^*$, which is isomorphic to $\mathbb{Z}_2 \times \mathbb{Z}_2$. The group H_8 acts on $\mathbb{P}(V)$ and[1] on $\mathrm{S}^2 V$; for each $\chi \in \widehat{H}_8$, we denote by $(\mathrm{S}^2 V)_\chi$ the eigensubspace of $\mathrm{S}^2 V$ with respect to χ, that is, the space of quadratic forms Q on $\mathbb{P}(V)$ such that $h \cdot Q = \chi(h)Q$ for all $h \in H_8$.

Theorem 1.1 *For each $\chi \in \widehat{H}_8$, let Q_χ be a general element of $(\mathrm{S}^2 V)_\chi$. The subvariety $\widetilde{X}$ of $\mathbb{P}(V)$ defined by the 4 equations*

$$Q_\chi = 0 \quad \text{for all } \chi \in \widehat{H}_8$$

is a smooth threefold, on which the group H_8 acts freely. The quotient $X := \widetilde{X}/H_8$ is a Calabi–Yau threefold with $\pi_1(X) = H_8$.

Let me observe first that the last assertion is an immediate consequence of the others. Indeed, since $\widetilde{X}$ is a Calabi–Yau threefold, we have $h^{1,0}(\widetilde{X}) = h^{2,0}(\widetilde{X}) = \chi(\mathcal{O}_{\widetilde{X}}) = 0$, hence $h^{1,0}(X) = h^{2,0}(X) = \chi(\mathcal{O}_X) = 0$. This implies

[*]Partially supported by the European HCM project "Algebraic Geometry in Europe" (AGE).

[1]I use Grothendieck's notation, that is, $\mathbb{P}(V)$ is the space of hyperplanes in V.

13

$h^{3,0}(X) = 1$, so there exists a nonzero holomorphic 3-form ω on X; since its pullback to $\widetilde{X}$ is everywhere nonzero, ω has the same property, hence X is a Calabi–Yau threefold. Finally $\widetilde{X}$ is a complete intersection in $\mathbb{P}(V)$, hence simply connected by the Lefschetz theorem, so the fundamental group of X is isomorphic to H_8.

So the problem is to prove that H_8 acts freely and $\widetilde{X}$ is smooth. To do this, we will need to write down explicit elements of $(S^2 V)_\chi$. As an H_8-module, V is the direct sum of the 4 one-dimensional representations of H_8 and twice the irreducible two-dimensional representation ρ. Thus there exists a system of homogeneous coordinates $(X_1, X_\alpha, X_\beta, X_\gamma; Y, Z; Y', Z')$ such that

$$g \cdot (X_1, X_\alpha, X_\beta, X_\gamma; Y, Z; Y', Z') =$$
$$(X_1, \alpha(g)X_\alpha, \beta(g)X_\beta, \gamma(g)X_\gamma; \rho(g)(Y, Z); \rho(g)(Y', Z')).$$

To be more precise, I denote by α (respectively β, γ) the nontrivial character which is $+1$ on i (respectively j, k), and I take for ρ the usual representation via Pauli matrices:

$$\rho(i)(Y, Z) = (\sqrt{-1}\,Y, -\sqrt{-1}\,Z), \quad \rho(j)(Y, Z) = (-Z, Y),$$
$$\rho(k)(Y, Z) = (-\sqrt{-1}\,Z, -\sqrt{-1}\,Y).$$

Then the general element Q_χ of $(S^2 V)_\chi$ can be written

$$Q_1 = t_1^1 X_1^2 + t_2^1 X_\alpha^2 + t_3^1 X_\beta^2 + t_4^1 X_\gamma^2 + t_5^1(YZ' - Y'Z),$$
$$Q_\alpha = t_1^\alpha X_1 X_\alpha + t_2^\alpha X_\beta X_\gamma + t_3^\alpha YZ + t_4^\alpha Y'Z' + t_5^\alpha(YZ' + ZY'),$$
$$Q_\beta = t_1^\beta X_1 X_\beta + t_2^\beta X_\alpha X_\gamma + t_3^\beta(Y^2 + Z^2) + t_4^\beta(Y'^2 + Z'^2) + t_5^\beta(YY' + ZZ'),$$
$$Q_\gamma = t_1^\gamma X_1 X_\gamma + t_2^\gamma X_\alpha X_\beta + t_3^\gamma(Y^2 - Z^2) + t_4^\gamma(Y'^2 - Z'^2) + t_5^\gamma(YY' - ZZ').$$

For fixed $\mathbf{t} := (t_i^\chi)$, let $\mathcal{X}_\mathbf{t}$ be the subvariety of $\mathbb{P}(V)$ defined by the equations $Q_\chi = 0$. Let us check first that the action of H_8 on $\mathcal{X}_\mathbf{t}$ has no fixed points for $\mathbf{t}$ general enough. Since a point fixed by an element h of H_8 is also fixed by h^2, it is sufficient to check that the element $-1 \in H_8$ acts without fixed point, that is, that $\mathcal{X}_\mathbf{t}$ does not meet the linear subspaces L_+ and L_- defined by $Y = Z = Y' = Z' = 0$ and $X_1 = X_\alpha = X_\beta = X_\gamma = 0$ respectively.

Let $x = (0, 0, 0, 0; Y, Z; Y', Z') \in \mathcal{X}_\mathbf{t} \cap L_-$. One of the coordinates, say Z, is nonzero; since $Q_1(x) = 0$, there exists $k \in \mathbb{C}$ such that $Y' = kY$, $Z' = kZ$. Substituting in the equations $Q_\alpha(x) = Q_\beta(x) = Q_\gamma(x) = 0$ gives

$$(t_3^\alpha + t_5^\alpha k + t_4^\alpha k^2)YZ = (t_3^\beta + t_5^\beta k + t_4^\beta k^2)(Y^2 + Z^2) =$$
$$(t_3^\alpha + t_5^\alpha k + t_4^\alpha k^2)(Y^2 - Z^2) = 0$$

which has no nonzero solutions for a generic choice of $\mathbf{t}$.

Now let $x = (X_1, X_\alpha, X_\beta, X_\gamma; 0, 0; 0, 0) \in \mathcal{X}_t \cap L_+$. As soon as the t_i^χ are nonzero, two of the X-coordinates cannot vanish, otherwise all the coordinates would be zero. Expressing that $Q_\beta = Q_\gamma = 0$ has a nontrivial solution in (X_β, X_γ) gives X_α^2 as a multiple of X_1^2, and similarly for X_β^2 and X_γ^2. But then $Q_1(x) = 0$ is impossible for a general choice of **t**.

Now we want to prove that $\mathcal{X}_t$ is smooth for **t** general enough. Let $\mathcal{Q} = \bigoplus_{\chi \in \widehat{H}_8} (S^2 V)_\chi$; then $\mathbf{t} := (t_i^\chi)$ is a system of coordinates on $\mathcal{Q}$. The equations $Q_\chi = 0$ define a subvariety $\mathcal{X}$ in $\mathcal{Q} \times \mathbb{P}(V)$, whose fibre above a point $\mathbf{t} \in \mathcal{Q}$ is $\mathcal{X}_t$. Consider the second projection $p: \mathcal{X} \to \mathbb{P}(V)$. For $x \in \mathbb{P}(V)$, the fibre $p^{-1}(x)$ is the linear subspace of $\mathcal{Q}$ defined by the vanishing of the Q_χ, viewed as linear forms in **t**. These forms are clearly linearly independent as soon as they do not vanish. In other words, if we denote by B_χ the base locus of the quadrics in $(S^2 V)_\chi$ and put $B = \bigcup B_\chi$, the map $p: \mathcal{X} \to \mathbb{P}(V)$ is a vector bundle fibration above $\mathbb{P}(V) \setminus B$; in particular $\mathcal{X}$ is nonsingular outside $p^{-1}(B)$. Therefore it is enough to prove that $\mathcal{X}_t$ is smooth at the points of $B \cap \mathcal{X}_t$.

Observe that an element x in B has two of its X-coordinates zero. Since the equations are symmetric in the X-coordinates we may assume $X_\beta = X_\gamma = 0$. Then the Jacobian matrix

$$\left(\frac{\partial Q_\chi}{\partial X_\psi}(x) \right) \quad \text{takes the form} \quad \begin{pmatrix} 2t_1^1 X_1 & 2t_2^1 X_\alpha & 0 & 0 \\ t_1^\alpha X_\alpha & t_1^\alpha X_1 & 0 & 0 \\ 0 & 0 & t_1^\beta X_1 & t_2^\beta X_\alpha \\ 0 & 0 & t_2^\gamma X\alpha & t_1^\gamma X_1 \end{pmatrix}.$$

For generic **t**, this matrix is of rank 4 except when all the X-coordinates of x vanish; but we have seen that this is impossible when **t** is general enough. $\square$

2   Some comments

As mentioned in the introduction, the construction is inspired by Reid's example [R] of a surface of general type with $p_g = 0$, $K^2 = 2$, $\pi_1 = H_8$. This is more than a coincidence. In fact, let $\widetilde{S}$ be the hyperplane section $X_1 = 0$ of $\widetilde{X}$. It is stable under the action of H_8 (so that H_8 acts freely on $\widetilde{S}$), and we can prove as above that it is smooth for a generic choice of the parameters. The surface $S := \widetilde{S}/H_8$ is a Reid surface, embedded in X as an ample divisor, with $h^0(X, \mathcal{O}_X(S)) = 1$. In general, let us consider a Calabi–Yau threefold X which contains a *rigid ample surface*, that is, a smooth ample divisor S such that $h^0(\mathcal{O}_X(S)) = 1$. Put $L := \mathcal{O}_X(S)$. Then S is a minimal surface of general type (because $K_S = L_{|S}$ is ample); by the Lefschetz theorem, the natural map $\pi_1(S) \to \pi_1(X)$ is an isomorphism. Because of the exact sequence

$$0 \to \mathcal{O}_X \longrightarrow L \longrightarrow K_S \to 0,$$

the geometric genus $p_g(S) := h^0(K_S)$ is zero.

We have $K_S^2 = L^3$; the Riemann–Roch theorem on X yields

$$1 = h^0(L) = \frac{L^3}{6} + \frac{L \cdot c_2}{12} \ .$$

Since $L \cdot c_2 > 0$ as a consequence of Yau's theorem (see for instance [B], Cor. 2), we obtain $K_S^2 \leq 5$.

For surfaces with $p_g = 0$ and $K_S^2 = 1$ or 2, we have a great deal of information about the *algebraic* fundamental group, that is the profinite completion of the fundamental group (see [B-P-V] for an overview). In the case $K_S^2 = 1$, the algebraic fundamental group is cyclic of order ≤ 5; if $K_S^2 = 2$, it is of order ≤ 9; moreover the dihedral group D_8 cannot occur. D. Naie [N] has recently proved that the symmetric group $\mathfrak{S}_3$ can also not occur; therefore the quaternion group H_8 is the only non-Abelian group which occurs in this range.

On the other hand, little is known about surfaces with $p_g = 0$ and $K_S^2 = 3, 4$ or 5. Inoue has constructed examples with $\pi_1 = H_8 \times (\mathbb{Z}_2)^n$, with $n = K^2 - 2$ (*loc. cit.*); I do not know if they can appear as rigid ample surfaces in a Calabi-Yau threefold.

Let us denote by $\widetilde{X}$ the universal cover of X, by $\widetilde{L}$ the pullback of L to $\widetilde{X}$, and by ρ the representation of G on $H^0(\widetilde{X}, \widetilde{L})$. We have $\text{Tr}\,\rho(g) = 0$ for $g \neq 1$ by the holomorphic Lefschetz formula, and $\text{Tr}\,\rho(1) = \chi(\widetilde{L}) = |G|\,\chi(L) = |G|$. Therefore ρ *is isomorphic to the regular representation*. Looking at the list in *loc. cit.* we get a few examples of this situation, for instance:

- $G = \mathbb{Z}_5$, $\widetilde{X}$ = a quintic hypersurface in $\mathbb{P}^4$;

- $G = (\mathbb{Z}_2)^3$ or $\mathbb{Z}_4 \times \mathbb{Z}_2$, $\widetilde{X}$ = an intersection of 4 quadrics in $\mathbb{P}^7$ as above;

- $G = \mathbb{Z}_3 \times \mathbb{Z}_3$, $\widetilde{X}$ = a hypersurface of bidegree $(3,3)$ in $\mathbb{P}^2 \times \mathbb{P}^2$.

Of course, when looking for Calabi-Yau threefolds with interesting π_1, there is no reason to assume that it contains an ample rigid surface. Observe however that if we want to use the preceding method, in other words, to find a projective space $\mathbb{P}(V)$ with an action of G and a smooth invariant linearly normal Calabi-Yau threefold $\widetilde{X} \subset \mathbb{P}(V)$, then the line bundle $\mathcal{O}_{\widetilde{X}}(1)$ will be the pullback of an ample line bundle L on X, and by the above argument the representation of G on V will be $h^0(L)$ times the regular representation. This leaves little hope to find an invariant Calabi-Yau threefold when the product $h^0(L)|G|$ becomes large.

References

[B-P-V] W. Barth, C. Peters, A. Van de Ven, *Compact complex surfaces.* Ergebnisse der Math., Springer Verlag (1984)

[B]   J.-P. Bourguignon, *Premières formes de Chern des variétés kählér-iennes compactes.* Séminaire Bourbaki 77–78, exp. 507; Lecture Notes in Math. **710**, Springer Verlag (1979)

[N]   D. Naie, *Numerical Campedelli surfaces cannot have* $\mathfrak{S}_3$ *as algebraic fundamental group.* Bull. London Math. Soc., to appear

[R]   M. Reid, *Surfaces with* $p_g = 0$, $K^2 = 2$. Unpublished manuscript and letters (1979)

Arnaud Beauville,
DMI – École Normale Supérieure (URA 762 du CNRS),
45 rue d'Ulm,
F75230 Paris Cedex 05
e-mail: beauville@dmi.ens.fr

Algebraic Gromov–Witten invariants

K. Behrend

Abstract

We present an introduction to the algebraic theory of Gromov–
Witten invariants, as developed in collaboration with Yu. Manin and
B. Fantechi in [4], [3] and [2]. We try to make these three articles
more accessible. Proofs are generally omitted and there is little new
material.

Contents

1   Introduction

Gromov–Witten invariants are the basic enumerative geometry invariants of
a (nonsingular projective) algebraic variety W. Given a family $\Gamma_1, \dots, \Gamma_n$ of
algebraic cycles on W, one asks how many curves of fixed genus and degree
(or homology class) pass through $\Gamma_1, \dots, \Gamma_n$. The answer is given by the
associated Gromov–Witten invariant. (If there is an infinite number of such
curves, the Gromov–Witten invariant is a cycle in the moduli space of marked
curves, rather than a number.) Isolating the properties satisfied by these
invariants (formulated here as Axioms I–VIII) has had tremendous impact on
enumerative geometry in recent years. Moreover, Gromov–Witten invariants
tell us the *correct* way of counting curves. In simple cases (for example,
$W = \mathbb{P}^n$) the Gromov–Witten invariant simply gives the actual number of
curves through $\Gamma_1, \dots, \Gamma_n$ if $\Gamma_1, \dots, \Gamma_n$ are moved into general position. But
such a naive interpretation of Gromov–Witten invariants is impossible in
general, and so one should think of Gromov–Witten invariants as the *ideal*
number of curves through $\Gamma_1, \dots, \Gamma_n$.

Gromov–Witten invariants are defined as certain integrals over moduli
spaces of maps from curves to W. Integrating over the usual fundamental
class of the moduli space is problematic and can give the wrong result, because the moduli space might be of higher dimension than expected. This
necessitates the construction of a so-called virtual fundamental class. This
is the key step in the definition of Gromov–Witten invariants. Before the
virtual fundamental classes were understood, Gromov–Witten invariants had
only been constructed in special cases.

It turns out that the Gromov–Witten invariants of W (over $\mathbb{C}$) only depend on the underlying symplectic structure of W. (The only aspect one does
not see from the symplectic point of view is the motivic nature of Gromov–
Witten invariants.) The history of Gromov–Witten invariants in symplectic
geometry is actually much older than in algebraic geometry. Classically, one
perturbed the almost complex structure on W, instead of constructing a virtual fundamental class. For an exposition of this theory and its development,
see the article by Siebert in this volume, which also explains the fact that the
invariants constructed in symplectic geometry equal the algebraic ones.

The necessity of virtual fundamental classes for the definition of Gromov–
Witten invariants in algebraic geometry was felt from the very beginning (see
the seminal papers [10] and [11]). Before the general construction, several
special cases had been studied in detail, usually in genus zero, or for W a
homogeneous space. For more information on the results obtained and the
history of this part of the subject, see the survey [7].

The theory of virtual fundamental classes explained in this article is due
to B. Fantechi and the author (see [3]). Our work was inspired by a talk

of J. Li at the Santa Cruz conference on algebraic geometry in the summer of 1995. In his talk, Li reported on work in progress with G. Tian on the subject of virtual fundamental classes. At the time, that approach relied on analytic methods, for example, the existence of the Kuranishi map. Our work [3] grew out of an attempt to understand Li and Tian's work, to construct virtual fundamental classes in an algebraic context, and, most of all, to give the construction as intrinsically as possible. But, of course, our construction owes its existence to theirs. See [12] for the approach of Li and Tian.

Full details of the theory explained here can be found in the series of papers [4], [3] and [2]. In this article, we stress the geometric meaning of Gromov–Witten invariants and skip most proofs.

Our approach uses graphs to keep track of the moduli spaces involved. The graph theory we use here is much simpler than that of [4], for two reasons. Firstly, we restrict to 'absolutely stable' graphs (in the terminology of [4]). We lose a lot of invariants this way, but we gain a high degree of simplification of the formalism. However, even this simplified formalism contains all invariants $I_{g,n}^V(\beta)$ envisioned in [11]. The other aspect we do not go into here is that graphs form a category. Using the full power of the categorical (or 'operad') approach, the number of axioms for Gromov–Witten invariants can be distilled down to two (from the eight we need here), but only at the cost of a lot of formalities.

Introducing graphs here has two purposes. Firstly, we believe that graphs (as presented here) actually *simplify* the theory of Gromov–Witten invariants, and make their properties more transparent. For example, the use of graphs splits the famous 'splitting axiom' into three much simpler axioms. We also hope that presenting a simplified graph theoretical approach here will make [4] and [2] more accessible.

Our approach also relies heavily on the use of stacks. Again, stacks are introduced to simplify the theory; still, a few remarks seem in order. There are two ways in which stacks appear here, and two different kinds of stacks that play a role.

First of all, the moduli stacks involved are *Deligne–Mumford* stacks, which are algebraic geometry analogues of orbifolds. Thus, if one works over $\mathbb{C}$ and uses the analytic topology, such stacks are locally given as the quotient of an analytic space by the action of a finite group (except that the stack 'remembers' these group actions in a certain sense). A good way to think of a Deligne–Mumford stack is as a space (of points) together with a finite group attached to each point. (So if the stack is the quotient of a space by a finite group, the points of the stack are the orbits, and the group attached to an orbit is the isotropy group of any element of the orbit.) If the stack is a moduli stack, its points correspond to isomorphism classes of the objects the stack classifies, and the group attached to such an isomorphism class is the automorphism group of any object in the isomorphism class. The space of isomorphism classes is called the *underlying coarse moduli space.*

Deligne–Mumford stacks behave in many respects just like schemes. For example, their cohomological and intersection theoretic properties are identical to those of schemes, at least if one uses rational coefficients. Care is needed a just one place, when integrating a cohomology class over a Deligne–Mumford stack (which is not a scheme). Then fractions may appear (even if one integrates integral cohomology classes). More generally, one has to use fractions when doing *proper pushforwards* of homology or Chow cycles, if the morphism one pushes forwards along is not representable (i.e., has fibers which are stacks, not schemes).

For example, if our Deligne–Mumford stack X has one point, with finite group G attached to it (notation $X = BG$; we can view it as the quotient of a point by the action of G), then the Euler characteristic of X (i.e., the integral of the top Chern class of the tangent bundle, in this case the integral of $1 \in H^*(X)$) is $\chi(X) = \int_X 1 = 1/\#G$.

To calculate such an integral $\int_X \omega$ over a Deligne–Mumford stack X, one has to find a proper scheme X' together with a generically finite morphism $f\colon X' \to X$, and then one has $\int_X \omega = (1/\deg f) \int_{X'} f^*\omega$. In the above example $X = BG$, we may take X' to be the one-point variety and then $X' \to X$ has degree $\#G$ and so $\int_{BG} 1 = (1/\#G) \int_{\mathrm{pt}} 1 = 1/\#G$.

When explaining the general theory, it is not necessary to calculate a nonrepresentable proper pushforward explicitly, and so for this purpose we may as well pretend that all moduli stacks are spaces (i.e., schemes). We do this often, so that even if it says moduli space somewhere, it is implicitly understood that moduli stack is meant.

One reason that we must work with moduli stacks to do things properly, is that the corresponding coarse moduli spaces do not have universal families over them, whereas the construction of Gromov–Witten invariants uses universal families in an essential way.

The second way in which stacks appear is in the construction of virtual fundamental classes. Of course, one could construct the virtual fundamental class without using stacks, but we believe that stacks is the natural language for formulating the construction. The stacks used in this theory are so-called *cone stacks*, which are Artin stacks of a particular type. Artin stacks are more general than Deligne–Mumford stacks in that the groups attached to the points of the stack can be arbitrary algebraic groups, not just finite groups. These groups are too big to sweep under the carpet so easily, so that it is better not to pretend that Artin stacks are spaces, and we therefore include a 'heuristic' definition of cone stacks. (Cone stacks are special, since their groups are always vector groups.) The most important cone stack is the 'intrinsic normal cone'. It is an invariant of any Deligne–Mumford stack, and even for schemes it is an interesting object, which is nontrivial as a stack.

2   What are Gromov–Witten invariants?

Let k be a field[1] and W a smooth projective variety over k. We shall define the Gromov–Witten invariants of W, taking values in the cohomology of moduli spaces of curves.

2.1   Cohomology theories

So before we can begin, we have to choose a cohomology theory,

$$H^* \colon (\text{smooth proper DM stacks}/k) \longrightarrow (\text{vector spaces over } \Lambda)$$
$$X \longmapsto H^*(X)$$

This needs to be a 'graded generalized cohomology theory with coefficients in a field Λ of characteristic zero, with cycle map, such that $\mathbb{P}^1$ satisfies epu'. It should be defined on the category of smooth and proper Deligne–Mumford stacks over k. The precise definition can be found in [8].

Remark (for pedants) In [8], the cohomology theory is of course defined on the category of smooth and proper varieties, but the generalization of the definitions in [8] to Deligne–Mumford stacks is not difficult. The only point is that, strictly speaking, the category of (smooth, proper) Deligne–Mumford stacks is a 2-category, and so the cohomology theory is a functor from a 2- to a 1-category (i.e., a usual category). This means that it factors through the associated 1-category of the 2-category of Deligne–Mumford stacks, i.e., the category in which one passes to isomorphism classes of morphisms. In other words, one pretends that the category of Deligne–Mumford stacks is a usual category.

Rather than recalling the precise definition of a generalized cohomology theory with the required properties, we give a few examples.

1.   If the ground field k is $\mathbb{C}$ and the coefficient field Λ is $\mathbb{Q}$, then let

$$H^*(X) = H_{\mathrm{B}}^*(X) = \text{Betti cohomology of } X.$$

This can be defined in several ways.

---

[1]Because the theory is somewhat limited in positive characteristic (see footnote 4) the most important case is char $k = 0$.

The easiest case is when X has a moduli space $\widetilde{X}$. Then we can simply set

$$H_{\mathrm{B}}^*(X) = H_{\mathrm{sing}}^*(\widetilde{X}(\mathbb{C}), \mathbb{Q}),$$

the usual (singular) cohomology of the underlying topological space with the analytic topology. All the X that we consider have moduli spaces.[2]

More generally, one can consider $[X(\mathbb{C})]$, the set of isomorphism classes of the groupoid $X(\mathbb{C})$, in other words, the set of isomorphism classes of the objects the stack classifies. It comes with a natural topology, because the quotient of any groupoid exists in the category of topological spaces. The space $[X(\mathbb{C})]$ is thus the quotient of the topological groupoid associated to any presentation of X (with the analytic topology). Then we have

$$H_{\mathrm{B}}^*(X) = H_{\mathrm{sing}}^*([X(\mathbb{C})], \mathbb{Q}).$$

The canonical definition is the following. To the algebraic $\mathbb{C}$-stack X we associate a topological stack X^{top} (a stack on the category of topological spaces with the usual Grothendieck topology). This has an associated site (or topos) of sheaves $X_{\mathrm{et}}^{\mathrm{top}}$. (By abuse of notation we denote the usual topology by the subscript ét.) The Betti cohomology of X is then the cohomology of this topos

$$H_{\mathrm{B}}^*(X) = H^*(X_{\mathrm{et}}^{\mathrm{top}}, \mathbb{Q}).$$

This can also be defined in terms of geometric realizations.

2. Let ℓ be a prime not equal to the characteristic of k and consider the coefficient field $\Lambda = \mathbb{Q}_\ell$. Then we may take the ℓ-adic cohomology of X:

$$H^*(X) = H_\ell^*(X) = H^*(\overline{X}_{\mathrm{et}}, \mathbb{Q}_\ell) = \varprojlim_n H^*(\overline{X}_{\mathrm{et}}, \mathbb{Z}/\ell^n).$$

Here $\overline{X} = X \times_k \overline{k}$ and $\overline{X}_{\mathrm{et}}$ denotes the étale site of $\overline{X}$.

3. In the case char $k = 0$, we may take $\Lambda = k$ and consider algebraic de Rham cohomology

$$H^*(X) = H_{\mathrm{DR}}^*(X) = \mathbb{H}^*(X_{\mathrm{et}}, \Omega_X^\bullet).$$

---

[2]Note, however, that the existence of the moduli spaces is a nontrivial, *additional* fact, that we never need.

4. We may also take Chow cohomology

$$H^*(X) = A^{*/2}(X),$$

where the coefficient field is $\Lambda = \mathbb{Q}$. The Chow rings one needs for this definition were constructed by Vistoli [13].

2.2   Moduli stacks of curves

Gromov–Witten invariants take values in the cohomology of moduli stacks of curves. To list the axioms of Gromov–Witten invariants efficiently, we need slightly more general moduli spaces than the well-known $\overline{M}_{g,n}$. These are indexed by modular graphs.

Definition 2.1 A *graph* τ is a quadruple $(F_\tau, V_\tau, j_\tau, \partial_\tau)$ where F_τ is a finite set, the set of *flags*, V_τ is another finite set, the set of *vertices*, $\partial \colon F_\tau \to V_\tau$ is a map and $j_\tau \colon F_\tau \to F_\tau$ an involution. We use the notation:

$$S_\tau = \{f \in F_\tau \mid jf = f\}, \quad \text{the set of } \textit{tails} \text{ of } \tau.$$
$$E_\tau = \{\{f_1, f_2\} \subset F_\tau \mid f_2 = jf_1, f_1 \neq f_2\}, \quad \text{the set of } \textit{edges}.$$

For every vertex $v \in V$, the set $F_\tau(v) = \partial_\tau^{-1}(v)$ is the set of flags of v and $\# F_\tau(v)$ the *valency* of v.

We draw graphs by representing a vertex as a dot, and edge as a curve connecting vertices, and a tail as a half open curve, connected only to a vertex at its closed end. (The map ∂ specifies which vertex a flag is connected to.) Drawing graphs in this manner suggests an obvious notion of *geometric realization* of a graph. This is the topological space obtained in the way just indicated. The geometric realization of a graph τ is denoted by $|\tau|$.

Definition 2.2 A *modular graph* is a pair (τ, g), where τ is a graph and $g \colon V_\tau \to \mathbb{Z}_{\geq 0}$ a map. We use the terminology:

$g(v)$ is the *genus* of the vertex v.

$\chi(\tau) = \chi(|\tau|) - \sum_{v \in V_\tau} g(v)$ is the *Euler characteristic* of the graph τ.

If the geometric realization $|\tau|$ of τ is nonempty and connected then

$$g(\tau) = \sum_{v \in V_\tau} g(v) + \dim H^1(|\tau|, \mathbb{Q}) = 1 - \chi(\tau)$$

is the *genus* of τ. A graph of genus zero is a *tree*, and a possibly disconnected graph all of whose connected components are trees is called a *forest*. A nonempty connected graph without edges is called a *star*. Note that a star has exactly one vertex.

The Deligne–Mumford moduli stacks we are interested in are indexed by modular graphs. But we do not associate a moduli stack to every modular graph, only to the *stable* modular graphs.

Definition 2.3 The modular graph τ is *stable* if each vertex is stable, i.e., if

$$2g(v) + \# F_\tau(v) \geq 3 \quad \text{for all } v \in V_\tau.$$

We are now ready to define the moduli stacks of curves. First, for a nonnegative integer g and a finite set S such that $2g + \# S \geq 3$, we define the stack $\overline{M}_{g,S}$ to be the moduli stack of stable curves of genus g with marked points indexed by S.

Thus each point of $\overline{M}_{g,S}$ corresponds to a pair (C, x), where C is a nodal curve of arithmetic genus g (i.e., a connected but possibly reducible curve with at worst nodes as singularities) and x is an injective map $x \colon S \to C$, which avoids all nodes. The pair (C, x) is moreover required to be *stable*, meaning that for every irreducible component C' of C we have

$$2g(C') + \#\{\text{special points of } C'\} \geq 3.$$

Here $g(C')$ is the geometric genus of C', and a point is *special* if it is in the image of x or is a node of C. If both branches of a node belong to C', then this node counts as *two* special points.

If we choose an identification $S = \{1, \ldots, n\}$, then we get an induced identification

$$\overline{M}_{g,S} = \overline{M}_{g,n},$$

where the $\overline{M}_{g,n}$ are the moduli stacks of stable marked curves introduced by Mumford and Knudsen [9], and for $n = 0$ the stacks of stable curves defined by Deligne and Mumford [5].

Definition 2.4 The moduli stack associated to a stable modular graph τ is now simply defined to be

$$\overline{M}_\tau = \prod_{v \in V_\tau} \overline{M}_{g(v), F_\tau(v)}.$$

It may seem surprising that the involution j_τ does not enter here. The usefulness of this definition will become clear later.

Let (C, x) be a stable marked curve. We obtain its associated modular graph by associating

- to each irreducible component C' of C a vertex of genus $g(C')$ (geometric genus, that is, the genus of its normalization),

- to each node of C an edge connecting the vertices corresponding to the two branches of the node,

- to each marked point of C a tail attached to the vertex corresponding to the component containing the marked point.

If τ is the modular graph associated to (C, x), we say that (C, x) is of *degeneration type* τ. Note that τ is connected and $g(\tau) = g(C)$ (arithmetic genus).

If τ is a stable modular graph which is nonempty and connected, there exists a morphism

$$\overline{M}_\tau \longrightarrow \overline{M}_{g(\tau), S_\tau},$$

defined by associating to a V_τ-tuple of stable marked curves $(C_v, (x_i)_{i \in F_v})_{v \in V_\tau}$ the single curve $(C, (x_i)_{i \in S_\tau})$ obtained by identifying any two marks x_i that correspond to an edge of τ. This morphism is finite and its image is the stack of curves *of degeneration type τ or worse*. It is of generic degree $\#\operatorname{Aut}'(\tau)$ onto the image; here $\operatorname{Aut}'(\tau)$ is the group of automorphisms of τ fixing the tails.

If one fixes g and n and considers all connected stable modular graphs τ such that $g(\tau) = g$ and $S_\tau = \{1, \dots, n\}$, then one gets in this way the stratification of $\overline{M}_{g,n}$ by degeneration type.

2.3 Systems of Gromov–Witten invariants

Fix a smooth projective variety W over k. We use the notation

$$H_2(W)^+ = \left\{ \varphi \in \operatorname{Hom}(\operatorname{Pic} W, \mathbb{Z}) \,\middle|\, \begin{array}{l} \varphi(L) \geq 0 \text{ for all ample} \\ \text{invertible sheaves } L \text{ on } W \end{array} \right\}$$

Of course, if $k = \mathbb{C}$, then $H_2(W)^+$ contains the semigroup of effective cycle classes in $H_2(W, \mathbb{Z})$ (or, in general, the semigroup of effective cycle classes in $A_1(W)$) and we would not lose anything by restricting to this subsemigroup.

Definition 2.5 A *system of Gromov–Witten invariants* for W is a collection of (multi-)linear maps[3]

$$I_\tau(\beta)\colon H^*(W)^{\otimes S_\tau} \longrightarrow H^*(\overline{M}_\tau), \tag{1}$$

---

[3]To take Tate twists into account, one has to twist in a certain way, explained below, in the context of the grading axiom. So what we say here is only true up to Tate twists. Of course in the most important case, the Betti case, this is of no concern.

for every stable modular graph τ and every[4] $H_2(W)^+$ marking $\beta: V_\tau \to H_2(W)^+$ of τ, satisfying the list of eight axioms below.

Before listing the axioms, we say a few words about the geometric interpretation of Gromov–Witten invariants. For this, assume that we are over $\mathbb{C}$ and use singular cohomology. For purposes of intuition, it is better to dualize. So using Poincaré duality we identify H^* with H_* and get

$$I_\tau^\vee(\beta): H_*(W)^{\otimes S_\tau} \longrightarrow H_*(\overline{M}_\tau)$$
$$\gamma_1 \otimes \cdots \otimes \gamma_n \longmapsto I_\tau^\vee(\beta)(\gamma_1, \ldots, \gamma_n).$$

Note that, as the notation suggests, we are thinking of the $I_\tau^\vee(\beta)$ as multilinear maps (and we have chosen an identification $S_\tau = \{1, \ldots, n\}$).

To explain what $I_\tau^\vee(\beta)(\gamma_1, \ldots, \gamma_n)$ should be, choose cycles $\Gamma_1, \ldots, \Gamma_n \subset W$ in sufficiently general position representing the homology classes $\gamma_1, \ldots, \gamma_n$. Consider all triples (C, x, f), where

- $C = (C_v)_{v \in V_\tau}$ is a family of connected curves.

- $x = (x_i)_{i \in F_\tau}$ is a family of 'marks', i.e., for each $i \in F_\tau$ the mark x_i is a point on the curve $C_{\partial(i)}$. We also demand that (C, x) be a family of stable marked curves.

- $f = (f_v)_{v \in V_\tau}$ is a family of maps $f_v: C_v \to W$ such that

   1. for each edge $\{i_1, i_2\}$ of τ we have $f_{\partial(x_1)}(x_{i_1}) = f_{\partial(i_2)}(x_{i_2})$,
   2. for all $v \in V_\tau$ we have $f_*[C_v] = \beta(v)$,
   3. for all $i \in S_\tau$ we have that $f_{\partial(i)}(x_i) \in \Gamma_i$.

Let T be the 'space' of all such triples up to isomorphism. (An isomorphism from a triple (C, x, f) to a triple (D, y, g) is a V_τ-tuple $\varphi = (\varphi_v)_{v \in V_\tau}$ of isomorphisms of curves $\varphi_v: C_v \to D_v$ such that $\varphi_{\partial(i)}(x_i) = y_i$ for all $i \in F_\tau$ and $g_v \circ \varphi_v = f_v$ for all $v \in V_\tau$.)

We have a morphism $\varphi: T \to \overline{M}_\tau$, which simply maps a triple (C, x, f) to the first two components (C, x). The 'naive' definition of $I_\tau(\beta)$ is then

$$I_\tau^\vee(\beta)(\gamma_1, \ldots, \gamma_n) = \varphi_*[T].$$

Remark For simplicity, assume that τ is connected. To a triple (C, x, f) we may associate, as above, a single marked curve $(\widetilde{C}, \widetilde{x})$ by identifying the

---

[4] If char $k > 0$, then choose a very ample invertible sheaf L on W and consider only $\beta \in H_2(W)^+$ such that $\beta(L) < $ char k. This ensures that all maps considered are separable, which is needed for all the arguments (as stated here) to go through. However, one does not get 'as many' Gromov–Witten invariants as in characteristic zero.

two marks corresponding to each edge of τ, obtaining a stable marked curve of degeneration type τ or worse. The V_τ-tuple of maps f induces a map $\widetilde{f}\colon \widetilde{C} \to W$.

Let $\widetilde{T}$ be the space of triples (D, y, g), where (D, y) is a stable marked curve of degeneration type τ and $g\colon D \to W$ a morphism such that $g(y_i) \in \Gamma_i$ for all $i = 1, \ldots, n$, and $g_*[D_v] = \beta(v)$ for all $v \in V_\tau$. (Here D_v is the component of D corresponding to v.) Then we have a rational map $T \to \widetilde{T}$ of degree $\#\operatorname{Aut}'(\tau)$. (It is not defined everywhere, as we do not allow worse degeneration types than τ in $\widetilde{T}$.)[5]

So a slightly more naive but less abstract definition of $I_\tau(\beta)$ would be

$$I_\tau^\vee(\beta)(\gamma_1, \ldots, \gamma_n) = \#\operatorname{Aut}'(\tau)\varphi_*[\widetilde{T}].$$

Note that in the most important case, where τ is a star, the factor $\#\operatorname{Aut}'(\tau)$ is equal to 1.

For example, assume that T is finite (usually the case of interest). Then

$$\#T = I_\tau^\vee(\beta)(\gamma_1, \ldots, \gamma_n) \in \mathbb{Q}$$

is the 'ideal' number of solutions to an enumerative geometry problem.[6]

More precisely, passing to $(\widetilde{C}, \widetilde{x}, \widetilde{f})$ as before and then to $\widetilde{f}(\widetilde{C})$, we get a curve in W passing through $\Gamma_1, \ldots, \Gamma_n$. If $\Gamma_1, \ldots, \Gamma_n$ are sufficiently generic, then one would hope that this process sets up a bijection between points of T and the curves of degeneration type τ (or worse) through $\Gamma_1, \ldots, \Gamma_n$. Thus (if the hope is justified) $\dfrac{1}{\#\operatorname{Aut}'(\tau)} I_\tau^\vee(\beta)(\gamma_1, \ldots, \gamma_n)$ is the number of such curves intersecting $\Gamma_1, \ldots, \Gamma_n$.

For example, let $W = \mathbb{P}^2$ be the projective plane. Then $H_2(W)^+ = \mathbb{Z}_{\geq 0}$, and one writes $d = \beta$. Assume $d \geq 2$ and let $n = 3d - 1$. Let τ be the star of genus zero with n tails: $S_\tau = F_\tau = \{1, \ldots, n\}$. So we have

$$I_\tau^\vee(\beta) = I_{0,n}^\vee(d)\colon H^*(\mathbb{P}^2)^{\otimes n} \longrightarrow H^*(\overline{M}_{0,n}).$$

If we consider the homology class of a point in $\mathbb{P}^2$, call it γ, and consider $I_{0,n}^\vee(d)(\gamma^{\times n})$, where $\gamma^{\times n}$ stands for the n-tuple $(\gamma, \ldots, \gamma)$, then the corresponding 'space' T is a discrete set of points (if the n points $\Gamma_1, \ldots, \Gamma_n$ representing γ are in sufficiently general position).

---

[5]Allowing more degenerate curves in $\widetilde{T}$ would not make sense, because D_v would no longer be well defined.

[6]The word 'ideal' is essential here. In many cases, the Gromov–Witten invariant differs from the actual curve count.

One sees easily, that T corresponds one-to-one to the rational curves of degree d through $\Gamma_1, \ldots, \Gamma_n$. Thus

$$I_{0,n}(d)(\gamma^{\times n}) = I_{0,n}^{\vee}(d)(\gamma^{\times n})$$

$$= \# \left\{ \begin{array}{c} \text{rational curves of degree } d \text{ through} \\ n \text{ points in general position} \end{array} \right\}.$$

For example, the number of conics through 5 points is $I_{0,5}(2)(\gamma^{\times 5}) = 1$, and the number of rational cubics through 8 points is $I_{0,8}(3)(\gamma^{\times 8}) = 12$.

In view of this 'intuitive definition' the following eight axioms that we require of Gromov–Witten invariants are all very natural. Note, however, that there are two problems with this definition. First of all, T must be compactified in order for φ_* in homology to make sense. This can be dealt with using stable maps (see below). A more serious problem is that in general it is not possible to put the Γ_i into sufficiently general position to assure that T is smooth and of the 'correct' dimension. This necessitates the construction of a 'virtual fundamental class' in T, which is a homology class in the correct degree, whose image in $\overline{M}_\tau$ is taken to be the Gromov–Witten invariant.

The reasons for using this axiomatic approach are largely historical. Kontsevich and Manin [11] introduced these axioms before Gromov–Witten invariants were rigorously defined. Today there are several natural constructions of invariants satisfying the axioms. We present one of these later.

One should note that the axioms do not determine the invariants uniquely. For example, one can set all the I equal to zero (except for those forced to be nonzero by the axiom concerning mapping to a point). Certain re-scalings are also possible.

The axioms do comprise all properties used to construct quantum cohomology out of the Gromov–Witten invariants, and certainly imply all characteristic properties of Gromov–Witten invariants that do not involve change of the variety W.

2.4   Axioms for Gromov–Witten invariants

I. The grading axiom

This says that

$$I_\tau(\beta) \colon H^*(V)^{\otimes S_\tau}[2\chi(\tau) \dim W] \;\longrightarrow\; H^*(\overline{M}_\tau)[2\beta(\tau)(\omega_W)]$$

respects the natural grading on both vector spaces.[7] Here $[\,\cdot\,]$ denotes shifts of grading: if $H^* = \bigoplus H^k$ is a graded vector space then $H^*[m]$ is the graded

---

[7]If one is concerned about Tate twists, one needs to also twist by $(\chi(\tau) \dim W)$ on the left and $(\beta(\tau)(\omega_W))$ on the right.

vector space such that $(H^*[m])^k = H^{k+m}$. In other words, $I_\tau(\beta)$ raises degrees by $2(\beta(\tau)(\omega_W) - \chi(\tau)\dim W)$. We use the notation

$$\beta(\tau) = \sum_{v \in V_\tau} \beta(v)$$

and ω_W is the canonical line bundle on W.

The idea behind this axiom is that the moduli 'space' T of triples, alluded to in the section on geometric intuition, has an expected dimension. It is computed using deformation theory (assuming that there are no obstructions). Even if there are obstructions, one still requires that $I_\tau(\beta)$ changes the grading by this expected dimension (minus $\sum_i \deg \gamma_i$), which is then called 'virtual dimension'.

The reasoning behind this is that one wants Gromov–Witten invariants to be invariant under continuous (or better, algebraic) deformations of the whole situation, like all good enumerative geometry numbers are. So one supposes that one could deform the situation into sufficiently general position for the obstructions to vanish and the space T to actually attain the expected dimension.

Note however, that in general it is not possible to deform the variety W algebraically to make it sufficiently generic in this sense.

For the computation of the expected dimension see Section 4.7. See also Remark 3.3

II. Isomorphisms

Let $\varphi \colon \sigma \to \tau$ be an isomorphism of $H_2(W)^+$-marked stable modular graphs. Then we get induced isomorphisms $V^{S_\sigma} \to V^{S_\tau}$ and $\overline{M}_\sigma \to \overline{M}_\tau$ and the isomorphism axiom requires the diagram

$$\begin{array}{ccc} H^*(V)^{\otimes S_\tau} & \xrightarrow{I_\tau(\beta)} & H^*(\overline{M}_\tau) \\ \downarrow & & \downarrow \\ H^*(V)^{\otimes S_\sigma} & \xrightarrow{I_\sigma(\beta)} & H^*(\overline{M}_\sigma) \end{array}$$

to commute.

This axiom leads to a covariant behavior of the $I_{g,n}(\beta)$ with respect to the action of the symmetric group on n letters. It is motivated by the expectation that the ideal number of curves through the cycles $\Gamma_1, \ldots, \Gamma_n$ should not depend on the labelling of the cycles.

III. Contractions

Let $\varphi \colon \sigma \to \tau$ be a contraction of stable modular graphs. This means that there exists an edge $\{f, \bar{f}\}$ of σ which, at the level of geometric realizations, is literally contracted to a vertex by φ. It also implies a certain compatibility between the genera of the vertices involved. There are two cases to distinguish.

Case (a) The edge $\{f, \overline{f}\}$ is a loop with vertex v. Then if v' is the corresponding vertex of τ to which our edge is contracted, we have $g(v') = g(v) + 1$.

Case (b) The edge $\{f, \overline{f}\}$ has two different vertices v_1 and v_2. In this case we let v be the vertex of τ obtained by merging the vertices v_1 and v_2 via the contraction φ. The requirement is that $g(v) = g(v_1) + g(v_2)$.

All other vertices of τ have the genus of the corresponding vertex of σ.

In either case we get an induced morphism $\Phi \colon \overline{M}_\sigma \to \overline{M}_\tau$, defined as follows.

Case (a)

$$\Phi \colon \overline{M}_\sigma \longrightarrow \overline{M}_\tau$$
$$(C, \dots) \longmapsto (C/x_f = x_{\overline{f}}, \dots)$$

Here C stands for the component of the V_σ-tuple of stable marked curves corresponding to the vertex v. This curve has two marked points, indexed by f and $\overline{f}$. The morphism Φ identifies them with each other, creating a node in the curve C and losing two marked points in the process. This curve obtained from C by creating an additional node we call C' and then C' is the component of the V_τ-tuple of stable marked curves corresponding to the index v'.

Case (b)

$$\Phi \colon \overline{M}_\sigma \longrightarrow \overline{M}_\tau$$
$$(C_1, C_2, \dots) \longmapsto (C_1 \amalg C_2/x_f = x_{\overline{f}}, \dots)$$

Here C_1 and C_2 are the components of the V_σ-tuple of stable marked curves corresponding to the vertices v_1 and v_2, respectively. On C_1 there is a marked point indexed by f, and on C_2 there is a marked point indexed by $\overline{f}$, and $C_1 \amalg C_2/x_f = x_{\overline{f}}$ refers to the curve obtained by identifying these two points in the disjoint union of these two curves. In the process one loses two marked points, which is OK, because the graph also lost two flags.

In either case, the image of Φ is a 'boundary' divisor in $\overline{M}_\tau$. Usually, Φ is a closed immersion. Only if exchanging the two flags f and $\overline{f}$ can be extended to an automorphism of σ inducing the identity on τ (i.e., always in Case (a), almost never in Case (b)) is Φ a degree two cover followed by an immersion. In Case (a) the image of Φ can also intersect itself.

The axiom now demands that for each $H_2(W)^+$ marking β on τ, the diagram

$$
\begin{array}{ccc}
H^*(W)^{\otimes S_\tau} & \xrightarrow{I_\tau(\beta)} & H^*(\overline{M}_\tau) \\
\downarrow & & \downarrow \Phi^* \\
H^*(W)^{\otimes S_\sigma} & \xrightarrow{\sum_{\beta'} I_\sigma(\beta')} & H^*(\overline{M}_\sigma)
\end{array}
$$

commutes. Here the vertical map on the left is the canonical isomorphism coming from the fact that the contraction φ does not affect the tails of the graphs involved.

The sum in the lower horizontal map is taken over all maps

$$\beta': V_\sigma \longrightarrow H_2(W)^+$$

that are compatible with β. This means

- in Case (a), that $\beta'(w) = \beta(w)$ for all $w \in V_\sigma$. In particular that $\beta'(v) = \beta(v')$,

- in Case (b), that $\beta'(w) = \beta(w)$ for all $w \neq v_1, v_2$, and $\beta'(v_1) + \beta'(v_2) = \beta(v)$.

Note that in Case (a) there is only one summand and in Case (b) there is a finite number of summands.

The meaning of this axiom is very simple. For example, in Case (b) it says that the number of curves in class β that have two components is the sum over all pairs (β_1, β_2) such that $\beta_1 + \beta_2 = \beta$ of the number of curves that have two components whose first component is of class β_1 and whose second component is of class β_2. (The invariant $I_\tau^\vee(\beta)$ might be a 1-cycle in $\overline{M}_\tau$ and Φ^* would intersect it with the boundary divisor $\overline{M}_\sigma$ and so count the number of curves in the family $I_\tau^\vee(\beta)$ that have two components, where the generic member has one.) In case Φ is generically two to one, Φ^* involves a multiplication by a factor of two, which reflects the ambiguity in marking the two points lying over the node.

IV. Glueing tails

Let τ be a stable modular graph and $\{f, \overline{f}\}$ an edge of τ. Let σ be the modular graph obtained from τ by 'cutting the edge' $\{f, \overline{f}\}$. This means that all the data describing σ is the same as the data describing τ, except for the involution j. In the case of τ, the set $\{f, \overline{f}\}$ is an orbit of j_τ and in the case of σ it is the union of two orbits of j_σ.

In this situation we have a 'morphism of stable modular graphs of type cutting edges' $\tau \to \sigma$ and an 'extended isogeny of type glueing tails' $\sigma \to \tau$. For the definitions of these terms see [4]. The direction of the arrow joining σ and τ depends on the context. Part I of [4] takes the 'morphism of stable modular graphs' approach to describe the morphisms between moduli spaces. In Part II, where Gromov–Witten invariants are given a graph theoretic treatment, the 'extended isogeny' viewpoint is needed.

Anyway, to state our axiom, the direction of the arrow joining σ and τ is not relevant. What is important to note is that σ has two tails more than τ, and therefore we have

$$H^*(W)^{\otimes S_\sigma} = H^*(W)^{\otimes S_\tau} \otimes H^*(W \times W).$$

The axiom now requires the diagram

$$
\begin{array}{ccccc}
H^*(W) \otimes H^*(W)^{\otimes S_\tau} & \xleftarrow{\;p^*\;} & H^*(W)^{\otimes S_\tau} & \xrightarrow{I_\tau(\beta)} & H^*(\overline{M}_\tau) \\
{\scriptstyle \Delta_*}\downarrow & & & & \downarrow{\scriptstyle \cong} \\
H^*(W \times W) \otimes H^*(W)^{\otimes S_\tau} & = & H^*(W)^{\otimes S_\sigma} & \xrightarrow{I_\sigma(\beta)} & H^*(\overline{M}_\sigma)
\end{array}
$$

to commute. Here $p\colon W \times W^{S_\tau} \to W^{S_\tau}$ is the projection onto the second factor, and $\Delta\colon W \to W \times W$ is the diagonal. This diagram is required to commute for any $H_2(W)^+$ marking β one can put on $V_\tau = V_\sigma$.

Note that the image of $\Gamma_1 \times \cdots \times \Gamma_n$ under $\Delta_* \circ p^*$ is $\Delta \times \Gamma_1 \times \cdots \times \Gamma_n$, where Δ takes up the two first components in W^{S_τ}. So $I_\sigma(\beta) \circ \Delta_* \circ p^*$ should count the number of marked curves whose first two marks map to the same point in W. These are exactly the curves that $I_\tau(\beta)$ should count.

V. Products

Let τ and τ' be two stable modular graphs and σ the stable modular graph whose geometric realization is the disjoint union of the those of τ and τ'. We write $\sigma = \tau \times \tau'$ (and not $\tau \amalg \tau'$). For $H_2(W)^+$ markings β on τ and β' on τ' we denote by $\beta \times \beta'$ the induced $H_2(W)^+$ marking on σ. The product axiom requires that under such conditions the diagram

$$
\begin{array}{ccc}
H^*(W)^{\otimes S_\tau} \otimes H^*(W)^{S_{\tau'}} & \xrightarrow{I_\tau(\beta) \otimes I_{\tau'}(\beta')} & H^*(\overline{M}_\tau) \otimes H^*(\overline{M}_{\tau'}) \\
\downarrow & & \downarrow \\
H^*(W)^{\otimes S_\sigma} & \xrightarrow{I_\sigma(\beta \times \beta')} & H^*(\overline{M}_\sigma)
\end{array}
$$

always commutes. Here the vertical maps are the isomorphisms induced by the isomorphisms $W^{S_\sigma} = W^{S_\tau} \times W^{S_{\tau'}}$ and $\overline{M}_\sigma = \overline{M}_\tau \times \overline{M}_{\tau'}$.

This axiom expresses the expectation that the number of solutions to the enumerative geometry problem (τ, β) multiplied by the number of solutions of the problem (τ', β') is the number of *pairs* solving the 'composite enumerative geometry problem'.

VI. Fundamental class

Let σ be a stable modular graph and $f \in S_\sigma$ a tail of σ. Let τ be the modular graph obtained by simply omitting f. We assume that τ is still stable.

Remark Since there is a canonical way of associating a stable modular graph to any modular graph (called its 'stabilization'), one might wonder if there is also an axiom that applies in the case that τ is not stable. The answer is that such an axiom would follow from the others and is therefore not necessary. To see this, assume that the stabilization of τ is not empty. Then the process of removing the tail from σ and stabilizing the graph thus obtained can also be described (albeit not uniquely) as an edge contraction followed by a tail omission that does not lead to an unstable graph.[8]

In this situation we get a morphism

$$\Phi \colon \overline{M}_\sigma \longrightarrow \overline{M}_\tau,$$

defined in the following way: take the curve corresponding to the vertex of the tail f, which has a marked point on it indexed by f. Omit this point x_f and stabilize the marked curve thus obtained. To stabilize means to contract (blow down) the component on which x_f lies, if it becomes unstable by omitting x_f. (This can only happen in case this component is rational.) It is proved in [9] that $\overline{M}_\sigma \to \overline{M}_\tau$ is the universal curve corresponding to the vertex of f. More on stabilization in the next section.

Our axiom requires that the diagram

$$
\begin{array}{ccc}
H^*(W)^{\otimes S_\tau} & \xrightarrow{I_\tau(\beta)} & H^*(\overline{M}_\tau) \\
{\scriptstyle p^*} \downarrow & & \downarrow {\scriptstyle \Phi^*} \\
H^*(W)^{\otimes S_\sigma} & \xrightarrow{I_\sigma(\beta)} & H^*(\overline{M}_\sigma)
\end{array}
$$

commutes for every $H_2(W)^+$ marking β one can put on $V_\sigma = V_\tau$. Note that σ has exactly one tail more than τ and that therefore we can identify $W^{S_\sigma} = W \times W^{S_\tau}$ and p is the projection onto the second factor.

The geometric meaning of this axiom is that if one of the homology classes $\gamma_1, \ldots, \gamma_n$, say γ_1, is $[W]$, then the space T_1 obtained for $\gamma_1, \ldots, \gamma_n$ is a curve over the corresponding space T for $\gamma_2, \ldots, \gamma_n$. This is because for x_1 to be in W is no condition, so it can move anywhere on C leading to $T_1 \to T$ being the universal curve.

---

[8]It is precisely for this reason that the notion of 'isogeny' of stable graphs is introduced in Part II of [4]. If one were to use only the morphisms defined in Part I, one would not be able to decompose a tail omission that necessitates stabilization in this way.

VII. Divisor

The setup is the same as in Axiom VI for the fundamental class. The divisor axiom says that for every line bundle $L \in \text{Pic}(W)$ (and every β) the diagram

$$
\begin{array}{ccc}
H^*(W)^{\otimes S_\tau} & \xrightarrow{\beta(L)I_\tau(\beta)} & H^*(\overline{M}_\tau) \\
c_1(L) \downarrow & & \uparrow \Phi_* \\
H^*(W)^{\otimes S_\sigma} & \xrightarrow{I_\sigma(\beta)} & H^*(\overline{M}_\sigma)
\end{array}
$$

commutes. Here the vertical map on the left is

$$
\begin{array}{ccc}
H^*(W)^{\otimes S_\tau} & \longrightarrow & H^*(W) \otimes H^*(W)^{\otimes S_\tau} \\
\gamma & \longmapsto & c_1(L) \otimes \gamma.
\end{array}
$$

This axiom expresses the expectation that modifying an enumerative problem by adding a divisor D (such that $L = \mathcal{O}(D)$) to the list $\Gamma_1, \ldots, \Gamma_n$ multiplies the number of solutions by $\beta(L)$, because for a curve C of class β to intersect D is no condition, and in fact the additional marking on C can be any of the $\beta(L)$ points of intersection of C with D.

VIII. Mapping to point

This axiom deals with the case that $\beta = 0$. Let τ be a nonempty connected stable modular graph. Over the moduli space $\overline{M}_\tau$ there are universal curves, one for each vertex of τ, obtained by pulling back the universal curves from the factors of $\overline{M}_\tau$. If $v \in V_\tau$ is a vertex of τ then the associated universal curve C_v has sections (x_f), one for each flag $f \in F_\tau(v)$. Now define a new curve $\widetilde{C}$ over $\overline{M}_\tau$ by identifying x_f with $x_{\bar{f}}$ for each edge $\{f, \bar{f}\}$ of τ. We call $\widetilde{C}$ the *universal curve* over $\overline{M}_\tau$. It has connected fibers since the geometric realization of τ is connected. Denote the structure morphism by $\pi \colon \widetilde{C} \to \overline{M}_\tau$.

Consider the direct product of $\overline{M}_\tau$ and W, with projections labelled as in the diagram

$$
\begin{array}{ccc}
\overline{M}_\tau \times W & \xrightarrow{p} & W \\
q \downarrow & & \\
\overline{M}_\tau & &
\end{array}
$$

We get an induced homomorphism

$$
\begin{array}{ccc}
\rho \colon H^*(W) & \longrightarrow & H^*(\overline{M}_\tau) \\
\gamma & \longmapsto & q_*(p^*(\gamma) \cup c_{\text{top}}(R^1\pi_*\mathcal{O}_{\widetilde{C}} \boxtimes T_W)).
\end{array}
$$

Here T_W stands for the tangent bundle of W and c_{top} for the highest Chern class, which in this case will be of degree $g(\tau) \dim W$.

The mapping to point axiom now states that

$$I_\tau(0)\colon H^*(W)^{\otimes S_\tau} \longrightarrow H^*(\overline{M}_\tau)$$

is given by

$$I_\tau(0)(\gamma_1,\ldots,\gamma_n) = \rho(\gamma_1 \cup \cdots \cup \gamma_n).$$

This axiom expresses the fact that in this case

$$T = \overline{M}_\tau \rtimes \Gamma_1 \cap \cdots \cap \Gamma_n,$$

since a constant map to W has to map to $\Gamma_1 \cap \cdots \cap \Gamma_n$. Note that for $g(\tau) \geq 1$ the factor $c_{\mathrm{top}}(R^1\pi_*\mathcal{O}_{\tilde{\mathcal{C}}}\boxtimes T_W)$ is put in to satisfy the grading axiom. It is a sort of excess intersection term coming from the fact that there are obstructions in this case. More on this later (see Section 4.6).

Remark Axioms I–VIII imply the axioms listed in [11], except for the motivic axiom. This follows from the construction we give below. Axioms III, IV and V (contractions, glueing tails and products) imply the splitting and genus reduction axioms.

3   Construction of Gromov–Witten invariants

3.1   Stable maps

Gromov–Witten invariants are constructed as integrals over moduli spaces, namely, moduli spaces of stable maps. The notion of stable map is a natural generalization, due to Kontsevich, of stable curve (Deligne and Mumford [5]) and stable marked curve (Knudsen and Mumford [9]). Let us recall the definition.

Definition 3.1 Fix a smooth projective k-variety W. A *stable map* (to W) over a k-scheme T, of genus $g \in \mathbb{Z}_{\geq 0}$, class $\beta \in H_2(W)^+$ and marked by a finite indexing set S is:

1. a proper flat curve $C \to T$ such that every geometric fiber C_t is connected, one dimensional, has only ordinary double points (i.e., nodes) as singularities and arithmetic genus $1 - \chi(\mathcal{O}_{C_t}) = g$;

2. a family $(x_i)_{i\in S}$ of sections $x_i\colon T \to C$ such that for every geometric point $t \in T$, the points $(x_i(t))_{i\in S}$ are distinct, and not equal to a node;

3. a morphism $f: C \to W$ such that for every geometric point $t \in T$, denoting the restriction of f to the fiber C_t by $f_t: C_t \to W$, we have $\beta(L) = \deg f_t^*(L)$ for all $L \in \operatorname{Pic} W$ (or written more suggestively, $f_{t*}[C_t] = \beta$);

such that, for every geometric point $t \in T$, the normalization C' of every irreducible component of C_t satisfies:

$$f_t(C') \text{ is a point} \implies 2g(C') + \#\{\text{special points of } C'\} \geq 3;$$

here a *special* point is one that that lies over a mark x_i or a node of C_t.

A *morphism* of stable maps $\varphi: (C, x, f) \to (C', x', f')$ over T is a T-isomorphism $\varphi: C \to C'$ such that $\varphi(x_i) = x_i'$ for all $i \in S$ and $f'(\varphi) = f$.

Let $\overline{M}_{g,S}(W, \beta)$ denote the k-stack of stable maps of type (g, S, β) to W. Just like an algebraic space, or a scheme, or a variety (all over k), a stack is defined by giving its set of T-valued points for every k-scheme T, except that the set of T-valued points is not a set, but a category, in fact a category in which all morphisms are isomorphisms, in other words a *groupoid*. So the moduli stack $\overline{M}_{g,S}(W, \beta)$ is given by

$$\overline{M}_{g,S}(W, \beta)(T) = \text{category of stable maps over } T \text{ of type } (g, S, \beta) \text{ to } W,$$

for every k-scheme T.

The concept of stable maps was invented to make the following theorem true.

Theorem 3.2 (Kontsevich) *The k-stack $\overline{M}_{g,S}(W, \beta)$ is a proper algebraic Deligne–Mumford stack.*[9]

The Deligne–Mumford property signifies that the 'points' of $\overline{M}_{g,S}(W, \beta)$ have finite automorphism groups. The properness says two things. First, that every one dimensional family in $\overline{M}_{g,S}(W, \beta)$ has a 'limit', and secondly that this limit is unique. This translates into two facts about stable maps, namely first of all that every stable map over $T - \{t\}$, where T is one dimensional, extends to a stable map over T. For this to be true one has to allow certain degenerate maps, namely those with singular curves. The amazing fact is that by including exactly the degenerate maps which are *stable*, one picks out exactly one extension to T from all the possible extensions of the stable map over $T - \{t\}$. This makes the 'limit' unique, and hence the stack $\overline{M}_{g,S}(W, \beta)$ proper.

For a proof of this theorem we refer to [7].

---

[9]At least if char $k = 0$ or if $\beta(L) < \operatorname{char}(k)$ for some ample invertible sheaf L on W.

Note For each $i \in S$ there is an evaluation morphism

$$\mathrm{ev}_i \colon \overline{M}_{g,S}(W, \beta) \longrightarrow W$$
$$(C, x, f) \longmapsto f(x_i).$$

These moduli stacks are now taken as building blocks to define moduli stacks of stable maps associated to graphs.

So let (τ, β) be a stable modular graph with an $H_2(W)^+$ marking. The associated moduli stack we construct is denoted by $\overline{M}(W, \tau)$, abusing notation by leaving out β. We list three conditions on these moduli stacks that determine them completely.

1. Stars If τ is a star, i.e., a graph with only one vertex v, and set of flags S, which are all tails, then

$$\overline{M}(W, \tau) := \overline{M}_{g(v),S}(W, \beta(v)).$$

2. Products If τ and σ are stable modular graphs with $H_2(W)^+$ markings, and $\sigma \times \tau$ denotes the obvious stable modular graph with $H_2(W)^+$ marking whose geometric realization is the disjoint union of the geometric realizations of σ and τ, then

$$\overline{M}(W, \tau \times \sigma) := \overline{M}(W, \tau) \times \overline{M}(W, \sigma).$$

3. Edges If τ has two tails i_1 and i_2, and σ is obtained from τ by glueing these two tails to an edge (so that conversely, τ is obtained from σ by cutting an edge), then $\overline{M}(W, \sigma)$ is defined to be the fibered product

$$
\begin{array}{ccc}
\overline{M}(W, \sigma) & \longrightarrow & W \\
\downarrow & & \downarrow \Delta \\
\overline{M}(W, \tau) & \overset{\mathrm{ev}_{i_1} \times \mathrm{ev}_{i_2}}{\longrightarrow} & W \times W
\end{array}
$$

It is not difficult to see that $\overline{M}(W, \tau)$ is well defined by these conditions for every stable modular graph τ with $H_2(W)^+$ marking. Moreover, all $\overline{M}(W, \tau)$ are proper Deligne–Mumford stacks.

Note There exists an evaluation morphism $\mathrm{ev}_i \colon \overline{M}(W, \tau) \to W$ for each tail $i \in S_\tau$, and the product of these is the *evaluation morphism* $\mathrm{ev} \colon \overline{M}(W, \tau) \to W^{S_\tau}$.

For future reference, we now construct the *universal curve* on $\overline{M}(W, \tau)$. Fix a vertex $v \in V_\tau$. By construction, there exists a projection morphism

$$\overline{M}(W, \tau) \longrightarrow \overline{M}_{g(v),F_\tau(v)}(W, \beta(v)),$$

and we can pull back the universal stable map. This gives us a curve C_v over $\overline{M}(W, \tau)$ together with a morphism $f_v \colon C_v \to W$, and sections $x_i \colon \overline{M}(W, \tau) \to C_v$ for all $i \in F_\tau(v)$.

We glue the C_v according to the edges of τ (i.e., identify x_i with $x_{j_\tau(i)}$) to get a curve $C \to \overline{M}(W, \tau)$ called the *universal curve*, even though its fibers are only connected if $|\tau|$ is. There are induced sections $(x_i)_{i \in S_\tau}$ of C and an induced morphism $f \colon C \to W$.

Stabilization

Let $\tau = (\tau, \beta)$ be a stable modular graph with $H_2(W)^+$ structure. There exists a morphism

$$\overline{M}(W, \tau) \longrightarrow \overline{M}_\tau$$

given by 'stabilization'. To define it, it suffices to consider the case that τ is a star. So we are claiming that there exists a morphism

$$\overline{M}_{g,S}(W, \beta) \longrightarrow \overline{M}_{g,S}$$
$$(C, x, f) \longmapsto (C, x)^{\text{stab}}.$$

In other words, we take a stable map (C, x, f) and forget about the map f, retaining only the marked curve (C, x). The problem is that (C, x) might not be stable, so to get a point in $\overline{M}_{g,S}$ we need to associate to (C, x) a stable marked curve, in a natural way.

This is done as follows. Let $\pi \colon C \to T$ be a curve with a family of sections $x \colon T \to C$, $x = (x_i)_{i \in S}$. Then its stabilization is defined to be the curve

$$C' = \mathrm{Proj}_T \left(\bigoplus_{\nu \geq 0} \pi_*(L^{\otimes \nu}) \right),$$

where

$$L = \omega_{C/T} \left(\sum_{i \in S} x_i \right).$$

Here $\omega_{C/T}$ is the relative dualizing sheaf twisted by the Cartier divisor given by the images of the sections x_i in C. Note that there is a natural map $C \to C'$ and so one gets induced sections in C'.

One proves that C' together with these induced sections is a stable marked curve, and one calls it $(C, x)^{\text{stab}}$. For details of this construction, see [9].

The morphism $C \to C'$ just contracts (blows down) all the unstable rational components any fiber of $C \to T$ might have.

3.2   The construction: an overview

As usual, let $\tau = (\tau, \beta)$ be a stable modular graph with vertices marked by elements of $H_2(W)^+$.

Consider the diagram

$$\overline{M}(W, \tau) \xrightarrow{\text{ev}} W^{S_\tau}$$
$$\downarrow \text{stab} \qquad\qquad\qquad (2)$$
$$\overline{M}_\tau$$

In what follows, we construct a rational equivalence class

$$[\overline{M}(W, \tau)]^{\text{virt}} \in A_{\dim(W,\tau)}(\overline{M}(W, \tau))$$

called the *virtual fundamental class* of $\overline{M}(W, \tau)$. Here A stands for the Chow group (with rational coefficients) of a separated Deligne–Mumford stack, constructed by Vistoli [13]. This class has degree

$$\dim(W, \tau) = \chi(\tau)(\dim W - 3) - \beta(\tau)(\omega_W) + \# S_\tau - \# E_\tau,$$

which is the 'expected dimension' of the moduli stack $\overline{M}(W, \tau)$. If $\overline{M}(W, \tau)$ happens to be of dimension $\dim(W, \tau)$, then $[\overline{M}(W, \tau)]^{\text{virt}} = [\overline{M}(W, \tau)]$ is just the usual fundamental class.

Then the Gromov–Witten invariant

$$I_\tau(\beta) \colon H^*(W)^{\otimes S_\tau} \longrightarrow H^*(\overline{M}_\tau)$$
$$\gamma \longmapsto I_\tau(\beta)(\gamma)$$

is defined by

$$I_\tau(\beta)(\gamma) \cap [\overline{M}_\tau] = \text{stab}_*(\text{ev}^*(\gamma) \cap [\overline{M}(W, \tau)]^{\text{virt}}). \qquad (3)$$

Note that this condition defines $I_\tau(\beta)(\gamma)$ uniquely, because of Poincaré duality on the smooth stack $\overline{M}_\tau$.

Alternatively, consider the morphism

$$\pi \colon \overline{M}(W, \tau) \longrightarrow \overline{M}_\tau \times W^{S_\tau}$$

(induced by Diagram 2) which is proper and so we may consider the pushforward

$$\pi_*[\overline{M}(W, \tau)]^{\text{virt}},$$

which is a correspondence $\overline{M}_\tau \rightsquigarrow W^{S_\tau}$, so we get $I_\tau(\beta)$ through pullback via this correspondence:

$$I_\tau(\beta)(\gamma) = p_{1*}(p_2^*(\gamma) \cup \text{cl}\,\pi_*[\overline{M}(W, \tau)]^{\text{virt}}).$$

Hence this construction implies the 'motivic axiom' of [11].

Now all the axioms required of $I_\tau(\beta)$ reduce to axioms for

$$J(W,\tau) := [\overline{M}(W,\tau)]^{\text{virt}}.$$

Before we list these, some more remarks.

As in the above discussion of the geometric interpretation of Gromov–Witten invariants, consider the situation where $S = \{1,\ldots,n\}$, and $\Gamma_1,\ldots,\Gamma_n$ are dual cycles to the cohomology classes $\gamma_1,\ldots,\gamma_n \in H^*(W)$. For ease of exposition, let us assume that the Γ_i are actually algebraic subvarieties of W. We can now give a more precise definition of the moduli space T mentioned above. It is defined to be the fibered product

$$\begin{array}{ccc} T & \longrightarrow & \Gamma_1 \times \cdots \times \Gamma_n \\ \downarrow & & \downarrow \\ \overline{M}(W,\tau) & \xrightarrow{\text{ev}} & W \times \cdots \times W \end{array}$$

This will in fact assure that T is proper, and thus we have solved the problem of compactifying the earlier T.

Now, if the Γ_i are in general position, then T should be smooth of the expected dimension, which is

$$\dim(W,\tau) - \sum_{i=1}^{n} \text{codim}_W \Gamma_i.$$

In fact, one could use this principle as a *definition* of general position, *defining* $\Gamma_1,\ldots,\Gamma_n$ to be in general position if T is smooth[10] of this dimension. As mentioned above, the problem is that one cannot always find Γ_i in general position.

But let us assume that $\Gamma_1,\ldots,\Gamma_n$ are in general position. Then

$$[T] = \text{ev}^*[\Gamma_1 \times \cdots \times \Gamma_n]$$

and

$$I_\tau^\vee(\beta)([\Gamma_1],\ldots,[\Gamma_n]) = \text{stab}_*[T],$$

because the virtual fundamental class agrees with the usual one in this case.

So defining Gromov–Witten invariants by (3) leads to the situation anticipated by our geometric interpretation detailed above. In particular, for the case that $\overline{M}(W,\tau)$ is of dimension $\dim(W,\tau)$ the above heuristic arguments explaining the motivations of the various axioms give proofs of the axioms.

---

[10]In fact, purely of the expected dimension would be enough, if one is willing to count components of T with multiplicities given by the scheme (or stack) structure. The difference is the same as that between transversal and proper intersection in intersection theory.

Remark 3.3 If one prefers cohomology, in the case that $\overline{M}(W,\tau)$ of smooth of the expected dimension $\dim(W,\tau)$, one can also think of $I_\tau(\beta)(\gamma_1,\ldots,\gamma_n)$ as obtained by pulling back $\gamma_1,\ldots,\gamma_n$ by the evaluation maps, taking the cup product of these pullbacks and then integrating over the fibers of the morphism stab: $\overline{M}(W,\tau) \to \overline{M}_\tau$. Thus $I_\tau(\beta)$ should lower the grading by twice the dimension of the fibers of this map, which is

$$\dim(W,\tau) - \dim \overline{M}_\tau = \chi(\tau)\dim W - \beta(\tau)(\omega_W).$$

This gives another interpretation of the grading axiom.

3.3   Axioms for $J(W,\tau)$

We now list the properties that the

$$J(W,\tau) = [\overline{M}(W,\tau)]^{\overline{\mathrm{virt}}}$$

must satisfy in order that the induced Gromov–Witten invariants $I_\tau(\beta)$ should satisfy their respective properties. This amounts to five axioms for $J(W,\tau)$, and we refer to them by the names given in [4].

I. Mapping to point

Assume that $|\tau|$ is nonempty connected and that $\beta(\tau) = 0$, so that in fact $\beta(v) = 0$ for all $v \in V_\tau$.

In this situation the universal map $f\colon C \to W$ factors through the structure map $\pi\colon C \to \overline{M}(W,\tau)$ of the universal curve, since a map of class zero maps to a single point in W. We call the resulting map ev: $\overline{M}(W,\tau) \to W$, since it is also equal to all the evaluation maps. Now the morphism

$$\overline{M}(W,\tau) \overset{\mathrm{stab}\times\mathrm{ev}}{\longrightarrow} \overline{M}_\tau \times W$$

is an isomorphism, since giving a stable map to a point in W is the same as giving a stable curve and a point in W.

The axiom is that

$$J(W,\tau) = c_{g(\tau)\dim W}(R^1\pi_*\mathcal{O}_{\widetilde{C}} \boxtimes T_W) \cap [\overline{M}(W,\tau)].$$

For future reference, let us give an alternative description of $R^1\pi_*\mathcal{O}_{\widetilde{C}} \boxtimes T_W$. Consider the vector bundle

$$\begin{aligned}
R^1\pi_*f^*T_W &= R^1\pi_*(\pi^*\,\mathrm{ev}^*\,T_W) \\
&= R^1\pi_*\mathcal{O}_C \otimes \mathrm{ev}^*\,T_W.
\end{aligned}$$

Note that we have a Cartesian diagram

$$\begin{array}{ccc} C & \xrightarrow{\pi} & \overline{M}(W,\tau) \\ \downarrow & & \downarrow \text{stab} \\ \widetilde{C} & \xrightarrow{\pi} & \overline{M}_\tau \end{array}$$

since in the case of mapping to a point, there is no need to stabilize and thus the pullback of the universal curve over the moduli space of curves is the universal curve over the moduli space of stable maps to a point. Thus we can write the above tensor product as an exterior tensor product:

$$R^1\pi_* f^* T_W = R^1\pi_* \mathcal{O}_{\widetilde{C}} \boxtimes T_W.$$

Note that rank $R^1\pi_* f^* T_W = g(C)\dim W = g(\tau)\dim W$.

II. Products

Let σ and τ be stable modular graphs with $H_2(W)^+$ marking. Recall that we have

$$\overline{M}(W, \tau \times \sigma) = \overline{M}(W,\tau) \times \overline{M}(W,\sigma).$$

Our axiom is that

$$J(W, \tau \times \sigma) = J(W,\tau) \times J(W,\sigma).$$

III. Glueing tails

Let τ be obtained from σ by cutting an edge. Recall that then we have a Cartesian diagram

$$\begin{array}{ccc} \overline{M}(W,\sigma) & \longrightarrow & W \\ \downarrow & & \downarrow \Delta \\ \overline{M}(W,\tau) & \longrightarrow & W \times W \end{array}$$

Since W is smooth, Δ is a regular immersion, and so the Gysin pullback $\Delta^!\colon A_*(\overline{M}(W,\tau)) \to A_*(\overline{M}(W,\sigma))$ exists. (See [6], Section 6.2 for Gysin pullbacks in the context of schemes, [13] in the context of stacks.) The axiom is that

$$\Delta^! J(W,\tau) = J(W,\sigma).$$

IV. Forgetting tails

Let σ and τ be as in the fundamental class axiom for Gromov–Witten invariants. Endow both σ and τ with the same $H_2(W)^+$ marking β. Then we get an induced morphism of moduli of stable maps

$$\Phi\colon \overline{M}(W,\sigma) \longrightarrow \overline{M}(W,\tau).$$

To construct it, it suffices to consider the case of stars. So let σ be a star with set of tails $F_\sigma = S = \{0,\dots\}$ and let τ have set of tails $F_\tau = S' = \{\dots\}$. Then Φ is defined by

$$\Phi\colon \overline{M}_{g,S}(W,\beta) \longrightarrow \overline{M}_{g,S'}(W,\beta)$$
$$(C,x_0,(x_i),f) \longmapsto (C,(x_i),f)^{\mathrm{stab}}.$$

The construction of the stabilization is as before. One chooses a very ample invertible sheaf M on W. Then stabilization replaces the curve $\pi\colon C \to T$ by

$$C' = \mathrm{Proj}\bigoplus_{k\geq 0}\pi_*(L^{\otimes k}),$$

where $L = \omega_{C/T}(\sum x_i)\otimes f^* M^{\otimes 3}$. As before, this amounts to contracting or blowing down any rational components that become unstable on leaving out the section x_0.

Now a slightly nontrivial fact is that

$$\Phi\colon \overline{M}_{g,S}(W,\beta) \longrightarrow \overline{M}_{g,S'}(W,\beta)$$

is isomorphic to the universal curve over $\overline{M}_{g,S'}(W,\beta)$ (see [4], Corollary 4.6).

In the case of general graphs σ and τ this translates into the fact that the morphism x_0 in the diagram

$$\begin{array}{ccc} \overline{M}(W,\sigma) & \xrightarrow{\;x_0\;} & C_v \\ {\scriptstyle \Phi}\searrow & & \downarrow \\ & \overline{M}(W,\tau) & \end{array}$$

is an isomorphism. Here C_v is the universal curve over $\overline{M}(W,\tau)$ corresponding to the vertex $v = \partial(0)$.

In particular, the morphism Φ is flat of constant fiber dimension 1. Therefore the flat pullback homomorphism

$$\Phi^!\colon A_*(\overline{M}(W,\tau)) \longrightarrow A_*(\overline{M}(W,\sigma))$$

exists. Our axiom is that

$$\Phi^! J(W,\tau) = J(W,\sigma).$$

V. Isogenies

This axiom is really four axioms in one. The name of the axiom comes from the fact that it deals with those operations on a graph that do not affect its genus.

So let σ be a stable modular graph and let τ be obtained from σ by contracting an edge or omitting a tail. Assume that τ is also stable. Then choose an $H_2(W)^+$ structure on τ. In each case, we construct a commutative diagram

$$
\begin{array}{ccc}
\coprod \overline{M}(W,\sigma) & \longrightarrow & \overline{M}(W,\tau) \\
\downarrow & & \downarrow \text{stab} \\
\overline{M}_\sigma & \stackrel{\Phi}{\longrightarrow} & \overline{M}_\tau
\end{array}
\qquad (4)
$$

where the disjoint sum is taken over certain $H_2(W)^+$ structures on σ.

Case I   This is the case where we contract a loop in σ to obtain τ. Here there is only one possible $H_2(W)^+$ structure on σ compatible with that on τ. So $\coprod \overline{M}(W,\sigma) = \overline{M}(W,\sigma)$ and the two horizontal maps in (4) are obtained by glueing two marked points (or sections), as described above. The two vertical maps are given by forgetting the map part of a triple and then stabilizing.

Case II   Here we contract a nonlooping edge of σ, i.e., an edge with two vertices. Let v be the edge of τ onto which this edge is being contracted and v_1, v_2 the two vertices of this edge in σ. For an ordered pair $\beta_1, \beta_2 \in H_2(W)^+$ such that $\beta_1 + \beta_2 = \beta(v)$, define a marking on σ by setting $\beta(v_1) = \beta_1$, $\beta(v_2) = \beta_2$ and for the other vertices of σ take the marking induced from τ. Then take the disjoint union over all such pairs (β_1, β_2) of the associated stack of stable maps $\overline{M}(W,\sigma)$. This is $\coprod \overline{M}(W,\sigma)$. The maps in (4) are now defined the same way as in Case I.

Case III   This is the case where τ is obtained from σ by forgetting a tail. The $H_2(W)^+$ structure on σ is induced in a unique way from τ, $\coprod \overline{M}(W,\sigma) = \overline{M}(W,\sigma)$ and all the maps in (4) have been explained already.

Case IV   In this case τ is obtained from σ by 'relabelling'. In other words there is given an isomorphism between σ and τ. The $H_2(W)^+$ structure on σ is induced via this isomorphism from τ and $\coprod \overline{M}(W,\sigma) = \overline{M}(W,\sigma)$. Moreover, the horizontal maps in (4) are isomorphisms.

Now in each case the commutative diagram (4) induces a morphism

$$
h \colon \coprod \overline{M}(W,\sigma) \longrightarrow \overline{M}_\sigma \times_{\overline{M}_\tau} \overline{M}(W,\tau)
$$

and our axiom states that

$$h_*\left(\sum J(W,\sigma)\right) = \Phi^! J(W,\tau).$$

Note that Φ is a local complete intersection morphism, since both $\overline{M}_\sigma$ and $\overline{M}_\tau$ are smooth. Therefore the Gysin pullback

$$\Phi^! \colon A_*(\overline{M}(W,\tau)) \longrightarrow A_*(\overline{M}_\sigma \times_{\overline{M}_\tau} \overline{M}(W,\tau))$$

exists. Since all the stacks involved are complete, the morphism h is proper and so the proper pushforward exists.

Proposition 3.4 *The five axioms for $J(W,\tau)$ imply the eight axioms for $I_\tau(\beta)$.*

Proof The grading axiom follows from the fact that the $J(W,\tau)$ have the correct degree. The product, glueing tails and mapping to point axiom for I follow from the axioms for J with the same name. The forgetting tails axiom for J implies the divisor axiom for I. Finally, the isomorphisms, contractions and fundamental class axioms for I all follow from the isogenies axiom (with the same proof). $\square$

Remark The part of the isogenies axiom dealing with omitting tails (Case III) follows from the forgetting tails axiom (as can be seen, for example, by examining the proof of the isogenies axiom in [2]). So technically, the axioms for the virtual fundamental classes $J(W,\tau)$ contain some redundancy.

The reason why the isogenies axiom is stated in this slightly redundant form is that in this formulation it characterizes an aspect of the operad nature of $J(W,\cdot)$. (By this, we mean its description as a natural transformation between functors from a category of graphs to a category of vector spaces.) The forgetting tails axiom does not feature in the operad picture, but it is still needed for the divisor axiom (which does not fit naturally into the operad framework).

3.4   The unobstructed case

In the unobstructed case there is no need for a virtual fundamental class. The usual fundamental class of the moduli stack will do the job.

Definition 3.5 We say that a stable map $f \colon (C,x) \to W$ is *trivially unobstructed*, if $H^1(C, f^*T_W) = 0$.

Definition 3.6 A smooth and projective variety W is *convex* if for every morphism $f\colon \mathbb{P}^1 \to W$ we have $H^1(\mathbb{P}^1, f^*T_W) = 0$.

Examples of convex varieties are projective spaces $\mathbb{P}^n$, generalized flag varieties G/P (were G is a reductive algebraic group and P a parabolic subgroup), and in fact any variety for which the tangent bundle is generated by global sections.

The following proposition is not difficult to prove.

Proposition 3.7 *If W is convex, then all stable maps of genus 0 to W are trivially unobstructed.*

Because of this, the 'tree level' system of Gromov–Witten invariants for convex varieties may be constructed without recourse to virtual fundamental classes. By the tree level system we mean all the invariants $I_\tau(\beta)$, where the graph τ is a forest.

Theorem 3.8 *Let W be convex. Then for every forest τ the stack $\overline{M}(W, \tau)$ is smooth of dimension*

$$\dim(W, \tau) = \chi(\tau)(\dim W - 3) - \beta(\tau)(\omega_W) + \# S_\tau - \# E_\tau.$$

Moreover, the system of fundamental classes (where τ runs over all stable forests with $H_2(W)^+$ marking)

$$J(W, \tau) = [\overline{M}(W, \tau)]$$

satisfies the above five axioms.[11]

Proof Details of the proof can be found in [4]. Essentially, what is going on is that the definition of trivially unobstructed is of course precisely the condition needed to assure that the obstructions vanish,[12] which implies that the moduli stack is smooth. (More about obstruction theory in a later section.) The first four axioms for J follow from basic properties of Chern classes and Gysin pullbacks. For the last axiom one has to note also that h is an isomorphism generically. $\square$

It is explained in [11] and [7] how to construct the quantum cohomology algebra of W from the tree level system of Gromov–Witten invariants.

If one wants to count rational curves through a number of points in general position on a convex variety, then the cycles $\Gamma_1, \ldots, \Gamma_n$ are all just points, and

---

[11]Of course only those instances of the axioms for which all graphs involved are forests.

[12]The obstructions may also vanish if $H^1(C, f^*T_W) \neq 0$, but $H^1(C, f^*T_W) = 0$ is the only 'general' condition that always assures vanishing of the obstructions.

it follows from generic smoothness, that (at least in characteristic zero) the points can be put into general position. Therefore these special Gromov–Witten invariants actually solve numerical geometry problems (i.e., they are *enumerative*).

For the case of generalized flag varieties G/P all cohomology is algebraic and so all Gromov–Witten invariants can be defined in terms of algebraic cycles $\Gamma_1, \ldots, \Gamma_n$. Using results of Kleiman one can then prove that it is possible to move $\Gamma_1, \ldots, \Gamma_n$ into general position. Therefore these tree level Gromov–Witten invariants are enumerative. For more details see [7].

Let us now give a few examples of stable maps that are not trivially unobstructed.

1. Consider stable maps to $W = \mathbb{P}^r$. If $f: (C, x) \to \mathbb{P}^r$ is such a map, then we may pull back the exact sequence

$$0 \longrightarrow \mathcal{O} \longrightarrow \mathcal{O}(1)^{r+1} \longrightarrow T_{\mathbb{P}^r} \longrightarrow 0$$

to C to get the surjection

$$f^*\mathcal{O}(1)^{r+1} \longrightarrow f^*T_{\mathbb{P}^r} \longrightarrow 0$$

and the surjection

$$H^1(C, f^*\mathcal{O}(1))^{r+1} \longrightarrow H^1(C, f^*T_{\mathbb{P}^r}) \longrightarrow 0.$$

So if C is irreducible and $\deg f = \deg f^*\mathcal{O}(1) > 2g(C) - 2$, then f is trivially unobstructed.

Thus the 'good' elements of $\overline{M}_{g,n}(\mathbb{P}^r, d)$ (i.e., those corresponding to irreducible C) are trivially unobstructed for sufficiently high degree $d > 2g - 2$. If we knew that $\overline{M}_{g,n}(\mathbb{P}^r, d)$ is irreducible, then its generic element would be trivially unobstructed, and the virtual fundamental class would be equal to the usual one. But whether $\overline{M}_{g,n}(\mathbb{P}^r, d)$ is irreducible is far from clear. Anyway, the Gromov–Witten axioms involve the boundary of $\overline{M}_{g,n}(\mathbb{P}^r, d)$ in an essential way, and so this unobstructedness result is not of much help.

2. For $g > 0$ already the constant maps are not trivially unobstructed. As we already saw in the two mapping to point axioms, the moduli stack $\overline{M}_{g,n}(W, 0)$ has higher dimension than expected. On the other hand, $\overline{M}_{g,n}(W, 0) = \overline{M}_{g,n} \times W$ is smooth, so there are no obstructions. Constant maps are unobstructed, but not trivially so.

The fact that $\overline{M}_{g,n}(W, 0)$ has higher dimension than expected, leads to boundary components of $\overline{M}_{g,n}(W, \beta)$ with $\beta \neq 0$ having higher dimension than expected. For example, consider $W = \mathbb{P}^r$ and the graph τ with two vertices v_0 and v_1, one edge connecting v_0 and v_1, and $g(v_0) = 0$, $g(v_1) = g$. Let $d(v_0) = d \neq 0$ and $d(v_1) = 0$. Then

$$\overline{M}(\mathbb{P}^r, \tau) = \overline{M}_{0,1}(\mathbb{P}^r, d) \times \overline{M}_{g,1}$$

and so

$$\dim \overline{M}(\mathbb{P}^r, \tau) = r + d(r+1) + 3g - 4$$

whereas the expected dimension is $r + d(r+1) + (3-r)g - 4$. The stack $\overline{M}(\mathbb{P}^r, \tau)$ is a boundary component in $\overline{M}_{g,0}(\mathbb{P}^r, d)$, whose 'good' component attains the expected dimension $r+d(r+1)+(3-r)g-3$ in the range $d > 2g-2$. So if $d > 2g - 2$ and $rg > 1$ this boundary component has larger dimension than the 'good' component.

3. Let W be a surface and $E \subset W$ a rational curve with negative self-intersection $E^2 = -n$. Let $f \colon \mathbb{P}^1 \to E \subset W$ be a morphism of degree $d \neq 0$. Pulling back the sequence

$$0 \longrightarrow T_E \longrightarrow T_W \longrightarrow N_{E/W} \longrightarrow 0$$

via f, we get the sequence

$$0 \longrightarrow f^*T_E \longrightarrow f^*T_W \longrightarrow f^*N_{E/W} \longrightarrow 0. \tag{5}$$

Now $\deg(N_{E/W}) = E^2 = -n$ and so $f^*N_{E/W} = \mathcal{O}(-dn)$. Moreover, $T_{\mathbb{P}^1} = \mathcal{O}(2)$ and so $f^*T_E = \mathcal{O}(2d)$. Therefore, we get from the long exact cohomology sequence associated to (5) that

$$
\begin{aligned}
\dim H^1(\mathbb{P}^1, f^*T_W) &= \dim H^1(\mathbb{P}^1, \mathcal{O}(-dn)) \\
&= \dim H^0(\mathbb{P}^1, \mathcal{O}(dn-2)) \\
&= dn - 1.
\end{aligned}
$$

So if $d > 1$ or $n > 1$, then f is not trivially unobstructed. Since the 'boundary' of the moduli space usually contains maps of degree larger than 1, we must deal with maps that are not trivially unobstructed as soon as the surface W has -1 curves.

4   Virtual fundamental classes

4.1   Construction of $J(W, \tau)$, an overview

We will give an overview of how to construct the virtual fundamental classes $J(W, \tau)$. Many points are discussed in greater detail in the following sections.

This is where Artin stacks appear for the first time, and so we have to stop pretending that stacks are just spaces. Otherwise many facts would seem counterintuitive. But the only Artin stacks involved are of a particularly simple type, namely quotient stacks associated to the action of a vector bundle on a scheme of cones.

As usual, let τ be a stable modular graph with an $H_2(W)^+$ marking on the vertices. Consider the universal stable map of type τ

$$C \xrightarrow{f} W$$
$$\pi \downarrow$$
$$\overline{M}(W, \tau)$$

From this diagram we get the complex $R\pi_* f^* T_W$, which is an object of $D(\overline{M}(W, \tau))$, the derived category of $\mathcal{O}$-modules on $\overline{M}(W, \tau)$. In fact we may realize $R\pi_* f^* T_W$ as a two-term complex $[E_0 \to E_1]$ of vector bundles on $\overline{M}(W, \tau)$. Then we have $\ker(E_0 \to E_1) = \pi_* f^* T_W$ and $\operatorname{coker}(E_0 \to E_1) = R^1 \pi_* f^* T_W$.

In this context, the basic facts of obstruction theory are that for a morphism $f \colon C \to W$, the vector space $H^0(C, f^* T_W)$ classifies the infinitesimal deformations of f and $H^1(f^* T_W)$ contains the obstructions to deformations of f. Thus the complex $R\pi_* f^* T_W$ is intimately related to the obstruction theory of $\overline{M}(W, \tau)$.

To $R\pi_* f^* T_W$ we get an associated *vector bundle stack* $\mathfrak{E}$, which is simply given by the stack quotient $\mathfrak{E} = [E_1/E_0]$ (but is an invariant of the isomorphism class of $R\pi_* f^* T_W$ in the derived category).

The next ingredient is the *intrinsic normal cone*. If X is any scheme (or algebraic space, or Deligne–Mumford stack), it has an associated intrinsic normal cone, which, as the name indicates, is an intrinsic invariant of X, but is constructed from the normal cones coming from various local embeddings of X. The intrinsic normal cone is denoted $\mathfrak{C}_X$ and it is a *cone stack*, i.e., a stack that is étale locally over X of the form $[C/E]$, where $E \to C$ is a vector bundle over X operating on a cone over X. (As above, $[C/E]$ denotes the associated stack quotient.)

The intrinsic normal cone $\mathfrak{C}_X$ is constructed as follows. We choose a local embedding $i \colon X \hookrightarrow M$, where M is smooth. Then we get an action of the vector bundle $i^* T_M$ on the normal cone $C_{X/M}$ and the essential observation is that the associated stack quotient $[C_{X/M}/i^* T_M]$ is independent of the choice of the local embedding $i \colon X \to M$. Thus the various $[C_{X/M}/i^* T_M]$ coming from local embeddings of X glue together to give the cone stack $\mathfrak{C}_X$ over X. A basic fact about $\mathfrak{C}_X$ is that it is always purely of dimension zero.

For our application we will use the *relative* intrinsic normal cone. This is an intrinsic invariant of a morphism $X \to Y$ and is denoted $\mathfrak{C}_{X/Y}$. It has the property that for every local embedding

$$X \xrightarrow{\ i\ } M$$
$$\searrow \quad \downarrow$$
$$Y$$

of X into a scheme which is smooth over Y, it is canonically isomorphic to

$$\mathfrak{C}_{X/Y} = [C_{X/M}/i^* T_{M/Y}].$$

In our case we consider the morphism $\overline{M}(W,\tau) \to \mathfrak{M}_\tau$, where $\mathfrak{M}_\tau$ has the same definition as $\overline{M}_\tau$, except that the stability requirement is waived. So $\overline{M}_\tau$ is an open substack of $\mathfrak{M}_\tau$, and $\mathfrak{M}_\tau$ is not of finite type, or Deligne–Mumford, or even separated, but still smooth. The map $\overline{M}(W,\tau) \to \mathfrak{M}_\tau$ is given by forgetting f in a triple (C, x, f), but not stabilizing. We define

$$\mathfrak{C} = \mathfrak{C}_{\overline{M}(W,\tau)/\mathfrak{M}_\tau}.$$

Finally, we remark that there is a natural closed immersion of the cone stack $\mathfrak{C}$ into the vector bundle stack $\mathfrak{E}$ over $\overline{M}(W,\tau)$. This is because $R\pi_* f^* T_W$ is what is called a *(relative) obstruction theory* for $\overline{M}(W,\tau)$ over $\mathfrak{M}_\tau$.

We now consider the pullback diagram

$$\overline{M}(W,\tau) \longrightarrow \mathfrak{C}$$
$$\downarrow \qquad\qquad \downarrow$$
$$\overline{M}(W,\tau) \xrightarrow{\ 0\ } \mathfrak{E}$$

where 0 is the zero section of the vector bundle stack $\mathfrak{E}$. We obtain the virtual fundamental class as the intersection of the cone stack $\mathfrak{C}$ with the zero section of $\mathfrak{E}$.

$$J(W,\tau) = [\overline{M}(W,\tau)]^{\text{virt}} = 0^![\mathfrak{C}].$$

We should point out, though, that lacking an intersection theory for Artin stacks, we cannot apply this construction directly. Therefore we choose as above a two-term complex of vector bundles $[E_0 \to E_1]$ representing $R\pi_* f^* T_W$. Then $\mathfrak{C} \subset \mathfrak{E}$ induces a cone $C \subset E_1$ and we define

$$0^!_{\mathfrak{E}}[\mathfrak{C}] = 0^!_{E_1}[C].$$

4.2   Cones and cone stacks

We explain the basics of the theory of cones and cone stacks. For proofs see [3]. Let X be a Deligne–Mumford stack (or algebraic space, or scheme) over k, where k is our ground field. Later, X will be our moduli stack.

Cones

Let us recall the definition of a cone over X.

Consider a graded quasicoherent sheaf of $\mathcal{O}_X$-algebras

$$S = \bigoplus_{i \geq 0} S^i,$$

such that $S^0 = \mathcal{O}_X$, S^1 is coherent and S is locally generated by S^1. Then the affine X-scheme[13] $C = \operatorname{Spec} S$ is a *cone* over X.

The augmentation $S \to S^0$ defines a section $0 \colon X \to C$, the *vertex* of the cone $C \to X$. The morphism of $\mathcal{O}_X$-algebras $S \to S[x]$ mapping a homogeneous element $s \in S^i$ of degree i to sx^i defines a morphism $\mathbb{A}^1 \times C \to C$, which we call the $\mathbb{A}^1$-*action* on C. It is an action in the sense that $(\lambda\mu) \cdot c = \lambda(\mu \cdot c)$, $1 \cdot c = c$ and $0 \cdot c = 0$. Another, longer, but more descriptive name for this map could be the 'multiplicative contraction onto the vertex'.

Example (Abelian cones) Let $\mathcal{F}$ be a coherent $\mathcal{O}_X$-module. Then we get an associated cone by

$$C(\mathcal{F}) = \operatorname{Spec} \operatorname{Sym} \mathcal{F}.$$

Note that for a k-scheme T we have $C(\mathcal{F})(T) = \operatorname{Hom}(\mathcal{F}_T, \mathcal{O}_T)$, so that $C(\mathcal{F})$ is a group scheme over X. A cone obtained in this way is called an *Abelian cone*.

If C is any cone, then $\operatorname{Sym} S^1 \to \bigoplus S^i$ defines a closed immersion $C \hookrightarrow C(S^1)$. We denote $C(S^1)$ by $A(C)$ and call it the *Abelian hull* of C. It contains C as a closed subcone and is the smallest Abelian cone with this property.

Example (Vector bundles) Let $E \to X$ be a vector bundle and $\mathcal{E}$ the corresponding $\mathcal{O}_X$-module of sections. Then $E \cong C(\mathcal{E}^\vee)$ is an Abelian cone. Note that a cone $C \to X$ is smooth if and only if it is a vector bundle.

Example (Normal cones) Let $i \colon X \to Y$ be a closed immersion (or more generally a local immersion), with ideal sheaf I. Then

$$C_{X/Y} = \operatorname{Spec}_X \left(\bigoplus_{n \geq 0} I^n / I^{n+1} \right)$$

is the *normal cone* of X in Y. Its Abelian hull,

$$N_{X/Y} = C(I/I^2)$$

is the *normal sheaf* of X in Y. Note that i is a regular immersion if and only if $C_{X/Y}$ is Abelian (i.e., $C_{X/Y} = N_{X/Y}$) which in turn is equivalent to $C_{X/Y}$ being a vector bundle.

---

[13]It should be noted that whenever we talk of X-schemes or schemes over a stack X, we actually mean stacks over X that are *relative* schemes over X.

Vector bundle cones

Now consider the following situation. Let E be a vector bundle and C a cone over X, and let $d\colon E \to C$ be a morphism of cones (i.e., an X-morphism that respects the vertices and the $\mathbb{A}^1$-actions). Passing to the Abelian hulls we get a morphism $E \to A(C)$ of cones over X, which is necessarily a homomorphism of group schemes over X, so that E acts on $A(C)$. If C is invariant under the action of E on $A(C)$, so that we get an induced action of E on C, then we say that C is an *E-cone*.

Example Let $i\colon X \to M$ be a closed immersion, where M is smooth (over k). Then $C_{X/M}$ is automatically an i^*T_M-cone.

We now come to a construction that may seem intimidating, if one is not familiar with the language of stacks. We will try to explain why it shouldn't be.

Whenever we have an E-cone C, we associate to it the stack quotient $[C/E]$. At this point it is not very important to know what $[C/E]$ *is*, only to understand the main property, in fact the defining property of $[C/E]$, namely that the diagram of stacks over X

$$
\begin{array}{ccc}
E \times C & \overset{\sigma}{\longrightarrow} & C \\
{\scriptstyle p}\downarrow & & \downarrow \\
C & \longrightarrow & [C/E]
\end{array}
\tag{6}
$$

is Cartesian and co-Cartesian.[14] Here σ and p are the action and projection, respectively.

Recall that for an action of a group (like E) on a space (like C), the quotient C/E is defined to be the object (if it exists) which makes the diagram (6) co-Cartesian, i.e., the pushout. (This applies to many categories, not just (schemes/X).) If it turns out that (6) is also Cartesian, then C/E is the best possible kind of quotient, since to say that (6) is Cartesian means that the quotient map $C \to C/E$ is a principal E-bundle (or torsor, in different terminology).

The construction of stacks like $[C/E]$ should be viewed as a purely formal process which supplies such ideal quotients if they do not exist. On a certain level, this is analogous to the construction of the rational numbers from the integers. If a certain division 'doesn't go', one formally adjoins a quotient.

Applying this viewpoint to our situation, where we are trying to divide cones by vector bundles, we may say that the division C/E 'goes' (or that

---

[14]Note that everything is happening *over* X; E is a relative group over X (so its fibers over X are groups), C is a relative cone over X (so its fibers over X are 'usual' cones), the action of E on C is relative to X and so in particular, the product $E \times C$ is a product over X.

E divides C) if there exists a cone such that when inserted for $[C/E]$ in (6) it makes (6) Cartesian and co-Cartesian. If E does not divide C, then we formally adjoin the quotient. Of course, one has to introduce an equivalence relation on these formal quotients. So if C is an E-cone and C' an E'-cone, and there exists a Cartesian diagram

$$
\begin{array}{ccc}
E' & \longrightarrow & C' \\
\downarrow & & \downarrow \\
E & \longrightarrow & C
\end{array}
\qquad (7)
$$

where $C' \to C$ is a smooth epimorphism, then we call the quotients $[C/E]$ and $[C'/E']$ isomorphic. This may be motivated by noting that if we have a diagram (7) then there exists a vector bundle F such that $C = C'/F$ and $E = E'/F$ and thus we should have

$$
[C/E] = \frac{[C'/F]}{[E'/F]} = [C'/E'].
$$

By this process one enlarges the category of cones over X, and obtains a category where quotients of cones by vector bundles always exist. The one convenience one has to give up in the process is that of having a *category* of objects. The stacks that we obtain in this way form a 2-category, where there are objects, morphisms, and isomorphisms of morphisms. But for the most part we ignore that effect, to keep things simple.

Of course one has to do some work to prove that one can still do geometry with these new objects $[C/E]$. If one does this, then the quotient map $C \to [C/E]$ turns out to be an honest principal E-bundle. So over each point x of X the fiber $[C/E]_x$ is the quotient $[C_x/E_x]$. This is a usual cone divided by a usual vector bundle, but the quotient map $C_x \to [C_x/E_x]$ is a principal E_x-bundle, which means that the fibers of $C_x \to [C_x/E_x]$ are all copies of E_x (but not canonically).

It also makes sense to speak of the dimension of $[C/E]$. Since the fibers of the morphism $C \to [C/E]$ are vector spaces of dimension rank E we have $\dim[C/E] = \dim C - \operatorname{rank} E$.

Two extreme cases might be worth pointing out: if $E = X$, then $[C/E] = C$. If $C = X$, then $[C/E] = [X/E]$ is the stack over X whose fiber over $x \in X$ is $[\{x\}/E_x]$, a point divided by a vector space. One also uses the notation $BE_x = [\mathrm{pt}/E_x]$ and $BE = [X/E]$. So in the naive picture of a stack as a collection of points with groups attached, BE has points $\{x \mid x \in X\}$ and groups $(E_x)_{x \in X}$. Note that $\dim BE = \dim X - \operatorname{rank} E$.[15]

---

[15]The appearance of negative dimensions for Artin stacks is completely analogous to the appearance of fractions when counting points of finite Deligne–Mumford stacks.

Cone stacks

We have to take one more step to get the category of cone stacks. We need to localize, meaning that we want to call objects cone stacks if they locally look like the $[C/E]$ just constructed. For the applications we have in mind, this step is not really necessary, since in the end all cone stacks we use will turn out to be of the form $[C/E]$. But for the general theory of the intrinsic normal cone it would be an awkward restriction to require cone stacks to be global quotients. So we make the following definition.

Definition 4.1 Let $\mathfrak{C} \to X$ be an algebraic stack with vertex $0\colon X \to \mathfrak{C}$ and $\mathbb{A}^1$-action $\gamma\colon \mathbb{A}^1 \times \mathfrak{C} \to \mathfrak{C}$. Then $\mathfrak{C}$ is a *cone stack* if, étale locally on X, there exists a vector bundle E over X and an E-cone C over X such that $\mathfrak{C} \cong [C/E]$ as stacks over X with $\mathbb{A}^1$-action and vertex.

Every such C is called a *local presentation* of $\mathfrak{C}$. If one can find local presentations C which are vector bundles (so that locally $\mathfrak{C} \cong [E_1/E_0]$, for a homomorphism of vector bundles $E_0 \to E_1$), then $\mathfrak{C}$ is called a *vector bundle stack*.

4.3   The intrinsic normal cone

As before, let X be a Deligne–Mumford stack over k.

A *local embedding* of X is a diagram

$$
\begin{array}{ccc}
U & \xrightarrow{\;f\;} & M \\
{\scriptstyle i}\downarrow & & \\
X & &
\end{array}
$$

where i is étale, f a closed immersion and M is smooth.

A *morphism of local embeddings* is a commutative diagram

$$
\begin{array}{ccc}
U' & \xrightarrow{\;f'\;} & M' \\
\downarrow & & \downarrow \\
U & \xrightarrow{\;f\;} & M
\end{array}
$$

where $U' \to U$ is an étale X-morphism and $M' \to M$ is smooth.

Given such a morphism of local embeddings we get a commutative diagram

$$
\begin{array}{ccccc}
f'^*T_{M'/M} & \longrightarrow & f'^*T_{M'} & \longrightarrow & f^*T_M|U' \\
\downarrow & & \downarrow & & \downarrow \\
f'^*T_{M'/M} & \longrightarrow & C_{U'/M'} & \longrightarrow & C_{U/M}|U'
\end{array}
$$

The rows are exact sequences of cones. The square on the right is Cartesian and $C_{U'/M'} \to C_{U/M}|U'$ is a smooth epimorphism. All these are basic properties of normal cones and tangent bundles.

As explained above, in this situation the quotients $[C_{U/M}/f^*T_M]|U'$ and $[C_{U'/M'}/f'^*T_{M'}]$ are canonically isomorphic. Thus all these locally defined cone stacks (one for each local embedding) glue together to give rise to a globally defined cone stack on X (note that X can be covered by étale $U \to X$ that are embeddable into smooth varieties). This cone stack is called *the intrinsic normal cone* of X and is denoted by $\mathfrak{C}_X$.

Proposition 4.2 *The stack $\mathfrak{C}_X$ is a cone stack of pure dimension[16] zero. For any local embedding we have*

$$\mathfrak{C}_X|U = [C_{U/M}/f^*T_M].$$

Proof It is a general property of normal cones that they always have the dimension of the ambient variety. So we have $\dim C_{U/M} = \dim M$ and therefore

$$\dim[C_{U/M}/f^*T_M] = \dim C_{U/M} - \operatorname{rank} f^*T_M = \dim M - \dim M = 0. \quad \square$$

Remark One can do the same construction with normal sheaves $N_{U/M}$ instead of normal cones $C_{U/M}$. Then one gets the *intrinsic normal sheaf* $\mathfrak{N}_X$ of X. Moreover, $\mathfrak{C}_X \subset \mathfrak{N}_X$ is a closed substack and $\mathfrak{N}_X$ is the Abelian hull of $\mathfrak{C}_X$ (this notion also makes sense for cone stacks).

Let L_X be the cotangent complex of X and $\tau_{\geq -1} L_X$ its truncation at -1. Again, this is nothing deep, over a local embedding it is simply given by the two term complex

$$(\tau_{\geq -1} L_X)|U = [I/I^2 \to f^*\Omega_M],$$

where I is the ideal sheaf, and the map is the map appearing in the second fundamental exact sequence of Kähler differentials.

One can prove that the stack $\mathfrak{N}_X$ only depends on the quasi-isomorphism class of $\tau_{\geq -1} L_X$, in other words it is an invariant of the object $\tau_{\geq -1} L_X \in \operatorname{ob} D_{\operatorname{coh}}^{[-1,0]}(\mathcal{O}_X)$. In fact, one can define for every $M^\bullet \in \operatorname{ob} D_{\operatorname{coh}}^{[-1,0]}(\mathcal{O}_X)$ an associated Abelian cone stack $\mathfrak{C}(M^\bullet)$. To do this, write (locally over X) $M^\bullet = [M^{-1} \to M^0]$ with M^0 free. Then pass to $C(M^0) \to C(M^{-1})$, the associated Abelian cones, and let $\mathfrak{C}(M^\bullet)$ be the stack quotient $\mathfrak{C}(M^\bullet) = [C(M^{-1})/C(M^0)]$. This construction globalizes and is functorial. Alternative notations are $\mathfrak{C}(M^\bullet) = h^1/h^0(M^{\bullet\vee})$, used in [3] or $\mathfrak{C}(M^\bullet) = \operatorname{ch}(M^{\bullet\vee})$, in [1], Exposé XVII.

---

[16] Absolute dimension over k, not dimension over X

The following are a few basic results on the intrinsic normal cone. None of them are deep or difficult to prove, they just reformulate known results about normal cones and tangent bundles.

Proposition 4.3 *The following are equivalent.*

1. *X is a local complete intersection,*

2. *$\mathfrak{C}_X$ is a vector bundle stack,*

3. *$\mathfrak{C}_X = \mathfrak{N}_X$.*

If X is smooth then $\mathfrak{C}_X = \mathfrak{N}_X = BT_X$.

Proposition 4.4 $\mathfrak{C}_{X \times Y} = \mathfrak{C}_X \times \mathfrak{C}_Y$ *(absolute product, over k).*

Proposition 4.5 *Let $f \colon X \to Y$ be a local complete intersection morphism. Then there is a short exact sequence of cone stacks*

$$\mathfrak{N}_{X/Y} \longrightarrow \mathfrak{C}_X \longrightarrow f^*\mathfrak{C}_Y.$$

Here $\mathfrak{N}_{X/Y} = \mathfrak{C}(L^{\bullet}_{X/Y})$, which is a vector bundle stack. The notion of short exact sequence of cone stacks is a straightforward generalization of the notion of short exact sequence of cones. What it means is that the cone stack on the right may be viewed as the quotient of the cone stack in the middle by the action of the vector bundle stack on the left.

For example, if f is smooth we have an exact sequence

$$BT_{X/Y} \longrightarrow \mathfrak{C}_X \longrightarrow f^*\mathfrak{C}_Y,$$

and if f is a regular immersion we have

$$N_{X/Y} \longrightarrow \mathfrak{C}_X \longrightarrow f^*\mathfrak{C}_Y.$$

4.4   The intrinsic normal cone and obstructions

We will now look at the 'fiber' of the intrinsic normal cone over a point of X. So let $p \colon \operatorname{Spec} k \to X$ be a geometric point of X (which just means that k is an algebraically closed field, not necessarily equal to the ground field, by abuse of notation). Pulling back the intrinsic normal cone $\mathfrak{C}_X$ via p, we get a cone stack over $\operatorname{Spec} k$.

If we look at cone stacks over an algebraically closed field, they are necessarily given as the quotient $[C/E]$ associated to an E-cone C, where E is just a vector space. In this case the quotient of C by the image of $d \colon E \to C$ exists, and choosing a complementary subspace for $\ker d$ in E, we get a Cartesian diagram

$$
\begin{array}{ccc}
E & \xrightarrow{\;d\;} & C \\
\downarrow & & \downarrow \\
\ker d & \xrightarrow{\;0\;} & C/\operatorname{im} d
\end{array}
$$

showing that, as cone stacks, $[C/E]$ is isomorphic to the the quotient of $C' = C/\operatorname{im} d$ by $\ker d$ acting trivially. So for studying this cone stack, we may as well replace $d\colon E \to C$ by $0\colon \ker d \to C'$ and assume from the start that E acts trivially on C, i.e., that the map $d\colon E \to C$ is the zero map. Then we have that $[C/E] \cong BE \times C$, where BE is the quotient of the point $\operatorname{Spec} k$ by the vector space E.

Considering such a cone stack $BE \times C$ over $\operatorname{Spec} k$, we may interpret the cone C as the 'coarse moduli space' of $BE \times C$. Any stack has a coarse moduli space associated to it; it *is* the set of isomorphism classes of whatever the objects are that the stack classifies. The vector space E is the common automorphism group of all the objects that the stack classifies.

Now let us determine what these objects and automorphisms are, for the case of $p^*\mathfrak{C}_X$. Before dealing with the intrinsic normal cone, though, let us consider the intrinsic normal sheaf. We have $p^*\mathfrak{N}_X = p^*\mathfrak{C}(\tau_{\geq -1}L_X) = \mathfrak{C}(p^*\tau_{\geq -1}L_X)$.

Recall the 'higher tangent spaces'

$$
T^i_{X,p} = \operatorname{Ext}^i(p^*L_X, k) = h^i(p^*L_X)^\vee
$$

of X at p. For example, $T^0_{X,p} = \operatorname{Hom}(\Omega_X, k)$ is the usual Zariski tangent space. It classifies first order deformations of p, i.e., (isomorphism classes of) diagrams

$$
\begin{array}{ccc}
\operatorname{Spec} k & \longrightarrow & \operatorname{Spec} k[\epsilon] \\
 & {}_{p}\searrow \quad & \downarrow p' \\
 & X &
\end{array}
$$

where $k[\epsilon]$ is the ring of dual numbers (meaning that $\epsilon^2 = 0$). The first higher tangent space $T^1_{X,p}$ is the *obstruction space*, and classifies obstructions.

Now

$$
p^*\tau_{\geq -1}L_X \cong [h^{-1}(p^*L_X) \xrightarrow{\;0\;} h^0(p^*L_X)]
$$

and so

$$
p^*\mathfrak{N}_X = BT^0_{X,p} \times T^1_{X,p}.
$$

Thus the intrinsic normal sheaf classifies obstructions, with deformations as automorphism group. Since the intrinsic normal cone is a closed substack of $\mathfrak{N}_X$, we get that

$$p^*\mathfrak{C}_X \cong BT^0_{X,p} \times C_{X,p},$$

where $C_{X,p} \subset T^1_{X,p}$ is some cone of obstructions.

To describe what kind of obstructions the intrinsic normal cone classifies, let us recall what an obstruction is. Let $A' \to A$ be an epimorphism of local Artinian k-algebras with kernel k (i.e., a *small extension*). Let $T = \operatorname{Spec} A$ and $T' = \operatorname{Spec} A'$ and assume given an extension x of p to A, i.e., a diagram

$$
\begin{array}{ccc}
\operatorname{Spec} k & \longrightarrow & T \\
& {}_{p}\searrow & \downarrow {}_{x} \\
& & X
\end{array}
$$

In this situation we get a canonical morphism $x^*L_X \to L_T$, by the contravariant nature of the cotangent complex. From the morphism $T \to T'$ we get a morphism $L_T \xrightarrow{+1} k$ of degree 1. It is essentially the morphism from L_T to $L_{T/T'}$. Composing, we get a morphism $x^*L_X \to k$ of degree 1, in other words an element of

$$\operatorname{Ext}^1(x^*L_X, k) \xrightarrow{\sim} \operatorname{Ext}^1(p^*L_X, k) = T^1_{X,p},$$

which is called the *obstruction* of $(A' \to A, x)$. The justification for this terminology is that it vanishes if and only if x extends to A', i.e., if and only if there exists $x' : T' \to X$ making the diagram

$$
\begin{array}{ccc}
T & \longrightarrow & T' \\
{}_{x}\searrow & & \downarrow {}_{x'} \\
& X &
\end{array}
$$

commute.

In more concrete terms the obstruction of $(A' \to A, x)$ can be described as follows. Choose a local embedding $f : U \to M$ of X at p, where U and M are affine. Let I be the corresponding sheaf of ideals, which we identify with an ideal in the affine coordinate ring of M. Then we have

$$T^1_{X,p} = \operatorname{coker}(p^*f^*T_M \to (p^*I/I^2)^\vee).$$

Now given $x : T \to X$ it is possible to choose $x'' : T' \to M$ such that

$$
\begin{array}{ccc}
T & \longrightarrow & T' \\
{}_{x}\downarrow & & \downarrow {}_{x''} \\
U & \xrightarrow{f} & M
\end{array}
$$

commutes, since M is smooth. This diagram of two closed immersions induces a morphism on the level of ideals, namely $I \to \ker(A' \to A) = k$. This element of $I^\vee$ induces the obstruction in $T^1_{X,p}$ (which is independent of the choice of f and M and x'').

The small extension $A' \to A$ is *curvilinear* if it is isomorphic to $k[t]/t^{s+1} \to k[t]/t^s$ for some $s \geq 1$. This notion gives the answer to our question of what are the obstructions classified by the intrinsic normal cone:

Proposition 4.6 *Every element of $T^1_{X,p}$ obstructs some small extension. It obstructs a small curvilinear extension if and only if it is in $C_{X,p} \subset T^1_{X,p}$.*

Proof See [3], Proposition 4.7. □

4.5 Obstruction theory

Let us start with an example. Consider the Cartesian diagram

$$
\begin{array}{ccc}
X & \xrightarrow{v} & V \\
\downarrow & & \downarrow f \\
\operatorname{Spec} k & \xrightarrow{w} & W
\end{array}
\qquad (8)
$$

where f is a morphism between smooth varieties. Thus X is a fiber of f. This is a typical intersection theory situation. One defines a cycle class on X by $[X]^{\text{virt}} = w^![V]$. This class is called the *specialization* of $[V]$ at w. The class $[X]^{\text{virt}}$ is first of all in the expected degree, namely $\dim V - \dim W$, even if X actually has larger dimension. Moreover, it leads to numerical data which is independent of the parameter $w \in W$. In the case that $\dim V = \dim W$ this means that the degree of the zero cycle $[X]^{\text{virt}}$ is independent of w. For an explanation of what it means if $\dim V > \dim W$, see [6], Chapter 10.

Because of this invariance of numerical data defined in terms of $[X]^{\text{virt}}$, this class is a sensible one to use for questions in enumerative geometry. Let us recall the construction of $w^![V]$ (or at least how the definition is reduced to the linear case). One replaces Diagram (8) by the following:

$$
\begin{array}{ccc}
X & \longrightarrow & C_{X/V} \\
\downarrow & & \downarrow \\
X & \xrightarrow{0} & T_W(w) \times X
\end{array}
\qquad (9)
$$

Here $C_{X/V}$ is the normal cone and the normal bundle of $w \colon \operatorname{Spec} k \to W$ pulled back to X is $T_W(w) \times X$, the tangent space to W at w times X. Then $[X]^{\text{virt}} = 0^![C_{X/V}]$.

Now we also have a Cartesian diagram

$$
\begin{array}{ccc}
X & \longrightarrow & [C_{X/V}/v^*T_V] \\
\downarrow & & \downarrow \\
X & \xrightarrow{\ 0\ } & [T_W(w)_X/v^*T_V]
\end{array}
\qquad\qquad (10)
$$

which is obtained from (9) simply by dividing through (in the stack sense) by v^*T_V. Now note that $[C_{X/V}/v^*T_V] = \mathfrak{C}_X$ is just the intrinsic normal cone of X and $\mathfrak{E} = [T_W(w)_X/v^*T_V]$ is a vector bundle stack on X into which $\mathfrak{C}_X$ is embedded.

If there was an intersection theory for Artin stacks (or just cone stacks), then certainly $0^![C_{X/V}] = 0^!_{\mathfrak{E}}[\mathfrak{C}_X]$, where the first 0 is the zero section from (9). So we can characterize the virtual fundamental class of X in terms of the intrinsic normal cone of X, which is completely intrinsic to X and the vector bundle stack $\mathfrak{E}$, which is, of course, not intrinsic to X, but has to do with the obstruction theory of X.

In fact, $\mathfrak{E} = v^*[f^*T_W/T_V]$, and because $[T_V \to f^*T_W] = T^\bullet_{V/W}$ is the tangent complex of f, we have that $\mathfrak{E} = v^*\mathfrak{C}(L_{V/W})$. So $\mathfrak{E}$ can be thought of as the linearization of f. Moreover, $h^0(v^*T_{V/W}) = T_X$ classifies the first order deformations of X, and $h^1(v^*T_{V/W})$ contains the obstructions to deforming X.

So we have replaced the ambient morphism f, which defined a virtual fundamental class on X, by this vector bundle stack $\mathfrak{E}$, which serves the same purpose. Now if X is a moduli space (or stack), then it might be hopeless to try to embed X globally into a smooth space (or stack) but such an $\mathfrak{E}$ can sometimes still be found; in fact, it comes naturally from the moduli problem that X solves.

The two essential properties of $\mathfrak{E}$ are

1. $\mathfrak{E}$ is a vector bundle stack, i.e., it is locally defined in term of a complex of two vector bundles $[E^{-1} \to E^0]$. Such a complex is referred to as being *perfect of amplitude contained in* $[-1, 0]$. In other words, it is an object of $D^{[-1,0]}_{\text{parf}}(\mathcal{O}_X)$.[17]

2. The intrinsic normal cone $\mathfrak{C}_X$ is embedded as a closed subcone stack into $\mathfrak{E}$.

This motivates the following definition.

---

[17]Note that the superscript $[-1, 0]$ does not refer to the object of the derived category having cohomology in the interval $[-1, 0]$, but to its perfect amplitude being in that interval. The latter is stronger than the former.

Definition 4.7 Let E be an object of $D_{\text{parf}}^{[-1,0]}(\mathcal{O}_X)$. A homomorphism $\varphi\colon E \to L_X$ (and by abuse of language also E itself) is called a *perfect obstruction theory* for X if

1. $h^0(\varphi)$ is an isomorphism,

2. $h^{-1}(\varphi)$ is surjective.

It is not difficult to prove that the two conditions on φ are equivalent to the morphism $\mathfrak{N}_X \to \mathfrak{C}$ (where $\mathfrak{C} = \mathfrak{C}(E)$) induced by φ being a closed immersion. Moreover, if $p\colon \operatorname{Spec} k \to X$ is a geometric point of X, then an obstruction theory induces an isomorphism

$$T_{X,p}^0 \xrightarrow{\sim} h^0(p^*E^\vee)$$

and a monomorphism

$$T_{X,p}^1 \hookrightarrow h^1(p^*E^\vee),$$

so, in a sense, E reflects the deformation theory of X and contains the obstructions of X.

As an example, let C be a pre-stable curve, W a smooth projective variety and $f\colon C \to W$ a morphism. Then $H^0(C, f^*T_W)$ classifies the infinitesimal deformations of f. The obstructions are contained in $H^1(C, f^*T_W)$. To see this, let U_α be an affine open cover of C. By the infinitesimal lifting property the morphism f can be extended over each U_α. Over the overlaps $U_{\alpha\beta}$ two extensions differ by an infinitesimal deformation, i.e., a section of f^*T_W over $U_{\alpha\beta}$. The vanishing of this Čech 1-cocycle with values in f^*T_W means extendability of f.

These observations can be translated into the following statement. Let $X = \operatorname{Mor}(C, W)$ be the scheme of morphisms from C to W. Then there is a perfect obstruction theory on X given by $(R\pi_*f^*T_W)^\vee \to L_X$, where

$$C \times X \xrightarrow{f} W$$
$$\pi \downarrow$$
$$X$$

is the universal map. Note that since $\pi\colon C \times X \to X$ has one dimensional fibers, the complex $(R\pi_*f^*T_W)^\vee$ is indeed perfect of amplitude 1.

This is in fact the obstruction theory we want to use to construct the virtual fundamental class on $\overline{M}(W,\tau)$. But a deformation of a stable map may deform the curve C as well as the map $f\colon C \to W$. So we note that the morphism $\overline{M}(W,\tau) \to \mathfrak{M}_\tau$ that forgets the map (and does not stabilize) has fibers of the form $\mathrm{Mor}(C, W)$. So we would like to adapt the above theory to this relative situation.

Working in the relative rather than the absolute setting has the advantage that the obstruction theory is much simpler. Also, many of the axioms we will have to check involve the relative setting of $\overline{M}(W, \tau)$ over $\overline{M}_\tau$. So the relative obstruction theory is better suited for proving the axioms of Gromov–Witten theory. (Note, however, the difference between $\overline{M}_\tau$ and $\mathfrak{M}_\tau$. It is the main difficulty in proving the axioms.)

The reason why the relative obstruction theory works, is that the base $\mathfrak{M}_\tau$ is smooth.

So we replace the base $\mathrm{Spec}\, k$ by Y, where Y is any smooth algebraic k-stack of constant dimension n. It does not even have to be of Deligne–Mumford type. Let $X \to Y$ be a morphism which makes X a *relative* Deligne–Mumford stack over Y. This just means that any base change to a base Y', where Y' is a scheme, makes the fibered product X' a Deligne–Mumford stack.

Embedding X locally into stacks that are smooth and relative schemes over Y, one defines just as in the absolute case the intrinsic normal cone $\mathfrak{C}_{X/Y}$ and its Abelian hull $\mathfrak{N}_{X/Y}$. A complex of $\mathcal{O}_X$-modules E that is locally quasi-isomorphic to a two term complex of vector bundles, together with a map in the derived category $E \to L_{X/Y}$ is called a perfect relative obstruction theory, if it induces a closed immersion of cone stacks $\mathfrak{C}_{X/Y} \to \mathfrak{C}(E)$.

It follows from [3], Proposition 2.7 that the relative intrinsic normal cone $\mathfrak{C}_{X/Y}$ is 'just' the quotient of the absolute intrinsic normal cone $\mathfrak{C}_X$ by the natural action of the tangent vector bundle stack $\mathfrak{T}_Y$ of Y. The same is true for the intrinsic normal sheaves. Moreover, in our application, the relationship between the vector bundle stacks given by the relative and absolute obstruction theories, respectively, is also the same. This implies that the virtual fundamental class defined in the relative setting is the same as that defined in the absolute setting.

Let us make more precise the sense in which a relative obstruction theory $E \to L_{X/Y}$ governs the obstructions of X over Y.

Let

$$
\begin{array}{ccc}
T & \xrightarrow{x} & X \\
\downarrow & & \downarrow \\
T' & \longrightarrow & Y
\end{array}
\tag{11}
$$

be a commutative diagram, where $T \to T'$ is a square zero extension of (affine) schemes, with ideal N (i.e., a closed immersion with ideal sheaf N such that $N^2 = 0$). Such a diagram induces an obstruction $o \in \mathrm{Ext}^1(x^*E, N)$, which vanishes if and only if a map $T' \to X$ completing Diagram (11) exists. Moreover, if the obstruction o vanishes, then all arrows $T' \to X$ completing (11) form a torsor under $\mathrm{Ext}^0(x^*E, N)$, i.e., there exists a natural action of $\mathrm{Ext}^0(x^*E, N)$ on the set of such arrows $T' \to X$, which is simply transitive.

The obstruction o is obtained as follows. A fundamental fact about the cotangent complex is that is classifies extensions of algebras, i.e., that

$$\mathrm{Extalg}_Y(T, N) = \mathrm{Ext}^1(L_{T/Y}, N).$$

Thus $T \to T'$ gives a morphism of degree one $L_{T/Y} \to N$. Composing with the natural maps $x^*L_{X/Y} \to L_{T/Y}$ and $x^*E \to x^*L_{X/Y}$ we get a morphism of degree one $x^*E \to N$, in other words an element $o \in Ext^1(x^*E, N)$.

Note In the case that $X \to Y$ is a morphism of smooth schemes, we can take the identity $L^\bullet_{X/Y} \to L^\bullet_{X/Y}$ as relative obstruction theory for X over Y. Pulling this relative obstruction theory back to a fiber of $X \to Y$, we get the absolute obstruction theory of the fiber described earlier.

4.6 Fundamental classes

Since the relative case is no more difficult than the absolute one, we assume from the start that we have a perfect relative obstruction theory E for X over Y. To define the associated virtual fundamental class we need to assume that E has *global resolutions*, i.e., that E is globally quasi-isomorphic to a two term complex $[E^{-1} \to E^0]$ of vector bundles over X. This condition is satisfied for the relative obstruction theory of $\overline{M}(W, \tau)$ (see [2], Proposition 5). Then the stack $\mathfrak{E} = \mathfrak{C}(E)$ associated to E is isomorphic to $[E_1/E_0]$, where E_i denotes the dual bundle of E^{-i}. Since $\mathfrak{C}_X$ is a closed subcone stack of $\mathfrak{E}$, it induces a closed subcone C of E_1 and we define

$$[X]^{\mathrm{virt}} = [X, E] = 0^!_{E_1}[C] \in A_{\dim Y + \mathrm{rank}\, E}(X),$$

which is a class in Vistoli's Chow group with rational coefficients. This class is independent of the global resolution chosen to define it. The fact that this class is in the expected degree $\dim Y + \mathrm{rank}\, E$ follows immediately from the fact that the relative intrinsic normal cone has pure dimension $\dim Y$ (which corresponds to the fact that the absolute intrinsic normal cone has pure dimension zero).

Example 4.8 If X is smooth over Y, then $h^0(E) \cong h^0(L_{X/Y}) \cong \Omega_{X/Y}$ is locally free. Hence $E^{-1} \to E^0$ has locally free h^0 and h^{-1}, and so the same holds for the dual $\varphi\colon E_0 \to E_1$. Also, $\mathfrak{C}_{X/Y} = BT_{X/Y} \hookrightarrow [E_1/E_0]$ identifies $BT_{X/Y}$ with $B\ker\varphi$, which is isomorphic to $[\mathrm{im}\,\varphi/E_0]$, and so the cone induced by $\mathfrak{C}_X$ in E_1 is equal to $\mathrm{im}\,\varphi$. Hence

$$\begin{aligned}[X]^{\mathrm{virt}} &= 0^!_{E_1}[\mathrm{im}\,\varphi] \\ &= c_{\mathrm{top}}(\mathrm{coker}\,\varphi) \cap [X] \\ &= c_{\mathrm{top}}(h^1(E^\vee)) \cap [X].\end{aligned}$$

In the above example of $\text{Mor}(C, W)$ we have

$$[X]^{\text{virt}} = c_{\text{top}}(R^1\pi_* T_W) \cap [X].$$

Proposition 4.9 *If X has the expected dimension $\dim Y + \text{rank}\, E$, then X is a local complete intersection and $[C, E] = [X]$, the usual fundamental class.*

Proof For simplicity, let us explain the absolute case $Y = \text{pt}$. Let k be algebraically closed and A a localization of a finite type k-algebra at a maximal ideal. Write $A = (k[x_1, \ldots, x_n]/(f_1, \ldots, f_r))_{(x_1, \ldots, x_n)}$, let $\mathfrak{m}$ be the maximal ideal of $k[x_1, \ldots, x_n]_{(x_1, \ldots, x_n)}$ and $I = (f_1, \ldots, f_r) \subset k[x_1, \ldots, x_n]_{(x_1, \ldots, x_n)}$. Then the truncation at -1 of the cotangent complex of A is $I/I^2 \to \Omega_{k[x_1, \ldots, x_n]} \otimes A$ and if we tensor it over A with k we get $I/\mathfrak{m}I \to \mathfrak{m}/\mathfrak{m}^2$ and so there is an exact sequence

$$0 \longrightarrow T^1(A)^\vee \longrightarrow I/\mathfrak{m}I \longrightarrow \mathfrak{m}/\mathfrak{m}^2 \longrightarrow T^0(A)^\vee \longrightarrow 0,$$

where $T^i(A)$ is the i-th tangent space of A at the maximal ideal.

After projecting $\text{Spec}\, A$ into its tangent space at the origin, which only changes A by an étale map, we may assume that $I \subset \mathfrak{m}^2$. This entails that $I/\mathfrak{m}I \to \mathfrak{m}/\mathfrak{m}^2$ is the zero map and hence $T^0(A)^\vee = \mathfrak{m}/\mathfrak{m}^2$ and $T^1(A)^\vee = I/\mathfrak{m}I$. Clearly, $\bar{x}_1, \ldots, \bar{x}_n$ is a basis of $\mathfrak{m}/\mathfrak{m}^2$. By Nakayama's lemma we may also assume that $\bar{f}_1, \ldots, \bar{f}_r$ form a basis of $I/\mathfrak{m}I$. Hence $n = \dim T^0(A)$ and $r = T^1(A)$.

Now clearly, $\dim A \geq n - r$. If equality holds, then $f_1, \ldots, f_r$ is a regular sequence for $k[x_1, \ldots, x_r]_{(x_1, \ldots, x_r)}$ and so A is Cohen–Macaulay and a local complete intersection.

Now assume given a perfect obstruction theory $E^\bullet$ for A. Then $T^0(A) = h^0(E^{\bullet\vee} \otimes k)$ and $T^1(A) \hookrightarrow h^1(E^{\bullet\vee} \otimes k)$. Hence $n - r \geq \text{rank}\, E^\bullet$ and so $\dim A \geq \text{rank}\, E^\bullet$. By the previous argument $\dim A = \text{rank}\, E^\bullet$ implies that A is a local complete intersection. Moreover, $\dim A = \text{rank}\, E^\bullet$ implies $n - r = \text{rank}\, E^\bullet$ and $T^1(A) = h^1(E^{\bullet\vee} \otimes k)$. This, in turn, implies that $E^\bullet \to L_A^\bullet$ is an isomorphism and so $[X, E^\bullet] = [X]$. $\square$

Corollary 4.10 *If the expected and actual dimension are both zero, then $[X]^{\text{virt}}$ counts the number of points of X with their scheme (or stack) theoretic multiplicity.*

Remark If X can be embedded into a smooth scheme M and $E^\bullet$ is an absolute obstruction theory for X then we have (in the notation above)

$$[X, E^\bullet] = (c(E_1) \cap s(C))_{\text{rank}\, E^\bullet},$$

where $s(C)$ is the Segre class of C and c denotes the total Chern class. The subscript denotes the component of degree $\operatorname{rank} E^\bullet$. (See [6], Chapter 6.) Now we have

$$
\begin{aligned}
c(E_1) \cap s(C) &= c(E_1)c(E_0)^{-1}c(E_0) \cap s(C) \\
&= c(E_\bullet)^{-1}c(E_0) \cap s(C) \\
&= c(E_\bullet)^{-1}c(i^*T_M) \cap s(C_{X/M}) \\
&= c(E^{\bullet\vee})^{-1}c_*(X),
\end{aligned}
$$

where $c_*(X)$ is the canonical class of X (see [ibid.]). Hence

$$
[X, E^\bullet] = (c(E^{\bullet\vee})^{-1}c_*(X))_{\operatorname{rank} E^\bullet}.
$$

Thus the intrinsic normal cone may be viewed as the geometric object underlying the canonical class. As such it glues (which cycle classes usually do not) and is thus also defined for nonembeddable X.

4.7 Gromov–Witten invariants

As always, let W be a smooth projective k-variety and τ a stable modular graph with an $H_2(W)^+$ marking β.

As indicated, we use the Artin stacks

$$
\mathfrak{M}_\tau = \prod_{v \in V_\tau} \mathfrak{M}_{g(v), F_\tau(v)},
$$

where $\mathfrak{M}_{g,S}$ is the stack of S-marked pre-stable curves of genus g. Pre-stable means that the singularities are at worst nodes and all marks avoid the nodes. Note that $\mathfrak{M}_\tau$ is smooth of dimension

$$
\dim(\tau) = \# S_\tau - \# E_\tau - 3\chi(\tau).
$$

We consider the morphism

$$
\begin{aligned}
\overline{M}(W,\tau) &\longrightarrow \mathfrak{M}_\tau \\
(C,x,f) &\longmapsto (C,x),
\end{aligned}
$$

where no stabilization takes place. Note that the fiber of this morphism over a point of $\mathfrak{M}_\tau$ corresponding to a curve C is an open subscheme of the scheme of morphisms $\operatorname{Mor}(C,W)$.

As before, let $\pi\colon C \to \overline{M}(W,\tau)$ be the universal curve and $f\colon C \to W$ the universal map. Then we have a perfect relative obstruction theory (even though this was only explained in the absolute case)

$$
(R\pi_* f^* T_W)^\vee \longrightarrow L_{\overline{M}(W,\tau)/\mathfrak{M}_\tau},
$$

and hence a virtual fundamental class

$$J(W,\tau) = [\overline{M}(W,\tau)]^{\text{virt}} = [\overline{M}(W,\tau),(R\pi_*f^*T_W)^\vee]$$

in $A_*(\overline{M}(W,\tau))$ of degree $\dim(\tau) + \operatorname{rank} R\pi_*f^*T_W$.

Let us check that this is the degree we claimed $[\overline{M}(W,\tau)]^{\text{virt}}$ to have:

$$
\begin{aligned}
&\dim(\tau) + \operatorname{rank} R\pi_*f^*T_W \\
&= \ \# S_\tau - \# E_\tau - 3\chi(\tau) + \chi(f^*T_W) \\
&= \ \# S_\tau - \# E_\tau - 3\chi(\tau) + \deg f^*T_W + \dim W\chi(\mathcal{O}_C) \\
&= \ \# S_\tau - \# E_\tau - 3\chi(\tau) - \beta(\omega_W) + \dim W\chi(\mathcal{O}_C) \\
&= \ \chi(\tau)(\dim W - 3) - \beta(\omega_W) + \# S_\tau - \# E_\tau \\
&= \ \dim(W,\tau).
\end{aligned}
$$

This calculation justifies the grading axiom for Gromov–Witten invariants.

Theorem 4.11 *The classes $J(W,\tau)$ satisfy all five axioms required.*

Proof The mapping to point axiom follows from the Example 4.8. For the proofs of the other axioms see [2]. One has to prove various compatibilities of virtual fundamental classes. These follow from the properties of normal cones proved by Vistoli [13]. □

Corollary 4.12 *The Gromov–Witten invariants $I_\tau(\beta)$ defined in terms of $J(W,\tau)$ satisfy all eight axioms required.*

4.8   Complete intersections

These ideas can easily be adapted to construct the tree level system of Gromov–Witten invariants for possibly singular complete intersections.

So let $W \in \mathbb{P}^n$ be a complete intersection, $i\colon W \to \mathbb{P}^n$ the inclusion morphism. Then $[i^*T_{\mathbb{P}^n} \to N_{W/\mathbb{P}^n}]$ is the tangent complex of W. So as obstruction theory for $\overline{M}(W,\tau) \to \mathfrak{M}_\tau$ we may take $(R\pi_*f^*[i^*T_{\mathbb{P}^n} \to N_{W/\mathbb{P}^n}])^\vee$. This will be a perfect obstruction theory if we restrict to the case where τ is a forest, because then the higher direct images under π of $f^*i^*T_{\mathbb{P}^n}$ and $f^*N_{W/\mathbb{P}^n}$ vanish. So we get the tree level system of Gromov–Witten invariants of W.

As an example, consider a cone over a plane cubic, which is a degenerate cubic surface in $\mathbb{P}^3$. There is a one dimensional family of lines on this cubic, namely the ruling of the cone. On the other hand, the expected dimension of the space of lines on a cubic in $\mathbb{P}^3$ is zero. Therefore the Gromov–Witten invariant $I_{0,0}(1)$ is a number, which turns out to be 27. So the 'ideal' number of lines on a cubic is 27, even in degenerate cases.

References

[1] M. Artin, A. Grothendieck, and J. L. Verdier. *Théorie des topos et cohomologie étale des schémas,* **SGA4**. Lecture Notes in Mathematics Nos. 269, 270, 305. Springer, Berlin, Heidelberg, New York, 1972, 73.

[2] K. Behrend. Gromov–Witten invariants in algebraic geometry. *Invent. Math.* 127, 1997, 601–617.

[3] K. Behrend and B. Fantechi. The intrinsic normal cone. *Invent. Math.,* 128, 1997, 45–88.

[4] K. Behrend and Yu. Manin. Stacks of stable maps and Gromov–Witten invariants. *Duke Mathematical J.,* 85:1–60, 1996.

[5] P. Deligne and D. Mumford. The irreducibility of the space of curves of given genus. *Publications Mathématiques, Institut des Hautes Études Scientifiques,* 36:75–109, 1969.

[6] W. Fulton. *Intersection Theory.* Ergebnisse der Mathematik und ihrer Grenzgebiete 3. Folge Band 2. Springer-Verlag, Berlin, Heidelberg, New York, Tokyo, 1984.

[7] W. Fulton and R. Pandharipande. Notes on stable maps and quantum cohomology. Algebraic geometry—Santa Cruz 1995. *Proc. Sympos. Pure Math.* 62, Part 2:45–96, 1997.

[8] S. Kleiman. Motives. In F. Oort, editor, *Algebraic Geometry, Oslo 1970,* pages 53–96. Wolters-Noordhoff, 1972.

[9] F. Knudsen. The projectivity of the moduli space of stable curves, II: The stacks $M_{g,n}$. *Math. Scand.,* 52:161–199, 1983.

[10] M. Kontsevich. Enumeration of rational curves via torus actions. In *The moduli space of curves,* Progress in Mathematics. 1995, 335–368.

[11] M. Kontsevich and Yu. Manin. Gromov–Witten classes, quantum cohomology, and enumerative geometry. *Communications in Mathematical Physics,* 164:525–562, 1994.

[12] J. Li and G. Tian. Virtual moduli cycles and Gromov–Witten invariants of algebraic varieties. *J. Amer. Math. Soc.,* 11, no. 1:119–174, 1998.

[13] A. Vistoli. Intersection theory on algebraic stacks and on their moduli spaces. *Inventiones mathematicae,* 97:613–670, 1989.

K. Behrend,
University of British Columbia,
Mathematical Department,
1984 Mathematics Road,
Vancouver, B.C. VGT 1Z2, Canada
behrend@math.ubc.ca

Kähler hyperbolicity and variations
of Hodge structures

Philippe Eyssidieux

There is a well-known duality, pervasive throughout the theory of Hermitian symmetric spaces, sending a symmetric space of compact type to its noncompact dual (see, for example, [9], Chap. VIII, §4 and Chap. XI, p. 354); whereas the former is a projective variety, the latter can be realized as a bounded domain in $\mathbb{C}^n$. For example, the Grassmannian $\mathrm{Grass}(p, p + q) = \mathrm{SU}(p+q)/\mathrm{S}\big(\mathrm{U}(p) \times \mathrm{U}(q)\big)$ has as its dual the quotient $\mathrm{SU}(p, q)/\mathrm{S}(\mathrm{U}(p) \times \mathrm{U}(q))$, which is the bounded symmetric domain $\{Z \in \mathbb{C}^{p,q} \mid {}^t\overline{Z}Z - I_q < 0\}$. A geometric statement concerning a Hermitian symmetric space of compact type often has an associated 'dualized' statement about its noncompact dual. I formulate such a dual pair of problems and discuss some aspects of the version on a Hermitian symmetric space of noncompact type.

Let Ω be a Hermitian symmetric space of noncompact type that has a compact dual (that is, a bounded symmetric domain). Let A_Ω be the evenly graded commutative real algebra generated by the Chern–Weil forms of homogenous holomorphic Hermitian vector bundles on Ω; then A_Ω is isomorphic to the real cohomology algebra of the compact dual of Ω. Setting $\deg c_i = 2i$ defines a grading on the polynomial algebra $B_\Omega = A_\Omega[c_1, \ldots, c_i, \ldots]$.

Let $T = (M, \rho, i)$ be a triple consisting of a connected n-dimensional complex compact manifold M, a representation ρ of $\pi_1(M)$ in the isometry group $\mathrm{Aut}\,\Omega$ and a ρ-equivariant holomorphic immersion $i\colon \widetilde{M} \hookrightarrow \Omega$ from the universal covering space $\widetilde{M}$ of M into Ω. Typical examples arise from complex submanifolds of Hermitian locally symmetric manifolds uniformized by Ω. Since every homogenous vector bundle V_Ω on Ω descends to a vector bundle V_M on M, there is a canonical morphism $i^*\colon B_\Omega \to H^*(M, \mathbb{R})$ defined by $i^*c_i(V_\Omega) = c_i(V_M), i^*c_i = c_i(M)$. Evaluation on the fundamental class of M defines a linear form $e_T\colon B_\Omega^{2n} \to \mathbb{R}$. I call such a triple an n-dimensional *positive cycle* of Ω.

Define the closed convex cone of universal Chern numbers inequalities to be $K_\Omega^n = \{\alpha \in B_\Omega^{2n}; e_T(\alpha) \geq 0 \text{ for all } T\}$. Consider the following problem:

Problem Describe completely the closed convex cone K_Ω^n. Construct optimal universal inequalities, and describe the cases of equality. A specialization of the latter question is of interest: describe the optimal inequalities whose cases of equality are the totally geodesic submanifolds of Ω/Γ for Γ a discrete cocompact subgroup of Aut Ω.

The dual problem is easy to describe. To the compact dual $\widehat{\Omega}$, we associate the algebra $A_{\widehat{\Omega}}$ generated by the Chern–Weil forms of homogenous vector bundles on $\widehat{\Omega}$ and we set $B_{\widehat{\Omega}} = A_{\widehat{\Omega}}[c_1, \ldots, c_n, \ldots]$, with $\deg c_i = 2i$. Then B_{Ω} and $B_{\widehat{\Omega}}$ are isomorphic. We define the n-dimensional *positive cycles* as the n-dimensional projective algebraic submanifolds of $\widehat{\Omega}$. The cone $K_{\widehat{\Omega}}^n$ is defined as before. The structure of $K_{\widehat{\Omega}}^n \cap A_{\widehat{\Omega}}^n$ is known when $\widehat{\Omega}$ is a Grassmannian: this is the content of the Fulton–Lazarsfeld theorem on universal inequalities for ample vector bundles [5].

However, there are also very significant differences between the problems of describing universal Chern numbers inequalities for Ω and $\widehat{\Omega}$. In the latter case, every projective algebraic manifold M (of sufficiently small dimension) may occur as the first term of a cycle (M, ρ, i). In the case of interest, the projective manifolds M that occur are of general type and have large fundamental groups.

I cannot formulate any reasonable conjecture describing K_{Ω}^n, and I only explain how to construct some interesting elements. There are various ways of constructing elements of K_{Ω}^n. For instance, because Chern–Weil theory enables us to compute Chern numbers as integrals of differential forms, starting from curvature tensors (whose explicit form is well known, see for instance [14]), explicit computations can be performed. The computational complexity of the problem grows so quickly with n that more subtle approaches are necessary.

In the situation of interest, Ω carries many homogeneous flat complex vector bundles which in addition underly a complex variation of Hodge structure. This structure pulls back to the universal covering $\widetilde{M}$ of M and then descends to M. The universal inequalities we will discuss in this article will be deduced from the study of these complex variations of Hodge structures. In the case that $\widetilde{M}$ is a totally geodesic complex submanifold of Ω, we call the resulting variation of Hodge structure a *locally homogeneous variation* and its period map a *homogeneous period map*.

More generally, we can consider complex variations of Hodge structures on M with their period map $\widetilde{M} \to D$, where D is the associated Griffiths period domain. See [3] for this construction and [7] for a thorough study of the homogenous domain D.

The manifold $\widetilde{M}$ is almost never compact. Fortunately, if the period map does not contract curves, the metric properties of the period map $\widetilde{M} \to D$ ensure that $\widetilde{M}$ admits a Kähler form $\widetilde{\omega}$ which is Lipschitz equivalent to the pullback of any given Kähler form on M and is the de Rham coboundary of a smooth 1-form which is bounded with respect to $\tilde{\omega}$. Such a Kählerian compact manifold M is called *weakly Kähler hyperbolic.*

This structure makes it possible to apply Gromov's methods from [8] to the L^2 cohomology with values in the variation of Hodge structure on the universal cover and the main result from [4] follows.

Main Theorem *Let M be a compact weakly Kähler hyperbolic manifold, and suppose that $(M, \nabla, F^\bullet, S)$ is a complex polarized variation of Hodge structures (VHS for short). Denote by K_P^r the holomorphic vector bundle on M given by $K_P^r = F^{P-r}/F^{P-r+1} \otimes \Omega^r$ and by $K_P^\bullet$ the complex*

$$K_P^0 \xrightarrow{\nabla'} K_P^1 \xrightarrow{\nabla'} \cdots \longrightarrow K_P^{\dim_{\mathbb{C}} M}.$$

Then $(-1)^{\dim_{\mathbb{C}}(M)} \chi(M, K_P^\bullet) \geq 0$.

Via the Riemann–Roch theorem, these give inequalities for the Chern classes, which I call *Arakelov type inequalities.* Going back to homogeneous variations over a bounded domain $D = \Omega$, one obtains nontrivial elements of K_Ω^n. In the general case, one has to work with A_D, the algebra generated by the Chern forms of homogeneous Hermitian vector bundles, and with $B_D = A_D[c_1, \ldots]$. I define a *positive n-dimensional cycle* of D to be a triple $T = (M, \rho, i)$ with M a connected n-dimensional complex compact manifold, $\rho \colon \pi_1(M) \to \mathrm{Aut}(D)$ a representation and $i \colon \widetilde{M} \to D$ a ρ-equivariant holomorphic horizontal immersion; the cone K_D^n is defined as the dual cone of the convex cone generated by irreducible positive cycles. The Arakelov type inequalities define nontrivial elements in this cone. An interesting feature is that, for a general Griffiths domain, these Arakelov type inequalities cannot be deduced from the explicit formulas for the curvature, even when $n = 1$. Another interesting phenomenon is that one can construct cases of equality for some of them that are locally homogenous. I conjecture that in fact every case of equality is locally homogenous (I call the corresponding positive cycles *geodesic* cycles). I can only prove this when the geodesic cycles satisfy some strong rigidity properties.

We describe the organization of this article. Section 1 reviews Gromov's notion of Kähler hyperbolicity, and vanishing theorems for L^2 cohomology. Section 2 recalls the definition of a VHS and sketches the proof of the Main Theorem; for more details, see [4], on which my talk at the Warwick Euro-Conference was based. The rest of the article is devoted to a systematic investigation of the cases of equality in the main theorem, which was lacking in [4]. Section 3 describes locally homogenous VHS on irreducible symmetric domains along the lines of Zucker's article [17] and gives a criterion derived from a formula of Kostant that identifies the cases of equality in the main theorem among them (Proposition 3.5.2). This criterion leads in principle to an algorithm producing the list of every locally homogenous VHS that gives rise to a case of equality in the main theorem. In this spirit, Appendix A.2 gives an algorithm to determine every such locally homogenous VHS uniformized by a classical domain. Section 4 establishes a part of the conjecture that every nontrivial case of equality of the universal inequalities among Chern numbers of Theorem 2.2.1 is actually a locally homogenous VHS, and describes the strong rigidity properties satisfied by certain of these examples (a list is given in Appendix A.3), which gives some more evidence for the gap phenomenon conjectured by N. Mok and the author (see [12], §4 and the introduction to [13]).

I thank the editors and referees for help in improving the exposition.

1   Harmonic bundles and Kähler hyperbolicity

1.1   Kähler hyperbolicity

Recall the following definition of Gromov [8]: let (M, ω) be a compact Kähler manifold and $\widetilde{M} \to M$ its universal covering. (M, ω) is *Kähler hyperbolic* if the Kähler form ω pulled back to $\widetilde{M}$ is the de Rham coboundary of a bounded smooth 1-form. This notion depends on the chosen Kähler class. The basic examples are Kähler manifolds admitting a Riemannian metric of strictly negative sectional curvature (for any Kähler form) and submanifolds of Hermitian locally symmetric manifolds of noncompact type (for the restriction of the symmetric Kähler form). I need the following variant: M is said to be *weakly Kähler hyperbolic* if $\widetilde{M}$ carries a bounded smooth 1-form α whose de Rham coboundary $d\alpha$ is a Kähler form Lipschitz equivalent to the pullback of some (and hence every) Kähler form on M.

1.2   The Lefschetz–Gromov vanishing theorem

Let (X, g) be a complete Riemannian manifold and V a flat vector bundle endowed with a flat connection D and a Hermitian metric h, not necessarily flat. Let $L^2 \, \mathrm{dR}^p(X, g, V, h)$ be the Hilbert space of square integrable p-forms on X with values in V whose distributional de Rham coboundary is still square integrable. Define the L^2 de Rham complex as the following complex of topological vector spaces:

$$L^2 \, \mathrm{dR}^0(X, g, V, h) \xrightarrow{\mathrm{d}} L^2 \, \mathrm{dR}^1(X, g, V, h) \xrightarrow{\mathrm{d}} \ldots \xrightarrow{\mathrm{d}} L^2 \, \mathrm{dR}^{\dim X}(X, g, V, h).$$

The cohomology of this complex is called the L^2 cohomology of (X, g) with values in (V, h), and denoted by $H^{\bullet}_{(2)}(X, g, V, h)$. It is obviously a bi-

Lipschitz invariant. Gromov [8], 1.2.B, p. 273, asserts that the L^2 cohomology of the universal covering of a compact Kähler hyperbolic manifold (M, ω) with values in the trivial flat vector bundle (or any unitary flat vector bundle) vanishes in degree $\neq \dim_{\mathbb{C}} M$.

As observed in [4], Théorème 1, Gromov's proof also extends to VHS (see below). In fact, the same proof also holds for Higgs bundles equipped with a harmonic metric. Recall ([16]) that a Higgs bundle on a complex manifold M is a holomorphic vector bundle V together with a holomorphic map $\theta : V \to V \otimes \Omega^1_M$ with $\theta \wedge \theta = 0$ in $\operatorname{End}(V) \otimes \Omega^2_M$. Not all such bundles admit harmonic metrics; one needs that the underlying flat bundle is semisimple. Such Higgs bundles are called *harmonic Higgs bundles*. Variations of Hodge structures provide examples, as we see below.

Proposition 1.2.1 *Let M be a compact weakly Kähler hyperbolic manifold. Then the L^2 cohomology of its universal covering with coefficients in the pullback of a harmonic Higgs bundle on M endowed with its harmonic metric (cf. [16]) vanishes in degrees $\neq \dim_{\mathbb{C}} M$.*

Note that on $\widetilde{M}$ the harmonic metric is Lipschitz equivalent to a Hermitian metric constructed by parallel transport of a Hermitian metric on some fibre if and only if the (real) monodromy group is compact, which is rarely the case.

2   Variations of Hodge structures

2.1   Definition

Definition 2.1.1 Let M be a complex manifold. A quadruple $(M, \mathbb{V}, F^\bullet, S)$ is called a *complex polarized variation of Hodge structure* if $\mathbb{V}$ is a flat bundle of finite dimensional complex vector spaces with a flat connection D, $F^\bullet$ a decreasing filtration of $\mathbb{V} \otimes \mathcal{O}_M$ by holomorphic subbundles indexed by the integers, and S a flat nondegenerate sesquilinear pairing such that

1. The C^∞ vector bundle V associated to $\mathbb{V}$ decomposes as a direct sum $V = \bigoplus_p H^p$ with $F^P = \bigoplus_{p \geq P} H^p$;

2. $S(H^p, H^r) = 0$ for $p \neq r$, and $(-1)^p S$ is positive definite on H^p;

3. $D^{1,0} F^p \subset F^{p-1} \otimes \Omega^{1,0}_M$.

The numerical vector $\{\dim_{\mathbb{C}} H^r\}_{r \in \mathbb{Z}}$ is called the *Hodge vector* of the VHS.

The subbundle H^p can be given a holomorphic structure by the isomorphism $H^p \to F^p / F^{p+1}$. Write d''_p for the corresponding Dolbeault operator, and set $\mathrm{d}'' = \bigoplus_p \mathrm{d}''_p$. Then $D^{1,0}$ induces a C^∞-linear map $\nabla'_p : H^p \to$

$H^{p-1} \otimes \Omega^1$ called the *Gauss–Manin connection*; we set $\nabla' = \bigoplus_p \nabla'_p$. The Hermitian metric $H = \bigoplus_p (-1)^p S_{H^p}$ is called the *Hodge metric*. The pair $(\mathbb{V} \otimes \mathcal{O}_M, \nabla')$ is a Higgs bundle and H is a harmonic metric on the associated flat bundle $\mathbb{V}$.

For our purposes, the weight of a VHS is not relevant; instead, we are interested in the *length* of a VHS, the number $p_{max} - p_{min}$, where p_{max} and p_{min} are the maximum and minimum of the (finite) set of integers p with $\dim H^p \neq 0$.

The Kähler hyperbolic assumption in Proposition 1.2.1 is well adapted to nondegenerate VHS by [4], Proposition 4.6.1:

Proposition 2.1.2 *Let M be a compact Kähler manifold and $(M, \mathbb{V}, F^\bullet, S)$ a VHS on M. Assume that for any connected smooth curve C mapping to M with 1-dimensional image, the induced VHS on C has a nonzero Gauss–Manin connection (equivalently, no curve on M is contracted by the Griffiths period map). Then M is a projective weakly Kähler hyperbolic manifold. Thus the L^2 cohomology $H^\bullet_{(2)}(\widetilde{M}, \mathbb{V})$ vanishes in degree $\neq \dim_{\mathbb{C}} M$.*

2.2   Arakelov type inequalities

Set $E^{P,Q} = \bigoplus_{p+r=P, s-p=Q} H^p \otimes \Omega^{rs}$ and $D'' = \mathrm{d}'' + \nabla'$. It follows that $D'' E^{P,Q} \subset E^{P,Q+1}$. Furthermore, Deligne proved (cf. [4], Proposition 2.2.1) that given any Kähler metric, if we take formal adjoints of differential operators with respect to this Kähler metric and the Hodge metric on $\mathbb{V}$, the usual Kähler identities hold, for instance $2(D''(D'')^* + (D'')^* D'') = DD^* + D^* D$. It follows that the L^2 cohomology on the universal covering $\widetilde{M}$ with coefficients in $\mathbb{V}$ has a Hodge decomposition.

Theorem 2.2.1 *Let M be a compact weakly Kähler hyperbolic manifold and $(M, \mathbb{V}, F^\bullet, S)$ a VHS. Denote by K_P^r the holomorphic vector bundle on M given by $K_P^r = H^{P-r} \otimes \Omega^r$ and by $K_P^\bullet$ the complex*

$$0 \to K_P^0 \xrightarrow{\nabla'} K_P^1 \xrightarrow{\nabla'} \cdots \to K_P^{\dim_{\mathbb{C}} M} \to 0.$$

Then $(-1)^{\dim_{\mathbb{C}}(M)} \chi(M, K_P^\bullet) \geq 0$. In terms of Chern numbers, this reads:

$$(-1)^{\dim_{\mathbb{C}}(M)} \sum_r (-1)^r \mathrm{ch}(H^{P-r}) \, \mathrm{ch}(\Omega_M^r) \, \mathrm{Todd}(T_M)[M] \geq 0.$$

Sketch of proof   Let G be a countable group, endowed with the counting measure, and write l or $r \colon G \to \mathrm{U}(L^2(G))$ for its left (respectively right) regular representation. The commutant of l is a von Neumann algebra $W_r^*(G)$. Every element a of $W_r^*(G)$ has a unique expression as $a = \sum_{g \in G} a_g r(g)$, with

$a_g \in \mathbb{C}$. Set $\tau(a) = a_e$. The linear form τ is a finite trace on $W_r^*(G)$, because for all a, b,

$$\tau(ab) = \tau(ba) \text{ and } a \neq 0 \implies +\infty > \tau(aa^*) > 0.$$

The orthogonal projection p onto a closed G-invariant subspace E of $L^2(G)$ belongs to $W_r^*(G)$ and we may try to define $\dim_G(E) = \tau(p)$. In fact, this definition can be extended to any Hilbert space with a left G-action, provided it can be realized as a G-stable closed linear subspace of $L^2(G)^{\oplus N}$. This provides a dimension theory for a certain class of unitary representations of G, similar to the dimension theory of linear algebra over $\mathbb{C}$ (the particular case $G = \{1\}$). For instance, $E = 0 \iff \dim_G E = 0$. The differences that have to be stressed in this article are that, when the group is infinite, the dimension function takes values in $\mathbb{R}^+$ and that an admissible G-module of nonzero G-dimension is infinite dimensional in the usual sense.

We can now explain Atiyah's L^2 index theorem (for details, see [1]). Let D be an elliptic differential operator on the compact connected C^∞-manifold X, and $\operatorname{Ind} D$ its index. Then D can be lifted to a differential operator $\widetilde{D}$ on the universal covering space $\widetilde{X}$ which is invariant under the natural action of $\pi_1(X)$. Then the $\pi_1(X)$-module $\ker_{L^2}(\widetilde{D}) = \{u \in L^2(\widetilde{X}); \widetilde{D}u = 0\}$ has a well-defined $\pi_1(X)$-dimension and

$$\dim_{\pi_1(X)} \ker_{L^2}(\widetilde{D}) - \dim_{\pi_1(X)} \ker_{L^2}(\widetilde{D}^*) = \operatorname{Ind} D.$$

This theorem is the analogue for infinite coverings of the multiplicative property of the usual index under finite coverings.

$\chi(M, K_P^\bullet)$ is the usual index of the elliptic complex $D'': E^{P,*} \to E^{P,*+1}$. By Atiyah's L^2 index theorem, it can be computed as the L^2 index of the lifted elliptic complex on $\widetilde{M}$ where the vanishing theorem Proposition 2.1.2 is available. $\square$

3 Locally homogenous variations of Hodge structures

3.1 Hermitian symmetric spaces of noncompact type

Let Ω be an irreducible Hermitian symmetric space of noncompact type. Its group of isometries is an almost simple adjoint Lie group $\operatorname{Aut}\Omega$, and we set $\mathfrak{g} = \operatorname{Lie}(\operatorname{Aut}\Omega)$. Let $G_{\mathbb{C}}$ be the simply connected complex semisimple algebraic group associated to $\mathfrak{g}_{\mathbb{C}} = \mathfrak{g} \otimes_{\mathbb{R}} \mathbb{C}$. Then $G_{\mathbb{C}}$ has a connected real form G with Lie algebra $\mathfrak{g}$, and $\operatorname{Aut}\Omega$ is the adjoint group G^{ad} of G. Let K be a maximal compact subgroup of G and observe that $\Omega = G/K$. Since Ω

is Hermitian symmetric, $\mathfrak{l} = \mathrm{Lie}\, K$ has a 1-dimensional center $z(\mathfrak{l})$ and splits as $\mathfrak{l} = z(\mathfrak{l}) \oplus [\mathfrak{l}, \mathfrak{l}]$. Thus $Z = \exp(z(\mathfrak{l}))$ is isomorphic to $U(1)$. We write μ for the degree of the covering map $Z \to Z^{\mathrm{ad}}$.

Now suppose that $z_0 \in Z^{\mathrm{ad}}$ is an element of order 4. Then $z_0^2 = \theta$ is a Cartan involution. Write $\mathfrak{g} = \mathfrak{l} \oplus \mathfrak{p}$ for the corresponding Cartan decomposition. Then z_0 induces on $\mathfrak{p} = T_{eK} G / K$ the tensor $\pm J$, where J is the almost complex structure of Ω. We may thus choose z_0 inducing J on $\mathfrak{p}$. Write accordingly $\mathfrak{p} \otimes \mathbb{C} = \mathfrak{p}^+ \oplus \mathfrak{p}^-$, where $\mathfrak{p}^{\pm} = \ker(\mathrm{ad}\, z_0 - \pm\sqrt{-1})$.

We can choose a Cartan subalgebra $\mathfrak{h}$ of $\mathfrak{g}$ contained in $\mathfrak{l}$. Let $\Delta_{\mathfrak{g}} \subset \mathfrak{h}_{\mathbb{C}}^*$ be the root system for $\mathfrak{g}_{\mathbb{C}}$. These roots take purely imaginary values on $\mathfrak{h}$. There exists a partition $\Delta_{\mathfrak{g}} = \Delta_{\mathfrak{l}} \cup \Delta_{\mathfrak{p}}$ for which $\mathfrak{p}_{\mathbb{C}} = \bigoplus_{\alpha \in \Delta_{\mathfrak{p}}} \mathfrak{g}^{\alpha}$ and $\mathfrak{l}_{\mathbb{C}} = \bigoplus_{\alpha \in \Delta_{\mathfrak{l}}} \mathfrak{g}^{\alpha}$. One can choose an ordering on $\sqrt{-1}\mathfrak{h}^*$ such that $\mathfrak{p}^+ = \bigoplus_{\alpha \in \Delta_{\mathfrak{p}}^+} \mathfrak{g}^{\alpha}$, where $\Delta_{\mathfrak{p}}^+ = \Delta^+ \cap \Delta_{\mathfrak{p}}$ (respectively, $\Delta_{\mathfrak{l}}^+ = \Delta^+ \cap \Delta_{\mathfrak{l}}$).

3.2 Homogenous holomorphic vector bundles

We say that a holomorphic vector bundle $V \to \Omega$ of finite rank is *homogenous* if the natural action of G on Ω lifts to an action on V which preserves the complex structure. Homogenous holomorphic vector bundles correspond 1-to-1 to finite dimensional K-modules. Under this correspondence the holomorphic tangent (or cotangent) bundle of Ω corresponds to $\mathfrak{p}^+$ (respectively $\mathfrak{p}^-$).

3.3 Homogenous VHS

There is another natural correspondence between flat homogenous vector bundles on Ω of finite rank and finite dimensional complex linear representations of G. As observed in [17], an irreducible flat homogenous complex vector bundle $\mathbb{V}$ on Ω can be given the structure of a homogenous VHS.

Let us describe Zucker's construction. $V = \mathbb{V}_{eK}$ is the representation space of the associated representation ρ of G. Let $\hat{Z} = \mathrm{Hom}(Z, U(1)) = \chi_0^{\mathbb{Z}}$ with χ_0^{μ} the character of Z acting on $\mathfrak{p}^+$. Decompose V according to the action of Z: $V = \bigoplus_{\chi \in \hat{Z}} V(\chi)$. Then $V(\chi)$ is K-invariant. Observe that $\rho(\mathfrak{p}^{\pm} V(\chi)) \subset V(\chi \cdot \chi_0^{\pm \mu})$. Let χ_V be the highest nontrivial character (with the convention that $\chi_0 > 0$). It follows from the irreducibility of ρ that there exists an integer N_{ρ} such that $V = \bigoplus_{l=0}^{N_{\rho}} V(\chi_V \chi_0^{-l\mu})$, and every summand is nonzero.

Define a decreasing filtration of V by setting $F^p V = \bigoplus_{l=p}^{N_{\rho}} V(\chi_V \chi_0^{-l\mu})$. The Lie subalgebra $\mathfrak{g}_c = \mathfrak{l} \oplus \sqrt{-1}\mathfrak{p} \subset \mathfrak{g}_{\mathbb{C}}$ is compactly embedded, and preserves a positive definite Hermitian sesquilinear pairing H (unique up to scalar multiple). The decomposition $V = \bigoplus_{l=0}^{N_\rho} V(\chi_V \chi_0^{-l\mu})$ is orthogonal and preserved by $\mathfrak{l}$.

Since $\rho(\mathfrak{p}) \bigoplus_{l \text{ even}} V(\chi_V \chi_0^{-l\mu}) = \bigoplus_{l \text{ odd}} V(\chi_V \chi_0^{-l\mu})$, it follows that $\mathfrak{g}$ preserves the Hermitian nondegenerate pairing $S = \bigoplus_l (-1)^l H|_{V(\chi_V \chi_0^{-l\mu})}$ which polarizes the filtration $F^\bullet$.

Let $x = gK \in \Omega$, and set $F^p \mathbb{V}_x = \rho(g) \cdot F^p V \subset \mathbb{V}_x$, $S_x = S \circ \rho(g)^{-1}$. The $F^\bullet_x$ and S_x glue together to yield a holomorphic filtration on $\mathbb{V}$ satisfying Griffiths transversality (because $\rho(\mathfrak{p}^+)F^p \subset F^{p-1}$) and a polarization S of this filtration. I henceforth refer to this VHS as the irreducible homogenous VHS associated to ρ. The number N_ρ is the length of the VHS (see Definition 2.1).

For the reader's convenience, Appendix A.1 describes all fundamental VHS on classical domains.

A simple modification of this construction works for reducible Hermitian symmetric spaces and yields:

Proposition 3.3.1 *Let Ω be a Hermitian locally symmetric space of non-compact type, $\mathfrak{g}$ the Lie algebra of its automorphism group and $\rho \colon \mathfrak{g} \to \mathfrak{gl}(V)$ a complex linear representation.*

There exists a $\mathfrak{g}$-invariant Hermitian form S_ρ that is nondegenerate (but indefinite unless ρ is trivial). Moreover, the constant local system $V \times \Omega \to \Omega$ underlies a VHS polarized by S_ρ which is invariant under a finite covering group G of the automorphism group of Ω, G acting on $\Omega \times V$ by $g \cdot (o, v) = (g \cdot o, g \cdot v)$.

For every $p \in \Omega$, we let $\varphi(p) = \{F^\bullet(V)\}$ be the corresponding Hodge flag. The group $U(S_\rho)$ acts transitively on Hodge flags and if we fix some origin $o \in \Omega$ and let U be the stabiliser of the flag defined by o, the set of Hodge flags can be described as a homogeneous domain $D_\rho = D(U(S_\rho), \varphi(o)) = U(S_\rho)/U$. This domain is called the *Griffiths domain*. The Griffiths period map corresponding to the VHS constructed above is the holomorphic horizontal map $f \colon \Omega \to D_\rho$ described by $f(p) := \varphi(p)$.

From a *homogenous* VHS, that is, a VHS on a bounded symmetric domain Ω invariant under some (finite) covering group G of the automorphism group of Ω, one associates a representation ρ of G in the obvious way, so that there is a 1-to-1 correspondance between homogenous VHS on a bounded symmetric domain and complex linear representations of the Lie algebra of its automorphism group.

3.4 Locally homogenous VHS

Let $M = \Omega/\Gamma$ be a Hermitian locally symmetric manifold of noncompact type with universal covering space the Hermitian symmetric space Ω. Say that a

VHS on M is *locally homogenous* if its lift to Ω is a homogenous VHS. It is proved in [4] that the partial converse to Theorem 2.2.1 holds:

Theorem 3.4.1 *Let M be a compact Hermitian locally symmetric manifold of noncompact type and $(M, \mathbb{V}, F^\bullet, S)$ a locally homogenous VHS. Recall that K_P^r is the holomorphic vector bundle on M given by $K_P^r = H^{P-r} \otimes \Omega_M^r$ and $K_P^\bullet$ the complex $0 \to K_P^0 \xrightarrow{\nabla'} K_P^1 \xrightarrow{\nabla'} \cdots \to K_P^{\dim_{\mathbb{C}} M} \to 0$. Then*

$$(-1)^{\dim_{\mathbb{C}}(M)} \chi(M, K_P^\bullet) \geq 0,$$

with equality if and only if $K_P^\bullet$ is acyclic.

Matsushima and Murakami ([11], Theorem 12.3, p. 33) prove that, if ρ is an irreducible complex representation of G associated to a *nonsingular* highest weight, $H^i(\pi_1(M), \rho) = H^i(M, \mathbb{V}_\rho) = 0$ unless $i = \dim_{\mathbb{C}}(M)$. This vanishing theorem is stronger than the L^2 vanishing theorem applied to a Hermitian locally symmetric space. Furthermore, Theorem 3.4.1 can be deduced, for a nonsingular weight, from their work, and should be viewed as a weak version of their theorem holding for singular weights.

3.5   A formula of Kostant

Let $\rho \colon G \to \mathrm{GL}(V_\rho)$ be an irreducible complex representation. The complex $K_P^\bullet$ associated to the homogenous VHS it defines on Ω is obviously the complex of homogenous vector bundles associated to the following complex of K-modules:

$$K_\chi^\bullet(V_\rho) \colon \cdots \to \Lambda^r \mathfrak{p}^- \otimes V(\chi\chi_0^{r\mu}) \xrightarrow{\nabla'_\rho} \Lambda^{r+1}\mathfrak{p}^- \otimes V(\chi\chi_0^{(r+1)\mu}) \to \cdots,$$

where $\chi = \chi_V \chi_0^{-P\mu}$ and $\nabla'_\rho = \sum_{\alpha \in \Delta_\mathfrak{p}^+} \rho(e_\alpha) \otimes e_{-\alpha} \wedge$. This complex computes the part of the K-module $H^\bullet(\mathfrak{p}^+, V_\rho)$ on which the central subgroup Z acts with character χ.

Write $W_\mathfrak{g}$ and $W_\mathfrak{l} = W_{[\mathfrak{l},\mathfrak{l}]}$ for the Weyl groups of $\mathfrak{g}$, respectively $\mathfrak{l}$, and $\Pi_\mathfrak{g} \in \mathfrak{h}$ and $\Pi_\mathfrak{l} \in \mathfrak{h}$ for the weight lattice of $\mathfrak{g}$, respectively $\mathfrak{l}$ in $\sqrt{-1}\mathfrak{h}^*$. The Grothendieck ring of the category of representations of $\mathfrak{g}$ and $\mathfrak{l}$ is isomorphic to the ring $\mathbb{Z}[\Pi_\mathfrak{g}]^{W_\mathfrak{g}}$, respectively $\mathbb{Z}[\Pi_\mathfrak{l}]^{W_\mathfrak{l}}$. The fact that the Weyl group is simply transitive on the Weyl chambers yields:

Lemma 3.5.1 ([10]) *Denote by $W_{\mathfrak{g},\mathfrak{l}}$ the right coset space $W_\mathfrak{l} \backslash W_\mathfrak{g}$. Then $W_0 = \{w \in W_\mathfrak{g} : w\Delta_\mathfrak{g}^- \cap \Delta_\mathfrak{g}^+ \subset \Delta_\mathfrak{p}^+\} \to W_{\mathfrak{g},\mathfrak{l}}$ is a set of representatives.*

Set $W_0 = \bigcup_q W_0(q)$ where $W_0(q) = \{w \in W_0 : |w\Delta_\mathfrak{g}^- \cap \Delta_\mathfrak{p}^+| = q\}$.

Let $C_\mathfrak{g}^+$ and $C_\mathfrak{l}^+$ be positive Weyl chambers for $\mathfrak{g}$ and $\mathfrak{l}$. For a dominant weight $\lambda \in \Pi_\mathfrak{g} \cap \overline{C}_\mathfrak{g}^+$ for $\mathfrak{g}$, write $E_\lambda^\mathfrak{g}$ for the irreducible $\mathfrak{g}$-module with highest weight λ; similarly, for a dominant weight $\mu \in \Pi_{\mathfrak{l}} \cap \overline{C}_{\mathfrak{l}}^+$ for $\mathfrak{l}$, write $E_\mu^{\mathfrak{l}}$ for the irreducible $\mathfrak{l}$-module with highest weight μ.

Let $\rho_{\mathfrak{g}}$, $\rho_{\mathfrak{l}}$ and $\rho_{\mathfrak{p}}$ be half the sum of the positive roots in $\Delta_{\mathfrak{g}}^+$ (respectively, $\Delta_{\mathfrak{l}}^+$ and $\Delta_{\mathfrak{p}}^+$).

The following is due to Kostant:

Proposition 3.5.2 (see [17],[10]) *Let* $\lambda \in \Pi_{\mathfrak{g}} \cap C_{\mathfrak{g}}^+$. *Then*

$$H^q(\mathfrak{p}^+, E_\lambda^{\mathfrak{g}}) = \sum_{w_0 \in W_0(q)} E_{w_0(\lambda+\rho_{\mathfrak{g}})-\rho_{\mathfrak{g}}}^{\mathfrak{l}}.$$

We need to decompose this formula further according to the action of Z. First of all, observe that:

Lemma 3.5.3 $w\rho_{\mathfrak{p}} = \rho_{\mathfrak{p}}$ *for all* $w \in W_{\mathfrak{l}}$.

Proof Let ζ be a generator of $z(\mathfrak{l})$. Clearly $(\rho_{\mathfrak{p}}, \zeta) = c|\Delta_{\mathfrak{p}}^+| \neq 0$. Observe that $w \cdot \Delta_{\mathfrak{p}} = \Delta_{\mathfrak{p}}$. It follows that

$$(w\rho_{\mathfrak{p}}, \zeta) = c(|w\Delta_{\mathfrak{p}}^+ \cap \Delta_{\mathfrak{p}}^+| - |w\Delta_{\mathfrak{p}}^+ \cap \Delta_{\mathfrak{p}}^-|).$$

The Weyl group $W_{\mathfrak{l}}$ acts orthogonally on $\mathfrak{h}$ (w.r.t. the Killing form of $\mathfrak{g}$), leaving $\mathfrak{h} \cap [\mathfrak{l}, \mathfrak{l}]$ invariant. Thus, $z(\mathfrak{l}) = [\mathfrak{l}, \mathfrak{l}]^\perp$ is also $W_{\mathfrak{l}}$-invariant. Since $\varepsilon(w)$, equal to -1 to the power the number of reflections in any decomposition of w as a product of reflections, does not depend on whether we view w as an element of $W_{\mathfrak{l}}$ or of $W_{\mathfrak{g}}$ and is the determinant of the linear transformation it induces on $\mathfrak{h}$ and on $[\mathfrak{l}, \mathfrak{l}]$, $z(\mathfrak{l})$ is fixed by $W_{\mathfrak{l}}$. It follows that $(w(\rho_{\mathfrak{p}}), \zeta) = (\rho_{\mathfrak{p}}, w^{-1}\zeta) = (\rho_{\mathfrak{p}}, \zeta)$, and thus $w\rho_{\mathfrak{p}} = \rho_{\mathfrak{p}}$. $\square$

The latter argument implies also that, by means of the identification of $\mathfrak{h}$ and its dual space via the Killing form, $\rho_{\mathfrak{p}}$ is identified with a positive generator of $z(\mathfrak{l})$, since $[\mathfrak{l}, \mathfrak{l}]^{W_{\mathfrak{l}}} = \{0\}$. Set $\overline{\omega} = \dim_{\mathbb{C}}(\mathfrak{p}^+)\rho_{\mathfrak{p}}/\mu(\rho_{\mathfrak{p}}, \rho_{\mathfrak{p}})$. Let $\Pi_{\mathfrak{g}} \subset \Pi_{\mathfrak{l}}^K \subset \Pi_{\mathfrak{l}}$ be the lattice consisting of those weights of $\mathfrak{l}$ arising from K-modules, and let $\lambda \in \Pi_{\mathfrak{l}}^K$ be a dominant weight. Observe that Z acts by the character χ_0^l on $E_\lambda^{\mathfrak{l}}$ if and only if $(\lambda, \overline{\omega}) = l$.

This implies the following:

Corollary 3.5.4 *Let* $\lambda \in \Pi_{\mathfrak{g}}$ *be a dominant weight. Then:*

$$H^q(K_{\chi_0^l}^\bullet(E_\lambda^{\mathfrak{g}})) = \sum_{\substack{w_0 \in W_0(q) \ s.t. \\ (w_0(\lambda+\rho_{\mathfrak{g}})-\rho_{\mathfrak{g}}, \overline{\omega})=l}} E_{w_0(\lambda+\rho_{\mathfrak{g}})-\rho_{\mathfrak{g}}}^{\mathfrak{l}}.$$

We apply this to $\rho = 0$, where $\nabla'_\rho = 0$ and $H^\bullet(K^\bullet_{\chi^l_0}(E^{\mathfrak{g}}_0)) = K^\bullet_{\chi^l_0}(E^{\mathfrak{g}}_0) = \Lambda^{-l/\mu}\mathfrak{p}^-$ ($l \leq 0$), and call $c_{\mathfrak{p}}$ the (unique) element of W_0 that corresponds to $\Lambda^{\dim_{\mathbb{C}} \mathfrak{p}^+}\mathfrak{p}^-$.

In particular $W_0(0) = \{\text{id}\}$, $W_0(\dim \Omega) = \{c_{\mathfrak{p}}\}$. For a possibly reducible Hermitian symmetric space, elements of $W_0(1)$ are in 1-to-1 correspondance with its irreducible factors.

It follows from Corollary 3.5.4 that for any $\lambda \in \Pi_{\mathfrak{g}} \cap \overline{C}^+_{\mathfrak{g}}$ that $H^0(\mathfrak{p}^+, E^{\mathfrak{g}}_\lambda)$ and $H^{\dim \Omega}(\mathfrak{p}^+, E^{\mathfrak{g}}_\lambda)$ are irreducible K-modules; these modules are easily identified. They correspond to the two complexes $K^\bullet_\chi(\lambda)$ with a single term; namely $K^\bullet_{\chi_V}(\lambda) = V(\chi_V)$, and

$$K^\bullet_{\chi_V \chi_0^{-(N_\lambda + \dim(\mathfrak{p}^+))\mu}}(\lambda) = V(\chi_V \chi_0^{-N_\lambda \mu}) \otimes \Lambda^{\dim(\mathfrak{p}^+)}\mathfrak{p}^-$$

correspond to the elements $w = e \in W_0$ respectively $w = c_{\mathfrak{p}}$. Therefore:

Lemma 3.5.5 *The length of the VHS associated to* $\lambda \in \Pi_{\mathfrak{g}} \cap C^+_{\mathfrak{g}}$ *is*

$$N_\lambda = (\lambda, (\text{id} - c_{\mathfrak{p}}^{-1})\overline{\omega}).$$

We say that the character $\chi \in \widehat{Z}$ is *critical* with respect to the dominant weight $\lambda \in \Pi_{\mathfrak{g}}$ if some $K^q_{\chi^l_0}(E^{\mathfrak{g}}_\lambda)$ is not 0 but $K^\bullet_{\chi^l_0}(E^{\mathfrak{g}}_\lambda)$ is acyclic. In view of Theorem 3.4.1, we get the following criterion:

Proposition 3.5.6 χ^l_0 *is critical with respect to* λ *if and only if the following hold:*

1. $l \equiv (\lambda, \overline{\omega}) \mod \mu$;

2. $l \notin \{(w_0(\lambda + \rho_{\mathfrak{g}}) - \rho_{\mathfrak{g}}, \overline{\omega})\}_{w_0 \in W_0}$;

3. $(c_{\mathfrak{p}}(\lambda + \rho_{\mathfrak{g}}) - \rho_{\mathfrak{g}}, \overline{\omega}) \leq l \leq (\lambda, \overline{\omega})$.

This provides us with many nontrivial examples. For example:

Corollary 3.5.7 *Let* Ω *be a Hermitian symmetric space of noncompact type. Set* $l_c(\Omega) = |W_0| - \dim_{\mathbb{C}} \Omega$. *Every dominant weight whose associated VHS has length* $\geq l_c(\Omega)$ *has critical characters. Moreover, every dominant weight in the series* $(n\rho_{\mathfrak{g}})_{n \geq 1}$ *or* $(n\rho_{\mathfrak{g}} + (n+1)\lambda)_{n \geq 1, \lambda \in \Pi_{\mathfrak{g}} \cap \overline{C}^+_{\mathfrak{g}}}$ *has critical characters.*

Since the length of the VHS associated to the dominant root λ is given by the positive linear form $l(E_\lambda) = (\lambda, (\text{id} - c_{\mathfrak{p}}^{-1})\overline{\omega})$, the corollary shows that, for each bounded symmetric domain Ω, only a finite number of weights have no critical character (I refer to these as the *special* weights of Ω).

Although I did not find a simple condition for a weight of a classical domain of type I, II or III to be special, the proposition gives in principle an algorithm to determine every special weight. See Appendix A.2 for details and specific examples.

We consolidate our gains in the following result.

Theorem 3.5.8 *Let Ω be an n-dimensional Hermitian symmetric space of noncompact type and λ a nonspecial weight with critical character $\chi_V \chi_0^{-P\mu}$. Let $E_\lambda^\mathfrak{g}$ be the highest weight module attached to λ. Let $\mathbb{G} = (\Omega, \mathbb{V}, F^\bullet, S)$ be the homogenous VHS attached to $E_\lambda^\mathfrak{g}$ (see Proposition 3.3.1).*

Then equality occurs in the universal inequality

$$(-1)^n \big(\mathrm{ch}(K_P^\bullet)\,\mathrm{Todd}(T_M)\big)_{n,n}[M] \geq 0$$

(which, in virtue of Theorem 2.2.1, is valid for every VHS with the same Hodge vector as $\mathbb{G}$ having an immersive period map) if the VHS is a locally homogenous VHS uniformized by $\mathbb{G}$.

4 Hyperrigid locally homogenous variations of Hodge structure

4.1 Scalar curvature for a Griffiths period map

Let $D = W/U$ be a Griffiths domain. The Lie algebra of W has a real Hodge structure of weight 0, polarized by the Killing form.

On D there is an invariant closed symplectic form ω which is not a Kähler form with respect to its complex structure. In fact, $h(X) = \omega(X, JX)$ is an indefinite nondegenerate form. However ω is positive definite on horizontal directions.

Let $T = (M, \rho, i)$ be an irreducible n-dimensional positive cycle. The form ω restricts to a Kähler metric ω_h on $\widetilde{M}$ which descends to M. In fact, it may also be defined as the curvature of the pullback of a certain equivariant holomorphic line bundle on D. Assume that the Griffiths period map i sends $x \in \widetilde{M}$ to $eU \in D(W, \varphi) = W/U$. Then $i_* T_x \widetilde{M}$ can be identified with a complex Abelian subspace of $\mathrm{Lie}(W)^{-1,1}$. Let $(e_i)_i$ be a unitary basis of this subspace. Using the Gauss equation and the formulas in [7] (for quick reference, see [14], Corollary 2.2), one easily computes:

Lemma 4.1.1 *The scalar curvature of M at x is:*

$$\mathrm{Scal}(M, x) = -\|\sum_i [e_i, \bar{e}_i]\|^2 - \Sigma^2 = S(\mathcal{A}) - \Sigma^2.$$

Viewing $\widetilde{TM}$ as a subbundle of the bundle associated to $\mathrm{Lie}(W)^{-1,1}$ (this is the horizontal tangent bundle) identifies the term Σ^2 with the square of the norm of the metric second fundamental form; thus it vanishes identically on M if and only if T is a geodesic cycle.

If ρ is a linear representation of W, one constructs the associated VHS $(M, \mathbb{V}_\rho, F^\bullet, S)$, and for every integer P, the complex $K_P^\bullet$:

$$K_P^\bullet: 0 \to F^P/F^{P-1} \xrightarrow{\nabla'} F^{P-1}/F^{P-2} \otimes \Omega_M^1 \xrightarrow{\nabla'}$$

$$\cdots \xrightarrow{\nabla'} F^{P-n}/F^{P-n-1} \otimes \Omega_M^n \to 0.$$

Lemma 4.1.2 *There exists a constant C_D^P and a function S_D^P on the space E_{ab} of n-dimensional complex Abelian subspaces of $\mathrm{Lie}(W)^{-1,1}$ such that, if $T = (M, \rho, i)$ denotes an n-dimensional positive cycle, then:*

$$c_1(K_P^\bullet)\omega_h^{n-1} = C_D^P\big(\mathrm{Scal}(M, x) + S_D^P(T_x M)\big)\omega_h^n \quad \text{at every } x \in M.$$

4.2 Hyperrigidity

4.2.1 Definitions

Definition 4.2.1 Let $D = W/U$ be a Griffiths domain and $\Omega = G/K$ a Hermitian symmetric space of noncompact type; we say that the homogeneous period map $\Omega \to D$ is 0- (respectively 1)-hyperrigid if it satisfies Condition 1 (respectively, Conditions 1–2) below:

1. $S(T\Omega)$ is the maximal value of the scalar curvature in the space on n-dimensional Abelian subspaces of $\mathrm{Lie}(W)^{-1,1}$.

2. There is a linear representation ρ of W and an integer P such that for the induced homogenous VHS on Ω, the complex $K_P^\bullet$ is nontrivial but acyclic, satisfies $C_D^P \neq 0$ and S_D^P is maximal at $T\Omega$. (This condition follows automatically if $\mathrm{Lie}(W)^{-1,1}$ happens to be an irreducible U-module.)

4.2.2 Rigidity phenomena

The following lemma is elementary:

Lemma 4.2.2 *Assume that there is some 0-hyperrigid period map $\Omega \to D$ with scalar curvature S, and $\dim \Omega = n$. Then, for each n-dimensional positive cycle (M, ρ, i) of D, one has $(S\omega_h^n - c_1(M)\omega_h^{n-1})[M] \geq 0$ and every case of equality is a geodesic cycle uniformized by some 0-hyperrigid locally homogenous period map.*

Now, assume that there is some 1-hyperrigid homogenous period map $\Omega \to D$, with $\dim \Omega = n$. Denote by $F^\bullet$ the Hodge filtration on the representation space of ρ. For each Abelian n-dimensional complex subspace $\mathcal{A} \subset \mathrm{Lie}(W)^{-1,1}$ one defines the complex:

$$K_P^\bullet(\mathcal{A}): 0 \to F^P/F^{P-1} \xrightarrow{\nabla'} F^{P-1}/F^{P-2} \otimes \mathcal{A}^* \xrightarrow{\nabla'}$$

$$\cdots \xrightarrow{\nabla'} F^{P-n}/F^{P-n+1} \otimes \Lambda^n \mathcal{A}^* \to 0$$

When $\mathcal{A}$ arises as the tangent space to a Griffiths period map, this complex is identified with the already defined complex $K_P^\bullet$ of the associated VHS. Let $U \subset E_{ab}$ be the (Zariski open) subset of the space of all Abelian n-dimensional complex subspaces $\mathcal{A}$ of $\mathrm{Lie}(W)^{-1,1}$ such that $K_P^\bullet(\mathcal{A})$ is acyclic.

One may build holomorphic homogenous fibre bundles $\mathbf{U} \to D$ and $\mathbf{E}_{ab} \to D$ with fibres at the origin U, respectively E_{ab}. The Gauss map of a positive cycle $T = (M, \rho, i)$ is the map $\gamma \colon M \to i^* \mathbf{E}_{ab}/\pi_1(M)$ sending each point to its embedded tangent space.

The following theorem follows almost immediately from the definition of 1-hyperrigidity:

Theorem 4.2.3 *Let Ω be an n-dimensional Hermitian symmetric space of noncompact type. Suppose that there is some 1-hyperrigid homogenous period map $\Omega \to D$. Let $T = (M, \rho, i)$ be a positive n-dimensional cycle of D. The Gauss map of T takes values in $i^* \mathbf{U}/\pi_1(M)$ if and only if T is a 0-hyperrigid geodesic cycle.*

Proof Theorem 3.4.1 tells us that for any cocompact fixed point free discrete subgroup Γ of $\mathrm{Aut}\,\Omega$, one can construct a positive cycle $T = (\Omega/\Gamma, \rho|_\Gamma, i)$ such that $\mathrm{ch}(K_P^\bullet) = 0$. In particular, $\int_{\Omega/\Gamma} c_1(K_P^\bullet) \omega_h^{n-1} = 0$. The integrand from Lemma 4.1.2 is constant along Ω/Γ. So the maximal value of S_D^P is exactly minus the scalar curvature of Ω. In particular, due to Definition 4.2.1, for any nongeodesic n-dimensional cycle $T = (M, \rho, i)$ (and also for any geodesic cycle of nonmaximal scalar curvature) $\int_M c_1(K_P^\bullet) \omega_h^{n-1} \neq 0$.

But, tautologically, to say that the Gauss map γ of the cycle T should take values in $i^* \mathbf{U}/\pi_1(M)$ means that $K_P^\bullet$ is an acyclic complex. Thus $\int_M c_1(K_P^\bullet) \omega_h^{n-1} = 0$, and T is a geodesic cycle. $\square$

On the other hand, the following theorem of [4] is more difficult to prove, and uses in a crucial way an improvement, due to Gromov [8], of Atiyah's L^2 index theorem.

Assume that there is some 1-hyperrigid period map $\Omega \to D = W/U$. Set $n = \dim \Omega$. Let ρ be the representation of W and P the integer provided by Definition 4.2.1. For a positive n-dimensional cycle of D, $T = (M, \rho, i)$, we write $(M, \mathbb{V}, F^\bullet, S)$ for the VHS on M associated to the representation ρ.

Theorem 4.2.4 *Assume that (M, ω_h) is a Kähler hyperbolic manifold in the strong sense. (This condition is automatic when D is Hermitian symmetric.) Then $\chi(M, K_P^{\bullet}) = 0$ if and only if T is a 0-hyperrigid geodesic cycle.*

Sketch of proof Let L be the homogenous holomorphic vector bundle on D such that $c_1(L, h) = \omega_h$. Since $i^*\omega_h$ is exact on $\widetilde{M}$, it follows that $i^*L \to \widetilde{M}$ is topologically trivial. Thus, one can construct, for each $\varepsilon \in \mathbb{R}$, a Hermitian holomorphic line bundle L^ε on $\widetilde{M}$ such that $c_1(L^\varepsilon) = \varepsilon\omega_h$. Furthermore, there is an extension Γ_ε of $\pi_1(M)$ by U(1) whose action on $\widetilde{M}$ lifts to L^ε. One can construct a variant of Atiyah's L^2 index theory for Γ_ε-invariant elliptic operators.

Set $E_\varepsilon^{P,Q} = \bigoplus_{p+r=P, s-p=Q} H^p \otimes \Omega^{rs} \otimes L^\varepsilon$ and $D_\varepsilon'' = \mathrm{d}_\varepsilon'' + \nabla_\varepsilon'$. In view of the higher index theory mentioned above, the L^2 D_ε''-complex with its Γ_ε-action has index given by the polynomial

$$P(\varepsilon) = \int_M \exp(\varepsilon\omega_h)\,\mathrm{ch}(K_P^{\bullet})\,\mathrm{Todd}(T_M).$$

Now by Proposition 1.2.1, $\chi(X, K_P^{\bullet}) = 0$ implies that the L^2 D''-complex is acyclic for $\varepsilon = 0$.

Due to the boundedness of the postulated primitive of ω_h, the L^2 D_ε''-complex is also acyclic for small values of ε. In particular:

$$\chi(M, K_P^{\bullet}) = 0 \implies a_k = \omega_h^{n-k}\big(\mathrm{ch}(K_p^{\bullet})\,\mathrm{Todd}(T_M)\big)_{k,k}[M] = 0 \quad \text{for all } k \leq n.$$

Now, for $k = n - 1$, $a_k = \int_M c_1(K_P^{\bullet})\omega_h^{n-1}$ (note that $a_n = 0$ since there is some n-dimensional geodesic cycle M' which satisfies $\chi(M', K_P^{\bullet}) = 0$). We are thus reduced to the situation of Theorem 4.2.3. $\square$

4.3   Constructing examples

The conditions in Definition 4.2.1 are so strong (and so artificial) that one might wonder if 1-hyperrigid homogenous period maps can exist at all. Since I have not found any general principle to classify 1-hyperrigid homogenous period maps, I only explain here how to construct all the nontrivial examples I know (see Appendix A.3 for a list of examples).

Lemma 4.3.1 *Let $\Omega = G/K$ be an irreducible Hermitian symmetric space of noncompact type and $\lambda \in \Pi_{\mathfrak{g}} \cap \overline{C}_{\mathfrak{g}}^+$ a weight whose associated VHS has length 1 (or 2 and is defined over the real numbers with Hodge vector (a, b, a)). Then:*

(1) *The Griffiths period map associated with the direct sum of k copies of $E_\lambda^{\mathfrak{g}}$, $\Omega \to D_{kp,kq}^{\mathrm{I}}$ (respectively,*

$$\Omega \to D_{\mathbb{R}}(ka, kb, ka) = \mathrm{SO}(2ka + kb)/\mathrm{S}\big(\mathrm{U}(ka) \times \mathrm{O}(kb)\big)$$

is 0-hyperrigid.

(2) Assume moreover that λ has a critical character, or that some multi-linear functor (for example, a symmetric or an exterior power) of $E_\lambda^{\mathfrak{g}}$ is an irreducible $\mathfrak{g}$-module whose highest weight has a critical character. If $C_D^P \neq 0$, then the homogenous period map is also 1-hyperrigid. In the case of a length 1 VHS with a critical character, this is always satisfied. In the length 2 case,

$$C_D^P = 0 \iff \dim \Omega = 2P \quad and \quad b/a = 2\dim \Omega/(\dim \Omega + 2).$$

Proof I give the proof of (1) only for VHS of length 1, the case of length 2 being similar. Recall the formula of Lemma 4.1.1 for the scalar curvature of a Griffiths period map. Observe that, for any locally homogenous VHS, $\sum_i [e_i, \bar{e}_i] \in \sqrt{-1}\mathfrak{s}(\mathfrak{u}(kp) \times \mathfrak{u}(kq))$ is the image under the representation of ρ_p. Thus it is a central element, that is, a matrix of the form (aI_{kp}, bI_{kq}), with $pa + bq = 0$.

For any n-dimensional subspace of $D_{kp,kq}^{\mathrm{I}}$, the element

$$\sum_i [e_i, \bar{e}_i] \in \sqrt{-1}\mathfrak{s}(\mathfrak{u}(kp) \times \mathfrak{u}(kq))$$

is a pair (A, B) of symmetric matrices, with A positive definite, B negative definite, $\mathrm{Tr}\, A = n$ and $\mathrm{Tr}\, A + \mathrm{Tr}\, B = 0$. Then $\|\sum_i [e_i, \bar{e}_i]\|^2 = \mathrm{Tr}(A^2) + \mathrm{Tr}(B^2)$ takes its minimal value at (aI_{kp}, bI_{kq}) with $pa + bq = 0$. This gives the maximality assumption on the scalar curvature.

For (2), the irreducibility of T_h^1 under the action of the isotropy group of the origin of the Griffiths domain D enables to conclude that S_D^P is constant on the space of Abelian subspaces. The main point to be verified for the statement on 1-hyperrigidity is thus $C_D^P \neq 0$. This follows from simple calculations. $\square$

A   Tables

The tables in this appendix were completed using the tables in [2], p. 251–276 and [9], p. 518.

A.1   Length of the fundamental representations

In Table 1, ω_i denotes the weight of the ith fundamental representation using the notation of [2].

The exceptional isomorphisms between classical domains are: $D_2^{\mathrm{II}} = \Delta$, $D_3^{\mathrm{II}} = B^3$, $D_4^{\mathrm{II}} = D_6^{\mathrm{IV}}$, $D_2^{\mathrm{III}} = D_3^{\mathrm{IV}}$, $D_{2,2}^{\mathrm{I}} = D_4^{\mathrm{IV}}$.

Ω	$\mathfrak{g}$ and $\mathfrak{l}$	$\dim_{\mathbb{C}}(\Omega)$	$|W_0|$	$l(E_{\omega_i})$
B^n	$\mathfrak{su}(n,1),$ $\mathfrak{s}(\mathfrak{u}(n) \times \mathfrak{u}(1))$	n	$n+1$	$l(E_{\omega_i}) = 1, 1 \leq i \leq n$
$D^I_{p,q}$, for $q \geq p \geq 2$	$\mathfrak{su}(p,q),$ $\mathfrak{s}(\mathfrak{u}(p) \times \mathfrak{u}(q))$	pq	$\binom{p+q}{p}$	$l(E_{\omega_i}) = i,\ 1 \leq i \leq p$ $l(E_{\omega_i}) = p,\ p \leq i \leq q$ $l(E_{\omega_i}) = p+q-i$ $q \leq i \leq p+q-1$
D^{II}_n, for $n \geq 4$	$\mathfrak{so}^*(2n), \mathfrak{u}(n)$	$\binom{n}{2}$	2^{n-1}	$l(E_{\omega_i}) = i, 1 \leq i \leq n-2$ $l(E_{\omega_{n-1}}) = [(n-1)/2]$ $l(E_{\omega_n}) = [n/2]$
D^{III}_n, for $n \geq 3$	$\mathfrak{sp}(n,\mathbb{R}), \mathfrak{u}(n)$	$\binom{n+1}{2}$	2^n	$l(E_{\omega_i}) = i, 1 \leq i \leq n$
D^{IV}_{2n}, for $n \geq 2$	$\mathfrak{so}(2n,2),$ $\mathfrak{so}(2n) \times \mathfrak{so}(2)$	$2n$	$2n+2$	$l(E_{\omega_i}) = 2, 1 \leq i \leq n-1$ $l(E_{\omega_n}) = 1, l(E_{\omega_{n+1}}) = 1$
D^{IV}_{2n+1}, for $n \geq 2$	$\mathfrak{so}(2n+1,2),$ $\mathfrak{so}(2n+1) \times \mathfrak{so}(2)$	$2n$	$2n+1$	$l(E_{\omega_i}) = 2, 1 \leq i \leq n$ $l(E_{\omega_{n+1}}) = 1$
D^V	$\mathfrak{e}_{6(-14)},$ $\mathfrak{so}(10) \times \mathfrak{so}(2)$	16	27	
D^{VI}	$\mathfrak{e}_{7(-25)}, \mathfrak{e}_6 \times \mathfrak{so}(2)$	27	56	

Table 1: Lengths of fundamental representations

In the case of type I domains, the weight ω_1 corresponds to the fundamental representation $\rho_1 \colon \mathfrak{su}(p,q) \to \mathfrak{gl}(p+q,\mathbb{C})$. $E^{\mathfrak{su}(p,q)}_k = \Lambda^k \rho_1$.

In the case of type III domains, the weight ω_1 corresponds to the fundamental representation $\rho_1 \colon \mathfrak{sp}(n,\mathbb{C}) \to \mathfrak{gl}(2n,\mathbb{C})$. $E^{\mathfrak{sp}(n,\mathbb{R})}_k$ is the primitive part of $\Lambda^k \rho_1$ (and can be seen as the primitive part of $R^k \pi_* \mathbb{C}$ where $\pi \colon A^{III}_n \to D^{III}_n$ is the universal family parametrizing marked principaly polarized n-dimensional Abelian varieties).

In the case of type II_n and IV_{2n-2} domains, E_{ω_1} is the fundamental representation $\rho_1 \colon \mathfrak{so}(2n) \to \mathfrak{gl}(2n,\mathbb{C})$. $E_{\omega_i} = \Lambda^i \rho_i$ for $i \leq n-2$, and $E_{\omega_{n-1}}$ and E_{ω_n} are the two halfspin representations.

In the case of type IV_{2n+1} domains, E_{ω_1} is the fundamental representation $\rho_1 \colon \mathfrak{so}(2n+1,2) \to \mathfrak{gl}(2n+3,\mathbb{C})$. $E_{\omega_i} = \Lambda^i \rho_i$ for $i \leq n$, and $E_{\omega_{n+1}}$ is the spin representation.

Lacking expertise on the two exceptional symmetric domains, I did not perform all the relevant computations. However, observing that length one homogenous VHS can be used to classify totally geodesic holomorphic embeddings on symmetric domains into classical ones (compare our table with the list of embeddings of a classical domain into another one given by [15], p. 188), the fact that D^V and D^{VI} have no such embedding ([15], p. 187) is equivalent to the fact that they do not carry any length one homogenous VHS.

A.2 Special weights for classical domains

For each classical domain, the combinatorial recipe for testing whether a dominant weight is special is given in Table 2:

Ω	$\lambda = \sum_i n_i \omega_i$ is special if and only if:
B^n	$\lambda = 0$
$D^I_{p,q}$, for $q \geq p \geq 2$	$\forall k \in \{0, \ldots, pq + \sum_{i=1}^{p} i(n_i + n_{p+q-i}) + p\sum_{p+1}^{p+q-1} n_i\}$, $\exists A \subset \{1, \ldots, p+q\}$ with $|A| = p$ such that $\sum_{j \in A}(1 + n_j)(\min(j, p) - |A \cap \{1, \ldots, j\}|) = k$
D^{II}_n, for $n \geq 4$	$\forall k \in \{0, \ldots, \binom{n}{2} + \sum_{i=1}^{n-2} i l_i + [\frac{n-1}{2}]l_{n-1} + [\frac{n}{2}]l_n\}$, $\exists A \subset \{1, \ldots, n\}$ with $|A| \equiv 0 \mod 2$ such that $\delta_{n \in A}\frac{l_n - l_{n-1}}{2} + \sum_{\substack{i \in A \\ i < n}}\left(n - i + \frac{l_{n-1} + l_n}{2} + \sum_{j=i}^{n-2} l_j\right) = k$
D^{III}_n, for $n \geq 3$	$\forall k \in \{0, \ldots, \frac{n(n+1)}{2} + \sum_{i=1}^{n} i n_i\}, \exists A \subset \{1, \ldots, n\}$ such that $\sum_{i \in A}(n - i + 1 + \sum_{l=i}^{n} n_l) = k$
D^{IV}_{2n}, for $n \geq 2$	$\lambda \in \{0, \omega_n, \omega_{n+1}\}$, where $E_{\omega_n}, E_{\omega_{n+1}}$ are the halfspin representations of $\mathfrak{so}(2n + 2)$
D^{IV}_{2n+1}, for $n \geq 2$	$\lambda = 0$

Table 2: Special weights

Except in the particular cases of type IV domains and of the complex ball or of some other small domains, I cannot give the whole list of special weights, but can give specific examples. In the following list, the weight λ is supposed to be the sum $\lambda = \sum_i l_i \omega_i$, with $l_i \in \mathbb{N}$:

Case $D_{p,q}^{I}$ for $q \geq p \geq 2$

- If $l_p \neq 0$ (or $l_q \neq 0$) then $\lambda = \sum l_i \omega_i$ is not a special weight.

- If $l_{p+1} \neq 0$ and $l_{p-1} \neq 0$ (or $l_{q\pm1} \neq 0$) then $\lambda = \sum l_i \omega_i$ is not a special weight.

- $r\omega_{p+q-1}$ (respectively $r\omega_1$) is special if and only if $r < 1 + (p-1)(q-1)$.

Case D_n^{III} for $n \geq 2$

- If $l_n \neq 0$ or $l_{n-1} \neq 0$ then λ is not special.

- $p\omega_1 + q\omega_2$ is special if and only if $p \leq \frac{(n-1)(n-2)}{2}$ and $q \leq \frac{(n-2)(n-3)}{2}$.

- $p\omega_1 + q\omega_2 + r\omega_3$ is special if and only if $r \leq \frac{(n-4)(n-3)}{2}$, $q \leq \frac{(n-2)(n-3)}{2}$ and $|n - 3 + r - p| \leq 1 + \frac{(n-3)(n-2)}{2}$.

In particular, for D_5^{III}, $7\omega_1 + 3\omega_2$ is not special whereas $7\omega_1 + 3\omega_2 + \omega_3$ is special; this shows that being nonspecial is not a monotonous property, that is, λ nonspecial does not necessarily imply that $\lambda + \mu$ is nonspecial.

Case D_n^{II} for $n \geq 4$

- If $l_n \neq 0$, λ is not special. When $n \equiv 1 \mod 2$, a special weight also has $l_{n-1} = 0$ and $l_n \leq 1$.

- The weight $\sum_{i=1}^{n-2} l_i \omega_i^{II}$ is a special weight for D_n^{II} iff $\sum_{i=1}^{n-2} l_i \omega_i^{III}$ is a special weight for D_{n-1}^{III}.

Ω	D	Nonspecial isotypic VHS
B^n	$D_{kC_n^{r-1}, kC_n^r}^{I}, 1 \leq r \leq n, k \in \mathbb{N}$	$(E_{\omega_r}^{I})^{\oplus k}$
D_{2n+1}^{IV}	$D_{k2^n, k2^n}^{I}, k \in \mathbb{N}$	$(E_{\omega_n}^{I})^{\oplus k}$
$D_n^{II}, n \geq 4$	$D_{n,n}^{I}$	$E_{\omega_{n-2}}^{II}(= \Lambda^{n-2} E_{\omega_1}^{II})$
$D_n^{III}, n \geq 3$ $n \equiv 0, 1[4]$	$D_{n,n}^{I}$	$E_{(1+d)\omega_1}^{III}(= S^{1+d} E_{\omega_1}^{III})$, $d = \binom{n-1}{2}$
B^n	$D_{\mathbb{R}}(ka_i, kb_i, ka_i)$	$E_{\omega_i + \omega_{n-i}}^{I}, i \leq \frac{n}{2}$
D_n^{IV}	$D_{\mathbb{R}}(kC_n^{r-1}, k(C_n^r + C_n^{r-2}), kC_n^{r-1})$	$(E_{\omega_r}^{IV})^{\oplus k} = (\Lambda^r E_{\omega_1}^{IV})^{\oplus k}, r \leq n$

Table 3: Hypperrigid homogeneous period maps

A.3 Examples of 1-hyperrigid period maps

Table 3 gives the examples of hyperrigid homogenous period maps $\Omega \to D$ constructed thanks to Lemma 4.3.1.

References

[1] M. Atiyah, Elliptic operators, discrete groups and von Neumann algebras, Soc. Math. Fr. Astérisque **32–33**, 1976, 43–72

[2] N. Bourbaki, Groupes et algèbres de Lie, Chap. 4, 5 et 6, Eléments de mathématique, fascicule XXXIV, Hermann, 1968

[3] P. Deligne, Travaux de Griffiths, Séminaire Bourbaki 1970, exp. 376, Springer Lecture Notes in Mathematics **180**

[4] P. Eyssidieux, La caractéristique d'Euler du complexe de Gauss–Manin, Journ. reine angew. Mathematik **490** (1997), 155-212

[5] W. Fulton and R. Lazarsfeld, Positive polynomials for ample line bundles, Ann. of Math. **118** (1983), 35–60

[6] P. Griffiths, Periods of integrals on algebraic varieties, III, Publ. Math. IHES **43** (1973), 125–180

[7] P. Griffiths and W.Schmid, Locally homogenous complex manifolds, Acta Mathematica **123** (1969), 253–301

[8] M. Gromov, Kähler hyperbolicity and L^2-Hodge theory, J. Diff. Geom. **33** (1991), 263–291

[9] S. Helgason, Differential geometry, Lie groups, and symmetric spaces, Academic Press, 1978

[10] B. Kostant, Lie algebra cohomology and the generalized Borel–Weil theorem, Ann. of Math. **74** (1961), 329–387

[11] Y. Matsushima and S. Murakami, On certain cohomology groups attached to Hermitian symmetric spaces, Osaka J. Math. **2** (1965), 1–35

[12] N. Mok, Aspects of Kähler geometry on arithmetic varieties in several complex variables and complex geometry, Part 2 (Santa Cruz, 1989) Proc. Symp. Pure math **52** Part 2, AMS, 1991, 335–394

[13] N. Mok and P. Eyssidieux, Characterization of certain totally geodesic cycles on Hermitian locally symmetric spaces of the noncompact type, in Modern methods in complex analysis (in honor of R. Gunning and J.J. Kohn), ed. T. Bloom, D. Catlin, J.P. D'Angelo and Y.T. Siu, Annals of Math. Studies **137** (1996), 79–117

[14] C. Peters, Curvature for period domains, Proc. Symp. in Pure Math. **53**, 1991, 261–268

[15] I. Satake, Algebraic structures of symmetric domains, Publications of the Mathematical Society of Japan **14**, Iwanami Shoten and Princeton University Press, 1980

[16] C. Simpson, Higgs bundles and local systems, Pub. Math. IHES **75** (1992), 5–93

[17] S. Zucker, Locally homogenous variations of Hodge structures, L'Enseign. Math. (2) **27** (1981), 243–276

Philippe Eyssidieux,
Laboratoire Emile Picard, CNRS UMR 5580, UFR MIG
Université Paul Sabatier, 118 route de Narbonne,
31062 Toulouse Cedex 4, France
e-mail: eyssi@picard.ups-tlse.fr

Algorithms for computing intersection numbers on moduli spaces of curves, with an application to the class of the locus of Jacobians

Carel Faber

The first purpose of this note is to explain how to compute intersection numbers of divisors on the moduli spaces $\overline{\mathcal{M}}_{g,n}$ of stable pointed curves. The Witten conjecture, proved by Kontsevich, gives a recipe to compute the intersection numbers of the n basic line bundles on $\overline{\mathcal{M}}_{g,n}$. As we will see, knowing these numbers allows one to compute all other intersection numbers of divisors as well. That this is possible was pointed out to me by Rahul Pandharipande. Earlier, Eduard Looijenga had made a remark that went a long way in the same direction.

After describing the various divisors on $\overline{\mathcal{M}}_{g,n}$, we proceed to discuss the algorithm computing their intersection numbers. We discuss our implementation of this algorithm and the results we obtain from it. For example, we compute all intersection numbers on $\overline{\mathcal{M}}_g$ for $g \leq 6$. (Copies of the program and some data computed with it are available from the author.)

A refined version of the algorithm requires us to take into account certain higher codimension classes, introduced by Mumford and Arbarello–Cornalba; it computes all intersection numbers of these classes and divisors. Recently, we realized that the Chern classes of the Hodge bundle can be taken along as well; hence all intersection numbers of Mumford's tautological classes and divisors can be computed. This has several applications. In §4 we discuss one application in detail: the calculation of the class of the locus of Jacobians in the moduli space of principally polarized abelian varieties of dimension g (projected in the tautological ring). This class was classically known for $g = 4$ and we computed it by ad hoc methods for $g = 5$; the new method allows us in principle to compute the class for all g, in practice currently for $g \leq 7$. Other applications may be found in the recent papers of Graber and Pandharipande [GP] and Kontsevich and Manin [KM].

1   Line bundles and divisors on $\overline{\mathcal{M}}_{g,n}$

For nonnegative integers g and n with $2g - 2 + n > 0$, denote by $\overline{\mathcal{M}}_{g,n}$ the moduli space of stable n-pointed curves of genus g, over an algebraically closed field k. This is the Deligne–Mumford compactification of the moduli space $\mathcal{M}_{g,n}$ of smooth n-pointed curves $(C; x_1, \ldots, x_n)$ of genus g (with $x_i \neq x_j$ if $i \neq j$). We consider certain classes in the rational Picard group of $\overline{\mathcal{M}}_{g,n}$. First, for $1 \leq i \leq n$, let ψ_i denote the first Chern class of the line bundle whose fiber at a stable n-pointed curve $(C; x_1, \ldots, x_n)$ is the cotangent space to C at x_i; that is, $\psi_i = c_1(\sigma_i^*(\omega_{\pi_{n+1}}))$, where $\pi_{n+1} \colon \overline{\mathcal{M}}_{g,n+1} \to \overline{\mathcal{M}}_{g,n}$ is the morphism obtained by forgetting the $(n + 1)$st marked point (the universal curve, cf. [Kn 1]), $\omega_{\pi_{n+1}}$ is the relative dualizing sheaf, and $\sigma_1, \ldots, \sigma_n$ are the natural sections of π_{n+1} (the image of a stable n-pointed curve under σ_i is the stable $(n + 1)$-pointed curve obtained by attaching a 3-pointed rational curve at the ith point and considering the remaining 2 points on that curve as the ith and $(n + 1)$st point). Next, following [AC 2], §1, we define $\kappa_1 = \pi_{n+1,*}(K^2)$, with $K = c_1(\omega_{\pi_{n+1}}(\sum_{i=1}^n D_i))$, where D_i is the divisor that is the image of the section σ_i, for $1 \leq i \leq n$. Note that it is a consequence of results of Harer (cf. [Ha 1], [Ha 2], [AC 1]) that over $\mathbb{C}$ the restrictions to $\mathcal{M}_{g,n}$ of the classes κ_1 and $\psi_1, \ldots, \psi_n$ generate the rational Picard group of $\mathcal{M}_{g,n}$.

To get generators for the rational Picard group of $\overline{\mathcal{M}}_{g,n}$, we have to add the fundamental classes of the boundary divisors. Exactly when $g > 0$, there is a boundary component whose generic point corresponds to an irreducible singular curve. It is the image of $\overline{\mathcal{M}}_{g-1,n+2}$ under the degree 2 map that identifies the $(n + 1)$st and $(n + 2)$nd point on each curve. Following [AC 2], we denote this locus by Δ_{irr} and its class in the Picard group by δ_{irr}. (For $g = 0$ this class is 0 by definition.)

The other boundary components parametrize reducible singular curves. The generic point of such a component corresponds to a curve with two irreducible components C_1 and C_2 of genus g_1 and g_2 with $g_1 + g_2 = g$, and labelled by subsets N_1 and N_2 of $\underline{n} = \{1, 2, \ldots, n\}$ satisfying $N_1 \coprod N_2 = \underline{n}$ that correspond to the marked points on the two components. All partitions of g and $\underline{n}$ that lead to a stable curve occur; this just translates as the condition $|N_i| \geq 2$ when $g_i = 0$. Such a boundary component is the image of $\overline{\mathcal{M}}_{g_1,|N_1|+1} \times \overline{\mathcal{M}}_{g_2,|N_2|+1}$ under the natural map that identifies the two 'extra' points and labels the $|N_i|$ remaining points on C_i with the labels from N_i. We choose to denote this boundary component in the case $n > 0$ by Δ_{g_i,N_i}, where N_i is the subset of $\underline{n}$ containing 1. In the case $n = 0$ the N_i are empty and we may drop them in the notation; note that $\Delta_{g_1} = \Delta_{g_2}$ and that this component is usually denoted by $\Delta_{\min(g_1,g_2)}$.

Although this plays no role in the sequel, we point out that the classes in the rational Picard group introduced so far are independent whenever

$g \geq 3$. For $g = 2$, there is one relation, arising from the fact that κ_1 on $\overline{\mathcal{M}}_2$ comes from the boundary. For $g = 1$, both κ_1 and the ψ_i come from the boundary; the boundary components are independent. For $g = 0$, the boundary components generate, but are not independent; the relations arise from the various projections to $\overline{\mathcal{M}}_{0,4}$ and the equivalence of its 3 boundary cycles. For proofs of these statements, see [Ke] and [AC 3].

There is one other divisor class which is most useful: λ_1, the first Chern class of the Hodge bundle. The Hodge bundle is the locally free rank g sheaf (on the moduli functor) whose fiber at a curve C is $H^0(C, \omega_C)$. So it is 0 in genus 0, while it is a pullback from $\overline{\mathcal{M}}_{1,1}$ or $\overline{\mathcal{M}}_g$ in case $g = 1$, respectively $g \geq 2$.

2   The idea of the algorithm

Suppose given a monomial of degree $3g-3+n$ in the divisor classes $\kappa_1, \psi_1, \dots,$ $\psi_n, \delta_{\mathrm{irr}}$ and the δ_{g_i, N_i} on $\overline{\mathcal{M}}_{g,n}$; we want to compute the corresponding intersection number. We interpret the divisor classes as classes on the moduli functor. In the case of a boundary divisor, this means that we divide the usual fundamental class by the order of the automorphism group of the generic curve parametrized by the divisor. We denote these divisor classes by $\delta_{...}$, to distinguish them from the actual boundary divisors $\Delta_{...}$.

The case in which the monomial involves only the ψ_i is of course covered by the Witten conjecture [Wi], proved by Kontsevich [Ko]. As explained in [AC 2], this also allows us to compute intersection numbers involving both κ_1 and the ψ_i.

It remains to compute the intersection numbers involving a boundary class. We think of such a number as the intersection of the remaining classes on the corresponding boundary component. A problem with this approach appears to be that most boundary components have singularities that are not quotient singularities, which means that one cannot properly do intersection theory on them. This problem is easily solved: we have seen that each boundary component is the image under a finite map of a moduli space of stable pointed curves or a product of two such spaces. (The map almost always has degree 1; the only exceptions are the degree 2 map from $\overline{\mathcal{M}}_{g-1,n+2}$ to Δ_{irr} and, in the case g even, the degree 2 map from $\overline{\mathcal{M}}_{g/2,1} \times \overline{\mathcal{M}}_{g/2,1}$ to $\Delta_{g/2}$.) So we wish to pull the remaining divisor classes back by means of this map. If we can express the pullbacks in terms of the basic classes on the new moduli space (or product of moduli spaces), we will be done, by induction on the dimension of the moduli space.

So the point is to understand the pullbacks of the basic divisor classes from $\overline{\mathcal{M}}_{g,n}$ to the moduli spaces occurring in its boundary components.

It is clear that the ψ_i pull back to ψ_i on the new moduli space(s): to the first n on $\overline{\mathcal{M}}_{g-1,n+2}$, and to the $|N_1|$ classes ψ_i on $\overline{\mathcal{M}}_{g_1,|N_1|+1}$ and the $|N_2|$ classes ψ_i on $\overline{\mathcal{M}}_{g_2,|N_2|+1}$ that correspond to the points not identified in the map to Δ_{g_i,N_i}.

As explained in [AC 2], the class κ_1 pulls back to κ_1 on $\overline{\mathcal{M}}_{g-1,n+2}$, respectively to the sum of the pullbacks of the κ_1 from the two factors on the product $\overline{\mathcal{M}}_{g_1,|N_1|+1} \times \overline{\mathcal{M}}_{g_2,|N_2|+1}$.

Pulling back a boundary divisor other than the one under consideration to the new moduli space(s) is not difficult. The main point is that two distinct boundary divisors intersect transversally in the universal deformation space (see [DM]). It remains to identify the boundary divisors on the new moduli space(s) that arise as the inverse image of the intersection of the boundary divisor under consideration with a distinct one.

For example, in the case Δ_{irr}, the pullback of the class $\delta_{h,M}$ to $\overline{\mathcal{M}}_{g-1,n+2}$ is the sum

$$\delta_{h-1,M\cup\{n+1,n+2\}} + \delta_{h,M},$$

with some exceptions: when $n = 0$ and $2h = g$, the two classes in the sum are equal and the pullback consists of that class just once; when $n = 0$ otherwise, $\delta_{h,\emptyset}$ was denoted $\delta_{g-1-h,\{1,2\}}$ above; when $h = 0$ or $h = g$, the first, respectively the second, summand is not defined, and should be omitted.

We now consider the pullbacks of boundary divisors to a product of the form $\overline{\mathcal{M}}_{g_1,|N_1|+1} \times \overline{\mathcal{M}}_{g_2,|N_2|+1}$. The pullback of δ_{irr} is the sum of the δ_{irr} on the two factors. It remains to find the pullbacks of the boundary divisors parametrizing reducible curves of type other than that under consideration.

We start with the case $n = 0$. Hence $N_1 = N_2 = \emptyset$. We may assume that $g_1 \leq g_2$, and we want to pull back a class δ_h, with $h \leq g - h$ and $h \neq g_1$. The general point in a component of the intersection of the two boundary divisors parametrizes a chain of 3 irreducible curves. There are *a priori* 4 possibilities for the genera of the 3 components:

(1) $[h, g_1 - h, g_2]$, occurring when $g_1 > h$;

(2) $[h, g_2 - h, g_1]$, occurring when $g_2 > h$;

(3) $[g_1, h - g_1, g - h]$, occurring when $h > g_1$;

(4) $[g_2, h - g_2, g - h]$, occurring when $h > g_2$.

Here the second entry refers to the genus of the middle component; note that $[a, b, c]$ and $[c, b, a]$ describe the same type of curves.

Note that $h \leq g/2 \leq g_2$. In fact $h < g_2$, since equality implies $h = g_1$, which we have excluded. So (4) never occurs, while (2) always occurs. We conclude that, with one exception, the pullback of δ_h equals

$$\begin{cases} \mathrm{pr}_1^* \, \delta_{g_1-h,\{1\}} + \mathrm{pr}_2^* \, \delta_{g_2-h,\{1\}} & \text{if } g_1 > h; \\ \mathrm{pr}_2^* \, \delta_{g_2-h,\{1\}} + \mathrm{pr}_2^* \, \delta_{h-g_1,\{1\}} & \text{if } h > g_1, \end{cases}$$

where pr_1, pr_2 are the projections to the two factors of $\overline{\mathcal{M}}_{g_1,|N_1|+1} \times \overline{\mathcal{M}}_{g_2,|N_2|+1}$. The exception is when $h = g - h$ (implying $h > g_1$), and the two summands in the second line are equal; then the actual pullback consists of that class just once.

Now for the case $n > 0$. We may assume that $1 \in N_1$ and we want to pull back a class $\delta_{h,M}$ other than δ_{g_1,N_1}. So $1 \in M$ and $(h,M) \neq (g_1,N_1)$, but we no longer have $g_1 \leq g_2$ or $h \leq g - h$. Again there are *a priori* 4 possibilities for the genera of the 3 components, where as before the second entry refers to the genus of the middle component:

(1) $[h, g_1 - h, g_2]$, occurring when $g_1 \geq h$ and $M \subset N_1$;

(2) $[h, g_2 - h, g_1]$, occurring when $g_2 \geq h$ and $M \subset N_2$;

(3) $[g_1, h - g_1, g - h]$, occurring when $h \geq g_1$ and $N_1 \subset M$;

(4) $[g_2, h - g_2, g - h]$, occurring when $h \geq g_2$ and $N_2 \subset M$.

Observe that (2) never occurs, since $1 \in M$ and $1 \notin N_2$. Note also that the other possibilities indeed yield stable curves in all cases: in (1) and (3) the necessary condition $M \neq N_1$ when $h = g_1$ is fulfilled, and in (4) the equality $M = N_2$ never occurs. Finally, note that types (1), (3) and (4) are mutually exclusive.

This means that the pullback of $\delta_{h,M}$ consists of the sum of 0, 1 or 2 of the classes from the following list, depending on which conditions hold:

(1) $\mathrm{pr}_1^* \, \delta_{h,M}$, pullback from $\overline{\mathcal{M}}_{g_1,N_1\cup\{*\}}$ when $g_1 \geq h$ and $M \subset N_1$;

(3) $\mathrm{pr}_2^* \, \delta_{h-g_1,M-N_1\cup\{1\}}$, pullback from $\overline{\mathcal{M}}_{g_2,N_2\cup\{1\}}$ when $h \geq g_1$ and $N_1 \subset M$;

(4) $\mathrm{pr}_1^* \, \delta_{h-g_2,M-N_2\cup\{*\}}$, pullback from $\overline{\mathcal{M}}_{g_1,N_1\cup\{*\}}$ when $h \geq g_2$ and $N_2 \subset M$.

Here we have identified the factors of the product of moduli spaces by means of sets of marked points instead of just their number of elements. In light of our convention to label a divisor parametrizing reducible curves by the genus of the component containing 1 and by the set of marked points on that component, it is natural to give the 'extra' point in case (3) the label 1, rather than $*$.

Finally, we have to deal with self-intersections of boundary divisors: we need to pull the class of a boundary divisor back to the corresponding moduli space (or product of moduli spaces). It is not difficult to deal with this directly, but it is easier to use the fundamental identity on $\overline{\mathcal{M}}_{g,n}$

$$\kappa_1 = 12\lambda_1 - \delta + \psi,$$

where δ is the sum of the functorial classes of the boundary divisors and ψ is the sum of the n classes ψ_i (see [Co]). Namely, every divisor class on $\overline{\mathcal{M}}_{g,n}$ discussed so far occurs in this identity. So a given boundary divisor class can be expressed as a linear combination of other divisor classes, and if we know how to pull back the other classes, we will also know how to deal with self-intersections. We have discussed the pullbacks of κ_1, the ψ_i and the other boundary divisor classes above, so we only need to determine the pullback of λ_1. Note that the pullback of the Hodge bundle to $\overline{\mathcal{M}}_{g-1,n+2}$ is an extension of a trivial line bundle by the Hodge bundle in genus $g - 1$, whereas the pullback of the Hodge bundle to $\overline{\mathcal{M}}_{g_1,|N_1|+1} \times \overline{\mathcal{M}}_{g_2,|N_2|+1}$ is the direct sum of the Hodge bundle in genus g_1 and the Hodge bundle in genus g_2 (see [Kn 2] or [Co].) Hence we find that the pullback of λ_1 to $\overline{\mathcal{M}}_{g-1,n+2}$ equals λ_1, whereas its pullback to $\overline{\mathcal{M}}_{g_1,|N_1|+1} \times \overline{\mathcal{M}}_{g_2,|N_2|+1}$ equals $\mathrm{pr}_1^* \lambda_1 + \mathrm{pr}_2^* \lambda_1$.

This determines the pullback of a boundary divisor class to its corresponding (product of) moduli space(s). We find that for $n = 0$, the pullback of δ_{irr} to $\overline{\mathcal{M}}_{g-1,2}$ equals

$$-\psi_1 - \psi_2 + \delta_{\mathrm{irr}} + \sum_{h=1}^{g-2} \delta_{h,\{1\}},$$

whereas in the case $n > 0$ the pullback of δ_{irr} to $\overline{\mathcal{M}}_{g-1,n+2}$ equals

$$-\psi_{n+1} - \psi_{n+2} + \delta_{\mathrm{irr}} + \sum_{\substack{0 \le h \le g-1, 1 \in M \subset \underline{n} \\ M \ne \underline{n} \text{ when } h = g-1}} (\delta_{h,M \cup \{n+1\}} + \delta_{h,M \cup \{n+2\}}).$$

In the case of a boundary divisor parametrizing reducible curves, an actual self-intersection is much rarer. We find that the pullback of δ_{g_1,N_1} to $\overline{\mathcal{M}}_{g_1,N_1 \cup \{*\}} \times \overline{\mathcal{M}}_{g_2,N_2 \cup \{1\}}$ equals

$$\begin{cases} -\,\mathrm{pr}_1^* \psi_{\{*\}} - \mathrm{pr}_2^* \psi_1 + \mathrm{pr}_2^* \delta_{g_2-g_1,\{1\}} & \text{if } n = 0 \text{ and } g_1 < g_2; \\ -\,\mathrm{pr}_1^* \psi_{\{*\}} - \mathrm{pr}_2^* \psi_1 + \mathrm{pr}_1^* \delta_{g_1-g_2,\underline{n} \cup \{*\}} & \text{if } n > 0, N_1 = \underline{n} \text{ and } g_1 \ge g_2 > 0; \\ -\,\mathrm{pr}_1^* \psi_{\{*\}} - \mathrm{pr}_2^* \psi_1 & \text{otherwise.} \end{cases}$$

This finishes the theoretical description of the algorithm.

3 Implementation and results

We have implemented the algorithm outlined in §2 in Maple [Ma]. This first requires an implementation of Witten's algorithm [Wi] to compute the intersection numbers of the ψ_i on $\overline{\mathcal{M}}_{g,n}$. For this, we gratefully use the results of Chris Zaal [Za] who, using such an implementation, computed a table containing all the intersection numbers

$$\langle \tau_{d_1} \tau_{d_2} \cdots \tau_{d_n} \rangle = \psi_1^{d_1} \psi_2^{d_2} \cdots \psi_n^{d_n}$$

on $\overline{\mathcal{M}}_{g,n}$ with $g \leq 9$ for which all $d_i \geq 2$ (hence $n \leq 24$ since $\sum_{i=1}^{n}(d_i - 1) = 3g-3$). The intersection numbers for which a d_i equals 0 or 1 can be computed from these by means of the string and dilaton equations [Wi]:

$$\begin{cases} \langle \tau_0 \tau_{d_1} \tau_{d_2} \cdots \tau_{d_n} \rangle = \sum_{i:d_i>0} \langle \tau_{d_1} \cdots \tau_{d_i-1} \cdots \tau_{d_n} \rangle & \text{(string equation)}, \\ \langle \tau_1 \tau_{d_1} \tau_{d_2} \cdots \tau_{d_n} \rangle = (2g - 2 + n) \langle \tau_{d_1} \tau_{d_2} \cdots \tau_{d_n} \rangle & \text{(dilaton equation)}. \end{cases}$$

From the intersection numbers of the ψ_i, one can determine the intersection numbers of the classes κ_i on $\overline{\mathcal{M}}_g$ introduced by Mumford [Mu], as outlined in [Wi]. Arbarello and Cornalba [AC 2] introduced classes κ_i on $\overline{\mathcal{M}}_{g,n}$ generalizing Mumford's classes, and they show that the intersection numbers of the ψ_i determine the intersection numbers of these κ_i as well as the 'mixed' intersection numbers of the divisor classes κ_i and ψ_i. So in particular the intersection numbers of κ_1 and the ψ_i are determined. It is easy to implement the calculation of these numbers from the intersection numbers of the ψ_i (especially so with a formula we learned from Dijkgraaf [Dij]), and we have for instance calculated the numbers κ_1^{3g-3} on $\overline{\mathcal{M}}_g$ for $g \leq 9$.

Naturally the various divisors on $\overline{\mathcal{M}}_{g,n}$ have to be ordered in some consistent way. On $\overline{\mathcal{M}}_g$ we start with κ_1, followed by δ_{irr} and then the 'reducible' boundary divisors, ordered by the minimum of the genera of the 2 components, for a total of $[g/2]+2$ classes. (The class λ_1 was introduced only to deal with self-intersections of boundary divisors and is not actually used in the program.) When $n > 0$, there are $(g+1)2^{n-1}+1$ classes: first $\psi_1, \ldots, \psi_n, \kappa_1$ and δ_{irr}, then the reducible classes, ordered first by the genus of the component that contains the point 1, then by the number of points on that component, and finally by the lexicographic ordering of subsets of $\underline{n}$ (of equal size and containing 1). (Recall that in genus 0 the class δ_{irr} is 0; it is included for convenience.)

In the case of a pullback to a product of moduli spaces, we need to renumber the indices from $N_1 \cup \{*\}$ as well as those from $\{1\} \cup N_2$. For this we just use the natural ordering of the elements of $\underline{n}$, taking $*$ as the $(n+1)$st point.

The implementation of the actual algorithm is now rather straightforward. Given a monomial M in the divisor classes involving at least one boundary divisor, we order the classes as above, and pull M back to the (product of) moduli space(s) corresponding to the last divisor that occurs. Note that whenever the monomial contains several distinct boundary divisors, we have a choice here; while the ordering we choose is probably optimal for $\overline{\mathcal{M}}_g$, it is probably not for $n > 0$: we have recently experimented with another ordering, which indeed appears to be better; the idea is that in the reducible case one should try to pull back to two moduli spaces parametrizing curves of approximately equal genus and number of points.

In case the last divisor is Δ_{irr}, we find a homogeneous polynomial of degree $3g - 4 + n$ in the divisor classes on $\overline{\mathcal{M}}_{g-1,n+2}$. After expanding, it is a sum of monomials (with coefficients) in the divisor classes; these are evaluated by means of the (heavily recursive) algorithm.

In the reducible case, we also find a homogeneous polynomial of degree $3g - 4 + n$, but this time in two sets of variables, the divisors on $\overline{\mathcal{M}}_{g_1,|N_1|+1}$ and those on $\overline{\mathcal{M}}_{g_2,|N_2|+1}$. Many of the monomials will not have the correct degree $3g_1 - 2 + |N_1|$ in the first set of variables and are 0 for trivial reasons. The others automatically have degree $3g_2 - 2 + |N_2|$ in the second set of variables; writing such a monomial as $c \cdot M_1 \cdot M_2$, where c is the coefficient of the monomial and M_i is the monic monomial in the ith set of variables, it contributes $c \cdot a(M_1) \cdot a(M_2)$, where $a(M_i)$ is the result of applying the algorithm to M_i on $\overline{\mathcal{M}}_{g_i,|N_i|+1}$.

Using this implementation of the algorithm, we have computed, for example, the 28 intersection numbers of κ_1, δ_{irr} and δ_1 on $\overline{\mathcal{M}}_3$, confirming the results of [Fa 1]. However, because of its heavily recursive character, the algorithm already becomes impracticable in the computation of certain intersection numbers on $\overline{\mathcal{M}}_4$. Most intersection numbers are still easy to compute, but especially the numbers $\kappa_1^{9-i}\delta_{\text{irr}}^i$ with i large take a long time. It is quite clear why: firstly, a pullback to $\overline{\mathcal{M}}_{g-1,n+2}$ is the 'worst case', since the changes in genus and in dimension of the moduli space are minimal, while the number of points increases by 2, so that the number of divisors increases by a factor of almost 4; secondly, as we saw in §2, the pullback of δ_{irr} to $\overline{\mathcal{M}}_{g-1,n+2}$ involves by far the highest number of terms.

Observe however that the class δ_{irr} is a pullback from $\overline{\mathcal{M}}_g$, respectively $\overline{\mathcal{M}}_{1,1}$. This first of all implies

$$\delta_{\text{irr}}^{m+1} = 0,$$

where $m = \max(g, 3g - 3)$, for all $g \geq 0$. Using this identity systematically already saves a considerable amount of time. Moreover, any product involving only κ_1, the ψ_i and δ_{irr} can be pushed down to $\overline{\mathcal{M}}_g$, respectively $\overline{\mathcal{M}}_{1,1}$, by using the projection formula and the formulas in [AC 2]. This leads to intersection numbers on those spaces of monomials in δ_{irr} and the higher κ_i mentioned before. As observed by Arbarello and Cornalba, the κ_i behave very well under pullback, and it is clear that all intersection numbers involving divisors as well as the κ_i can be computed by means of an algorithm almost identical to that in §2. Even the implementation is easy to adapt. The point is that this greatly simplifies the calculation of the numbers involving only κ_1, the ψ_i and δ_{irr}: in the naive implementation, the complexity of a calculation increases at least exponentially with the number of points, and one would have to calculate certain numbers on $\overline{\mathcal{M}}_{0,2g}$ to get all numbers on $\overline{\mathcal{M}}_g$; but now many of the hardest numbers can be computed using at most 2-pointed curves at all stages of the computation.

With these simple changes implemented, the calculation of many more numbers becomes practical. We have calculated all intersection numbers of divisors on $\overline{\mathcal{M}}_g$ for $g \leq 6$ as well as on $\overline{\mathcal{M}}_{3,1}$ and $\overline{\mathcal{M}}_{4,1}$. To obtain the 2 numbers $\kappa_1 \delta_{\mathrm{irr}}^{14}$ and $\delta_{\mathrm{irr}}^{15}$ on $\overline{\mathcal{M}}_6$, we used the relations $\lambda_1^{13} \delta_{\mathrm{irr}} \kappa_1 = 0$ and $\lambda_1^{13} \delta_{\mathrm{irr}}^2 = 0$, consequences of the relation $\lambda_1^{13} \delta_{\mathrm{irr}} = 0$, which is geometrically obvious.

This section of the paper would hardly be complete without some actual intersection *numbers*. Here are a few:

On $\overline{\mathcal{M}}_4$: $\quad \delta_{\mathrm{irr}}^9 = \dfrac{-251987683}{4320}$ and $\lambda_1^9 = \dfrac{1}{113400}$;

On $\overline{\mathcal{M}}_5$: $\quad \delta_{\mathrm{irr}}^{12} = \dfrac{-1766321028967}{6048}$ and $\lambda_1^{12} = \dfrac{31}{680400}$;

On $\overline{\mathcal{M}}_6$: $\quad \delta_{\mathrm{irr}}^{15} = \dfrac{-32467988437272065977}{7257600}$ and $\lambda_1^{15} = \dfrac{431}{481140}$.

We computed the number λ_1^9 on $\overline{\mathcal{M}}_4$ in [Fa 2] by a completely different ad hoc method. Calculating it with the algorithm amounts to calculating all 220 intersection numbers of divisors on $\overline{\mathcal{M}}_4$, which provides a nice check of the implementation.

4   The class of the locus of Jacobians

We used the calculation of λ_1^9 on $\overline{\mathcal{M}}_4$ in [Fa 2] to obtain the well-known result that the class of the locus $\mathcal{J}_4$ of Jacobians of curves of genus 4 in the moduli space $\mathcal{A}_4$ of principally polarized abelian varieties of dimension 4 equals $8\lambda_1$. Using the computation of λ_1^{12} on $\overline{\mathcal{M}}_5$ as well as some computations in the tautological ring of $\mathcal{M}_5$ as in [Fa 3], we could determine the class of $\mathcal{J}_5$ in $\mathcal{A}_5$, as we explain in a moment. Recently we realized that Mumford's formula [Mu] for the Chern character of the Hodge bundle on $\overline{\mathcal{M}}_g$, together with an algorithm similar to that discussed in §2, enables one to compute the class of $\mathcal{J}_g$ in $\mathcal{A}_g$ for *all* g, at least in principle. In practice, we have carried this out for $g \leq 7$. We also discuss these results here.

We first recall the set-up, and explain our restricted interpretation of the "class of the locus of Jacobians" in the moduli space of principally polarized abelian varieties: by this, we mean a class in the *tautological ring* of $\widetilde{\mathcal{A}}_g$, the $\mathbb{Q}$-subalgebra of the cohomology ring of a toroidal compactification $\widetilde{\mathcal{A}}_g$ of $\mathcal{A}_g$ generated by the Chern classes λ_i of the Hodge bundle $\mathbb{E}$ on $\widetilde{\mathcal{A}}_g$. From [Mu], §5 we know the relation

$$\big(1 - \lambda_1 + \lambda_2 - \lambda_3 + \cdots + (-1)^g \lambda_g\big)\big(1 + \lambda_1 + \lambda_2 + \lambda_3 + \cdots + \lambda_g\big) = 1;$$

equivalently, $\mathrm{ch}_{2k}(\mathbb{E}) = 0$ for all $k \geq 1$. The tautological ring is in fact the quotient of $\mathbb{Q}[\lambda_1, \ldots, \lambda_g]$ by the ideal generated by the homogeneous components of the relation above and is thus a complete intersection ring. A detailed description of the tautological ring may be found in [vdG]. In particular, the relation with the cohomology ring of the *compact dual* of the Siegel upper half space via Hirzebruch's proportionality principle is explained there. This includes the fundamental identity

$$\prod_{i=1}^{g} \lambda_i = \prod_{i=1}^{g} \frac{|B_{2i}|}{4i}$$

that enables one to compute intersection numbers in the tautological ring of $\widetilde{\mathcal{A}}_g$: the monomial of top degree on the left, interpreted as an intersection number, equals the number on the right. Here B_m are the Bernoulli numbers, defined as in [BS] via $t/(e^t - 1) = 1 + \sum_{m=1}^{\infty}(B_m/m!)t^m$. So for $g = 1, 2, 3$ respectively,

$$\lambda_1 = \frac{1}{24}, \quad \lambda_1 \lambda_2 = \frac{1}{5760}, \quad \lambda_1 \lambda_2 \lambda_3 = \frac{1}{2903040}.$$

Denoting by $t \colon \overline{\mathcal{M}}_g \to \widetilde{\mathcal{A}}_g$ the extended Torelli morphism and its image by $\widetilde{\mathcal{J}}_g$, we are after the functorial class $[\widetilde{\mathcal{J}}_g]_Q$ of the locus of (generalized) Jacobians, which is one half its usual fundamental class. In other words, we wish to determine $\frac{1}{2}t_*1$, since a generic curve of genus at least 3 has no nontrivial automorphisms, while the generic p.p.a.v. and the generic Jacobian of dimension at least 3 have two automorphisms.

It is important to point out that this is *not* what we actually compute. We do not know whether, modulo boundary classes, the class of the locus of Jacobians lies in the tautological ring; our feeling is that this is probably false for g large enough, but we cannot even envisage a method to decide this. Instead, we compute the *projection* $[\widetilde{\mathcal{J}}_g]_T$ of this class in the tautological ring; this is well defined by the perfect pairing in the tautological ring and the cohomology ring of $\widetilde{\mathcal{A}}_g$. In other words, we compute the class $[\widetilde{\mathcal{J}}_g]_Q$ modulo a class X that has zero pairing with all tautological classes of the complementary dimension.

The method to compute $\left[\widetilde{\mathcal{J}}_g\right]_T$ is the following. It is a class in the tautological ring of $\widetilde{\mathcal{A}}_g$ of dimension $3g-3$, hence of codimension $c = \binom{g-2}{2}$. Write it as a linear combination with unknown coefficients a_i of the elements s_i of a basis of the degree c part of the tautological ring:

$$\left[\widetilde{\mathcal{J}}_g\right]_T = a_1 s_1 + a_2 s_2 + \cdots + a_k s_k, \quad \left[\widetilde{\mathcal{J}}_g\right]_Q = \left[\widetilde{\mathcal{J}}_g\right]_T + X.$$

(A natural basis is for instance the collection of square free monomials of degree c in $\lambda_1, \ldots, \lambda_g$.) To compute the coefficients a_i, we have to evaluate the k monomials Λ_i of a basis of the degree $3g-3$ part of the tautological ring on this class. The evaluation of expressions $\Lambda_i s_j$ in the tautological ring of $\widetilde{\mathcal{A}}_g$ uses the relations between the λ_i and the proportionality relation stated above. So the expressions $\Lambda_i(a_1 s_1 + \cdots + a_k s_k)$ yield k linearly independent rational linear combinations of the unknowns a_i. The values of these expressions can be determined as $\frac{1}{2}t_*(t^*\Lambda_i) = (\frac{1}{2}t_*1) \cdot \Lambda_i$, provided we know how to evaluate $t^*\Lambda_i$ on $\overline{\mathcal{M}}_g$.

The simplest nontrivial example is $g=4$. Here $c=1$, a basis in codimension 1 is λ_1, a basis in codimension 9 is λ_1^9 (or any nonzero monomial in the λ_i), so to compute the class of $\widetilde{\mathcal{J}}_4$ in $\widetilde{\mathcal{A}}_4$ we only need λ_1^9 on $\overline{\mathcal{M}}_4$. We have seen above that this can be evaluated, for example, by means of the implementation of the algorithm; we find the well-known result that $[\widetilde{\mathcal{J}}_4]_T = 8\lambda_1$.

The situation for $g=5$ is more interesting. Here $c=3$ and a basis in codimension 3 is given by $\lambda_1\lambda_2(= \frac{1}{2}\lambda_1^3)$ and λ_3. We need to evaluate 2 independent monomials of degree 12 in the λ_i on $\overline{\mathcal{M}}_5$. The algorithm will naturally yield only the number λ_1^{12} (whose value we gave at the end of §3). However, we can use the simple observation that the class $\lambda_g\lambda_{g-1}$ vanishes on the boundary $\overline{\mathcal{M}}_g - \mathcal{M}_g$ (see [Fa 3] or [Fa 4]). As a corollary, the numbers $\lambda_1\lambda_2\lambda_4\lambda_5$ and $\lambda_3\lambda_4\lambda_5$ on $\overline{\mathcal{M}}_5$ satisfy the same relation as the classes $\lambda_1\lambda_2$ and λ_3 in the degree 3 part of the *tautological ring* $R^*(\mathcal{M}_5)$ of $\mathcal{M}_5$, which is 1-dimensional. This relation was worked out in [Fa 3]: $10\lambda_3 = 3\lambda_1\lambda_2$. A quick calculation in the tautological ring of $\widetilde{\mathcal{A}}_5$ shows that this implies that the class of the locus of Jacobians satisfies

$$[\widetilde{\mathcal{J}}_5]_T = a(3\lambda_1\lambda_2 - 2\lambda_3).$$

Another such calculation, using $\lambda_1^{12} = \frac{31}{680400}$ on $\overline{\mathcal{M}}_5$, then shows that $a = 24$, hence

$$[\widetilde{\mathcal{J}}_5]_T = 72\lambda_1\lambda_2 - 48\lambda_3.$$

For higher genus, this method only provides some of the coefficients, not all of them. We start with a general formula:

Conjecture 1 *In the basis of monic square free monomials in* $\lambda_1, \ldots, \lambda_g$ *of degree* $\binom{g-2}{2}$*, the coefficient* $C_{1,2,\ldots,g-3}$ *of* $\lambda_1 \lambda_2 \cdots \lambda_{g-3}$ *in the (projected) class* $[\tilde{\mathcal{J}}_g]_T$ *equals*

$$\frac{1}{2g-2} \prod_{i=1}^{g-2} \frac{2}{(2i+1)|B_{2i}|}.$$

Thus for $g = 3, \ldots, 7$ *it equals* $1, 8, 72, 384, 768$ *respectively, while it is not an integer for any larger value of* g.

The conjecture holds for $g \leq 15$: it is a consequence of a conjectural formula in [Fa 3] for the number λ_{g-1}^3 on $\overline{\mathcal{M}}_g$ that is proved for $g \leq 15$. To derive it from that formula, note simply that $\lambda_{g-1}^3 = 2\lambda_{g-2}\lambda_{g-1}\lambda_g$ and that $\lambda_1 \lambda_2 \cdots \lambda_{g-3}$ is the unique monomial in the basis that has nonzero pairing with $\lambda_{g-2}\lambda_{g-1}\lambda_g$.

Note that we could have used this formula instead of that for λ_1^{12} to compute the class of $[\tilde{\mathcal{J}}_5]_T$.

For $g = 6$ we find the coefficient of $\lambda_2\lambda_4$ using the relation between $\lambda_1\lambda_3$ and λ_4 in $R^*(\mathcal{M}_6)$:

$$C_{2,4} = -3C_{1,2,3} = -1152.$$

The knowledge of λ_1^{15} (see §3) gives a nontrivial relation between the remaining coefficients of $\lambda_1\lambda_5$ and λ_6:

$$C_6 + 16C_{1,5} = \frac{7336704}{691}.$$

A new ingredient will be required to solve for these coefficients. It is provided by Mumford's formula [Mu] for the Chern character of the Hodge bundle on $\overline{\mathcal{M}}_g$:

$$\mathrm{ch}(\mathbb{E}) = g + \sum_{i=1}^{\infty} \frac{B_{2i}}{(2i)!} \left[\kappa_{2i-1} + \frac{1}{2} \sum_{h=0}^{g-1} i_{h,*} \left(K_1^{2i-2} - K_1^{2i-3}K_2 + \cdots + K_2^{2i-2} \right) \right].$$

(Note that Mumford's convention for the Bernoulli numbers differs from the one we use: $B_2 = \frac{1}{6}$, $B_4 = \frac{-1}{30}$ etc.) Here $i_0 \colon \overline{\mathcal{M}}_{g-1,2} \to \Delta_{\mathrm{irr}} \subset \overline{\mathcal{M}}_g$ and $i_h \colon \overline{\mathcal{M}}_{h,1} \times \overline{\mathcal{M}}_{g-h,1} \to \Delta_h \subset \overline{\mathcal{M}}_g$ are the natural maps, and K_i is the first Chern class of the relative cotangent line bundle at the ith point.

The formula is ideally suited for a recursive computation of the intersection numbers of the κ_i and the $\mathrm{ch}_{2j-1}(\mathbb{E})$. Namely, suppose given a monomial in those classes of degree $3g - 3 + n$ on $\overline{\mathcal{M}}_{g,n}$. If only κ_i occur, we can proceed as explained in §2, using the Witten conjecture (Kontsevich's theorem). In any case, as the Hodge bundle is a pullback from $\overline{\mathcal{M}}_g$ or $\overline{\mathcal{M}}_{1,1}$, we can push down the expression to $\overline{\mathcal{M}}_g$, respectively $\overline{\mathcal{M}}_{1,1}$, and obtain a sum of similar expressions. In genus 1, we only need the well-known equalities

$$\operatorname{ch}_1(\mathbb{E}) = \lambda_1 = \kappa_1 = \psi_1 = \langle \tau_1 \rangle = \frac{1}{24}.$$

So assume the genus is at least 2 and the monomial on $\overline{\mathcal{M}}_g$ contains at least one $\operatorname{ch}_{2j-1}(\mathbb{E})$. Take the highest odd Chern character component that occurs, and expand it using Mumford's formula. In the first term, that ch_{2k-1} is replaced by a κ_{2k-1} (up to a factor), so it is determined inductively. The other terms involve expressions in the K_i pushed forward via the maps i_h. The point is that these can be written as pushforwards from $\overline{\mathcal{M}}_{g-1,2}$ or $\overline{\mathcal{M}}_{h,1} \times \overline{\mathcal{M}}_{g-h,1}$ of intersection numbers of the classes K_i, κ_j and Chern character components of the Hodge bundles in genus $g - 1$, respectively genus h and $g - h$. This is clear from the Arbarello–Cornalba formulas for $i_h^* \kappa_j$ and the fact that $i_0^*(\mathbb{E})$ is the extension of a trivial line bundle by the Hodge bundle in genus $g - 1$, while for h positive $i_h^*(\mathbb{E})$ is the direct sum of the Hodge bundles in genera h and $g - h$.

After expanding and omitting the terms that are 0 for dimension reasons, we find an expression in intersection numbers of the classes just mentioned on spaces of 1- or 2-pointed curves. These can be pushed down again to intersection numbers of κ_i and $\operatorname{ch}_{2j-1}(\mathbb{E})$ on $\overline{\mathcal{M}}_h$ (with $h < g$) or $\overline{\mathcal{M}}_{1,1}$. By induction on the genus, these numbers are known.

Having discussed the implementation of the divisor algorithm in some detail, we content ourselves with saying that the implementation of the new algorithm proceeds along similar lines and is considerably easier.

Because the λ_i can be expressed in the $\operatorname{ch}_j(\mathbb{E})$, this means that all the intersection numbers of the λ_i on $\overline{\mathcal{M}}_g$ can be computed recursively. By the discussion above, it follows that for *all* g the projection in the tautological ring of $\widetilde{A}_g$ of the class of the locus of Jacobians can be computed, at least in principle.

Currently, we have carried this out for $g \leq 7$. For $g = 6$, only one more relation was required. Either one of the two relations following from

$$\lambda_2 \lambda_3 \lambda_4 \lambda_6 = \frac{1697}{2988969984000} \quad \text{and} \quad \lambda_1 \lambda_2 \lambda_3 \lambda_4 \lambda_5 = \frac{150719}{15692092416000}$$

suffices to solve for C_6 and $C_{1,5}$ (the first relation involves $C_{1,5}$ only). The result is

$$\begin{aligned}
[\widetilde{\mathcal{J}}_6]_T &= 384 \lambda_1 \lambda_2 \lambda_3 - 1152 \lambda_2 \lambda_4 + \frac{474048}{691} \lambda_1 \lambda_5 - \frac{248064}{691} \lambda_6 \\
&= 2^7 3 \left(\lambda_1 \lambda_2 \lambda_3 - 3 \lambda_2 \lambda_4 + \frac{2469}{1382} \lambda_1 \lambda_5 - \frac{646}{691} \lambda_6 \right).
\end{aligned}$$

It may be worthwhile to point out the relation $15C_6 + 28C_{1,5} = 2^9 3^3$.
   In genus 7 the result is:

$$[\widetilde{\mathcal{J}}_7]_T = 768\lambda_1\lambda_2\lambda_3\lambda_4 - 6912\lambda_2\lambda_3\lambda_5 + \frac{2209152}{691}\lambda_1\lambda_4\lambda_5$$
$$+ \frac{7522176}{691}\lambda_1\lambda_3\lambda_6 - \frac{8842752}{691}\lambda_4\lambda_6 + \frac{968832}{691}\lambda_3\lambda_7 - \frac{3276672}{691}\lambda_1\lambda_2\lambda_7.$$

As stated, this result is not very pretty; perhaps the class looks better in a different basis. We would like to point out though that the class can be computed with any choice of 7 independent monomials of degree 18 in the λ_i; once this is accomplished, the values of all other such monomials can be determined by means of an easy calculation in the tautological ring of $\widetilde{\mathcal{A}}_7$. In particular, one can choose 'easy' monomials to compute the class, and get the 'hard' ones for free. In this way, for instance, we computed λ_1^{18} on $\overline{\mathcal{M}}_7$:

$$\lambda_1^{18} = \frac{32017001}{638512875}.$$

   Finally, we have another general formula:

Conjecture 2 *In the basis of monic square free monomials in* $\lambda_1, \ldots, \lambda_g$ *of degree* $\binom{g-2}{2}$, *the coefficient* $C_{2,3,\ldots,g-4,g-2}$ *of* $\lambda_2\lambda_3 \cdots \lambda_{g-4}\lambda_{g-2}$ *in the (projected) class* $[\widetilde{\mathcal{J}}_g]_T$ *equals*

$$\left(\frac{g(2g-2)}{12} - 2^{g-3}\right) C_{1,2,\ldots,g-3}$$

$$= \frac{g}{12}\prod_{i=1}^{g-2}\frac{2}{(2i+1)|B_{2i}|} - \frac{1}{4g-4}\prod_{i=1}^{g-2}\frac{4}{(2i+1)|B_{2i}|}.$$

Thus for $g = 5, 6, 7$ *it equals* $-48, -1152, -6912$ *respectively, while it is not an integer for any larger value of* g.

   The conjecture again holds for $g \le 15$; it is a consequence of Conjecture 1 and of the conjectural formula $\kappa_1\lambda_{g-3} = g(2g-2)\lambda_{g-2}$ in $R^{g-2}(\mathcal{M}_g)$ that can be proved for $g \le 15$ using the results of [Fa 3] (note that the only 2 monomials in the basis that have nonzero pairing with $\lambda_1\lambda_{g-3}\lambda_{g-1}\lambda_g$ are $\lambda_2\lambda_3 \cdots \lambda_{g-4}\lambda_{g-2}$ and $\lambda_1\lambda_2\ldots\lambda_{g-3}$).

5   Acknowledgments and final remarks

The research in the first 3 sections and the determination of the classes of $\widetilde{\mathcal{J}}_4$ and $\widetilde{\mathcal{J}}_5$ was carried out at the Universiteit van Amsterdam and was made possible by a fellowship of the Royal Netherlands Academy of Arts and Sciences. Rahul Pandharipande's remark that led directly to the algorithm in §2 was made during a visit of the author to the University of Chicago in April/May 1995. The author would like to thank Bill Fulton for the invitation to visit Chicago, and the University of Chicago for partial support. Both this remark and an analogous remark by Eduard Looijenga half a year earlier (that I didn't get around to working out during that half year, unfortunately) were sparked by talks in which I discussed the problem of determining the intersection numbers of divisors on $\overline{\mathcal{M}}_g$ and the partial results I had obtained (cf. [Fa 2], §5). I am most grateful to them for these remarks. I also want to thank Chris Zaal whose results I used heavily in my computations.

The research in the remaining part of §4 was carried out at the Institut Mittag-Leffler of the Royal Swedish Academy of Sciences during the academic year 1996/97 devoted to *Enumerative geometry and its interaction with theoretical physics*. My stay at the Mittag-Leffler Institute was supported by a grant from the Göran Gustafsson Foundation for Mathematics, Physics and Medicine to Torsten Ekedahl. I would like to thank the organizers of the program and the staff of the institute for making it a wonderful year.

Finally, some remarks about possible uses of the implemented algorithms. The algorithm for computing intersection numbers involving Chern classes of the Hodge bundle (described in §4) has already found applications in the recent work of Graber and Pandharipande [GP] and Kontsevich and Manin [KM]. For moduli spaces $\overline{\mathcal{M}}_{g,n}$ of small dimension and with not too many divisors, one can use the divisor algorithm to determine the part of the cohomology ring generated by divisors. Recently, Getzler [Ge] did this for $\overline{\mathcal{M}}_{2,2}$; he shows in fact that the divisors generate the cohomology ring. As in that paper, such calculations may have applications to computing Gromov–Witten invariants. Also, the algorithms involving the κ_i and the ch_{2j-1} allow one to do calculations that include those classes; this covers, for example, the Chow ring of $\overline{\mathcal{M}}_3$. It is even possible to include arbitrary boundary strata as *module* generators, by pulling back all other classes to the corresponding product of moduli spaces via a sequence of maps, each identifying a single pair of points. Writing algorithms that can handle *intersections* of arbitrary boundary strata will be considerably more difficult, however.

The various algorithms, and some tables of intersection numbers computed with it, are available from the author by e-mail, although only the initial version of the divisor algorithm is currently available in a user friendly format.

Note added in proof Conjecture 1 is now proved: Rahul Pandharipande and I have proved the formula for λ_{g-1}^3 that implies it [FaP].

References

[AC 1] E. Arbarello, M. Cornalba, *The Picard groups of the moduli spaces of curves*, Topology **26** (1987), 153–171.

[AC 2] E. Arbarello, M. Cornalba, *Combinatorial and algebro-geometric cohomology classes on the moduli spaces of curves*, J. Alg. Geom. **5** (1996), 705–749.

[AC 3] E. Arbarello, M. Cornalba, *Calculating cohomology groups of moduli spaces of curves via algebraic geometry*, math.AG/9803001.

[BS]   Z.I. Borevich, I.R. Shafarevich, *Number theory*, Academic Press, New York, 1966.

[Co]   M. Cornalba, *On the projectivity of the moduli spaces of curves*, J. reine angew. Math. **443** (1993), 11–20.

[DM]   P. Deligne, D. Mumford, *The irreducibility of the space of curves of given genus*, Publ. Math. IHES **36** (1969), 75–109.

[Dij]   R. Dijkgraaf, *Some facts about tautological classes*, private communication (November 1993).

[Fa 1] C. Faber, *Chow rings of moduli spaces of curves I: The Chow ring of* $\overline{\mathcal{M}}_3$, Ann. of Math. **132** (1990), 331–419.

[Fa 2] C. Faber, *Intersection-theoretical computations on* $\overline{\mathcal{M}}_g$, in *Parameter Spaces* (Editor P. Pragacz), Banach Center Publications, Volume 36, Warszawa 1996.

[Fa 3] C. Faber, *A conjectural description of the tautological ring of the moduli space of curves*, preprint 1996 (available from http://www.math. okstate.edu/preprint/1997.html).

[Fa 4] C. Faber, *A non-vanishing result for the tautological ring of* $\mathcal{M}_g$, preprint 1995 (available from http://www.math.okstate.edu/preprint /1997.html).

[FaP]   C. Faber and R. Pandharipande, *Hodge integrals and Gromov-Witten theory*, math.AG/9810173, 24 pp.

[vdG]   G. van der Geer, *Cycles on the moduli space of abelian varieties*, alg-geom/9605011.

[Ge]   E. Getzler, *Topological recursion relations in genus 2*, math.AG /9801003.

[GP] T. Graber, R. Pandharipande, *Localization of virtual classes*, alg-geom/9708001.

[Ha 1] J. Harer, *The second homology group of the mapping class group of an orientable surface*, Invent. Math. **72** (1983), 221–239.

[Ha 2] J. Harer, *The cohomology of the moduli space of curves*, in *Theory of moduli*, 138–221, LNM 1337, Springer, Berlin-New York 1988.

[Ke] S. Keel, *Intersection theory of moduli space of stable N-pointed curves of genus zero*, Trans. A.M.S. **330** (1992), 545–574.

[Kn 1] F. Knudsen, *The projectivity of the moduli spaces of stable curves, II: The stacks $\mathcal{M}_{g,n}$*, Math. Scand. **52** (1983), 161–199.

[Kn 2] F. Knudsen, *The projectivity of the moduli spaces of stable curves, III: The line bundles on $\mathcal{M}_{g,n}$ and a proof of the projectivity of $\overline{\mathcal{M}}_{g,n}$ in characteristic 0*, Math. Scand. **52** (1983), 200–212.

[Ko] M. Kontsevich, *Intersection theory on the moduli space of curves and the matrix Airy function*, Comm. Math. Phys. **147** (1992), 1–23.

[KM] M. Kontsevich, Yu. I. Manin, *Relations between the correlators of the topological sigma-model coupled to gravity*, alg-geom/9708024.

[Ma] Maple V Release 3, computer algebra package, copyright by Waterloo Maple Software and the University of Waterloo. Documentation: B.W Char and others, Maple V, Springer-Verlag, 1997.

[Mu] D. Mumford, *Towards an enumerative geometry of the moduli space of curves*, in *Arithmetic and geometry* (Editors M. Artin and J. Tate), Part II, Progress in Math., Volume 36, Birkhäuser, Basel 1983.

[Wi] E. Witten, *Two dimensional gravity and intersection theory on moduli space*, Surveys in Diff. Geom. **1** (1991), 243–310.

[Za] C. Zaal, Maple procedures for computing the intersection numbers of the ψ_i and a table of the numbers for $g \leq 9$. Universiteit van Amsterdam 1992. (Available by e-mail from zaal@wins.uva.nl or from the author.)

Carel Faber,
Department of Mathematics, Oklahoma State University,
Stillwater, Oklahoma 74078–1058, U.S.A.
e-mail: cffaber@math.okstate.edu

On some tensor representations of the Cremona group of the projective plane

Marat Gizatullin

Intellectus est universalium et non singularium.
Thomas Aquinas, from *Summa contra gentiles* (1264)

0 Introduction

The Cremona group $\mathrm{Cr} = \mathrm{Cr}(2, K)$ is the group of birational automorphisms $\mathrm{Bir}\,\mathbb{P}_2$ of the projective plane; it is (anti-)isomorphic to the automorphism group $\mathrm{Aut}\,K(x, y)$ of the rational function field in two variables.

Let W be a 3-dimensional vector space (over an algebraically closed field K of characteristic zero), and $\mathbb{P}_2 = \mathbb{P}(W)$ its projectivization; $\mathcal{P} = \mathrm{Aut}(\mathbb{P}_2) = \mathrm{PGL}(W) = \mathrm{PGL}(3, K)$ is the collineation group of $\mathbb{P}_2$, that is, the group of projective linear transformations. Thus $\mathcal{P} \subset \mathrm{Cr}$ is a subgroup of the Cremona group. For a linear representation $r\colon \mathrm{GL}(W) \to \mathrm{GL}(V)$, consider the projectivization

$$\rho = \mathbb{P}(r)\colon \mathcal{P} = \mathrm{PGL}(W) \to \mathrm{PGL}(V) = \mathrm{Aut}\,\mathbb{P}(V).$$

An *extension* of ρ is a homomorphism

$$\tilde{\rho}\colon \mathrm{Cr} \to \mathrm{Bir}\,\mathbb{P}(V)$$

which restricts to ρ on $\mathcal{P}$, that is, $\tilde{\rho}_{|\mathcal{P}} = \rho\colon \mathcal{P} \to \mathrm{Aut}\,\mathbb{P}(V)$; in other words, there is a commutative diagram

$$
\begin{array}{ccc}
\mathcal{P} & \xrightarrow{\ \rho\ } & \mathrm{Aut}\,\mathbb{P}(V) \\
\downarrow & & \downarrow \\
\mathrm{Cr} & \xrightarrow{\ \tilde{\rho}\ } & \mathrm{Bir}\,\mathbb{P}(V)
\end{array}
$$

where the vertical arrows are the natural inclusions.

Question *Given the projectivization ρ of a linear representation r, does there exist an extension $\tilde{\rho}$ of ρ to the whole Cremona group Cr?*

111

We shall see that the answer is yes if $r = S^m(r_0)$ is the mth symmetric power of the natural representation r_0 of $\mathrm{GL}(W)$ in the vector space W^* of linear forms, and $m = 2, 3, 4$. In other words, the Cremona group of the plane has an action on the spaces of plane conics, cubics or quartics, extending the actions of the group of plane collineations.

A first approach to the above question was proposed by Igor Artamkin in his thesis [1], [2], where he constructed an action of the Cremona group of the plane on moduli spaces of stable vector bundles over the projective plane, and deduced an action on the curves of jumping lines of the bundles. A drawback of his approach is that the generic curve of degree > 3 is not realized as the curve of jumping lines of a vector bundle. Moreover, Artamkin's action applies to curves with the additional structure of an even theta characteristic.

Our approach is more algebraic, although we believe that at a deeper level, the reasons underlying Artamkin's constructions and ours are the same. A rough outline of our constructions is as follows.

The group $\mathrm{Cr}(2, K)$ is known to be generated by the collineations $\mathcal{P}$ and the standard quadratic transformation. Given a variety and an action of $\mathcal{P}$, we can obtain the required extension by choosing an action of the standard quadratic transformation with the lucky property that all the relations holding between the collineations and the standard quadratic transformation are satisfied. Of course, to realize this approach, one needs to find a handy and explicit way to verify the list of relations. Section 1 of this paper carries out this program. In a sense, this section complements the main theorem of [11]; it was omitted from [11] in view of the length of the paper.

Section 2 contains a series of general definitions of some objects connected with natural actions of the group $\mathrm{Cr}(n, K)$, or of a more general group $\mathrm{UCr}(n, K)$ (Definition 2.7), which we call the *universal Cremona group*. Our definitions are perhaps too general for applications, but we hope that this philosophy will clarify our constructions. Section 2 ends with a series of verifications of relations as just explained.

Section 3 describes actions of the Cremona group of the plane on the spaces of curves of degrees 2, 3 and 4. We present the first two actions in some detail, but only sketch the treatment for quartics; we hope to return to this case on another occasion.

As an introduction to these ideas, we describe the effect of the standard quadratic transformation s_0 on a generic conic C, following Artamkin [1]. We write $\widehat{C}$ for the dual conic of C; let P_0, P_1, P_2 be the three fundamental points of s_0 and $Q_0', Q_0'', Q_1', Q_1'', Q_2', Q_2''$ the six points of intersection of $\widehat{C}$ with the sides of the fundamental triangle, with Q_i', Q_i'' on the side L_i opposite the vertex P_i for $0 \le i \le 2$. Write R_i' and R_i'' for the intersection points of L_i and the proper transforms of the lines $P_i Q_i'$ and $P_i Q_i''$ under s_0; then all six points R_i', R_i'' lie on a conic D, and the dual conic $\widehat{D}$ is the image of C by the action of the standard quadratic transformation on the space of conics. About the same time, a letter from Dolgachev [7] contained the formulas

$$x'_0 = x_1 x_2, \ \ x'_1 = x_2 x_0, \ \ x'_2 = x_0 x_1, \ \ x'_3 = x_3 x_0, \ \ x'_4 = x_4 x_1, \ \ x'_5 = x_5 x_2$$

for a quadratic transformation of $\mathbb{P}_5$, which he considered as an analog for $\mathbb{P}_5$ of the standard quadratic transformation s_0. These formulas express the relation between the coefficients of $\widehat{C}$ and of D in Artamkin's construction (compare (3.0) below).

Acknowledgments

It is a pleasure to thank Igor Dolgachev who made possible our discussions of this topic in Ann Arbor in Summer 1995. Our discussion about Artamkin's ideas and actions of the Cremona group on spaces of plane curves of small degree was the original motivation for this work. The author also wishes to thank the Department of Mathematics and the Shapiro Library of the University of Michigan for hospitality, as well as the organizers of the Warwick conference, especially Miles Reid.

The author takes this opportunity of expressing his gratitude to the American Mathematical Society for a grant during 1993–1995 and the Russian Fund of Fundamental Researches (project 96-01-01230) for support during 1996–97.

After revising the initial text, I would like to thank the referees and the editor, whose numerous improvements and corrections (mathematical, linguistic and typographical) amount to the contribution of a joint author.

1 Generators and defining relations for Cr

1.1 Generators of the Cremona group

We suppose that the ground field K is algebraically closed; let $(x_0 : x_1 : x_2)$ be homogeneous coordinates on the projective plane $\mathbb{P}_2$ over K. A rational transformation of $\mathbb{P}_2$ can be written

$$x'_0 = f_0(x_0, x_1, x_2), \ \ x'_1 = f_1(x_0, x_1, x_2), \ \ x'_2 = f_2(x_0, x_1, x_2), \qquad (1.0)$$

where f_0, f_1, f_2 are either homogeneous polynomials of the same degree, or quotients of homogeneous polynomials having the same degrees of homogeneity. The image of a point $(a_0 : a_1 : a_2) \in \mathbb{P}_2$ under such a transformation is

$$\big(f_0(a_0, a_1, a_2) : f_1(a_0, a_1, a_2) : f_2(a_0, a_1, a_2)\big).$$

Let $\mathrm{Cr} = \mathrm{Cr}(2, K)$ denote the set of all invertible rational transformations of $\mathbb{P}_2$ over K.

Let $\mathcal{P} = \mathrm{Aut}\,\mathbb{P}_2 = \mathrm{PGL}(3, K)$ be the set of all projective transformations

$$
\begin{aligned}
x_0' &= c_{00}x_0 + c_{01}x_1 + c_{02}x_2, \\
x_1' &= c_{10}x_0 + c_{11}x_1 + c_{12}x_2, \quad \text{with } c_{ij} \in K \text{ and } \det(c_{ij}) \neq 0. \\
x_2' &= c_{20}x_0 + c_{21}x_1 + c_{22}x_2,
\end{aligned}
\tag{1.1}
$$

We write $\mathcal{Q}$ for the set of all quadratic transformations, that is, invertible rational transformations of the form (1.0), where f_0, f_1, f_2 are homogeneous polynomials of degree 2 with no common linear factor. The set of quadratic transformations splits into three double cosets under the group of collineations (more precisely, with respect to the natural two-sided action of $\mathcal{P} \times \mathcal{P}$ on $\mathcal{Q}$):

$$
\mathcal{Q} = \mathcal{P}s_0\mathcal{P} \sqcup \mathcal{P}s_1\mathcal{P} \sqcup \mathcal{P}s_2\mathcal{P},
\tag{1.2}
$$

(here $\sqcup$ means disjoint union), where s_0 is the so-called standard quadratic transformation, given by

$$
\begin{aligned}
s_0 : x_0' &= x_1x_2, \quad x_1' = x_0x_2, \quad x_2' = x_0x_1, \tag{1.3} \\
\text{or} \quad x_0' &= x_0^{-1}, \quad x_1' = x_1^{-1}, \quad x_2' = x_2^{-1}. \tag{1.4}
\end{aligned}
$$

Next, s_1 is the first degeneration of the standard quadratic transformation, where two of the three fundamental points of s_0 come together, and is given by

$$
\begin{aligned}
s_1 : x_0' &= x_1^2, \quad x_1' = x_0x_1, \quad x_2' = x_0x_2, \tag{1.5} \\
\text{or} \quad x_0' &= x_1x_0^{-1}x_1, \quad x_1' = x_1, \quad x_2' = x_2. \tag{1.6}
\end{aligned}
$$

Finally, s_2 is the second degeneration of s_0 (or a further degeneration of s_1), where all the fundamental points of s_0 come together to one point, and is given in formulas by

$$
\begin{aligned}
s_2 : x_0' &= x_0^2, \quad x_1' = x_0x_1, \quad x_2' = x_1^2 - x_0x_2, \tag{1.7} \\
\text{or} \quad x_0' &= x_0, \quad x_1' = x_1, \quad x_2' = x_1x_0^{-1}x_1 - x_2. \tag{1.8}
\end{aligned}
$$

Note that (1.4), (1.6), (1.8) are destined for future noncommutative generalizations (see (2.9)).

Remark 1.1 The third double coset $\mathcal{P}s_2\mathcal{P}$ of (1.2) contains all nonunit elements of the following one-parameter subgroup σ_t (with parameter $t \in K$)

$$
\begin{aligned}
\sigma_t : x_0' &= x_0^2, \quad x_1' = x_0x_1, \quad x_2' = x_0x_2 + tx_1^2 \\
\text{or} \quad x_0' &= x_0, \quad x_1' = x_1, \quad x_2' = x_2 + tx_1x_0^{-1}x_1.
\end{aligned}
$$

We can write these elements as composites $\sigma_t = p_2 \circ s_2 \circ p_1$ in terms of s_2 and the collineations $p_1, p_2 \in \mathcal{P}$ given by

$$p_1: x_0' = x_0, \quad x_1' = -tx_1, \quad x_2' = -tx_2,$$
$$\text{and} \quad p_2: x_0' = x_0, \quad x_1' = -t^{-1}x_1, \quad x_2' = t^{-1}x_2.$$

Theorem 1.1 (Max Noether) *The Cremona group* $\mathrm{Cr}(2, K)$ *is generated by* $\mathcal{P} \cup \mathcal{Q}$.

Note that one can, as usual, replace $\mathcal{P} \cup \mathcal{Q}$ by the more economic set of generators $\mathcal{P} \cup \{s_0\}$, writing the transformations s_1, s_2 in terms of s_0 and collineations. More precisely,

$$s_1 = g_0 \circ s_0 \circ g_0 \circ s_0 \circ g_0, \tag{1.9}$$
$$\text{where} \quad g_0: x_0' = x_1 - x_0, \quad x_1' = x_1, \quad x_2' = x_2; \tag{1.10}$$

and

$$s_2 = s_1 \circ g_1 \circ s_1, \tag{1.11}$$
$$\text{where} \quad g_1: x_0' = x_0, \quad x_1' = x_1, \quad x_2' = x_0 - x_2 \tag{1.12}$$

Remark 1.2 The identities (1.9) and (1.11) have interesting analogs in the Cremona group $\mathrm{Cr}(3, K)$ of 3-space. The involution

$$S_0: x_0' = x_0^{-1}, \quad x_1' = x_1^{-1}, \quad x_2' = x_2^{-1}, \quad x_3' = x_3^{-1} \tag{1.4a}$$

is the standard *cubic* transformation of $\mathbb{P}_3$. Take the projective transformation

$$G_0: x_0' = x_1 - x_0, \quad x_1' = x_1, \quad x_2' = x_2 \quad x_3' = x_3 \tag{1.10a}$$

as an analog of (1.10). Then the composite

$$S_1 = G_0 \circ S_0 \circ G_0 \circ S_0 \circ G_0, \tag{1.9a}$$

is the *quadratic* space transformation

$$S_1: x_0' = x_1^2, \quad x_1' = x_0 x_1, \quad x_2' = x_0 x_2, \quad x_3' = x_0 x_3. \tag{1.5a}$$

Moreover, if we take the collineation

$$G_1: x_0' = x_0, \quad x_1' = x_1, \quad x_2' = x_2, \quad x_3' = x_0 - x_3 \tag{1.12a}$$

as an analog of g_1 in (1.12), then the composite

$$S_2 = S_1 \circ G_1 \circ S_1, \tag{1.11a}$$

is the *quadratic* transformation

$$S_2\colon x'_0 = x_0^2,\; x'_1 = x_0 x_1,\; x'_2 = x_0 x_2,\; x'_3 = x_1^2 - x_0 x_3 \qquad (1.7a)$$

or (in affine coordinates $x = x_1/x_0, y = x_2/x_0, z = x_3/x_0$)

$$S_2\colon x' = x,\quad y' = y,\quad z' = x^2 - z.$$

The composites (1.9a) and (1.11a) contradict some propositions of Dolgachev and Ortland [9], p. 93 (which are comparatively lucid paraphrases of some claims of S. Kantor [13], A. Coble [6], H. Hudson [12], and P. Du Val [10]).

More precisely, write $\mathrm{Cr}_{\mathrm{reg}}(3, K)$ for the subgroup of $\mathrm{Cr}(3, K)$ generated by S_0 and the subgroup of collineations; the elements of $\mathrm{Cr}_{\mathrm{reg}}(3, K)$ are the "regular" transformations in the sense of Coble. Let $\mathrm{Punct}(3, K)$ be the set of Cremona transformations of $\mathbb{P}_3$ without fundamental curves of the first kind, that is, transformations without curves whose proper image in the projective space is a surface, see [12], [9]. The authors listed above start by asserting (sometimes with some provisos) that "one can prove that all punctual transformations form a subgroup of the Cremona group". This is false, because each factor of the right-hand side of (1.9a) is a punctual transformation, whereas the composite S_1 has $x_0 = 0, x_1 = 0$ as a fundamental line of the first kind (maybe, more precisely, a curve infinitely near to this line is a fundamental curve of the first kind); at any rate, no blowups of $\mathbb{P}_3$ at a finite sets of points can reduce the transformations S_1 and S_2 to pseudoisomorphisms in the sense of Dolgachev and Ortland [9]. The identity (1.9a) also refutes the conjectured equality $\mathrm{Punct}(n, K) = \mathrm{Cr}_{\mathrm{reg}}(n, K)$, or even the inclusion $\supset$. Note that the fact that the composites (1.9a) and (1.11a) have even degree also contradicts Coble's formulas, according to which the degree of a "regular" transformation of $\mathbb{P}_d$ is of the form $(d - 1)m + 1$.

1.2   Defining relations between the generators $\mathcal{P} \cup \mathcal{Q}$

We now reproduce and comment on the main theorem of [11], Theorem 10.7, with some changes of formulation. If $a, b, c, \ldots$ are finitely many elements of the set $\mathcal{P} \cup \mathcal{Q}$ of generators of $\mathrm{Cr}(2, K)$ (see Theorem 1.1), we write $abc \cdots$ to mean a *word* over the *alphabet* $\mathcal{P} \cup \mathcal{Q}$, whereas the expression $a \circ b \circ c \cdots$ means the ordinary *composite* in Cr, that is, a birational transformation of $\mathbb{P}_2$. The theorem on relations is as follows.

Theorem 1.2 *Every relation between the generators* $\mathcal{P} \cup \mathcal{Q}$ *that holds in* $\mathrm{Cr}(2, K)$ *is a consequence of the 3-term relations of the form*

$$g_1 g_2 g_3 = 1, \qquad (1.13)$$

where $\{g_1, g_2, g_3\}$ is an ordered triple of elements of $\mathcal{P} \cup \mathcal{Q}$ for which the corresponding composite $g_1 \circ g_2 \circ g_3$ of rational maps equals the identity transformation of $\mathbb{P}_2$.

We pull out some special relations from the above large family (1.13), and each relation of the family will be a consequence of the marked special ones.

1.2.1 The first family of special relations: the multiplication law of the projective group

They are relations of the form

$$p_1 p_2 p_3 = 1, \tag{1.14}$$

where $p_1, p_2, p_3 \in \mathcal{P}$ are collineations, and $p_1^{-1} = p_2 \circ p_3$.

1.2.2 Generalities on edge relations

The *edge relations* arise from the two-sided action of the collineations on the set of quadratic transformations, and are of the form

$$p_1 q_1 p_2 = q_2, \tag{1.15}$$

where $p_1, p_2 \in \mathcal{P}$, $q_1, q_2 \in \mathcal{Q}$, and $p_1 \circ q_1 \circ p_2 = q_2$. More precisely, each relation (1.15) gives three 3-term relations for use in Theorem 1.2:

$$p_1 q_1 (p_2 \circ q_2^{-1}) = 1, \quad p_1(q_1 \circ p_2) q_2^{-1} = 1,$$
$$\text{and} \quad (p_1 \circ q_1) p_2 q_2^{-1} = 1. \tag{1.15a}$$

We picture a relation (1.15) as follows:

$$p_1 \overset{\overset{q_1}{\longrightarrow}}{\underset{\underset{q_2^{-1}}{\longleftarrow}}{\bigcirc\!\!-\!\!-\!\!-\!\!-\!\!\bigcirc}}\, p_2$$

This describes a relation (1.15) as a loop of length 2 going out along an edge and back along the same edge; our term "edge relation" arises from this. The family of all relations (1.15) is still too large and cumbersome, but in 1.2.3, 1.2.4, 1.2.5 below, we distinguish three special edge relations which, together with (1.14), imply all the edge relations. Note that in any relation (1.15), the quadratic transformations q_1 and q_2 both belong to the same double coset of (1.2); this leads us to separate and classify the edge relations according to the subscript n of the representative s_n of the double coset $\mathcal{P} s_n \mathcal{P}$ for $n \in \{0, 1, 2\}$. We call the corresponding relation an *(n)-edge relation*.

1.2.3   The (0)-edge relations

Our second family of special relations are edge relations arising from a loop of length 2 obtained by going out and back along an edge corresponding to the standard quadratic transformation s_0. The loop in question is a marked

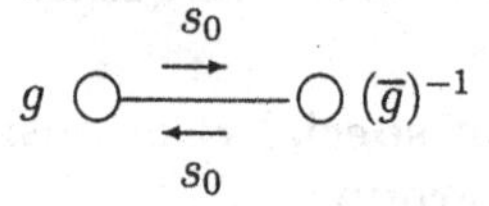

$$g \; \underset{\displaystyle s_0}{\overset{\displaystyle s_0}{\longleftrightarrow}} \; (\bar{g})^{-1}$$

Figure 1: (0)-edge relation

path in the graph Γ_1 (see Figure 1 and compare [11], 4.1, 4.4, 4.5 and 10.5.3). Let G_0 be the collineation group consisting of the transformations

$$g\colon x_0' = t_0 x_i, \quad x_1' = t_1 x_j, \quad x_2' = t_2 x_k, \tag{1.16}$$

where $\{i, j, k\}$ is a permutation of $\{0, 1, 2\}$ and $t_0, t_1, t_2 \in K^*$. In other words, $G_0 = \operatorname{Aut} V_0$, where $V_0 \to \mathbb{P}_2$ is the blowup of $\mathbb{P}_2$ in the three points

$$\{(1:0:0), (0:1:0), (0:0:1)\}.$$

Let $g \mapsto \bar{g}$ be the involutive automorphism of G_0 taking (1.16) to

$$\bar{g}\colon x_0' = t_0^{-1} x_i, \quad x_1' = t_1^{-1} x_j, \quad x_2' = t_2^{-1} x_k. \tag{1.17}$$

Our second family consists of the relations of the form

$$s_0 g s_0 = \bar{g} \quad \text{for } g \in G_0 \subset \mathcal{P}. \tag{1.18}$$

More precisely, each relation (1.18) provides three 3-term relations for use in Theorem 1.2:

$$s_0 g (s_0 \circ (\bar{g})^{-1}) = 1, \quad s_0 (g \circ s_0)(\bar{g})^{-1} = 1,$$
$$\text{and} \quad (s_0 \circ g) s_0 (\bar{g})^{-1} = 1; \tag{1.18a}$$

compare (1.15) and (1.15a). Note that one of the simplest consequences of (1.18) is $s_0^2 = 1$ (take $g = 1$ in (1.18)).

1.2.4   The (1)-edge relations

Our third family of special relations are edge relations arising from a loop of length 2 obtained by going out and back along an edge corresponding to the first degeneration s_1 of the standard quadratic transformation (see (1.5), (1.6)). See Figure 2, where we omit arrows that can be deduced by analogy with Figure 1.

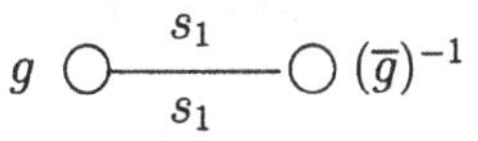

$$g \; \bigcirc\!\!\xrightarrow[\;s_1\;]{\;s_1\;}\!\!\bigcirc \; (\overline{g})^{-1}$$

Figure 2: (1)-edge relation

Let G_1 be the collineation group consisting of the transformations

$$g\colon x_0' = t_0 x_0, \quad x_1' = t_1 x_1, \quad x_2' = t_2 x_2 + r x_0, \tag{1.19}$$

where $t_0, t_1, t_2 \in K^*$ and $r \in K$. The group G_1 is $\operatorname{Aut} V_1$, where $V_1 \to \mathbb{P}_2$ is the minimal resolution of the indeterminacy of the rational map s_1 of (1.5). Let $g \mapsto \overline{g}$ be the involutive automorphism of G_1 sending (1.19) to

$$\overline{g}\colon x_0' = t_1^2 t_0^{-1} x_1, \quad x_1' = t_1 x_1, \quad x_2' = t_2 x_2 + r x_1. \tag{1.20}$$

Our third family consists of the relations of the form

$$s_1 g s_1 = \overline{g}, \tag{1.21}$$

where $g \in G_1$. More precisely, as in (1.15a) and (1.18a), (1.21) provides three 3-term relations. As before, $s_1^2 = 1$ is a consequence of (1.18).

1.2.5 The (2)-edge relations

Our fourth family of special relations are edge relations arising from a loop of length 2 obtained by going out and back along an edge corresponding to the second degeneration s_2 of the standard quadratic transformation (see (1.7), (1.8)). See Figure 3, where we omit arrows that can be deduced by analogy with Figure 1, and, here and below, we label the edges by n in place of s_n.

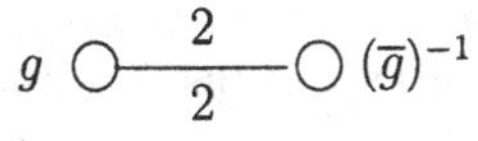

$$g \; \bigcirc\!\!\xrightarrow[\;2\;]{\;2\;}\!\!\bigcirc \; (\overline{g})^{-1}$$

Figure 3: (2)-edge relation

Let G_2 be the collineation group consisting of the transformations

$$g\colon x_0' = x_0, \quad x_1' = t x_1, \quad x_2' = t^2 x_2 + r x_1 + s x_0, \tag{1.22}$$

where $t \in K^*$ and $r, s \in K$. The group G_2 is isomorphic to the group $\operatorname{Aut} V_2$ of automorphisms of the surface V_2, the minimal resolution of the indeterminacy of the rational map s_2. Let $g \mapsto \overline{g}$ be the involutive automorphism of G_1 sending (1.22) to

$$\overline{g}\colon x_0' = x_0, \quad x_1' = t x_1, \quad x_2' = t^2 x_2 - r x_1 - s x_0. \tag{1.23}$$

Our fourth family consists of the relations of the form

$$s_2 g s_2 = \overline{g}, \tag{1.24}$$

where $g \in G_2$. As in (1.15)–(1.15a), (1.18)– (1.18a), (1.24) provides three 3-term relations. As before, $s_2^2 = 1$ is a consequence of (1.24).

1.2.6   Generalities on triangular relations

The relations (1.18), (1.21) and (1.24) were pictured as walks around the edge in Figures 1, 2 and 3. Our remaining special relations are pictured as marked loops around the triangle of Figure 4 (clockwise, as for the above edge relations). Here the vertices are marked by collineations $p_1, p_2, p_3 \in \mathcal{P}$, and

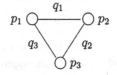

Figure 4: Triangular relations

the edges by quadratic transformations $q_1, q_2, q_3 \in \mathcal{Q}$. The marked triangular loop of Figure 4 gives a relation of the form

$$p_1 q_1 p_2 q_2 p_3 q_3 = 1, \tag{1.25}$$

whenever the composite of rational maps $p_1 \circ q_1 \circ p_2 \circ q_2 \circ p_3 \circ q_3$ is the identity. Although as it stands (1.25) has six terms, it actually reduces to a three-term relation (1.13) if we set $g_i = p_i \circ q_i$. We call (1.25) a *triangular relation*.

All the triangular relations follow from the special triangular relations written down in 1.2.7–1.2.12 below, together with the special relations already listed above. Each special triangular relation is of the form (1.25) with q_1, q_2, q_3 taken from the quadratic involutions s_0, s_1 or s_2 of (1.3)–(1.8). If (1.25) holds with

$$q_1 = s_{n(1)}, \quad q_2 = s_{n(2)} \quad \text{and} \quad q_3 = s_{n(3)}, \tag{1.26}$$

we say that Figure 4 is an $(n(1), n(2), n(3))$-*triangle* and that the relation (1.25) is an $(n(1), n(2), n(3))$-*triangular relation*. As in 1.2.5, we label edges with the number n instead of s_n.

1.2.7 The special (0,0,0)-triangular relation

This is the following relation:

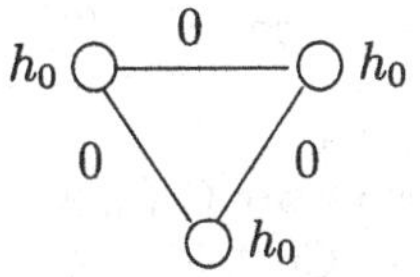

$$h_0 s_0 h_0 s_0 h_0 s_0 = 1, \tag{1.27}$$

where s_0 is the standard quadratic transformation, and h_0 the involutive collineation given by

$$h_0 \colon x_0' = x_0, \quad x_1' = x_0 - x_1, \quad x_2' = x_0 - x_2. \tag{1.28}$$

1.2.8 The special (1,0,0)-triangular relation

This is the identity (1.9) written down as the relation

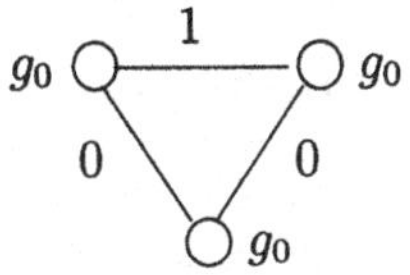

$$g_0 s_1 g_0 s_0 g_0 s_0 = 1, \tag{1.29}$$

where g_0 is the projective involution (1.10).

1.2.9 The special (2,1,1)-triangular relation

This is the identity (1.11) written down as the relation:

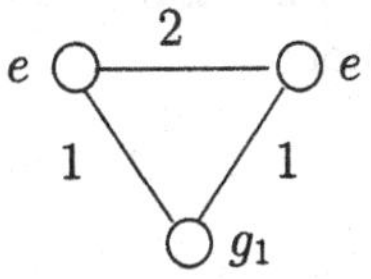

$$s_2 s_1 g_1 s_1 = 1, \tag{1.30}$$

where g_1 is the projective involution (1.12) and $e \in \mathcal{P}$ the identity.

1.2.10 The special (0,1,1)-triangular relation

This is the following relation:

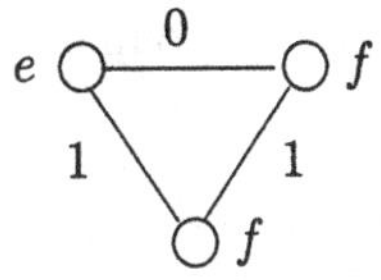

$$s_0 f s_1 f s_1 = 1, \tag{1.31}$$

where f is the collineation

$$f\colon \quad x_0' = x_2, \quad x_1' = x_1, \quad x_2' = x_0.$$

Remark 1.3 It is interesting to note in passing that the relation (1.31) yields as a corollary:

the set $\mathcal{P} \cup \{s_1\}$ *generates the group* $\mathrm{Cr}(2, K)$

(if K is an algebraically closed field, of course).

In contrast, the set $\mathcal{P} \cup \{s_2\}$ does not generate $\mathrm{Cr}(2, K)$. Indeed, $\mathcal{P} \cup \{s_2\}$ is contained in the subgroup $\mathrm{Cr}^{(3)}(2, K) \subset \mathrm{Cr}(2, K)$ consisting of Cremona transformations (f_0, f_1, f_2) (in the notation of (1.0)) having Jacobian determinant a perfect cube; this is a proper subgroup because, for example, $s_0 \notin \mathrm{Cr}^{(3)}(2, K)$.

1.2.11 The special (1,1,1)-triangular relation

This is the relation:

$$h_1 s_1 h_1 s_1 h_1 s_1 = 1, \qquad (1.32)$$

where h_1 is the projective transformation

$$h_1 \colon x_0' = x_1 - x_0, \quad x_1' = x_1, \quad x_2' = x_2.$$

1.2.12 The special (2,2,2)-triangular relations

Our final family of relations depends on a parameter $t \in K$, with $t \neq 0, 1$. Write p_t for the projective transformation:

$$p_t \colon x_0' = x_0, \quad x_1' = -tx_1, \quad x_2' = tx_2,$$

and s_2 for the second degeneration of the standard quadratic transformation as in (1.7). Then our final special triangular relations are:

$$p_{t''} s_2 p_{t'} s_2 p_t s_2 = 1, \quad \text{where } t' = 1 - \frac{1}{t} \text{ and } t'' = \frac{1}{1 - t}. \qquad (1.33)$$

Remark 1.4 There is a more natural and convenient form of (1.33), namely, the multiplication law for the one-parameter group σ_t of Remark 1.1, that is, the relation

$$\sigma_t \sigma_s = \sigma_{t+s},$$

where $s, t \in K$, with $s, t \neq 0$ and $s + t \neq 0$. Note that the last equality is of the form presented by (1.25) with $p_1 = p_2 = p_3 = 1$, $q_1 = \sigma_{t+s}$, $q_2 = \sigma_{-s}$, $q_3 = \sigma_{-t}$.

1.2.13   Theorem 1.2 revisited

The more detailed statement of the theorem on relations is as follows.

Theorem 1.3 *Every relation holding between the generators* $\mathcal{P} \cup \{s_0, s_1, s_2\}$ *of the Cremona group* $\mathrm{Cr}(2, K)$ *is a consequence of the special relations (1.14), (1.18), (1.21), (1.24), (1.27), (1.29), (1.30), (1.31), (1.32), (1.33).*

For the proof, see [11], 10.6–10.7.

2   The universal Cremona group

2.1   Admissible triples, their spaces and maps

Definition 2.1 An *admissible triple* is a triple (R, A, M), where:

1. R is a commutative ring with a unit.

2. A is an R-algebra, not necessarily commutative or associative, but at least *alternative*; this means that R is contained in the centre of A and the subring of A generated by *any two* elements is associative.

3. $M \subset A$ is an R-submodule such that
$$mMm \subset M \quad \text{for every } m \in M.$$

4. If $m \in M$ has a total inverse m^{-1} in A, then $m^{-1} \in M$; here a *total inverse* of a (Mal'tsev [15], Chap. II, 4.3) means an element a^{-1} such that
$$a^{-1}(ax) = (xa)a^{-1} = x \quad \text{for every } x \in A.$$

Let $\mathbb{G}(M)$ denote the set of *units* or totally invertible elements of M.

Definition 2.2 Let R be a commutative ring having an involutive automorphism $r \mapsto \bar{r}$; by default, $\bar{}$ is the identity map if no involution is specified.

An *R-algebra with involution* is an R-algebra A with a semilinear involutive anti-automorphism $a \mapsto a^*$; that is, $*$ is an involution satisfying the identities
$$(ra + sb)^* = \bar{r}a^* + \bar{s}b^* \quad \text{and} \quad (ab)^* = b^*a^*.$$
We write
$$A^+ = \{a \in A \,|\, a^* = a\}, \quad A^- = \{a \in A \,|\, a^* = -a\}$$
for the set of $*$-invariant (respectively $*$-anti-invariant) elements of A.

For an R-algebra A with involution, a triple (R, A, M) is *admissible* if it is admissible in the sense of Definition 2.1, and $M^* = M$.

Remark 2.1 If A is an R-algebra with involution, then both (R, A, A^+) and (R, A, A^-) are admissible triples in the sense of Definition 2.2.

Let R be a commutative ring and $n \geq 0$ an integer. We construct a functor $\mathbb{S}_n$ from the category of admissible triples to the category of sets.

Definition 2.3 If (R, A, M) is an admissible triple, we say that an $(n + 1)$-tuple $\mathbf{m} = (m_0, \ldots, m_n) \in M^{n+1}$ is *invertible* if $\lambda_0 m_0 + \cdots + \lambda_n m_n$ is invertible in A for some $\lambda_0, \ldots, \lambda_n \in R$; in other words, if the components of $\mathbf{m}$ generate an R-submodule of M having nonempty intersection with $\mathbb{G}(M)$.

On invertible $(n+1)$-tuples, we introduce the equivalence relation $\sim$ which is generated by the elementary relation

$$\exists g \in \mathbb{G}(M) \quad \text{such that} \quad (m'_0, \ldots, m'_n) = (gm_0 g, \ldots, gm_n g).$$

(This is the point at which we need A to be alternative.) In other words, two $(n + 1)$-tuples $\mathbf{m}$ and $\mathbf{m}'$ are equivalent if and only if there are elements $g_1, \ldots, g_k \in \mathbb{G}(M)$ such that

$$m'_i = g_1(\cdots(g_{k-1}(g_k(m_i)g_k)g_{k-1})\cdots)g_1 \quad \text{for each } 0 \leq i \leq n.$$

We write $(m_0 : \cdots : m_n)$ (or sometimes simply $\mathbf{m}$) for the equivalence class of $\mathbf{m} = (m_0, \ldots, m_n)$, and define $\mathbb{S}_n(R, A, M)$ as the set of equivalence classes of invertible $(n + 1)$-tuples under $\sim$. We define the functors $\mathbb{S}_n^+$ and $\mathbb{S}_n^-$ on the category of R-algebras with involution by

$$\mathbb{S}_n^+(A) = \mathbb{S}_n(R, A, A^+) \quad \text{and} \quad \mathbb{S}_n^-(A) = \mathbb{S}_n(R, A, A^-).$$

Remark 2.2 The main example in what follows is the functor $\mathbb{S}_n^+$, especially its value $\mathbb{S}_n^+(\mathrm{Mat}_p(K))$ on the K-algebra $\mathrm{Mat}_p(K)$ of $p \times p$ matrices with entries in an algebraically closed field K, where the involution $*$ is matrix transposition.

In general, $\mathbb{S}_n^+(A)$ is the set of $(n + 1)$-tuples $(a_0, \ldots, a_n)$ with $a_i \in A^+$ and with an invertible R-linear combination $\sum \lambda_i a_i \in \mathbb{G}(A)$, modulo the equivalence relation:

$$(a_0, \ldots, a_n) \sim (a'_0, \ldots, a'_n) \iff ba_0 b = a'_0, \ldots, ba_n b = a'_n,$$

where $b \in \mathbb{G}(A)$ is a product of elements of $\mathbb{G}(A^+)$. $\mathbb{S}_n^+(A)$ is called the *spherical n-space* or the *n-sphere* over A. This is partly justified by the fact that if A is an algebra with involution $*$ over $\mathbb{R}$ or over $\mathbb{C}$, such that $A^+ = \mathbb{R}$, then $\mathbb{S}_n^+(A)$ is in natural one-to-one correspondence with the unit sphere $S_n \subset \mathbb{R}^{n+1}$.

If K is an algebraically closed field, then the set $\mathbb{G}(\mathrm{Mat}_p(K)^+)$ of invertible symmetric matrices generates the whole group $\mathrm{GL}(p, K)$, hence the spherical space $\mathbb{S}_n^+(\mathrm{Mat}_p(K))$ coincides with the "noncommutative projective space" of Tyurin and Tyurin [17].

Under certain conditions, we define the polynomial $\Delta(K, A, M)(\mathbf{m})$ and some other polynomials $\Gamma(\Delta)$, usually considered up to proportionality. These polynomials depend on dual variables $(u_0, \ldots, u_n)$ and $(x_0, \ldots, x_n)$, and define hypersurfaces in $\mathbb{P}_n$ and the dual $\check{\mathbb{P}}_n$.

Let A be a finite dimensional associative algebra over the ground field K, and $N_{A/K}\colon A \to K$ its norm (see Bourbaki, Algèbre, [4], Livre II, Chap. VII; for our purposes, we can use the so-called principal norm in the sense of the exercise in [4], *loc. cit.*, or the reduced norm if A is a semisimple algebra). Write p for the degree of the characteristic polynomial of A ([4], *loc. cit.*), and let $K[u_0, \ldots, u_n]_d$ be the vector space of homogeneous polynomials of degree d. We assume that the restriction of the norm $N_{A/K}$ to the subspace $M \subset A$ is the exact qth power of a polynomial: $N_{A/K}|_M = (N^0)^q$, where $N_{A/K}^0 \in K[M]$.

Definition 2.4 For a fixed $\mathbf{m} = (m_0 : \cdots : m_n) \in \mathbb{S}_n(M)$, we set

$$\Delta(K, A, M)(\mathbf{m})(u_0, \ldots, u_n) = N_{A/K}^0(u_0 m_0 + \cdots + u_n m_n);$$

thus $\Delta(K, A, M) \in K[u_0, \ldots, u_n]$ is a homogeneous polynomial of degree p/q.

If $\Gamma\colon K[u_0, \ldots, u_n]_{p/q} \to K[x_0, \ldots, x_n]$ is a contravariant and has nonzero value at $\Delta(K, A, M)(\mathbf{m})$, then $\Gamma(\Delta(K, A, M)(\mathbf{m}))$ is an equivariant like defined polynomial in the sense of Remark 2.5 below.

Remark 2.3 If $A = \mathrm{Mat}_p(K)$ is a matrix algebra over a field K, with involution matrix transposition, then $\Delta(K, A, A^+)(a) = \det(a)$.

Remark 2.4 If $\dim_K A < \infty$, the hypersurface $\Delta(M)(\mathbf{m})(u_0, \ldots, u_n) = 0$ coincides with the so-called *spectrum set* of $\mathbf{m}$, that is, with the set of all points $(y_0 : \cdots : y_n) \in \mathbb{P}_n(K)$ for which the linear combination $y_0 m_0 + \cdots + y_n m_n$ is a noninvertible element of M. Indeed, by [4], Chap. VII, Proposition 12, an element $x \in A$ is invertible in A if and only if its norm $N_{A/K}(x)$ is invertible in K.

Remark 2.5 The group $\mathrm{PGL}(n + 1, K)$ of projective transformations

$$g\colon x_i' = \sum_{j=0}^{n} g_{ij} x_j \quad \text{for } i = 0, \ldots, n$$

acts (on the left) on $\mathbb{P}_n$ and (on the right) on $\check{\mathbb{P}}_n$ by the transpose map

$$g^{\mathrm{T}}\colon u_i' = \sum_{j=0}^{n} g_{ji} u_j \quad \text{for } i = 0, \ldots, n.$$

This group acts (on the right) on $\mathbb{P}(K[u_0, \ldots, u_m]_d)$: if $F(u) \in K[u]_d$, then $g(F)(u) = F(g^{\mathrm{T}}(u))$. It also acts (on the left) on $\mathbb{S}_n(K, M, A)$:

$$g(\mathbf{m}) = (m_0' : \cdots : m_n'), \quad \text{where} \quad m_i' = \sum_{j=0}^n g_{ij} m_j.$$

Note that

$$\Delta(K, A, M)(g(\mathbf{m})) = g(\Delta(K, A, M)(\mathbf{m})),$$
$$\Gamma(\Delta(K, A, M)(g(\mathbf{m}))) = g^{-1}(\Gamma(\Delta(K, A, M)(\mathbf{m}))).$$

The last equality means that the map

$$\Gamma(\Delta(K, A, M)) \colon \mathbb{S}_n(K, A, M) \to \mathbb{P}(K[x_0, \ldots, x_n])$$

is equivariant with respect to $\mathrm{PGL}(n + 1, K)$, because the correspondence $g \mapsto (g^{-1})^{\mathrm{T}}$ is an automorphism of this group.

In the following two definitions, we now construct an analog of homogeneous rational functions, specially adapted to the noncommutative case; these are certain expressions in variables which are either letters, or elements of a K-algebra A. The pattern of our construction follows that of the well-formed formulas in the calculus of mathematical logic (for example, see Church's book [5]); our functions are always derived from well-formed rational expressions. Moreover, for a well-formed homogeneous rational expression f, we define at the same time its *domain of definition* $\mathrm{dom}\, f \subset M^{n+1}$ (here M is the third component of an admissible triple (K, A, M)), its *value* $f(\mathbf{m}) \in M$ at point $\mathbf{m} = (m_0, \ldots, m_n) \in \mathrm{dom}\, f$, and the notion of the *domain of invertibility* $\mathrm{dom}\, f^{-1} \subset M^{n+1}$ of such an expression. The degrees of homogeneity of our functions $f(x_0, \ldots, x_n)$ are the numbers $+1$ and -1; in what follows, ε stands for an element $\varepsilon \in \{+1, -1\}$. The ground field K is fixed, and its elements are called constants.

Definition 2.5 (i) For each $i \in \{0, \ldots, n\}$, we define the coordinate function $f(x_0, \ldots, x_n) = x_i$ to be an expression of degree 1. Its domain of definition $\mathrm{dom}\, x_i$ is the whole of M^{n+1}, its value at $\mathbf{m} = (m_0, \ldots, m_n)$ is equal to m_i, its domain of invertibility is

$$\{\mathbf{m} = (m_0, \ldots, m_i, \ldots, m_n) \mid m_i \in \mathbb{G}(M)\}.$$

(ii) If f is an expression of degree $\varepsilon = \pm 1$ and λ a nonzero constant, λf is an expression of degree ε by definition. Its domain of definition (or invertibility) coincides with that of f, and its value at $\mathbf{m}$ is equal to $\lambda f(\mathbf{m})$.

(iii) If f and g are expressions of the same degree $\varepsilon = \pm 1$, the sum $f + g$ is an expression of degree ε. Its domain of definition is the intersection of the corresponding domains for f and g, its value $(f + g)(\mathbf{m})$ equals $f(\mathbf{m}) + g(\mathbf{m})$ and its domain of invertibility is

$$\{\mathbf{m} \mid \mathbf{m} \in \operatorname{dom}(f + g) \text{ and } (f + g)(\mathbf{m}) \in \mathbb{G}(M)\}.$$

(iv) If f is an expression of degree ε, then $f^{-1} = 1/f$ is an expression of degree $-\varepsilon$ by definition. The domain of definition and the domain of invertibility of the expression f^{-1} coincide with the domain of invertibility of f. The value $f^{-1}(\mathbf{m})$ is equal to $(f(\mathbf{m}))^{-1}$.

(v) If f, g are expressions of degree ε then $fg^{-1}f$ and $f^{-1}gf^{-1}$, are expressions of degree ε and $-\varepsilon$ respectively by definition. The domain of definition or invertibility of each product is the intersection of the corresponding domains for all the three factors, and

$$(fg^{-1}f)(\mathbf{m}) = f(\mathbf{m})g^{-1}(\mathbf{m})f(\mathbf{m}),$$
$$(f^{-1}gf^{-1})(\mathbf{m}) = f^{-1}(\mathbf{m})g(\mathbf{m})f^{-1}(\mathbf{m}).$$

The smallest set of expressions satisfying the above conditions (i)–(v) is the set of well-formed homogeneous K-rational expressions of variables $(x_0, \ldots, x_n)$. Every K-rational expression f has a definite degree $\deg f \in \{+1, -1\}$.

Remark 2.6 If $f(\mathbf{x}) = f(x_0, \ldots, x_n)$ is a K-rational expression of degree δ and $g_0(\mathbf{y}), \ldots, g_n(\mathbf{y})$ are n expressions in variables $\mathbf{y} = (y_0, \ldots, y_m)$ of degree ε, then the composite $f(g_0, \ldots, g_n)$ is obviously a rational expression in $\mathbf{y}$ of degree $\varepsilon\delta$.

If f is a well-formed expression of degree ε, and $\mathbf{m} \in M^{n+1}$ belongs to $\operatorname{dom} f$, then $bmb \in \operatorname{dom} f$ for any $b \in \mathbb{G}(M)$, and $f(bmb) = b^\varepsilon f(\mathbf{m})b^\varepsilon$.

Definition 2.6 If f and g are two well-formed homogeneous expressions in $(n + 1)$ variables, each with nonempty domain of definition in some M^{n+1}, we say that f and g are equivalent (and write $f \equiv g$) if for every admissible triple (K, A, M) and for every element $\mathbf{m} \in \operatorname{dom} f \cap \operatorname{dom} g \subset M^{n+1}$, the equality $f(\mathbf{m}) = g(\mathbf{m})$ holds. A *homogeneous K-rational function* is defined as an equivalence class of well-formed homogeneous K-rational expressions $f(x_0, \ldots, x_n)$ with a nonempty domain of definition in some M^{n+1}; the set of these is denoted by $\operatorname{Rat}(n+1, K)$. Note that nonzero homogeneous K-rational functions fall into two sets according to their degree.

Remark 2.7 We have the following identities in two variables x, y:

$$(x^{-1})^{-1} \equiv x, \qquad (x^{-1}yx^{-1})^{-1} \equiv xy^{-1}x,$$

$$(x^{-1} - y^{-1})^{-1} \equiv x(x - xy^{-1}x)^{-1}x$$

$$xy^{-1}x \equiv x - (x^{-1} - (x - y)^{-1})^{-1}, \tag{2.1}$$

$$x(x - xy^{-1}x)^{-1}x \equiv x - x(x - y)^{-1}x \tag{2.2}$$

$$(x^{-1} - y^{-1})^{-1} \equiv x - x(x - y)^{-1}x. \tag{2.3}$$

See Mal'tsev [16], Chap. 2, 4.3 for (2.1); (2.3) follows from (2.1) on substituting $y \mapsto x - y$, and (2.2) is similar.

Definition 2.7 A well-formed K-rational map from projective n-space to projective p-space is given by a $(p + 1)$-tuple of K-rational functions of the same degree in variables $(x_0, \ldots, x_n)$:

$$\widetilde{f}(\mathbf{x}) = (f_0(\mathbf{x}) : \cdots : f_p(\mathbf{x})); \tag{2.4}$$

$$\text{or} \quad \widetilde{f} : x'_0 = f_0(x_0, \ldots, x_n), \ldots, x'_p = f_p(x_0, \ldots, x_n).$$

Two $(p + 1)$-tuples give the same map if they are equivalent under the equivalence relation generated by the following primitive relation: another $(p+1)$-tuple $(g_0(\mathbf{x}) : \cdots : g_p(\mathbf{x}))$ is equivalent to (2.4) if there exists a K-rational function $h(\mathbf{x})$ with $\deg(h) = -\deg(g_i)$ and with nonempty domain of invertibility, such that we have equivalences (in the sense of Definition 2.6) $f_i \equiv hg_ih$ for each $0 \leq i \leq p$.

The identity map is the transformation given by $x'_0 = x_0, \ldots, x'_n = x_n$.

The map $\widetilde{f}(\mathbf{x})$ (2.4) induces a family of partially defined maps $\mathbb{S}_n(M) \rightarrow \mathbb{S}_p(M)$, one for every admissible triple (K, A, M); it follows from Remark 2.6 that these maps are well defined. The domain of definition of a $(p + 1)$-tuple (2.4) consisting of rational expressions f_i is

$$\{\mathbf{m} \mid \mathbf{m} \in \text{dom } f_i, \text{ for } 0 \leq i \leq p\},$$

and on it $\widetilde{f}(\mathbf{m})$ defines a point of $\mathbb{S}_p(M)$ in the sense of Definition 2.3.

A well-formed K-birational transformation of projective n-space is a well-formed rational map F of this space to itself such that there is an inverse map G with the property that both composites $F \circ G$ and $G \circ F$ are equal (more precisely, equivalent) to the identity map. We call such a map F a *universal Cremona transformation*; the group of all these is called the *universal Cremona group*, and is denoted $\mathrm{UCr}(n, K)$.

A partially defined map $f : \mathbb{S}_n(M) \rightarrow \mathbb{S}_n(M)$ with a nonempty domain of definition, where M is the third component of an admissible triple (K, A, M), is a *Cremona transformation* if there is an element $F \in \mathrm{UCr}(n, K)$ inducing f. We identify two such maps if they coincide on some nonempty intersection of the domains of definition of some well-formed representatives for both maps. The group of these maps will be denoted by $\mathrm{Cr}(n, M)$.

Thus the universal Cremona group $\mathrm{UCr}(n, K)$ is endowed with a family of epimorphisms

$$\Pi(n, M)\colon \mathrm{UCr}(n, K) \to \mathrm{Cr}(n, M). \tag{2.5}$$

Our immediate goal is to construct (for the case of an algebraically closed ground field K) a section $\Sigma(2, K)$ of the epimorphism $\Pi(2, K)$.

Remark 2.8 If A is a finite dimensional associative algebra with involution over an algebraically closed field K and (K, A, M) an admissible triple such that the set $\mathbb{G}(M)$ generates a semisimple algebraic subgroup G of $\mathbb{G}(A)$, then, at least birationally, one may view $\mathbb{S}_n(K, A, M)$ as a geometric quotient of M^{n+1} with respect to the two-sided diagonal action of G. Thus a generic element of $\mathbb{S}_n(K, A, M)$ may be viewed as a generic point of some algebraic variety over K; and moreover, we may view the transformations of the $\mathbb{S}_n(K, A, M)$ induced by elements $\mathrm{UCr}(n, K)$ as birational transformations of the variety.

2.2 An action of the Cremona group of the plane on the 2-spaces $\mathbb{S}_2(K, A, M)$

Let A be a K-algebra over an algebraically closed field K and (K, A, M) an admissible triple. The collineation group $\mathcal{P} = \mathrm{PGL}(3, K)$ acts on the set $\mathbb{S}_2(K, A, M)$. Our goal is to extend the action to the whole Cremona group (see the Introduction), making it act on $\mathbb{S}_2(K, A, M)$ by birational transformations. The group $\mathrm{Cr}(2, M)$ acts on $\mathbb{S}_2(K, A, M)$ and, according to equation (2.5) (see (2.6) below), we have the epimorphism $\Pi(2, M)$ of the universal Cremona group $\mathrm{UCr}(2, K)$ onto $\mathrm{Cr}(2, M)$, hence this universal group acts on $\mathbb{S}_2(M)$. A special case of (2.5) is

$$\Pi(2, K)\colon \mathrm{UCr}(2, K) \to \mathrm{Cr}(2, K). \tag{2.6}$$

If the homomorphism (2.6) admits a section

$$\Sigma(2, K)\colon \mathrm{Cr}(2, K) \to \mathrm{UCr}(2, K), \tag{2.7}$$

(of course, by definition, so that the composite $\Pi(2, K) \circ \Sigma(2, K)$ is the identity of $\mathrm{Cr}(2, K)$), then this section provides the required extension.

In the rest of this section, our plan is as follows. First, we already have a natural inclusion

$$\Sigma_{\mathcal{P}}\colon \mathcal{P} \to \mathrm{UCr}(2, K). \tag{2.8}$$

of the collineation group $\mathcal{P}$ into the universal Cremona group.

Next, we find three universal birational maps $S_0, S_1, S_2 \in \mathrm{UCr}(2, K)$ that map to the quadratic transformations s_0, s_1, s_2 under $\Pi(2, K)$. Finally, we check that all the relations mentioned in Theorem 1.3 hold in $\mathrm{UCr}(2, K)$, or more precisely, the relations obtained from those by substituting S_0, S_1, S_2 respectively for s_0, s_1, s_2.

Let $\mathbf{x} = (x_0 : x_1 : x_2)$. We define the effect of the action on $\mathbf{x}$ of the quadratic maps S_0, S_1, S_2 (compare (1.4), (1.6), (1.8)) in the following natural way:

$$
\begin{aligned}
S_0(\mathbf{x}) &= (x_0^{-1} : x_1^{-1} : x_2^{-1}), \\
S_1(\mathbf{x}) &= (x_1 x_0^{-1} x_1 : x_1 : x_2), \\
S_2(\mathbf{x}) &= (x_0 : x_1 : x_1 x_0^{-1} x_1 - x_2).
\end{aligned}
$$

Note that, to be correct, we should perhaps write "$\equiv$" instead of "$=$" in all the verifications below, but we neglect to do it.

2.2.1 Verifying the relations (1.14)

These relations hold because the natural inclusion (2.8) of the collineation group of the plane into the universal Cremona group is a homomorphism.

2.2.2 Verifying the relations (1.18)

If G is the collineation $G(\mathbf{x}) = (t_0 x_i : t_1 x_j : t_2 x_k)$ (more precisely, the image of the collineation (1.16) under the inclusion (2.8)), then

$$GS_0(\mathbf{x}) = (t_0 x_i^{-1} : t_1 x_j^{-1} : t_2 x_k^{-1}), \quad S_0 GS_0(\mathbf{x}) = (t_0^{-1} x_i : t_1^{-1} x_j : t_2^{-1} x_k),$$

that is, $S_0 G S_0 = \overline{G}$, where $\overline{G} = \Sigma_{\mathcal{P}}(\overline{g})$. Thus (1.18) is satisfied here.

2.2.3 Verifying the relations (1.21)

Set $G = \Sigma_{\mathcal{P}}(g)$, where g is the collineation (1.19) and $\overline{G} = \Sigma_{\mathcal{P}}(\overline{g}))$, where $\overline{g}$ is (1.20); then

$$
\begin{aligned}
GS_1(\mathbf{x}) &= (t_0 x_1 x_0^{-1} x_1 : t_1 x_1 : t_2 x_2 + r x_0), \\
S_1 GS_1(\mathbf{x}) &= (t_1 x_1 (t_0 x_1 x_0^{-1} x_1)^{-1} t_1 x_1 : t_1 x_1 : t_2 x_2 + r x_0) \\
&= (t_1^2 t_0^{-1} x_0 : t_0 t_1 x_1 : t_2 x_2 + r x_0) = \overline{G}(\mathbf{x}).
\end{aligned}
$$

Thus (1.21) is satisfied here.

2.2.4   Verifying the relations (1.24)

Similarly, set $G = \Sigma_{\mathcal{P}}(g)$, where g is the collineation (1.22), and $\overline{G} = \Sigma_{\mathcal{P}}(\overline{g})$, where $\overline{g}$ is (1.23); then

$$
\begin{aligned}
GS_2(\mathbf{x}) &= (x_0 : t_1x_1 : t^2(x_1x_0^{-1}x_1 - x_2) + rx_1 + sx_0), \\
S_2GS_2(\mathbf{x}) &= (x_0 : tx_1 : -t^2(x_1x_0^{-1}x_1 - x_2) - rx_1 - sx_0 + tx_1x_0^{-1}tx_1) \\
&= (x_0 : tx_1 : t^2x_2 - rx_1 - sx_0) = \overline{G}(\mathbf{x}).
\end{aligned}
$$

Thus (1.24) is satisfied here.

2.2.5   Verifying the relation (1.27)

Let $H_0 = \Sigma_{\mathcal{P}}(h_0)$, where h_0 is the collineation (1.28). We have to check that

$$
S_0H_0S_0(\mathbf{x}) = H_0S_0H_0(\mathbf{x}).
$$

First, on the left-hand side,

$$
\begin{aligned}
S_0(\mathbf{x}) &= (x_0^{-1} : x_1^{-1} : x_2^{-1}), \\
H_0S_0(\mathbf{x}) &= (x_0^{-1} : x_0^{-1} - x_1^{-1} : x_0^{-1} - x_2^{-1}), \\
S_0H_0S_0(\mathbf{x}) &= (x_0 : (x_0^{-1} - x_1^{-1})^{-1} : (x_0^{-1} - x_2^{-1})^{-1}).
\end{aligned}
$$

Similarly, on the right-hand side,

$$
\begin{aligned}
S_0H_0(\mathbf{x}) &= (x_0^{-1} : (x_0 - x_1)^{-1} : (x_0 - x_2)^{-1}), \\
H_0S_0H_0(\mathbf{x}) &= (x_0^{-1} : x_0^{-1} - (x_0 - x_1)^{-1} : x_0^{-1} - (x_0 - x_2)^{-1}) \\
&= (x_0 : x_0 - x_0(x_0 - x_1)^{-1}x_0 : x_0 - x_0(x_0 - x_2)^{-1}x_0).
\end{aligned}
$$

Thus the required relation follows from the identity (2.3).

2.2.6   Verifying the relation (1.29)

Let $G_0 = \Sigma_{\mathcal{P}}(g_0)$, where g_0 is the collineation (1.10). We have to check that

$$
S_1(\mathbf{x}) = G_0S_0G_0S_0G_0(\mathbf{x}).
$$

We build the following pyramid of equivalences:

$$
\begin{aligned}
G_0(\mathbf{x}) &= (x_1 - x_0 : x_1 : x_2), \\
S_0G_0(\mathbf{x}) &= ((x_1 - x_0)^{-1} : x_1^{-1} : x_2^{-1}), \\
G_0S_0G_0(\mathbf{x}) &= (x_1^{-1} - (x_1 - x_0)^{-1} : x_1^{-1} : x_2^{-1}), \\
S_0G_0S_0G_0(\mathbf{x}) &= ((x_1^{-1} - (x_1 - x_0)^{-1})^{-1} : x_1 : x_2), \\
G_0S_0G_0S_0G_0(\mathbf{x}) &= (x_1 - (x_1^{-1} - (x_1 - x_0)^{-1})^{-1} : x_1 : x_2).
\end{aligned}
$$

By virtue of the identity (2.1), the first component of the last triple coincides with $x_1x_0^{-1}x_1$. Hence the required relation is established.

2.2.7 Verifying the relation (1.30)

Let $G_1 = \Sigma_{\mathcal{P}}(g_1)$, where g_1 is the collineation (1.12). We have to check that

$$S_2(\mathbf{x}) = S_1 G_1 S_1(\mathbf{x}).$$

This is easy; indeed,

$$
\begin{aligned}
S_1(\mathbf{x}) &= (x_1 x_0^{-1} x_1 : x_1 : x_2), \\
G_1 S_1(\mathbf{x}) &= (x_1 x_0^{-1} x_1 : x_1 : x_1 x_0^{-1} x_1 - x_2), \\
S_1 G_1 S_1(\mathbf{x}) &= (x_1 (x_1 x_0^{-1} x_1)^{-1} x_1 : x_1 : x_1 x_0^{-1} x_1 - x_2) \\
&= (x_0 : x_1 : x_1 x_0^{-1} x_1 - x_2) = S_2(\mathbf{x}).
\end{aligned}
$$

2.2.8 Verifying the relation (1.31)

Let $F = \Sigma_{\mathcal{P}}(f)$, where f is the collineation of (1.31). We have to check that

$$S_0(\mathbf{x}) = F S_1 F S_1(\mathbf{x}).$$

As before, this is easy; indeed,

$$
\begin{aligned}
S_1(\mathbf{x}) &= (x_1 x_0^{-1} x_1 : x_1 : x_2), \\
F S_1(\mathbf{x}) &= (x_2 : x_1 : x_1 x_0^{-1} x_1), \\
S_1 F S_1(\mathbf{x}) &= (x_1 x_2^{-1} x_1 : x_1 : x_1 x_0^{-1} x_1) = (x_2^{-1} : x_1^{-1} : x_0^{-1}), \\
F S_1 F S_1(\mathbf{x}) &= (x_0^{-1} : x_1^{-1} : x_2^{-1}).
\end{aligned}
$$

2.2.9 Verifying the relation (1.32)

Let $H_1 = \Sigma_{\mathcal{P}}(h_1)$, where h_1 is the collineation participating in (1.32). We have to check that

$$S_1 H_1 S_1(\mathbf{x}) = H_1 S_1 H_1(\mathbf{x}).$$

First, on the left-hand side,

$$
\begin{aligned}
S_1(\mathbf{x}) &= (x_1 x_0^{-1} x_1 : x_1 : x_2), \\
H_1 S_1(\mathbf{x}) &= (x_1 - x_1 x_0^{-1} x_1 : x_1 : x_2), \\
S_1 H_1 S_1(\mathbf{x}) &= (x_1 (x_1 - x_1 x_0^{-1} x_1)^{-1} x_1 : x_1 : x_2).
\end{aligned}
$$

Next, on the right-hand side,

$$
\begin{aligned}
H_1(\mathbf{x}) &= (x_1 - x_0 : x_1 : x_2), \\
S_1 H_1(\mathbf{x}) &= (x_1 (x_1 - x_0)^{-1} x_1 : x_1 : x_2), \\
H_1 S_1 H_1(\mathbf{x}) &= (x_1 - x_1 (x_1 - x_0)^{-1} x_1 : x_1 : x_2).
\end{aligned}
$$

Thus the required relation now follows from the identity (2.2).

2.2.10 Verifying the relation (1.33)

Let p_t be the collineation of 1.2.12, and $P_t = \Sigma_P(p_t)$ its image; as in (1.33), write $t' = 1 - \frac{1}{t}$ and $t'' = \frac{1}{1-t}$. Then

$$
\begin{aligned}
S_2(\mathbf{x}) &= (x_0 : x_1 : x_1 x_0^{-1} x_1 - x_2), \\
P_t S_2(\mathbf{x}) &= (x_0 : -tx_1 : t(x_1 x_0^{-1} x_1 - x_2)), \\
S_2 P_t S_2(\mathbf{x}) &= (x_0 : -tx_1 : (t^2 - t)(x_1 x_0^{-1} x_1) + tx_2), \\
P_{t'} S_2 P_t S_2(\mathbf{x}) &= (x_0 : (t-1)x_1 : (t-1)^2(x_1 x_0^{-1} x_1) + (t-1)x_2), \\
S_2 P_{t'} S_2 P_t S_2(\mathbf{x}) &= (x_0 : (t-1)x_1 : -(t-1)x_2), \\
P_{t''} S_2 P_{t'} S_2 P_t S_2(\mathbf{x}) &= (x_0 : x_1 : x_2).
\end{aligned}
$$

All our verifications are now completed.   Q.E.D.

3   Conics, cubics and quartics

3.1   Left and right actions of the Cremona group of the plane on the space of plane conics

Let $A = \mathrm{Mat}_2(K)$ be the 2×2 matrix algebra over an algebraically closed field K of characteristic $\neq 2$, with involution given by transposition $* = {}^{\mathrm{T}}$; thus A^+ is the set of symmetric matrices. We write $D(P, Q)$ for the mixed determinant of two 2×2-matrices P, Q; in other words, if

$$
P = \begin{pmatrix} p_{11} & p_{12} \\ p_{21} & p_{22} \end{pmatrix}, \quad Q = \begin{pmatrix} q_{11} & q_{12} \\ q_{21} & q_{22} \end{pmatrix},
$$

then

$$
D(P, Q) = \frac{1}{2} \left(\begin{vmatrix} p_{11} & q_{12} \\ p_{21} & q_{22} \end{vmatrix} + \begin{vmatrix} q_{11} & p_{12} \\ q_{21} & p_{22} \end{vmatrix} \right).
$$

The spherical 2-space $\mathbb{S}_2^+(A)$ consists of triples (m_0, m_1, m_2) of symmetric matrices $m_i \in A^+$, such that some K-linear combination $\lambda_0 m_0 + \lambda_1 m_1 + \lambda_2 m_2$ is invertible, modulo the equivalence relation: $(m_0, m_1, m_2) \sim (n_0, n_1, n_2)$ if $n_i = C m_i C^{\mathrm{T}}$ for some invertible matrix $C \in A$. Let $\mathbf{m} = (m_0 : m_1 : m_2)$ denote the equivalence class of (m_0, m_1, m_2). The collineation group $\mathcal{P}$ acts (on the left) on $\mathbb{S}_2^+(A)$. The $\mathcal{P}$-anti-equivariant map

$$
\Delta \colon \mathbb{S}_2^+(\mathrm{Mat}_2(K)) \to \mathbb{P}_5(K) = \mathbb{P}(K[u_0, u_1, u_2]_2)
$$

(compare Definition 2.4) sends each triple $\mathbf{m} = (m_0 : m_1 : m_2)$ to the ternary quadratic form

$$
\Delta(\mathbf{m})(u_0, u_1, u_2) = \det(u_0 m_0 + u_1 m_1 + u_2 m_2) = \sum_{ij} a_{ij}(\mathbf{m}) u_i u_j,
$$

considered up to proportionality; here $a_{ij}(\mathbf{m}) = D(m_i, m_j)$. The Cremona group of the plane acts (on the left) on $\mathbb{S}_2^+(A)$. It is possible to define a natural (right) action of this group on the space of conics in such a way that Δ is a $\mathrm{Cr}(2, K)$-anti-equivariant map. Indeed, we can use the identities

$$\det(P^{-1}) = (\det(P))^{-1}, \quad D(P, P) = \det(P),$$

$$D(P^{-1}, Q^{-1}) = \frac{D(P, Q)}{\det(P)\det(Q)},$$

$$D(QP^{-1}Q, Q) = \frac{D(P, Q)\det(Q)}{\det(P)},$$

$$D(QP^{-1}Q, P) = 2D(P, Q)^2(\det(P))^{-1} - \det(Q),$$

$$D(QP^{-1}Q, R) = \frac{2D(P, Q)D(Q, R) - D(P, R)\det(Q)}{\det(P)}.$$

For s_0, we get

$$a_{ij}(S_0(\mathbf{m})) = a_{ij}(\mathbf{m})(a_{ii}(\mathbf{m})a_{jj}(\mathbf{m}))^{-1} \quad \text{for } 0 \le i, j \le 2;$$

to write down explicit formulas for the actions of the three quadratic transformations s_0, s_1, s_2 on conics. That is, in other expressions, the right action of the standard quadratic transformation on the space of plane conics is described by the formulas

$$
\begin{aligned}
a'_{00} &= a_{11}a_{22}, & a'_{11} &= a_{22}a_{00}, & a'_{22} &= a_{00}a_{11}, \\
a'_{12} &= a_{12}a_{00}, & a'_{02} &= a_{02}a_{11}, & a'_{01} &= a_{01}a_{22}.
\end{aligned}
\tag{3.0}
$$

Similarly, we get the following formulas for s_1 and s_2:

$$
\begin{aligned}
a_{00}(S_1(\mathbf{m})) &= (a_{00}(\mathbf{m}))^{-1}(a_{11}(\mathbf{m}))^2, & a_{11}(S_1(\mathbf{m})) &= a_{11}(\mathbf{m}), \\
a_{22}(S_1(\mathbf{m})) &= a_{22}(\mathbf{m}), & a_{12}(S_1(\mathbf{m})) &= a_{12}(\mathbf{m}), \\
a_{01}(S_1(\mathbf{m})) &= a_{01}(\mathbf{m})a_{11}(\mathbf{m})(a_{00}(\mathbf{m}))^{-1}, \\
a_{02}(S_1(\mathbf{m})) &= (a_{00}(\mathbf{m}))^{-1}(2a_{01}(\mathbf{m})a_{12}(\mathbf{m}) - a_{02}(\mathbf{m})a_{11}(\mathbf{m}))
\end{aligned}
\tag{3.1}
$$

and

$$
\begin{aligned}
a_{00}(S_2(\mathbf{m})) &= a_{00}(\mathbf{m}), \quad a_{11}(S_2(\mathbf{m})) = a_{11}(\mathbf{m}), \quad a_{01}(S_2(\mathbf{m})) = a_{01}(\mathbf{m}), \\
a_{12}(S_2(\mathbf{m})) &= a_{11}(\mathbf{m})a_{01}(\mathbf{m})((a_{00}(\mathbf{m}))^{-1} - a_{12}(\mathbf{m}), \\
a_{02}(S_2(\mathbf{m})) &= 2(a_{01}(\mathbf{m}))^2(a_{00}(\mathbf{m}))^{-1} - a_{02}(\mathbf{m}) - a_{11}(\mathbf{m}), \\
a_{22}(S_2(\mathbf{m})) &= a_{22}(\mathbf{m}) + (a_{00}(\mathbf{m}))^{-1}((a_{11}(\mathbf{m}))^2 \\
&\qquad + 2a_{02}(\mathbf{m})a_{11}(\mathbf{m}) - 4a_{01}(\mathbf{m})a_{12}(\mathbf{m})).
\end{aligned}
\tag{3.2}
$$

An alternative way of writing the action of S_1 is as follows:

$$a'_{00} = a_{11}^2, \qquad a'_{11} = a_{11}a_{00}, \qquad\qquad a'_{22} = a_{22}a_{00},$$
$$a'_{12} = a_{12}a_{00}, \qquad a'_{02} = 2a_{01}a_{12} - a_{02}a_{11}, \qquad a'_{01} = a_{01}a_{11}, \tag{3.1a}$$

and similarly for the action of S_2:

$$a'_{00} = a_{00}^2, \qquad\qquad a'_{11} = a_{11}a_{00}, \qquad a'_{01} = a_{01}a_{00},$$
$$a'_{12} = a_{11}a_{01} - a_{12}a_{00}, \qquad a'_{02} = 2a_{01}^2 - (a_{02} + a_{11})a_{00}, \tag{3.2a}$$
$$a'_{22} = a_{22}a_{00} + a_{11}^2 + 2a_{02}a_{11} - 4a_{01}a_{12}.$$

All the special relations (where, of course, we replace each collineation g by its transpose g^{T}, and reverse the order of terms in products) are satisfied here.

If we want a left action of the Cremona group of the plane on the space of plane conics, then we must pass to the dual conic. Let $(A_{ij})_{0 \leq i,j \leq 2}$ be the adjoint matrix of (a_{ij}); then the left action of s_0 is given by the formulas

$$a'_{ij} = a_{ii}a_{jj}A_{ij} \tag{3.0b}$$

The left action of s_1 is defined by

$$a'_{00} = a_{00}^2 A_{00}, \qquad\qquad a'_{11} = a_{11}^2 A_{11} - 4a_{02}a_{12}A_{02},$$
$$a'_{22} = a_{11}^2 A_{22}, \qquad\qquad a'_{12} = a_{01}a_{11}A_{02} - a_{11}a_{12}A_{22}, \tag{3.1b}$$
$$a'_{02} = -a_{00}a_{11}A_{02}, \qquad a'_{01} = a_{00}a_{12}A_{02} - a_{00}a_{01}A_{00}.$$

The left action of s_2 is defined by

$$a'_{00} = a_{00}^2 A_{00} - 2a_{00}a_{11}A_{02} + a_{11}^2 A_{22},$$
$$a'_{11} = a_{00}^2 A_{11} + 4a_{00}a_{01}A_{12} + 4a_{01}^2 A_{22}, \qquad a'_{22} = a_{00}^2 A_{22},$$
$$a'_{01} = a_{00}^2 A_{01} + 2a_{00}a_{01}A_{02} - a_{00}a_{11}A_{12} - 2a_{01}a_{11}A_{22}, \tag{3.2b}$$
$$a'_{02} = -a_{00}^2 A_{02} + a_{00}a_{11}A_{22}, \qquad a'_{12} = -a_{00}^2 A_{12} - 2a_{00}a_{01}A_{22}.$$

More precisely, if a matrix $g \in \mathrm{PGL}(3, K) = \mathcal{P}$, viewed up to scalar multiples, acts on the space of symmetric 3×3 matrices (also viewed up to scalar multiples)

$$a = (a_{ij}) \in \check{\mathbb{P}}_5(K) = \mathbb{P}(S^2(W^*))$$

according to the rule

$$g(a) = (g^{-1})^{\mathrm{T}} a(g^{-1}),$$

and the quadratic transformations s_0, s_1, s_2 act on $\check{\mathbb{P}}_5(K)$ according to formulas (3.0b), (3.1b), (3.2b) respectively, then we have a well-defined (left) action of $\mathrm{Cr}(2, K)$ on the space of plane conics.

Note that our good luck in the case of conics is based on the fact that the map Δ is a birational isomorphism (for some analogs of this fact see below, 3.2.5, Theorem 3.3 and 3.5, formulas (3.39)-(3.40)).

3.2   Left and right actions of the Cremona group of the plane on the space of plane cubics

Now let $A = \mathrm{Mat}_3(K)$ be the 3×3 matrix algebra over an algebraically closed field K of characteristic $\neq 2, 3$, with the involution given by transposition $* = {}^{\mathrm{T}}$; thus A^+ is again the set of symmetric 3×3 matrices. Let $D(P, Q, R)$ denote the mixed determinant of three 3×3 matrices; that is, if

$$P = \begin{pmatrix} p_{11} & p_{12} & p_{13} \\ p_{21} & p_{22} & p_{23} \\ p_{31} & p_{32} & p_{33} \end{pmatrix}, \quad Q = \begin{pmatrix} q_{11} & q_{12} & q_{13} \\ q_{21} & q_{22} & q_{23} \\ q_{31} & q_{32} & q_{33} \end{pmatrix}, \quad R = \begin{pmatrix} r_{11} & r_{12} & r_{13} \\ r_{21} & r_{22} & r_{23} \\ r_{31} & r_{32} & r_{33} \end{pmatrix},$$

then $6D(P, Q, R)$ equals

$$\begin{vmatrix} p_{11} & q_{12} & r_{13} \\ p_{21} & q_{22} & r_{23} \\ p_{31} & q_{32} & r_{33} \end{vmatrix} + \begin{vmatrix} p_{11} & r_{12} & q_{13} \\ p_{21} & r_{22} & q_{23} \\ p_{31} & r_{32} & q_{33} \end{vmatrix} + \begin{vmatrix} q_{11} & p_{12} & r_{13} \\ q_{21} & p_{22} & r_{23} \\ q_{31} & p_{32} & r_{33} \end{vmatrix}$$

$$+ \begin{vmatrix} q_{11} & r_{12} & p_{13} \\ q_{21} & r_{22} & p_{23} \\ q_{31} & r_{32} & p_{33} \end{vmatrix} + \begin{vmatrix} r_{11} & p_{12} & q_{13} \\ r_{21} & p_{22} & q_{23} \\ r_{31} & p_{32} & q_{33} \end{vmatrix} + \begin{vmatrix} r_{11} & q_{12} & p_{13} \\ r_{21} & q_{22} & p_{23} \\ r_{31} & q_{32} & p_{33} \end{vmatrix}.$$

The spherical 2-space $\mathbb{S}_2^+(A)$ consists of triples of matrices (m_0, m_1, m_2) in A^+ for which some K-linear combination $\lambda_0 m_0 + \lambda_1 m_1 + \lambda_2 m_2$ is invertible, and $(m_0, m_1, m_2) \sim (n_0, n_1, n_2)$ if there exists an invertible matrix $C \in A$ such that $n_i = C m_i C^{\mathrm{T}}$. We write $\mathbf{m} = (m_0 : m_1 : m_2)$ for the equivalence class. The collineation group $\mathcal{P}$ acts (on the left) on $\mathbb{S}_2^+(A)$. The $\mathcal{P}$-anti-equivariant map (see Definition 2.4)

$$\Delta \colon \mathbb{S}_2^+(\mathrm{Mat}_3(K)) \to \mathbb{P}_9(K) = \mathbb{P}(K[u_0, u_1, u_2]_3)$$

associates with each triple $\mathbf{m} = (m_0 : m_1 : m_2)$ the ternary cubic form

$$\Delta(\mathbf{m})(u_0, u_1, u_2) = \det(u_0 m_0 + u_1 m_1 + u_2 m_2) = \sum_{ij} a_{ijk}(\mathbf{m}) u_i u_j u_k$$

(up to proportionality), where $a_{ijk}(\mathbf{m}) = D(m_i, m_j, m_k)$. The cubic curve $\Delta(\mathbf{m})(u_0, u_1, u_2) = 0$ inherits an additional structure from the matrix triple $\mathbf{m}$, namely, an even theta characteristic, that is, a nonzero 2-torsion point; we now treat these relations more explicitly.

3.2.1   Invariants, covariants, contravariants and 2-torsion of plane cubic curves

We take the three variables x_0, x_1, x_2 to be homogeneous coordinates on $\mathbb{P}_2$, and normalize the coefficients a_{ijk} of $x_i x_j x_k$ in a cubic form F as follows:

$$
\begin{aligned}
F = {} & a_{000}x_0^3 + a_{111}x_1^3 + a_{222}x_2^3 + 6a_{012}x_0x_1x_2 \\
& + 3a_{001}x_0^2x_1 + 3a_{002}x_0^2x_2 + 3a_{110}x_0x_1^2 \\
& + 3a_{112}x_1^2x_2 + 3a_{220}x_0x_2^2 + 3a_{221}x_1x_2^2.
\end{aligned} \tag{3.3}
$$

Let u_0, u_1, u_2 be dual homogeneous coordinates on the projective plane $\check{\mathbb{P}}_2$. The *Hessian form* $\mathrm{He}(F)$ of the cubic (3.3) is defined by

$$
\mathrm{He}(F) = \frac{1}{36}\det(\mathrm{HE}(F)),
$$

where $\mathrm{HE}(F)$ is the Hessian matrix

$$
\mathrm{HE}(F) = \begin{pmatrix}
\dfrac{\partial^2 F}{\partial x_0^2} & \dfrac{\partial^2 F}{\partial x_0 \partial x_1} & \dfrac{\partial^2 F}{\partial x_0 \partial x_2} \\[2mm]
\dfrac{\partial^2 F}{\partial x_1 \partial x_0} & \dfrac{\partial^2 F}{\partial x_1^2} & \dfrac{\partial^2 F}{\partial x_1 \partial x_2} \\[2mm]
\dfrac{\partial^2 F}{\partial x_2 \partial x_0} & \dfrac{\partial^2 F}{\partial x_2 \partial x_1} & \dfrac{\partial^2 F}{\partial x_2^2}
\end{pmatrix}. \tag{3.4}
$$

Note that our Hessian form differs slightly from that of Salmon's book [16] or Dolgachev and Kanev [8] (ours is multiplied by 6). Normalized coefficients by monomials $x_i x_j x_k$ of the Hessian are written down in [16], N° 218. The *Cayley form* $\mathrm{Ca}(F)$ of F is

$$
\mathrm{Ca}(F) = 3 \times \begin{vmatrix}
a_{000} & a_{110} & a_{220} & a_{012} & a_{002} & a_{001} \\
a_{001} & a_{111} & a_{221} & a_{112} & a_{012} & a_{011} \\
a_{002} & a_{112} & a_{222} & a_{122} & a_{022} & a_{012} \\
2u_0 & 0 & 0 & 0 & u_2 & u_1 \\
0 & 2u_1 & 0 & u_2 & 0 & u_0 \\
0 & 0 & 2u_2 & u_1 & u_0 & 0
\end{vmatrix}.
$$

There is a well-defined natural scalar product (or convolution) (F, G) of two ternary forms $F(x_0, x_1, x_2)$ and $G(u_0, u_1, u_2)$ of the same degree in dual variables. For example, if F and G are ternary cubic forms (where F is (3.3), and G has normalized coefficients b_{ijk}), then

$$
\begin{aligned}
(F,G) = {} & a_{000}b_{000} + a_{111}b_{111} + a_{222}b_{222} + 6a_{012}b_{012} \\
& + 3a_{001}b_{001} + 3a_{002}b_{002} + 3a_{110}b_{110} \\
& + 3a_{112}b_{112} + 3a_{220}b_{220} + 3a_{221}b_{221}.
\end{aligned}
\tag{3.5}
$$

The Aronhold invariants $S = S(F), T = T(F)$ and $R = R(F)$ of a cubic form F are defined by

$$
S(F) = -(F, \mathrm{Ca}(F)), \quad T(F) = -(\mathrm{He}(F), \mathrm{Ca}(F)), \quad R(F) = T(F)^2 - S(F)^3
$$

(compare [16], N°s 220–221). It is convenient to use the following contravariant cubic form

$$
D(F) = \frac{1}{3}\big(T(F)\,\mathrm{Ca}(F) - \mathrm{Ca}(\mathrm{He}(F))\big). \tag{3.6}
$$

The operation D is an analog of passing to the dual of a quadratic form. Indeed,

$$
D(D(F)) = -32R(F)^6 S(F)^2 F, \tag{3.7}
$$

or in other words, D iterated twice yields the initial cubic form (up to a factor). We may consider D as a "birational null-correlation", because the contravariant $D(F)$ defines a hyperplane in the space of cubic curves, and F belongs to this hyperplane: $(F, D(F)) = 0$, where $(\ ,\ )$ is the scalar product (3.5). The operation D interchanges the Hessian and the Cayley forms up to a factor, in the sense that

$$
\begin{aligned}
\mathrm{He}(D(F)) &= 2R(F)^2\,\mathrm{Ca}(F) \\
\text{and} \quad \mathrm{Ca}(D(F)) &= -4R(F)^2\,\mathrm{He}(F).
\end{aligned}
\tag{3.8}
$$

We refer to the pencil of cubic forms

$$
uF(x_0, x_1, x_2) + v\,\mathrm{He}(F)(x_0, x_1, x_2),
$$

as the *Hessian pencil* (an alternative term *syzygetic pencil* is due to L. Cremona), and

$$
u\,\mathrm{Ca}(F)(u_0, u_1, u_2) + vD(F)(u_0, u_1, u_2)
$$

as the *Cayley pencil*. The Hessian operation preserves both these pencils, giving rise to the following actions (compare [16], N° 225). On the Hessian pencil:

$$
\begin{aligned}
\mathrm{He}(uF + v\,\mathrm{He}(F)) = {} & 3v\Big(u^2 S + 2uvT + v^2 S^2\Big)\cdot F \\
& + \Big(u^3 - 3Suv^2 - 2Tv^3\Big)\cdot \mathrm{He}(F),
\end{aligned}
\tag{3.9}
$$

in particular $\mathrm{He}(\mathrm{He}(F)) = 3S^2 F - 2T\,\mathrm{He}(F).$ (3.10)

(Here and below, we write $S = S(F), T = T(F), R = R(F)$.) On the Cayley pencil:

$$\mathrm{He}(u\,\mathrm{Ca}(F) + vD(F)) = 6u\left(u^2 - 2Tuv + Rv^2\right)\cdot D(F) \qquad (3.11)$$
$$+ 2\left(2Tu^3 - 3Ru^2v + R^2v^3\right)\cdot \mathrm{Ca}(F),$$

in particular $\quad \mathrm{He}(\mathrm{Ca}(F)) = 6T(F)\,\mathrm{Ca}(F) - 2\,\mathrm{Ca}(\mathrm{He}(F)). \qquad (3.12)$

The Cayley operation takes the Hessian pencil into the Cayley pencil; namely,

$$S(F)\,\mathrm{Ca}(uF + v\,\mathrm{He}(F)) = 3v\left(u^2 - S(F)v^2\right)\cdot D(F)$$
$$+ \left(S(F)u^3 + 3T(F)u^2v + 3S(F)^2uv^2 + T(F)S(F)v^3\right)\cdot \mathrm{Ca}(F);$$

The operation D acts in a similar way. Furthermore,

$$\mathrm{Ca}(u\,\mathrm{Ca}(F) + vD(F)) = 12S(F)^2u\left(Rv^2 - u^2\right)\cdot F$$
$$+ 4\left(T(F)u^3 - 3R(F)u^2v + 3R(F)T(F)uv^2 - R(F)^2v^3\right)\cdot \mathrm{He}(F), \quad (3.13)$$

and

$$D(u\,\mathrm{Ca}(F) + vD(F)) = 16S^2R\Phi(u,v)^2[2uS\,\mathrm{He}(F) - (Tu + 2Rv)F],$$

where

$$\Phi(u,v) = 9u^4 - 8T(F)u^3v + 6R(F)u^2v^2 - v^4R(F)^2.$$

In particular,

$$D(\mathrm{Ca}(F)) = 288R(F)S(F)^2\left(T(F)F - S(F)\,\mathrm{He}(F)\right). \qquad (3.14)$$

Evaluating the Aronhold operations $S(\cdot), T(\cdot)$, and $R(\cdot)$ on our two pencils gives the following: on the Hessian pencil,

$$S(uF + v\,\mathrm{He}(F)) = u^4S + 4u^3vT + 6u^2v^2S^2$$
$$+ 4uv^3ST + v^4(4T^2 - 3S^3), \qquad (3.15)$$

in particular $\quad S(\mathrm{He}(F)) = 4T^2 - 3S^3 = T^2 + 3R. \qquad (3.16)$

Also,

$$T(uF + v\,\mathrm{He}(F)) = u^6T + 6u^5vS^2 + 15u^4v^2ST$$
$$+ 20u^3v^3T^2 + 15u^2v^4S^2T \qquad (3.17)$$
$$+ 6uv^5(3S^3 - 2T^2)S + v^6(9S^3 - 8T^2)T,$$

in particular $\quad T(\mathrm{He}(F)) = (9S^3 - 8T^2)T = T^3 - 9RT. \qquad (3.18)$

Further, (3.15) and (3.17) give

$$R(uF + v\,\mathrm{He}(F)) = (u^4 - 6Su^2v^2 - 8Tuv^3 - 3S^2v^4)^3R,$$
$$\text{in particular}\quad R(\mathrm{He}(F)) = -27S^6R(F).$$

On the Cayley pencil, we get

$$S(u\,\mathrm{Ca}(F) + vD(F)) = 4 \times \Big((4T^2 - 3R)u^4 - 4RTu^3v$$
$$+ 6RT^2u^2v^2 - 4R^2Tuv^3 + R^3v\Big),\quad (3.19)$$

in particular, $S(\mathrm{Ca}(F)) = 4(4T^2 - 3R)$ and $S(D(F)) = 4R(F)^3.$

Also,

$$T(u\,\mathrm{Ca}(F) + vD(F)) = 8 \times \Big(-T(9R - 8T^2)u^6 + 6R(3R - 2T^2)u^5v$$
$$- 15R^2Tu^4v^2 + 20R^2T^2u^3v^3 - 15R^3Tu^2v^4 + 6R^4uv^5 - TR^4v^6\Big),$$

in particular

$$T(\mathrm{Ca}(F)) = 8T(8T^2 - 9R)\ \text{and}\ T(D(F)) = -8R(F)^4T(F). \quad (3.20)$$

Finally,

$$R(\mathrm{Ca}(F)) = (-12S(F))^3\big(T(F)^2 - S(F)^3\big)^2 = -12^3S(F)^3R(F)^2,$$
$$R(D(F)) = 64R(F)^8S(F)^3.$$

3.2.2   The space of marked cubics

An even theta characteristic of a nonsingular plane cubic curve is a nonzero 2-torsion point on the Jacobian curve of this cubic. The right parameter space for marked cubics (that is, cubics with a marked 2-torsion point) is the weighted projective space $\mathbb{P}(1^{10}; 2)$ with coordinates

$$(F; \theta) = (a_{000}, a_{111}, a_{222}, a_{001}, a_{002}, a_{110}, a_{112}, a_{220}, a_{221}, a_{012}; \theta).$$

A similar statement holds for the spherical 2-space $\mathbb{S}_2^+(\mathrm{Mat}_3(K))$, compare Theorem 3.3 below.

Definition 3.1 The space of marked cubics is the hypersurface $V \subset \mathbb{P}(1^{10}; 2)$ defined by the equation

$$\theta^3 - 3S(F)\theta - 2T(F) = 0. \quad (3.21)$$

Remark 3.1 The affine equation

$$\frac{3}{2}y^2 + x^3 - 3S(F)x - 2T(F) = 0 \qquad (3.22)$$

defines the Jacobian curve of the generic cubic curve $F = 0$, where F is the form (3.3); hence a 2-torsion point of the Jacobian corresponds to a zero of the left-hand side of (3.22) of the form $(x, 0)$; this justifies the above definition. Here θ is an "irrational invariant" of a ternary cubic form, and its degree equals 2. The fact that θ is invariant ensures that the action of $\mathcal{P}$ on the space of cubic forms extends to V. A point of V is a cubic curve with a marked 2-torsion point. The hypersurface V is birationally equivalent to the projective space of bare (unmarked) plane cubics (compare Dolgachev and Kanev [8], who attribute this result to G. Salmon [16]). We give two constructive proofs of the Salmon–Dolgachev–Kanev theorem (see Theorem 3.1, Claims (A) and (B) below).

Example 3.1 Let F be a generic cubic form and $\mathrm{He}(F)$ its Hessian; then twice the value of the Aronhold T-invariant of F defines a 2-torsion point of $\mathrm{He}(F)$. That is, $\theta = 2T(F)$ is a root of the equation

$$\theta^3 - 3S(\mathrm{He}(F))\theta - 2T(\mathrm{He}(F)) = 0.$$

This follows from (3.18) and (3.16). Hence we get a map

$$\mathrm{he} \colon \mathbb{P}_9 \to V \quad \text{defined by} \quad \mathrm{he}(F) = \big(\mathrm{He}(F); 2T(F)\big) \qquad (3.23)$$

from the space $\mathbb{P}_9$ of ternary cubic forms to the space V of cubics with a marked 2-torsion point.

Example 3.2 Let F be a generic cubic form and $\mathrm{Ca}(F)$ its Cayley form; then $-4T(F)$ defines a 2-torsion point of $\mathrm{Ca}(F)$. That is, $\theta = -4T(F)$ is a root of the equation

$$\theta^3 - 3S(\mathrm{Ca}(F))\theta - 2T(\mathrm{Ca}(F)) = 0.$$

This follows from (3.20) and (3.19). Hence we get a map

$$\mathrm{ca} \colon \mathbb{P}_9 \to V \quad \text{defined by} \quad \mathrm{ca}(F) = \big(\mathrm{Ca}(F); -4T(F)\big) \qquad (3.24)$$

from the space $\mathbb{P}_9$ of ternary cubic forms to the space V of cubics with a marked 2-torsion point.

The next theorem shows that each of (3.23) and (3.24) is a birational equivalence.

Theorem 3.1   *(A) The map* $g\colon V \to \mathbb{P}_9$ *defined by*

$$g((F;\theta)) = \theta F + \mathrm{He}(F) \qquad (3.25)$$

is a birational inverse of the map he *of (3.23).*

(B) The map $d\colon V \to \mathbb{P}_9$ *defined by*

$$d((F;\theta)) = R(F)\,\mathrm{Ca}(F) + (\theta S(F) + T(F))D(F)$$

is a birational inverse of the map ca *of (3.24).*

Proof of (A)   This follows from (3.9), (3.10) and (3.25):

$$g(\mathrm{he}(F)) = g((\mathrm{He}(F); 2T(F))) = 2T(F)\,\mathrm{He}(F) + \mathrm{He}(\mathrm{He}(F)) = 3S^2(F)F,$$

hence $g \circ \mathrm{he} = \mathrm{id}_{\mathbb{P}_9}$. The right hand side of (3.17) equals:

$$\frac{9}{2}(Su^2 + 2Tuv + S^2v^2)^2uv + T(u^3 - 3Suv^2 - 2Tv^3)^2 +$$

$$+ \frac{3}{2}(u^3 - 3Suv^2 - 2Tv^3)((Su + Tv)^2 + 3Rv^2).$$

Using this, together with (3.15), (3.17), (3.23), (3.25), we get

$$\mathrm{he}(g((F;\theta))) = \mathrm{he}(\theta F + \mathrm{He}(F))$$

$$= \Big(\mathrm{He}(\theta F + \mathrm{He}(F)); 2T(\theta F + \mathrm{He}(F))\Big)$$

$$= \Big(3(S(F)\theta^2 + 2T(F)\theta + S(F)^2)F; 9(S(F)\theta^2 + T(F)\theta + S(F)^2)^2\theta\Big).$$

This point of $\mathbb{P}(1^{10}; 2)$ coincides with $(F;\theta)$, hence $\mathrm{he} \circ g = \mathrm{id}_V$.

Proof of (B)   Substituting from (3.14), (3.20), (3.19) gives

$$d(\mathrm{ca}(F)) = d((\mathrm{Ca}(F); -4T(F)))$$

$$= R(\mathrm{Ca}(F))\,\mathrm{Ca}(\mathrm{Ca}(F)) + (T(\mathrm{Ca}(F)) - 4T(F)S(\mathrm{Ca}(F)))D(\mathrm{Ca}(F))$$

$$= -12^3 R(F)^2 S(F)^2 S(\mathrm{Ca}(F))F,$$

that is, $d(\mathrm{ca}(F))$ is proportional to F, hence $d \circ \mathrm{ca} = \mathrm{id}_{\mathbb{P}_9}$.

   To study the inverse composite ca $\circ d$, and for some further comments on the theorem (Remark 3.2), we need an additional series of identities.

Lemma 3.1 *Suppose that θ satisfies (3.21), and set*

$$\tau = S(F)\theta + T(F). \tag{3.26}$$

Then the coefficient of $\mathrm{He}(F)$ in (3.13) vanishes at $u = R(F)$, $v = \tau$:

$$\tau^3 - 3T(F)\tau^2 + 3R(F)\tau - T(F)R(F) = 0. \tag{3.27}$$

Furthermore,

$$\begin{aligned}
T(R\,\mathrm{Ca}(F) + \tau D(F)) &= 72R^4 S^6(-9T\tau^2 + 8R\tau - 3TR), \\
(\tau^2 - R)^2 \theta &= -2S^2(-9T\tau^2 + 8R\tau - 3TR), \qquad (3.28) \\
-4T(R\,\mathrm{Ca}(F) + \tau D(F)) &= (12R^2 S^2(\tau^2 - R))^2 \theta.
\end{aligned}$$

Moreover, if we set

$$\Lambda(\tau) = T(F)\tau^2 - 2R(F)\tau + T(F)R(F), \tag{3.29}$$

then

$$\Lambda(\tau)^3 = S^6(9T\tau^2 - 8R\tau - 3TR)\tau^2, \tag{3.30}$$

$$\mathrm{He}(R\,\mathrm{Ca}(F) + \tau D(F)) = 6R^2 \Lambda(\tau)\tau^{-1}(\tau\,\mathrm{Ca}(F) + D(F)). \tag{3.31}$$

These identities can be checked directly, but we omit the details.

Proof of (B), continued   Applying (3.14), (3.27) and (3.28) yields:

$$\begin{aligned}
\mathrm{ca}(d(F)) &= \mathrm{Ca}\Big(R\,\mathrm{Ca}(F) + \tau D(F); \; -4T(\mathrm{Ca}(R\,\mathrm{Ca}(F) + \tau D(F)))\Big) \\
&= \Big(12S^2 R^2(\tau^2 - R)F; \; (12S^2 R^2)^2(\tau^2 - R)^2 \theta)\Big).
\end{aligned}$$

This point of $\mathbb{P}(1^{10}; 2)$ coincides with $(F; \theta)$, hence $\mathrm{ca} \circ d = \mathrm{id}$.

Remark 3.2 Comparing the two assertions (A) and (B) gives new infor-mation concerning two birational transformations: (1) the transformation $D \colon \mathbb{P}_9 \to \mathbb{P}_9$ of (3.6) of the projective space of plane cubics, and (2) an in-volutive transformation $E \colon V \to V$ described below of the space of marked cubics.

First, D equals the composite $g \circ \mathrm{ca}$ (and the composite $d \circ \mathrm{he}$): for

$$g(\mathrm{ca}(F)) = g(\mathrm{Ca}(F); -4T(F)) = \mathrm{He}(\mathrm{Ca}(F)) - 4T(F)\,\mathrm{Ca}(F) = 6D(F),$$

by (3.12) and (3.6). Our second map E is the composite $\mathrm{he} \circ d$ (which is equal to $\mathrm{ca} \circ g$). We claim that

$$E(F; \theta) = \Big((\theta S + T)\,\mathrm{Ca}(F) + D(F); \; -4S^2(T\theta^2 + 2S^2\theta + TS)\Big),$$

or in terms of the notation (3.26), (3.29),

$$E(F; \theta) = (\tau \, \mathrm{Ca}(F) + D(F); -4\Lambda(\tau)).$$

For, applying (3.28), (3.30), (3.11), (3.31), we get

$$
\begin{aligned}
\mathrm{he}(d(F)) &= \mathrm{he}\Big(R\,\mathrm{Ca}(F) + \tau D(F); 2T(R\,\mathrm{Ca}(F) + \tau D(F)\Big) \\
&= \Big(6R^2\Lambda(\tau)\tau^{-1}(\tau\,\mathrm{Ca}(F) + D(F); -4(6R^2\Lambda(\tau)\tau^{-1})^2\Lambda(\tau)\Big) \\
&= \Big(\tau\,\mathrm{Ca}(F) + D(F); -4\Lambda(\tau)\Big).
\end{aligned}
$$

3.2.3   A birational transformation of the space of marked cubics

We describe a birational transformation Σ_0 of the variety V and of the ambient weighted projective space $\mathbb{P}(1^{10}; 2)$. This transformation is an analog of the action of the standard quadratic transformation on the space of conics. It is convenient to make a coordinate change $(F; \theta) \mapsto (F; \eta)$ in $\mathbb{P}(1^{10}; 2)$, replacing the final coordinate θ by

$$\eta = -\frac{1}{4}(\theta + 2P), \quad (\text{so that } \theta = -4\eta - 2P),$$

where $P = a_{012}^2 - G$,   and $G = a_{110}a_{220} + a_{001}a_{221} + a_{002}a_{112}$.

In the new variables, the hypersurface $V \subset \mathbb{P}(1^{10}; 2)$ of (3.21) is now defined by the equation:

$$32\eta^3 + 48P\eta^2 + 6(4P^2 - S)\eta + T + 4P^3 - 3SP = 0. \qquad (3.32)$$

We introduce the monomial birational transformation Σ_0 of $\mathbb{P}(1^{10}; 2)$, given by $(a; \eta) \mapsto (a^*, \eta^*)$, where:

$$
\begin{aligned}
a_{000}^* &= a_{111}a_{222}, & a_{111}^* &= a_{000}a_{222}, & a_{222}^* &= a_{000}a_{111}, \\
a_{001}^* &= a_{110}a_{222}, & a_{002}^* &= a_{220}a_{111}, & a_{110}^* &= a_{001}a_{222}, \\
a_{112}^* &= a_{221}a_{000}, & a_{220}^* &= a_{002}a_{111}, & a_{221}^* &= a_{112}a_{000}, \\
a_{012}^* &= \eta; & & & \eta^* &= a_{000}a_{111}a_{222}a_{012}.
\end{aligned}
\qquad (3.33)
$$

Theorem 3.2 *(A) The map Σ_0 is an involutive birational transformation of $\mathbb{P}(1^{10}; 2)$.*

(B) It preserves the hypersurface V.

Proof It is obvious that Σ_0 is an involution, because on double application of Σ_0, each weight 1 coordinate is multiplied by

$$M = a_{000}a_{111}a_{222}, \tag{3.34}$$

and the final weight 2 coordinate η is multiplied by M^2.

Let $A = K[a_{000}, a_{111}, a_{222}, a_{001}, a_{002}, a_{110}, a_{112}, a_{220}, a_{221}]$ be the polynomial ring generated by all the coefficients of cubics except a_{012}. The first nine equations of (3.33) define an endomorphism $*$ of A; write f^* for the image of $f \in A$ under $*$. Expanding the terms in the defining equation (3.32) of V in powers of a_{012} gives:

$$4P^2 - S = 4Ba_{012} + 4C,$$
$$T + 4P^3 - 3SP = 32Ma_{012}^3 - 48G^*a_{012}^2 + 24Ea_{012} + D,$$

where $M, B, C, D, E \in A$ (here M is the multiplier spelt out in (3.34)),

$$\begin{aligned} M^* &= M^2, & B^* &= MB, & D^* &= M^2D, \\ C^* &= ME, & E^* &= M^2C, & G^{**} &= M^2G. \end{aligned} \tag{3.35}$$

We can rewrite (3.32) as

$$32(\eta^3 + Ma_{012}^3) + 48(\eta a_{012})^2 - 48(G\eta^2 + G^*a_{012}^2)$$
$$+ 24B\eta a_{012} + 24(C\eta + Ea_{012}) + D = 0. \tag{3.36}$$

Using this, we see that applying formulas (3.33) defining Σ_0 (see especially the last two formulas of (3.33) and (3.35)) to the left hand side of (3.32) or (3.36) multiplies it by M^2.   Q.E.D.

3.2.4 A birational map of the spherical space of symmetric 3×3 matrices onto the space V of marked cubics

We write $\widehat{m}$ for the adjoint matrix of a symmetric 3×3-matrix m and $D(\cdot, \cdot, \cdot)$ for the mixed discriminant of three symmetric 3×3 matrices. For a triple $\mathbf{m} = (m_0, m_1, m_2)$, we define a ternary cubic by

$$F_{\mathbf{m}} = 6 \det(x_0 m_0 + x_1 m_1 + x_2 m_2),$$

and a number $\theta(\mathbf{m})$ by

$$\begin{aligned} \theta(\mathbf{m}) = {} & 2(2D(\widehat{m_0}, \widehat{m_1}, \widehat{m_2}) - (D(m_0, m_1, m_2))^2 \\ & + D(m_0, m_1, m_1)D(m_0, m_2, m_2) \\ & + D(m_1, m_0, m_0)D(m_1, m_2, m_2) + D(m_2, m_0, m_0)D(m_2, m_1, m_1)). \end{aligned} \tag{3.37}$$

Theorem 3.3 *For a triple* $\mathbf{m} = (m_0 : m_1 : m_2) \in \mathbb{S}_2^+(\mathrm{Mat}_3(K))$, *the point* $(F_\mathbf{m}; \theta(\mathbf{m}))$ *belongs to* V, *and*

$$\alpha: \mathbb{S}_2^+(\mathrm{Mat}_3(K)) \to V \subset \mathbb{P}(1^{10}; 2) \quad \textit{given by } \alpha(\mathbf{m}) = (F_\mathbf{m}; \theta(\mathbf{m})) \quad (3.38)$$

is a well-defined birational map, having the inverse

$$\beta: V \to \mathbb{S}_2^+(\mathrm{Mat}_3(K)), \quad \textit{given by } \beta(F; \theta) = (m_0 : m_1 : m_2),$$

where $x_0 m_0 + x_1 m_1 + x_2 m_2 = \theta \, \mathrm{HE}(F) + \mathrm{HE}(\mathrm{He}(F))$; *see (3.4) for the Hessian matrix* $\mathrm{HE}(F)$.

The map α *is* $\mathcal{P}$-*anti-equivariant, and has the following compatibility with the action of the standard quadratic transformation*

$$\Sigma_0(\alpha(\mathbf{m})) = \alpha(S_0(\mathbf{m})).$$

Remark 3.3 Formula (3.37) is borrowed from the end of Salmon, Conic sections [15]. Salmon gives a different formula for $\theta(\mathbf{m})$, which he attributes to Burnside. Namely, write $[\cdot, \cdot, \cdot]$ for the determinant made up of three ternary linear forms, and

$$\det(x_0 m_0 + x_1 m_1 + x_2 m_2) = \begin{vmatrix} A_{00} & A_{01} & A_{02} \\ A_{10} & A_{11} & A_{12} \\ A_{20} & A_{21} & A_{22} \end{vmatrix}$$

where $A_{ij} = A_{ij}(x_0, x_1, x_2)$ are linear forms and $A_{ij} = A_{ji}$ for $i, j = 0, 1, 2$. Then

$$\begin{aligned}
\theta(\mathbf{m}) = 2 \times \big(&-8[A_{01}, A_{12}, A_{20}]^2 + [A_{00}, A_{11}, A_{22}]^2 \\
&+ 4[A_{21}, A_{10}, A_{02}][A_{00}, A_{11}, A_{22}] + 4[A_{00}, A_{11}, A_{12}][A_{00}, A_{22}, A_{12}] \\
&+ 4[A_{11}, A_{22}, A_{02}][A_{11}, A_{00}, A_{02}] + 4[A_{22}, A_{00}, A_{01}][A_{22}, A_{11}, A_{01}] \\
&+ 8[A_{11}, A_{02}, A_{01}][A_{22}, A_{02}, A_{01}] + 8[A_{00}, A_{12}, A_{10}][A_{22}, A_{12}, A_{10}] \\
&+ 8[A_{00}, A_{21}, A_{20}][A_{11}, A_{21}, A_{20}] - 8[A_{00}, A_{02}, A_{01}][A_{11}, A_{22}, A_{12}] \\
&- 8[A_{11}, A_{10}, A_{12}][A_{22}, A_{00}, A_{20}] - 8[A_{22}, A_{10}, A_{12}][A_{00}, A_{11}, A_{01}] \big).
\end{aligned}$$

Salmon [15] also sketches a proof that α is well defined.

Proof of Theorem 3.3 We introduce some notation. Let m and m' be two symmetric 3×3-matrices and $\widehat{[m, m']}$ their mixed adjoint matrix, that is,

$$(\widehat{um + vm'}) = u^2 \widehat{m} + uv \widehat{[m, m']} + v^2 \widehat{m'}, \quad \text{in particular} \quad \widehat{[m, m]} = 2\widehat{m}.$$

For six ternary quadratic forms A, B, C, D, E, F (or the corresponding symmetric matrices), we write $[A, B, C, D, E, F]$ for the 6×6 determinant whose columns are these forms, written out as normalized coefficients in the order $00, 11, 22, 12, 02, 01$. Salmon's (or Burnside's) second invariant $M = M(\mathbf{m})$ is

$$M = [[\widehat{m_0, m_0}], [\widehat{m_1, m_1}], [\widehat{m_2, m_2}], [\widehat{m_1, m_2}], [\widehat{m_0, m_2}], [\widehat{m_0, m_1}]].$$

The Aronhold invariants of the above symmetric determinant $F = F_{\mathbf{m}}$ are the following expressions (see Conic sections, [15], loc. cit.)

$$S(F) = \theta^2 - 24M, \quad T(F) = \theta^3 - 36\theta M, \quad R(F) = 432M^2(32M - \theta^2),$$

where $\theta = \theta(\mathbf{m})$ and $M = M(\mathbf{m})$. Hence $(F; \theta)$ satisfies (3.21) and belongs to $V \subset \mathbb{P}(1^{10}, 2)$. Thus the map α is well defined. Further, if $(F; \theta) \in V$, and if we identify $\mathbf{m}$ and the corresponding linear form with their matrix coefficients, then

$$\begin{aligned}
\alpha(\beta(F; \theta)) &= \alpha(\theta \mathrm{HE}(F) + \mathrm{HE}(\mathrm{He}(F))) \\
&= \big(\mathrm{He}(\theta F + \mathrm{He}(F)); 2T(\theta F + \mathrm{He}(F)) = (F; \theta)
\end{aligned}$$

by the proof of Theorem 3.1, (A). Because they map between varieties of the same dimension, it is now obvious that α and β are birational. That α is compatible with the standard quadratic transformation follows from the observation that the mixed determinant of adjoint matrices in formula (3.37) corresponds to the η of Theorem 3.1 and from the behaviour of mixed determinants of the third order when the matrices involved are replaced by their inverses (or adjoints).   Q.E.D.

3.2.5 An action of the Cremona group on the space of cubics

Consider the following two composite maps from the space of plane cubics to the spherical 2-space over 3×3-matrices:

$$\mathbb{P}(S^3(W^*)) \xrightarrow{\text{he}} V \xrightarrow{\beta} \mathbb{S}_2^+(\mathrm{Mat}_3(K)),$$

$$\mathbb{P}(S^3(W^*)) \xrightarrow{\text{ca}} V \xrightarrow{\beta} \mathbb{S}_2^+(\mathrm{Mat}_3(K)).$$

Each of these maps leads to an action of the Cremona group on the space of plane cubics, the first on the right, the second on the left. If C is a plane cubic defined by a form F, and $g \in \mathrm{Cr}(2, K)$ (or $g \in \mathrm{UCr}(2, K)$), then we may define

$$g(C) = (\beta \circ \text{he})^{-1}(g((\beta \circ \text{he})(F))) \quad \text{or} \quad g(C) = (\beta \circ \text{ca})^{-1}(g((\beta \circ \text{ca})(F))).$$

3.3   An action of the Cremona group of the plane on the space of quartics

Let K be an algebraically closed field of characteristic zero. Every ordered triple of symmetric 4×4-matrices $m_0, m_1, m_2 \in \mathrm{Mat}_4^+(K)$ defines a net of quadrics $x_0 M_0 + x_1 M_1 + x_2 M_2 = 0$ in $\mathbb{P}^3$; here M_i is a quaternary quadratic form with matrix m_i, and the x_i are parameters in the net. $\mathrm{GL}(4, K)$-equivalence classes of stable nets correspond one-to-one to points $\mathbf{m} = (m_0 : m_1 : m_2) \in \mathbb{S}_2^+(\mathrm{Mat}_4(K))$. The discriminant curve $C(\mathbf{m})$ of such a point is well defined and also has degree 4. This curve has a marked even theta characteristic $\theta(\mathbf{m})$ (at least, provided that it is nonsingular, see [3]). Thus a point of spherical 2-space defines a point of the variety M_4^{ev} of plane quartics with a marked even theta characteristic. By results of Barth [3], the map

$$\gamma \colon \mathbb{S}_2^+(\mathrm{Mat}_4(K)) \to M_4^{\mathrm{ev}} \quad \text{given by } \mathbf{m} \mapsto (C(\mathbf{m}); \theta(\mathbf{m})) \tag{3.39}$$

is one-to-one on some open subset, and hence birational.

Moreover, every ternary quartic $F \in S^4(W^*)$ defines a pair $(S(F); \theta(F))$, where $S(F)$ is the Clebsch covariant of degree 4 for F, and $\theta(F)$ is an even theta characteristic of the plane quartic S whose equation is $S(F) = 0$ (at least, provided that F is weakly nondegenerate, see [8] for details). This map

$$\mathrm{Sc} \colon \mathbb{P}(S^4(W^*)) \to M_4^{\mathrm{ev}} \quad \text{given by } F \mapsto (S(F); \theta(F)) \tag{3.40}$$

is the Scorza map. By a theorem of Scorza (see [8], 7.8), Sc is a $\mathcal{P}$-equivariant birational isomorphism. Thus, we get the following possibility to define a (right) action of the Cremona group on the space of plane quartics: if $F \in \mathbb{P}(S^4(W^*))$, and $g \in \mathrm{Cr}(2, K)$ (or $g \in \mathrm{UCr}(2, K)$), then we may define

$$g(F) = \mathrm{Sc}^{-1}(\gamma(g(\gamma^{-1}(\mathrm{Sc}(F))))).$$

Remark 3.4 Let $X \subset \mathbb{S}_2^+(A)$ be the subset defined by the equations

$$\det(m_i \widehat{m_j} m_k - m_k \widehat{m_j} m_i) = 0,$$

where (i, j, k) is an arbitrary permutation of $(0, 1, 2)$. Equivalently, X is the subvariety whose generic point $\mathbf{x} = (x_0 : x_1 : x_2)$ satisfies the equations

$$\det(x_i x_j^{-1} x_k - x_k x_j^{-1} x_i) = 0.$$

In other words, Barth's commutators (see [3]) for $\mathbf{x}$ have rank ≤ 2.

The variety X is preserved by collineations and the standard quadratic transformation; this is clear for collineations. As for the standard quadratic transformation, S_0 substitutes $x_i \mapsto x_i^{-1}$, and

$$\det(x_i^{-1} x_j x_k^{-1} - x_k^{-1} x_j x_i^{-1}) = \det(x_i^{-1}(x_j x_k^{-1} x_i - x_i x_k^{-1} x_j) x_i^{-1})$$
$$= (\det(x_i))^{-2} \det(x_j x_k^{-1} x_i - x_i x_k^{-1} x_j).$$

Therefore the action of the Cremona group we have just constructed on the space of quartics with an even theta characteristic extends Artamkin's action (see the Introduction) on the space of special marked quartics corresponding to certain vector bundles.

References

[1] I. V. Artamkin, Thesis: Vector bundles over the projective plane and the Cremona group, 1989, Leningrad

[2] I. V. Artamkin, Action of biregular automorphisms of the affine plane on pairs of matrices, Izv. Akad. Nauk SSSR Ser. Mat. 52 (1988), 1109–1115, 1120; translation in Math. USSR–Izv. 33 (1989)

[3] W. Barth, Moduli of vector bundles on the projective plane, Invent. Math., 42, 1977, 63–91

[4] N. Bourbaki, Algèbre, Livre II, Paris, 1958 (Russian translation 1968)

[5] A. Church, Introduction to mathematical logic, 1, Princeton, New Jersey, 1956 (Russian translation 1960)

[6] A. Coble, Point sets and allied Cremona group. II, Transactions of Amer. Math. Soc., 17, 1916, 345–388

[7] I. V. Dolgachev, Private letter

[8] I. V. Dolgachev and V. Kanev, Polar covariants of planes cubics and quartics, Advances in Math, 98, No.2, 1993, 216–301

[9] I. Dolgachev and D. Ortland, Point sets in projective space and theta functions, Astérisque, 165, Société Math. de France

[10] P. Du Val, Applications des idées cristallographiques à l'étude des groupes crémonniennes, Troisième Colloque de Géometrie Algébrique, Bruxelles 1959, 1960, Paris, Gauthier–Villars

[11] M. Kh. Gizatullin, Defining relations for the Cremona group of the plane, Izv. Akad. Nauk SSSR, Ser. Mat., 46, 1982, 909–970 (English translation in Math. USSR Izvestiya, 21 (1983), p.211–268)

[12] H. P. Hudson, Cremona transformations in plane and space, 1927, Cambridge University Press

[13] S. Kantor, Theorie der Transformationen im R_3 welche keine Fundamentalkurven 1. Art besitzen und ihrer endlichen Gruppen, Acta Mathematica, 1897, 21, 1–77

[14] A. I. Mal'tsev, Algebraic systems, "Nauka", Moscow, 1978 (in Russian)

[15] G. Salmon, A treatise on conic sections, London, 1878 (Russian translation 1900)

[16] G. Salmon, A treatise on the higher plane curves, 1879, Dublin

[17] A. N. Tyurin and N. A. Tyurin, Algebraic geometry today and tomorrow. Unpublished manuscript

Marat Gizatullin,
Department of Mathematics and Mechanics,
Samara State University,
Academician Pavlov Street, 1,
Samara, 443011, Russia
e-mail gizam@ssu.samara.ru

Hilbert schemes and simple singularities

Y. Ito and I. Nakamura*

Abstract

The first half of this article is expository; it contains a brief survey
of the famous ADE classification, and how it applies to six kinds of
objects, some old and some relatively new. The second half is a research article, discussing the two dimensional McKay correspondence
from the new point of view of Hilbert schemes.

Contents

*The first author is a JSPS Research Fellow and partially supported by the Fujukai
Foundation and JAMS. The second author is partially supported by the Grant-in-aid (No.
06452001) for Scientific Research, the Ministry of Education.

Mathematics subject classification: Primary 14-02, 14B05, 14J17; Secondary 01-02;
15A66; 16G20; 17B10, 17B67, 17B68, 17C20; 20C05, 20C15; 46L35

Key words: Simple singularity, ADE, Dynkin diagram, Simple Lie algebra, Finite
group, Quiver, Conformal field theory, von Neumann algebra, Hilbert scheme, Quotient
singularity, McKay correspondence, Invariant theory, Coinvariant algebra

0   Introduction

There is a whole series of apparently unrelated phenomena that are governed by the so-called ADE Dynkin diagram scheme. It is widely believed that, despite the diverse nature of the objects concerned, there must be some hidden reasons for these coincidences. The ADE Dynkin diagrams provide a classification of the following types of objects (among others):

(a) simple singularities (rational double points) of complex surfaces (Du Val, Artin, Brieskorn),

(b) finite subgroups of SL(2, $\mathbb{C}$),

(c) simple Lie groups and simple Lie algebras (Elie Cartan, Dynkin),

(d) quivers of finite type ([Gabriel72]),

(e) modular invariant partition functions in two dimensions (Capelli, Itzykson and Zuber [CIZ87]),

(f) pairs of von Neumann algebras of type II$_1$ ([Ocneanu88]).

0.1

The present article consists of two halves, an expository part and a research part. The expository part occupies the first six sections. In Sections 1–4, we recall briefly the above ADE classifications. Sections 2–3 report in some detail on the relatively new subjects of modular invariant partition functions and type II_1 von Neumann algebras (II_1 factors). In Section 4 we recall the two dimensional McKay correspondence. Section 5 summarizes some of the missing links between the six objects and related problems. We would like to say that while much is known about these, much remains unknown.

Next, in Section 6, we recall some basic facts about Hilbert schemes for use in the research part, and give a quick review on three dimensional quotient singularities in Section 7. Section 7 is not directly related to the rest of the paper, but it provides motivation for further study in the same direction as Sections 8–16. For instance, a natural three dimensional generalization of the McKay correspondence, quite different from that of Theorem 7.2, can be obtained by applying similar ideas. This direction is the subject of current research and we simply mention [Reid97], [INkjm98] and [Nakamura98] as available references for it.

In the second half of the article we discuss the two dimensional McKay correspondence from a somewhat new point of view, namely by applying the technique of Hilbert schemes. Any of the known explanations for the classical McKay correspondence enables each irreducible component of the exceptional set E to correspond naturally to an irreducible representation of a finite subgroup G. In the present article we do a little more. In fact, to any point of the exceptional set, we associate in a natural way a G-module, irreducible or otherwise, whose equivalence class is constant along each irreducible component of E. We discuss this in outline in Section 8, and in detail in Sections 8–16. Some new progress and related problems are mentioned in Section 17.

0.2

There are a number of excellent reports on the first four topics (a)–(d), for example: Hazewinkel, Hesselink, Siersma and Veldkamp [HHSV77] and [Slodowy95]. See [Slodowy90] and [Gawedzki89] for the topic (e). See also [Ocneanu88], Goodman, de la Harpe and Jones [GHJ89], [Jones91] and Evans and Kawahigashi [EK98], Section 11 for the last topic (f). The authors hope that the reader will also study or at least have a glance at these reports.

We have in mind both specialists in algebraic geometry and nonspecialists as readers of the expository part. Therefore we have tried to include elementary examples and algebraic calculations, though they are not completely self-contained.

Acknowledgments   We wish to thank many mathematical colleagues for assistance and discussions during the preparation of the expository part. We are very much indebted, among others, to Professors Y. Kawahigashi and T. Yamanouchi for the report on von Neumann algebras. Our thanks are also due to Professors A. Kato, H. Nakajima, K. Shinoda, T. Shioda and H. Yamada for their various support. Last but not least we also thank Professor M. Reid for his numerous suggestions for improving the manuscript, both in English and mathematics.

1   Simple singularities and ADE classification

1.1   Simple singularities (1)

We first recall the definition of simple singularities. A germ of a two dimensional isolated hypersurface singularity is called a *simple singularity* if one of the following equivalent conditions holds:

1. It is isomorphic to one of the following germs at the origin

$$
\begin{aligned}
A_n : \quad & x^{n+1} + y^2 + z^2 = 0 \quad && \text{for } n \geq 1, \\
D_n : \quad & x^{n-1} + xy^2 + z^2 = 0 \quad && \text{for } n \geq 4, \\
E_6 : \quad & x^4 + y^3 + z^2 = 0, \\
E_7 : \quad & x^3 y + y^3 + z^2 = 0, \\
E_8 : \quad & x^5 + y^3 + z^2 = 0.
\end{aligned}
$$

2. It is isomorphic to a germ of a weighted homogeneous hypersurface of $(\mathbb{C}^3, 0)$ of total weight one such that the sum of weights (w_1, w_2, w_3) of the variables is greater than one. The possible weights are $(\frac{1}{n+1}, \frac{1}{2}, \frac{1}{2})$, $(\frac{1}{n-1}, \frac{n-2}{2n-2}, \frac{1}{2})$, $(\frac{1}{4}, \frac{1}{3}, \frac{1}{2})$, $(\frac{2}{9}, \frac{1}{3}, \frac{1}{2})$ and $(\frac{1}{5}, \frac{1}{3}, \frac{1}{2})$.

3. It has a minimal resolution of singularities with exceptional set consisting of smooth rational curves of selfintersection -2 intersecting transversally.

4. It is a quotient of $(\mathbb{C}^2, 0)$ by a finite subgroup of $SL(2, \mathbb{C})$ ([Klein]).

5. Its (semi-)universal deformation contains only finitely many distinct isomorphism classes ([Arnold74]).

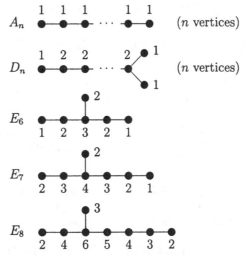

Figure 1: The Dynkin diagrams ADE

Many other characterizations of the singularities are given in [Durfee79]. The third characterization of a simple singularity classifies the exceptional set explicitly. In fact, the dual graph of the exceptional set is one of the Dynkin diagrams of simply connected complex Lie groups shown in Figure 1.

1.2   Simple singularities (2)

Let $(S, 0)$ be a germ of simple singularities, $\pi\colon X \to S$ its minimal resolution, $E := \pi^{-1}(0)_{\mathrm{red}}$ and E_i for $1 \leq i \leq r$ the irreducible components of E. It is known that $E_i \simeq \mathbb{P}^1$ and $(E_i^2)_X = -2$. Let Irr E be the set $\{E_i; 1 \leq i \leq r\}$. We see that $H_2 = H_{2,\mathrm{SING}}(S) := H_2(X, \mathbb{Z}) = \bigoplus_{1 \leq i \leq r} \mathbb{Z}[E_i]$. Then H_2 has a negative definite intersection pairing $(\ ,\ )_{\mathrm{SING}} \colon H_2 \times H_2 \to \mathbb{Z}$. Since $(E_i E_j)_{\mathrm{SING}} = 0$ or 1 for $i \neq j$, the pairing $(\ ,\ )_{\mathrm{SING}}$ can be expressed by a finite graph with simple edges. We rephrase this as follows: we associate a vertex $v(E')$ to any irreducible component E' of E, and join two vertices $v(E')$ and $v(E'')$ if and only if $(E'E'')_{\mathrm{SING}} = 1$. Thus we have a finite graph with simple edges, from which in turn the bilinear form $(\ ,\ )_{\mathrm{SING}}$ can be recovered in the obvious manner. We call this graph the *dual graph* of E, and denote it by $\Gamma(E)$ or $\Gamma_{\mathrm{SING}}(S)$. Let $H^2 = H^2_{\mathrm{SING}}(S) := H^2(X, \mathbb{Z})$.

There exists a unique divisor E_{fund}, called the *fundamental cycle* of X, which is the minimal nonzero effective divisor such that $E_{\mathrm{fund}} E_i \leq 0$ for all i. Let $E_{\mathrm{fund}} := \sum_{i=1}^{r} m_i^{\mathrm{SING}} E_i$ and $E_0 := -E_{\mathrm{fund}}$. For the simple singularities we have $E_0 E_i = 0$ or 1 for any $E_i \in \mathrm{Irr}\, E$, except for the case A_1, when $E_0 E_1 = 2$. Therefore we can draw a new graph $\widetilde{\Gamma}_{\mathrm{SING}}$ by adding the vertex

$v(E_0)$ to $\Gamma_{\mathrm{SING}}(S)$. By a little abuse of notation we denote Irr $E \cup \{E_0\}$ by Irr$_*$ E.

For instance let us consider the D_5 case. Then $E = \sum_{i=1}^{5} E_i$ with $E_i^2 = -2$ and

$$-E_0 = E_{\mathrm{fund}} = E_1 + 2E_2 + 2E_3 + E_4 + E_5.$$

Then $E_0 E_2 = E_1 E_2 = E_2 E_3 = E_3 E_4 = E_3 E_5 = 1$, and all other $E_i E_j = 0$. Hence $(m_1^{\mathrm{SING}}, \ldots, m_5^{\mathrm{SING}}) = (1, 2, 2, 1, 1)$, as indicated in Figure 2.

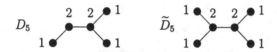

Figure 2: The Dynkin diagrams D_5 and $\widetilde{D}_5$

There are various ways of computing E. We do it starting from the fact that D_5 is the quotient singularity of $\mathbb{A}^2$ by the binary dihedral group $\mathbb{D}_3$ of order 12. The binary dihedral group $G := \mathbb{D}_3$ is generated by σ and τ:

$$\sigma = \begin{pmatrix} \varepsilon & 0 \\ 0 & \varepsilon^{-1} \end{pmatrix}, \quad \tau = \begin{pmatrix} 0 & 1 \\ -1 & 0 \end{pmatrix},$$

where $\varepsilon := e^{2\pi\sqrt{-1}/6}$. We have $\sigma^6 = \tau^4 = 1$, $\sigma^3 = \tau^2$ and $\tau\sigma\tau^{-1} = \sigma^{-1}$. The ring of G-invariants in $\mathbb{C}[x, y]$ is generated by three elements $F := x^6 + y^6$, $H := xy(x^6 - y^6)$ and $I := x^2 y^2$. The quotient $\mathbb{A}^2/G$ is isomorphic to the hypersurface $4I^4 + H^2 - IF^2 = 0$. Since G has a normal subgroup $N := \langle \sigma \rangle$ of order 6, we first take the quotient $\mathbb{A}^2/N$ and its minimal resolution X_N.

Since $P := x^6$, $Q := y^6$ and $R := xy$ are N-invariants, $\mathbb{A}^2/N$ is the hypersurface $PQ = R^6$. Hence X_N has an exceptional set consisting of a chain of 5 smooth rational curves $C_1 + \cdots + C_5$. The action of τ on $\mathbb{A}^2$ induces an action on X_N, which maps C_i into C_{5-i}, so in particular takes C_3 to itself. The action of τ on X_N has exactly two fixed points $\mathfrak{p}_+$ and $\mathfrak{p}_-$ on C_3, which give rise to all the singularities of $X_N/\langle \tau \rangle$.

The images of $\mathfrak{p}_\pm$ give smooth rational curves E_4 and E_5 on the minimal resolution X of $\mathbb{A}^2/G$ by resolving the singularities of $X_N/\langle \tau \rangle$ at $\mathfrak{p}_\pm$. Thus on X we have the images E_i of C_i for $i = 1, 2, 3$ and two new rational curves E_4 and E_5. This gives the exceptional set E of X. We see easily that $(E_i)_{\mathrm{SING}}^2 = -2$. The intersection pairing $(\ ,\ )_{\mathrm{SING}}$ is expressed with respect to the basis E_i for $0 \leq i \leq 5$ as a 6×6 symmetric matrix with diagonal entries equal to

-2. We write it down multiplied by -1 for convenience:

$$(-1) \cdot (E_i E_j)_{\text{SING}} = \begin{pmatrix} 2 & 0 & -1 & 0 & 0 & 0 \\ 0 & 2 & -1 & 0 & 0 & 0 \\ -1 & -1 & 2 & -1 & 0 & 0 \\ 0 & 0 & -1 & 2 & -1 & -1 \\ 0 & 0 & 0 & -1 & 2 & 0 \\ 0 & 0 & 0 & -1 & 0 & 2 \end{pmatrix}.$$

Let $v_i := v(E_i)$ for $0 \le i \le 5$. Then we obtain the Dynkin diagram D_5 from v_i for $1 \le i \le 5$ and the extended Dynkin diagram $\tilde{D}_5$ from v_i for $0 \le i \le 5$, as in Figure 2.

1.3 Simple singularities and simple Lie algebras (1)

Let $\mathfrak{G}$ be a simply laced simple Lie algebra and $\mathfrak{H}$ a Cartan subalgebra of $\mathfrak{G}$. We fix a lexicographical order of the roots of $\mathfrak{H}$ and let Δ (respectively Δ_+, Δ_{simple}) be the set of roots (respectively, positive roots, positive simple roots) of $\mathfrak{G}$ with respect to T. (See [Bourbaki] for more details.) Let r be the rank of $\mathfrak{G}$ $(= \dim \mathfrak{H})$ and $\Delta_{\text{simple}} = \{\alpha_i; 1 \le i \le r\}$.

Let Q be the root lattice, namely the lattice spanned by Δ over $\mathbb{Z}$ endowed with the Cartan–Killing form $(\ ,\ )_{\text{LIE}}$ and $P := \text{Hom}_{\mathbb{Z}}(Q, \mathbb{Z})$ the dual lattice of Q (the weight lattice):

$$Q := \bigoplus_{\alpha \in \Delta} \mathbb{Z}\alpha = \bigoplus_{\alpha \in \Delta_{\text{simple}}} \mathbb{Z}\alpha.$$

The Cartan–Killing form $(\ ,\ )_{\text{LIE}}$ with respect to the basis Δ_{simple} is a positive definite integral symmetric bilinear form with $(\alpha, \alpha) = 2$ for all $\alpha \in \Delta_{\text{simple}}$. Since $(\alpha, \beta)_{\text{LIE}} = 0$ or -1 for $\alpha \ne \beta \in \Delta_{\text{simple}}$, we can express the bilinear form by a finite graph with simple edges Γ_{LIE} as we did for the dual graph of the set of exceptional curves of simple singularities.

There is a maximal root in Δ with respect to the given order, called the *highest root* of Δ. (This name is justified by the fact that it is the highest root of the adjoint representation of $\mathfrak{G}$. See Table 1.) Let the highest root be $\alpha_0 := \alpha_{\text{highest}} = \sum_{i=1}^{r} m_i^{\text{LIE}} \alpha_i$. Then $(\alpha_0, \beta) = 0$ or -1 for any $\beta \in \Delta_{\text{simple}}$ (expect for the case A_1, when $(\alpha_0, \beta) = 2$), so that we can draw a new graph $\tilde{\Gamma}_{\text{LIE}}(\mathfrak{G})$ (called the *extended Dynkin diagram* of $\mathfrak{G}$) by adding the vertex α_0 to $\Gamma_{\text{LIE}}(\mathfrak{G})$.

Let us consider the D_5 case as an example. The Lie algebra $\mathfrak{G} := \mathfrak{G}(D_5)$ is given by $\mathfrak{o}(10) := \{X \in M_{10}(\mathbb{C}); {}^t X + X = 0\}$. Its Cartan subalgebra $\mathfrak{H}$ is spanned by $H_i := E_{i,i+5} - E_{i+5,i}$ for $1 \le i \le 5$ where E_{ij} is the matrix with (i,j)th entry equal to 1 and 0 elsewhere. We define $\varepsilon_i \in \text{Hom}_{\mathbb{C}}(\mathfrak{H}, \mathbb{C})$ by

Type	r	(m_0)	$m_1, m_2, m_3, \ldots, m_{r-1}; m_r$
A_n	n	1	$1, 1, \ldots, 1, 1$
D_n	n	1	$1, 2, 2, \ldots, 2, 1, 1$
E_6	6	1	$1, 2, 3, 2, 1; 2$
E_7	7	1	$1, 2, 4, 3, 2, 1; 2$
E_8	8	1	$2, 4, 6, 5, 4, 3, 2; 3$

Table 1: Multiplicities of the highest root

$\varepsilon_i(H) := t_i$ for all $H = \sum_{i=1}^{5} t_i H_i \in \mathfrak{H}$. Then we can choose simple roots α_i with order $\alpha_1 > \alpha_2 > \cdots > \alpha_5$ as follows:

$$\alpha_i := \varepsilon_i - \varepsilon_{i+1}, \quad \alpha_5 := \varepsilon_4 + \varepsilon_5 \quad \text{for } 1 \leq i \leq 4.$$

The highest root α_0 is $\varepsilon_1 + \varepsilon_2 = \alpha_1 + 2\alpha_2 + 2\alpha_3 + \alpha_4 + \alpha_5$. For each α_i we define an element $\widetilde{H}_i \in \mathfrak{H}$ by $\alpha_i(H) = -\frac{1}{2}\operatorname{Tr}(\widetilde{H}_i H)$ for all $H \in \mathfrak{H}$. We see that $\widetilde{H}_i = H_i - H_{i+1}$ for $1 \leq i \leq 4$, and $\widetilde{H}_5 = H_4 + H_5$. We define $(\alpha_i, \alpha_j) := \alpha_i(\widetilde{H}_j) = \alpha_j(\widetilde{H}_i)$. Then we have $(\alpha_i, \alpha_j) = -(E_i, E_j)$ for $0 \leq i \leq j \leq 5$ in the notation of 1.1–1.2. This shows that $\Gamma_{\mathrm{SING}}(D_5) = \Gamma_{\mathrm{LIE}}(\mathfrak{G}(D_5))$ and $\widetilde{\Gamma}_{\mathrm{SING}}(D_5) = \widetilde{\Gamma}_{\mathrm{LIE}}(\mathfrak{G}(D_5))$.

We note that $P = \sum_{i=1}^{5} \mathbb{Z}\varepsilon_i$ and $Q = \sum_{i=1}^{5} \mathbb{Z}\alpha_i$.

The first theorem to mention is the following:

Theorem 1.4 *Let S be a simple singularity and $\operatorname{Lie}(S)$ a simple Lie algebra of the same type as S. Then there is an isomorphism*

$$i \colon H^2_{\mathrm{SING}}(S) \simeq P(\operatorname{Lie}(S))$$

such that

1. $i(H_{2,\mathrm{SING}}(S)) = Q(\operatorname{Lie}(S))$;

2. $i(\operatorname{Irr}(E(S))) = \Delta_{\mathrm{simple}}(\operatorname{Lie}(S))$;

3. $i(E_{\mathrm{fund}}(S)) = -\alpha_{\mathrm{highest}}(\operatorname{Lie}(S))$;

4. $(\ ,\ )_{\mathrm{SING}} = -i^*(\ ,\ )_{\mathrm{LIE}}$;

5. $\Gamma_{\mathrm{SING}}(S) = \Gamma_{\mathrm{LIE}}(\operatorname{Lie}(S))$ *and* $\widetilde{\Gamma}_{\mathrm{SING}}(S) = \widetilde{\Gamma}_{\mathrm{LIE}}(\operatorname{Lie}(S))$.

1.5 Simple singularities and simple Lie algebras (2)

There are two kinds of similar constructions of simple singularities from simple Lie algebras: first of all, the Grothendieck–Brieskorn–Springer construction and second, the Knop construction. Good references for this topic are for instance [Slodowy80], [Slodowy95] and [Knop87].

1.6 Finite reflection groups and Coxeter exponents

Let V be a vector space over $\mathbb{R}$ endowed with a positive definite bilinear form $(\ ,\ )$. A linear automorphism s of V is called a *reflection* if there is a vector $\alpha \in V$ and a hyperplane H_α orthogonal to α such that $s(\alpha) = -\alpha$, and the restriction of s to H_α is trivial: $s|_{H_\alpha} = \mathrm{id}_{H_\alpha}$. There is a simple formula

$$s(v) = v - \frac{2(v, \alpha)}{(\alpha, \alpha)}\alpha. \qquad (1)$$

A finite group generated by reflections is called a finite reflection group. For instance, let Q be the root lattice of a simple Lie algebra $\mathfrak{G}$ over $\mathbb{C}$, $(\ ,\ )_{\mathrm{LIE}}$ its Cartan–Killing form, and set $V = Q \otimes \mathbb{C}$. For any simple root $\alpha_i \in \Delta_{\mathrm{simple}}$, we define a reflection $s_i := s_{\alpha_i}$ of V by the formula (1). The group W generated by all reflections s_α for $\alpha \in \Delta_{\mathrm{simple}}$ is finite, and is called the *Weyl group* of $\mathfrak{G}$. The Weyl group W acts on the polynomial ring $\mathbb{C}[V^*]$ generated by $V^* := \mathrm{Hom}_{\mathbb{Z}}(V, \mathbb{Z})$, the dual of V.

The product $s = \prod_{i=1}^r s_i$ of reflections for all the simple roots is called a *Coxeter element* of W. All s defined in this way for different choices of lexicographical order of the roots are conjugate in W. Therefore the order of s in W is uniquely determined, and we denote it by h and we call it the *Coxeter number* of $\mathfrak{G}$.

Theorem 1.7 ([Chevalley55]) *Let W be the Weyl group of a simple Lie algebra $\mathfrak{G}$ over $\mathbb{C}$, and r the rank of $\mathfrak{G}$. Then*

1. *The invariant ring $\mathbb{C}[V^*]^W$ is generated by r algebraically independent homogeneous polynomials $f_1, f_2, \ldots, f_r$. We order the f_i so that $\deg f_i$ is monotonically increasing.*

2. *For any choice of the generators f_i as above, the sequence of degrees $(\deg f_1, \ldots, \deg f_r)$ is uniquely determined.*

Definition 1.8 We define the *Coxeter exponents* e_i by $e_i := \deg f_i - 1$ for $1 \le i \le r$.

Theorem 1.9 *Let $\mathfrak{G}$ be a simple Lie algebra, h its Coxeter number, and e_i its Coxeter exponents. Then we have*

1. $e_i + e_{r-i} = h$ *for all* i;

2. $|W| = \prod_{i=1}^{r}(e_i + 1)$.

For the proof, see [Humphreys90], Orlik and Terao [OT92] and [Bourbaki].

Let us look at the D_5 case. From the root system given in 1.2–1.3 we see easily that the Weyl group $W(D_5)$ is a group of order $2^4 \cdot 5! = 1920$ fitting in the exact sequences

$$1 \to W(D_5) \to G \xrightarrow{\psi} \mathbb{Z}/2\mathbb{Z} \to 1$$

and

$$1 \to (\mathbb{Z}/2\mathbb{Z})^{\oplus 5} \to G \xrightarrow{\varphi} S_5 \to 1.$$

The group G, and hence the Weyl group $W(D_5)$ as a subgroup of G, acts on $\mathbb{C}[\mathfrak{H}(D_5)^*] \simeq \mathbb{C}[x_1, \ldots, x_5]$ by

$$\sigma^*(x_i) = \varepsilon_i x_{\varphi(\sigma)(i)},$$

where $\sigma \in G$, $\varepsilon_i = \pm 1$ and $\psi(\sigma) = \varepsilon_1 \cdots \varepsilon_5$. Write f_j for the jth elementary symmetric function of 5 variables. Then $\mathbb{C}[\mathfrak{H}(D_5)^*]^{W(D_5)}$ is generated by $g_j := f_j(x_1^2, \ldots, x_5^2)$ for $j = 1, 2, 3, 4$ and $g_5 := f_5 = x_1 \cdots x_5$. It follows that $\{\deg g_j\} = (2, 4, 6, 8, 5)$ so that the Coxeter exponents are $1, 3, 5, 7, 4$. Since the Coxeter number $h(D_5)$ equals 8, we have $8 = 1 + 7 = 3 + 5 = 4 + 4$. Moreover $|W(D_5)| = 1920 = 2 \cdot 4 \cdot 6 \cdot 8 \cdot 5$.

Type	r	$e_1, e_2, e_3, \ldots, e_{r-1}, e_r$	h
A_n	n	$1, 2, \ldots, n-1, n$	$n+1$
D_n	n	$1, 3, 5, \ldots, 2n-3, n-1$	$2n-2$
E_6	6	$1, 4, 5, 7, 8, 11$	12
E_7	7	$1, 5, 7, 9, 11, 13, 17$	18
E_8	8	$1, 7, 11, 13, 17, 19, 23, 29$	30

Table 2: Coxeter exponents and Coxeter numbers

1.10   Quivers (= oriented graphs) of finite type

Let Γ be a connected oriented graph. It consists of a finite set of vertices and (simple) oriented edges joining two vertices. Write $v(\Gamma)$ and $e(\Gamma)$ for the set of vertices and edges of Γ.

For an edge ℓ, we define $\partial(\ell) = \beta(\ell) - \alpha(\ell)$, where $\alpha(\ell)$ and $\beta(\ell)$ are the starting and end points of ℓ.

Definition 1.11 ([Gabriel72]) A *representation* $\mathbf{V} := \{V_\alpha, \varphi_\ell\}$ of Γ is a set of finite dimensional vector spaces V_α, one for each $\alpha \in v(\Gamma)$, coupled with a set of homomorphisms $\varphi_\ell \colon V_{\alpha(\ell)} \to V_{\beta(\ell)}$, one for each $\ell \in e(\Gamma)$. We define the *dimension vector* of a representation $\mathbf{V}$ to be $\mathbf{v} = \dim \mathbf{V} := \{\dim V_\alpha; \alpha \in v(\Gamma)\}$.

Two representations $\mathbf{V} = \{V_\alpha, \varphi_\ell\}$ and $\mathbf{W} = \{W_\alpha, \psi_\ell\}$ are *equivalent* if there are isomorphisms $f_\alpha \colon V_\alpha \to W_\alpha$ such that $\psi_\ell \cdot f_{\alpha(\ell)} = f_{\beta(\ell)} \cdot \varphi_\ell$ for any $\ell \in e(\Gamma)$. Two equivalent representations have the same dimension vector.

We say that Γ is a *quiver of finite type* if there are only finitely many equivalence classes of representations of Γ for any fixed dimension vector. This notion is independent of the choice of orientation of Γ.

Theorem 1.12 ([Gabriel72]) *Let Γ be a quiver of finite type. Then Γ with orientation forgotten is one of A_n, D_n and E_n. Conversely, if Γ is one of these types, it is a quiver of finite type.*

Proof (Outline) Suppose that Γ is of finite type. Let $\mathbf{v} = (n_\alpha)_{\alpha \in v(\Gamma)}$ be a vector with positive integer coefficients n_α. We choose and fix a representation $\mathbf{V} := \{V_\alpha, \varphi_\ell\}$ of Γ. Hence $n_\alpha = \dim V_\alpha$. Then the set of representations of Γ is the set $M := \prod_{\ell \in e(\Gamma)} \mathrm{Hom}(V_{\alpha(\ell)}, V_{\beta(\ell)})$. Let $G := \prod_{\alpha \in v(\Gamma)} \mathrm{End}(V_\alpha)$. Then G acts on M by

$$(\varphi_\ell) \mapsto (g_{\beta(\ell)} \cdot \varphi_\ell \cdot g_{\alpha(\ell)}^{-1}) \quad \text{for } g_\alpha \in \mathrm{End}(V_\alpha).$$

The set of equivalence classes of representations of Γ with fixed $\dim \mathbf{V} = \mathbf{v}$ is the quotient of M by the action of G. Since Γ is connected, the centre of G consists of scalar matrices. Therefore $\dim M \le \dim G - 1$ by assumption. It follows that $\sum_{\ell \in e(\Gamma)} n_\alpha n_\beta \le \sum_{\alpha \in v(\Gamma)} n_\alpha^2 - 1$. Since this holds for any $\mathbf{v} \in (\mathbb{Z}_+)^{\mathrm{Card}(v(\Gamma))}$, the bilinear form $\sum_{\alpha \in v(\Gamma)} x_\alpha^2 - \sum_{\ell \in e(\Gamma)} x_{\alpha(\ell)} x_{\beta(\ell)}$ is positive definite. It follows from the same argument as in the classification of simple Lie algebras that the graph Γ is one of ADE. $\square$

Theorem 1.13 ([Gabriel72]) *Let Γ be a quiver of finite type. Then the map $\mathbf{V} \mapsto \dim \mathbf{V}$ is a bijective correspondence between the set of equivalence classes of indecomposable representations and the set of positive roots of the root system corresponding to Γ.*

2 Conformal field theory

2.1 Background from physics

In the study of conformal field theories, systems fitting into an ADE classification arise on considering field theories satisfying certain physically natural assumptions on a real two dimensional torus (periodic in one space and one time direction).

We start by telling in very rough terms a story that physicists take for granted. Suppose we are given an infinite dimensional vector space $\mathcal{H}$ and a finite set of operators A_j on $\mathcal{H}$. The space $\mathcal{H}$ is supposed to be a realization of various physical states. The operators A_j are supposed to be selfadjoint in so far as they correspond to actual physical operators or "observables". In this sense, the vector space $\mathcal{H}$ is required to have a Hermitian inner product, namely, we require $\mathcal{H}$ to be unitary. Rather surprisingly, we will soon see that the unitary assumption picks up mathematically interesting objects.

If we have a kind of Hamiltonian operator in the algebra $\mathcal{A}$, the eigenvalue of the operator would be the energy of the (eigen)-state, and in general any state is an infinite linear combination of eigenstates, like a Fourier series expansion. The operators A_j are supposed to correspond to physical observables such as the energy of particles in the system, and they correspond in mathematical terms to irreducible representations of some algebra $\mathcal{A}$ on $\mathcal{H}$, where the system is said to admit $\mathcal{A}$-symmetry.

The system $\{\mathcal{A}, A_j, \mathcal{H}\}$ is called a *conformal field theory* if the algebra $\mathcal{A}$ contains a Virasoro algebra acting nontrivially on $\mathcal{H}$.

The distribution of various energy levels is captured by the so-called *partition function* of the system, which in mathematical terms is the generating function of $\mathcal{H}$ weighted by the values of energy. If the system has space-time symmetry, one proves by a physical argument that the partition function is $SL(2, \mathbb{Z})$-invariant.

The problem is to determine all possible systems admitting space-time symmetry; hence, as a first step, we consider the problem of classifying all possible modular invariant partition functions, namely $SL(2, \mathbb{Z})$-invariant partition functions in certain restricted situations. In the situations we are interested in, the algebra $\mathcal{A}$ is either the affine Lie algebra $A_1^{(1)}$ or the minimal unitary series of Virasoro algebras with central charge $c = 1 - 6/m(m + 1)$ for $m \geq 3$. Although the minimal unitary series is more interesting, the partition function for $A_1^{(1)}$ is easier to write down and more coherent to the ADE classification. Therefore we limit ourselves to $A_1^{(1)}$. It is not known whether the modular invariant partition functions in the subsequent table (Table 3) are partition functions of some conformal field theory admitting space-time symmetry.

We now rephrase all this in more mathematically rigorous terms.

Definition 2.2 Write

$$e = \begin{pmatrix} 0 & 1 \\ 0 & 0 \end{pmatrix}, \quad f = \begin{pmatrix} 0 & 0 \\ 1 & 0 \end{pmatrix}, \quad h = \begin{pmatrix} 1 & 0 \\ 0 & -1 \end{pmatrix}$$

for the standard generators of $\mathfrak{sl}_2(\mathbb{C})$. The Cartan–Killing form of $\mathfrak{sl}_2(\mathbb{C})$ is given by $(x, y)_{\mathrm{LIE}} = \mathrm{Tr}(xy)$. The affine Lie algebra $A_1^{(1)}$ is an infinite dimensional Lie algebra $\mathcal{A}$ over $\mathbb{C}$ spanned by $\mathfrak{sl}_2(\mathbb{C}) \otimes \mathbb{C}[t, t^{-1}]$, together with a central element c, subject to the relations

$$[x(m), y(n)] = [x, y](m + n) + mc\delta_{m+n,0}(x, y)_{\mathrm{LIE}} \quad \text{and} \quad [c, x(m)] = 0,$$

for all $m, n \in \mathbb{Z}$; here t is an indeterminate, and we write $x(m) := x \otimes t^m$ for $x \in \mathfrak{sl}_2(\mathbb{C})$.

Theorem 2.3 *Let k be a positive integer and s an integer with $0 \le s \le k$. We define an $A_1^{(1)}$-module $V(s, k) := A_1^{(1)} \cdot v(s, k)$ by*

$$x(n)v(s, k) = 0, \quad e(0)v(s, k) = 0 \quad \text{for } x \in \mathfrak{sl}_2(\mathbb{C}) \text{ and } n \ge 1,$$
$$h(0)v(s, k) = sv(s, k), \quad cv(s, k) = kv(s, k).$$

Then $V(s, k)$ is a unitary integrable irreducible $A_1^{(1)}$-module having highest weight vector $v(s, k)$. Conversely, any unitary irreducible integrable highest weight $A_1^{(1)}$-module V is isomorphic to $V(s, k)$ for some pair (s, k) as above.

By convention, we write $v(s, k)$ as the ket $|s, k\rangle$. The integer k is called the *level* of the $A_1^{(1)}$-module $V(s, k)$. By the Kac–Weyl character formula, we have

Theorem 2.4 *The character of $V(s, k)$ is given by*

$$\chi_{s,k}(q, \theta) = \sum_{m \in \mathbb{Z}} q^{(k+2)m^2 + (s+1)m} \left(e^{\sqrt{-1}\theta((k+2)m + \frac{s}{2})} - e^{-\sqrt{-1}\theta((k+2)m + \frac{s}{2}+1)} \right) / D,$$

where the denominator is $D = (1 - e^{-\sqrt{-1}\theta})\varphi(\tau)\varphi_+(\tau)\varphi_-(\tau)$, and

$$\varphi(q) = \prod_{n \ge 1}(1 - q^n), \quad \varphi_\pm(q, \theta) = \prod_{n \ge 1}(1 - e^{\pm\sqrt{-1}\theta}q^n).$$

Although this may look different from the usual form of the Kac–Weyl formula, the above form of the character is adjusted to the expression used by physicists to write down partition functions. In Kac's notation ([Kac90], Chapter 6 and p. 173) and the notation in 2.6

$$\chi_{s,k} = \chi_{L((k-s)\Lambda_0 + s\Lambda_1)}$$
$$= \mathrm{Tr}_{L((k-s)\Lambda_0 + s\Lambda_1)}\left(q^{(k+2)L_0} e^{\sqrt{-1}(k+2)\theta h(0)/2} \right).$$

We note that $L_0 = -d$ and $c = K$ in the notation of [Kac90], Chapters 6–7.

Definition 2.5 The *Virasoro algebra* Vir_c with central charge c is the infinite dimensional Lie algebra over $\mathbb{C}$ generated by L_n for $n \in \mathbb{Z}$ and c, subject to the following relations

$$[L_m, L_n] = (m - n)L_{m+n} + \frac{c}{12}(m^3 - m)\delta_{m+n,0},$$

$$[L_n, c] = 0 \quad \text{for all } n, m.$$

There is a way of constructing L_n from the affine Lie algebra $A_1^{(1)}$, called the *Segal–Sugawara construction*:

$$L_n = \frac{1}{2(k+2)} \sum_{m \in \mathbb{Z}} \Big(:e(n-m)f(m): + :f(n-m)e(m): + \tfrac{1}{2}:h(n-m)h(m): \Big).$$

Here : : is the *normal ordering* defined by

$$:x(m)y(n): = \begin{cases} x(m)y(n) & \text{if } m < n, \\ \frac{1}{2}(x(m)y(n) + y(n)x(m)) & \text{if } m = n, \\ y(n)x(m) & \text{if } m > n. \end{cases}$$

Then we infer the relations

$$[L_m, L_n] = (m - n)L_{m+n} + \frac{1}{12} \cdot \frac{3k}{k+2}(m^3 - m)\delta_{m+n,0},$$

$$[L_m, x(n)] = -nx(m+n) \quad \text{and} \quad [L_0, x(-n)] = nx(-n)$$

for all $m, n \in \mathbb{Z}$ and $x \in \mathfrak{sl}_2(\mathbb{C})$.

Thus given a system having $A_1^{(1)}$ symmetry of level k, the system admits a Virasoro algebra Vir_c symmetry with $c = 3k/(k+2)$. Write $v := x(-n_1)x(-n_2)\cdots x(-n_p)|s, k\rangle$; note that $V(s, k)$ is spanned by vectors v of this form for various $n_i > 0$. The element L_0 acts on v by

$$L_0(v) = \left\{ \frac{1}{4(k+2)}(s^2 + 2s) + (n_1 + n_2 + \cdots + n_p) \right\} v.$$

This shows that L_0 behaves as if it measures the *energy* of the state v.

2.6   Modular invariant partition functions

Write $\mathcal{A}$ for the affine Lie algebra $A_1^{(1)}$, and $\mathcal{A}^*$ for its complex conjugate. We fix the level k, and consider only unitary irreducible integrable $\mathcal{A}$ or $\mathcal{A}^*$-modules of level k. We consider the following particular $\mathcal{A} \otimes \mathcal{A}^*$-module:

$$\mathcal{H} = \bigoplus_{\ell, \ell'} m_{\ell, \ell'} V(\ell, k) \otimes (V(\ell', k))^*,$$

where $m_{\ell,\ell'}$ is the multiplicity of the copy $V(\ell, k) \otimes (V(\ell', k))^*$.

This is what physicists call Hilbert spaces in such a situation, without further qualifications. We only need to take the completion of $\mathcal{H}$ in order to be mathematically rigorous. Mathematicians might guess why we have to choose $\mathcal{H}$ as above. This is a special case of the factorization principle widely accepted by physicists. Now L_0 is supposed to play the same role as the Hamiltonian operator of the system, and therefore the eigenvalues of L_0 should express the energies. For the (physical) theory it is always important to know the energy level distribution inside the system. Thus it is important to know the eigenvalues of L_0 and to *count* the dimension of the eigenspaces, in other words to determine the *partition function Z* of the system. The partition function Z of the system (= the $A_1^{(1)}$-module) $\mathcal{H}$ is defined by

$$Z(q, \theta, \bar{q}, \bar{\theta}) := \mathrm{Tr}_{\mathcal{H}}\left(q^{(k+2)L_0} e^{\sqrt{-1}(k+2)\theta h(0)/2} \bar{q}^{(k+2)\bar{L}_0} e^{-\sqrt{-1}(k+2)\bar{\theta}\bar{h}(0)/2}\right)$$

$$= \sum_{\ell,\ell'} m_{\ell,\ell'}\, \chi_{\ell,k}\, \chi_{\ell',k}^* \,,$$

where $q = e^{2\pi\sqrt{-1}\tau}$ with τ in the upper half plane, and θ is a real parameter; when τ is purely imaginary, $-i\tau$ equals the ratio of sizes of time and one dimensional space. For more details see [Cardy88] and [EY89].

In this situation, the physicists assume

1. $m_{0,0} = 1$;

2. $Z(q, \theta, \bar{q}, \bar{\theta})$ is SL$(2, \mathbb{Z})$-invariant.

Condition (1) means that the system has a unique state of lowest energy, usually called the *vacuum*. This is one of the principles that physicists take for granted. We therefore follow the physicists' tradition, *doing as the Romans do*. Next, (2) is the condition of discrete space-time symmetry. It means that Z is invariant under the transformations $\tau \mapsto -1/\tau$ and $\theta \mapsto \theta + 1$. See [Cardy86] and [Cardy88] for more details. These assumptions have very surprising consequences.

Theorem 2.7 *Modular invariant partition functions are classified as in Table 3. We write the partition function $Z = \sum a_{ij}\chi_i\chi_j^*$ in terms of $A_1^{(1)}$-characters. Then the indices i with nonzero a_{ii} are Coxeter exponents of the Lie algebra of the same type. Moreover the value $k+2$ is equal to the Coxeter number.*

For example, for $k = 6$ there are two modular invariant partition functions:

$$Z(A_7) = |\chi_1|^2 + |\chi_2|^2 + \cdots + |\chi_6|^2 + |\chi_7|^2,$$

$$Z(D_5) = \sum_{\lambda} |\chi_{2\lambda-1}|^2 + (\chi_2\chi_6^* + \chi_2^*\chi_6) + |\chi_4|^2,$$

Type	$k+2$	partition function $Z(q, \theta, \bar{q}, \bar{\theta})$
A_n	$n+1$	$\sum_{\lambda=1}^{n} |\chi_\lambda|^2$
D_{2r}	$4r-2$	$\sum_{\lambda=1}^{r-1} |\chi_{2\lambda-1} + \chi_{4r+1-2\lambda}|^2 + 2|\chi_{2r-1}|^2$
D_{2r+1}	$4r$	$\sum_{\lambda=1}^{2r} |\chi_{2\lambda-1}|^2 + \sum_{\lambda=1}^{r-1} (\chi_{2\lambda}\bar{\chi}_{4r-2\lambda} + \bar{\chi}_{2\lambda}\chi_{4r-2\lambda}) + |\chi_{2r}|^2$
E_6	12	$|\chi_1 + \chi_7|^2 + |\chi_4 + \chi_8|^2 + |\chi_5 + \chi_{11}|^2$
E_7	18	$|\chi_1 + \chi_{17}|^2 + |\chi_5 + \chi_{13}|^2 + |\chi_7 + \chi_{11}|^2$
		$+|\chi_9|^2 + (\chi_3 + \chi_{15})\bar{\chi}_9 + \chi_9(\bar{\chi}_3 + \bar{\chi}_{15})$
E_8	30	$|\chi_1 + \chi_{11} + \chi_{19} + \chi_{29}|^2 + |\chi_7 + \chi_{13} + \chi_{17} + \chi_{23}|^2$

Table 3: Modular invariant partition functions

where A_7 (respectively D_5) has Coxeter exponents $\{1, 2, \ldots, 6, 7\}$ (respectively $\{1, 3, 5, 7, 4\}$). Note that the indices $2, 6$ are not among the Coxeter exponents of D_5. For $k = 10$, there are three types of modular invariant partition functions $Z(A_{11})$, $Z(D_7)$ and $Z(E_6)$.

For more details, see Capelli, Itzykson and Zuber [CIZ87], Kato [Kato87], Gepner and Witten [GW86] and Kac and Wakimoto [KW88]. Compare also [Slodowy90]. Pasquier [Pasquier87a] and [Pasquier87b] used Dynkin diagrams to construct some lattice models and rediscovered a series of associative algebras (called the *Temperly–Lieb algebras*) which are expected to appear as some algebra of operators on the Hilbert space in the continuum limit of the models. See also Section 3.4 and [GHJ89], p. 87, p. 259. Although the relation of the models with modular invariant partition functions remains obscure, the partition function of Pasquier's model is expected to coincide in some sense with those classified in Table 3. See [Zuber90]. The connection of CFT with graphs is studied by Petkova and Zuber [PZ96].

2.8 $N = 2$ superconformal field theories

There are other series of conformal field theories – the $N = 2$ superconformal field theories or (induced) topological conformal field theories, which are more intimately related to the theory of ADE singularities. However, these are a priori close to the theory of singularities. See Blok and Varchenko [BV92].

The following result might be worth mentioning here.

Theorem 2.9 *Suppose that there exists an irreducible unitary* Vir_c*-module, namely an irreducible* Vir_c*-module admitting a* Vir_c*-invariant Hermitian inner product. Then* $c \geq 1$ *or* $c = 1 - 6/m(m+1)$ *for some* $m \in \mathbb{Z}, m \geq 3$.

2.10   The minimal unitary series

Virasoro algebras of the second type are called the minimal $c < 1$ unitary series of Virasoro algebras. They attract attention because of their exceptional characters. There is a series of von Neumann algebras with indices equal to similar values $4\cos^2(\pi/h)$ for $h = 3, 4, \ldots$, where h is the Coxeter number in a suitable interpretation. Conjecturally, the minimal unitary $c < 1$ series of CFTs are deeply related to the class of subfactors which will be introduced in Section 3. Much is already known about this topic. See [GHJ89], [Jones91], [EK98].

3   Von Neumann algebras

3.1   Factors and subfactors

We give a brief explanation of von Neumann algebras, II_1 factors of finite type, and subfactors. The reader is invited to refer, for instance, to [GHJ89], [Jones91], [EK98]. Let $\mathcal{H}$ be a Hilbert space over $\mathbb{C}$ and $B(\mathcal{H})$ the space of all bounded $\mathbb{C}$-linear operators on $\mathcal{H}$ endowed with an operator seminorm in some suitable sense. A *von Neumann algebra* M is by definition a closed subalgebra of $B(\mathcal{H})$ containing the identity and stable under conjugation $x \mapsto x^*$. This is equivalent to saying that M is $*$-stable and is equal to its bicommutant. This is von Neumann's bicommutant theorem. See [Jones91], p. 2. The commutant of a subset S of $B(\mathcal{H})$ is by definition the centralizer of S in $B(\mathcal{H})$. The bicommutant of M is the commutant of the commutant of M. If M is a $*$-stable subset of $B(\mathcal{H})$, then the bicommutant of M is the smallest von Neumann algebra containing M.

A *factor* is defined to be a von Neumann algebra M with centre Z_M consisting only of constant multiples of the identity. Let M be a factor. A factor N is called a *subfactor* of M if it is a closed $*$-stable $\mathbb{C}$-subalgebra of M. A *II_1 factor* is by definition an infinite dimensional factor M which admits a $\mathbb{C}$-linear map $\mathrm{tr}\colon M \to \mathbb{C}$ (called the *normalized trace*) such that

1. $\mathrm{tr}(\mathrm{id}) = 1$,

2. $\mathrm{tr}(xy) = \mathrm{tr}(yx)$ for all $x, y \in M$,

3. $\mathrm{tr}(x^*x) > 0$ for all $0 \neq x \in M$.

We note that the above normalized trace is unique. Let $L^2(M)$ be the Hilbert space obtained by completing M with respect to the inner product $\langle x \mid y \rangle := \mathrm{tr}(x^*y)$ for $x, y \in M$. The normalized trace induces a trace (not necessarily normalized) $\mathrm{Tr}_{M'}$ on the commutant M' of M in $B(\mathcal{H})$, called the *natural trace.* If $\mathcal{H} = L^2(M)$, then $\text{Tr}_{M'}(JxJ) = \text{tr}_M(x)$ for all $x \in M$ where J is the extension to $L^2(M)$ of the conjugation $J(z) = z^*$ of M.

A *finite factor* M is either a II_1 factor or $B(\mathcal{H})$ for a finite dimensional Hilbert space $\mathcal{H}$. Let M be a finite factor, and N a subfactor of M. Then the *Jones index* $[M : N]$ is defined to be $\dim_N L^2(M) := \text{Tr}_{N'}(\text{id}_{L^2(M)})$, where N' is the commutant of N. In general $[M : N] \in [1, \infty]$ is a (possibly irrational) positive number.

For instance, $M = \text{End}_{\mathbb{C}}(W)$ is a factor (a simple algebra) for any finite dimensional $\mathbb{C}$-vector space W. If $N = \text{End}_{\mathbb{C}}(V)$ is a subfactor of M, then we have a representation of $N = \text{End}_{\mathbb{C}}(V)$ on W, in other words, W is an $\text{End}_{\mathbb{C}}(V)$-module. We recall that

1. any $\text{End}_{\mathbb{C}}(V)$-module is completely reducible, and

2. V is a unique nontrivial irreducible $\text{End}_{\mathbb{C}}(V)$-module up to isomorphism.

Therefore $W \simeq V \otimes_{\mathbb{C}} U$ for some $\mathbb{C}$-vector space U. Hence $\dim_{\mathbb{C}} W$ is divisible by $\dim_{\mathbb{C}} V$. Since M is complete with respect to the inner product, we have $[M : N] = \dim_N L^2(M) = \dim_N M = (\dim_{\mathbb{C}} M)(\dim_{\mathbb{C}} N)^{-1} = (\dim_{\mathbb{C}} U)^2$, a square integer. See [GHJ89], p. 38.

The importance of the index $[M : N]$ is explained by the following result:

Theorem 3.2 ([GHJ89], p. 138) *Suppose that M is a finite factor, and let H and H' be M-modules which are separable Hilbert spaces. Then*

1. $\dim_M H = \dim_M H'$ *if and only if H and H' are isomorphic as M-modules.*

2. $\dim_M H = 1$ *if and only if $H = L^2(M)$.*

3. $\dim_M H$ *is finite if and only if $\text{End}_M(H)$ is a finite factor.*

Theorem 3.3 ([GHJ89], p. 186) *Suppose that $N \subset M$ is a pair of II_1 factors whose principal graph is finite.*

1. *If $[M : N] < 4$ then $[M : N] = 4\cos^2(\pi/h)$ for some integer $h \geq 3$.*

2. *If $[M : N] = 4\cos^2(\pi/h) < 4$, the principal graph of the pair $N \subset M$ is one of the Dynkin diagrams A_n, D_n and E_n with Coxeter number h. (Only A_n, D_{2n}, E_6 and E_8 can appear, see [Izumi91], p. 972. This was proved independently by Kawahigashi and Izumi.)*

3. *If $[M : N] = 4$ then the principal graph of the pair $N \subset M$ is one of the extended Dynkin diagrams $\tilde{A}_n$, $\tilde{D}_n$ and $\tilde{E}_n$.*

4. *Conversely for any value* $\lambda = 4$ *or* $4\cos^2(\pi/h)$, *there exists a pair of* II_1 *factors* $N \subset M$ *with* $[M : N] = \lambda$.

See [GHJ89], [Jones91], p. 35. See [GHJ89], p. 186 for *principal graphs*. See also 3.8–3.10 where to each tower of finite dimensional semisimple algebras we associate a finite graph Γ analogous to a principal graph for a pair of factors. This will help us to guess the principal graphs for factors.

3.4   The fundamental construction and Temperly–Lieb algebras

Why do the constants $4\cos^2(\pi/h)$ appear? Let us explain this briefly.

Given a pair of finite II_1 factors $N \subset M$ with $\beta := [M : N] < \infty$, there exists a tower of finite II_1 factors M_k for $k = 0, 1, 2, \ldots$ such that

1. $M_0 = N$, $M_1 = M$,

2. for any $k \geq 1$, the algebra $M_{k+1} := \mathrm{End}_{M_{k-1}} M_k$ is obtained from M_k by taking the von Neumann algebra of operators on $L^2(M_k)$ generated by M_k and an orthogonal projection $e_k \colon L^2(M_k) \to L^2(M_{k-1})$, where M_k is viewed as a subalgebra of M_{k+1} under right multiplication.

By Theorem 3.2, (3), M_{k+1} is a finite II_1 factor. The sequence $\{e_k\}_{k=1,2,\ldots}$ of projections on $M_\infty := \bigcup_{k \geq 0} M_k$ satisfies the relations

$$e_i^2 = e_i, \quad e_i^* = e_i,$$
$$e_i = \beta e_i e_j e_i \quad \text{for} \quad |i - j| = 1,$$
$$e_i e_j = e_j e_i \quad \text{for} \quad |i - j| \geq 2.$$

We define $A_{\beta,k}$ to be the $\mathbb{C}$-algebra generated by $1, e_1, \ldots, e_{k-1}$ subject to the above relations, and $A_\beta := \bigcup_{k=1}^\infty A_{\beta,k}$. The algebra A_β is called the *Temperly–Lieb algebra*. Compare also [GHJ89], p. 259.

Thus given a pair of II_1 factors, the fundamental construction gives rise to a unitary representation of the Temperly–Lieb algebra. However, the condition that the representation is unitary restricts the possible values of β, as Theorem 3.5 shows.

Theorem 3.3, (1) follows from the following result

Theorem 3.5 ([Wenzl87]) *Suppose given an infinite sequence* $\{e_k\}_{k=1,2,\ldots}$ *of projections on a complex Hilbert space satisfying the following relations:*

$$e_i^2 = e_i, \quad e_i^* = e_i,$$
$$e_i = \beta e_i e_j e_i \quad \text{for} \quad |i - j| = 1,$$
$$e_i e_j = e_j e_i \quad \text{for} \quad |i - j| \geq 2.$$

If $e_1 \neq 0$, *then* $\beta \geq 4$ *or* $\beta = 4\cos^2(\pi/\ell)$ *for an integer* $\ell \geq 3$.

Proof We give an idea of the proof of Theorem 3.5. Suppose we are given a homomorphism $\varphi\colon A_\beta \to B(\mathcal{H})$ for some Hilbert space $\mathcal{H}$, that is, a unitary representation of A_β. For simplicity we identify $\varphi(x)$ with x for $x \in A_\beta$.

First we see that $0 \le e_1^* e_1 = e_1^2 = e_1 = \beta e_1 e_2 e_1 = \beta (e_2 e_1)^*(e_2 e_1)$. Hence $\beta \ge 0$. If $\beta = 0$ then $e_1 = 0$, contradicting the assumption. Hence $\beta > 0$.

Next we assume $0 < \beta < 1$ to derive a contradiction by using $A_{\beta,3}$. Let $\delta_2 := 1 - e_1$. Then the assumptions of Theorem 3.5 imply $\delta_2^* = \delta_2$, $\delta_2^2 = \delta_2$. Hence

$$0 \le (\delta_2 e_2 \delta_2)^*(\delta_2 e_2 \delta_2) = (\delta_2 e_2 \delta_2)^2 = (1 - \beta^{-1})(\delta_2 e_2 \delta_2) \le 0,$$

because $\delta_2 e_2 \delta_2 = (e_2 \delta_2)^*(e_2 \delta_2) \ge 0$. Thus $e_2 \delta_2 = 0$. It follows that $e_2 = e_1 e_2$, and $e_2 = e_2^2 = e_2 e_1 e_2 = \beta^{-1} e_2$, so that $e_2 = 0$. Therefore $e_1 = \beta e_1 e_2 e_1 = 0$, contradicting the assumption. If $4\cos^2(\pi/\ell) < \beta < 4\cos^2(\pi/(\ell+1))$, then we derive a contradiction by using $A_{\beta,\ell+1}$. See [GHJ89], pp. 272–273. □

3.6 Bipartite graphs

A *bipartite graph* Γ with multiple edges is a (finite, connected) graph with black and white vertices and multiple edges such that any edge connects a white and black vertex, starting from a white one (see, for example, Figure 3). If any edge is simple, then Γ is an oriented graph (a quiver) in the sense of Section 1. Let Γ be a connected bipartite finite graph with multiple oriented edges. Let $w(\Gamma)$ (respectively $b(\Gamma)$) be the number of white (respectively black) vertices of Γ. We define the *adjacency matrix* $\Lambda := \Lambda(\Gamma)$ of size $b(\Gamma) \times w(\Gamma)$ by

$$\Lambda_{b,w} = \begin{cases} m(e) & \text{if there exists } e \text{ such that } \partial e = b - w; \\ 0 & \text{otherwise,} \end{cases}$$

where $m(e)$ is the multiplicity of the edge e.

We define the norm $\|\Gamma\|$ as follows,

$$\|X\| = \max\{\|Xx\|_{\text{EUCL}};\, \|x\|_{\text{EUCL}} \le 1\};$$

$$\|\Gamma\| = \|\Lambda(\Gamma)\| = \left\| \begin{pmatrix} 0 & \Lambda(\Gamma) \\ \Lambda(\Gamma)^t & 0 \end{pmatrix} \right\|,$$

where X is a matrix, x a vector and $\| \ \|_{\text{EUCL}}$ the Euclidean norm. We note that when X is a square matrix, $\|X\|$ is the maximum of the absolute values of eigenvalues of X.

Lemma 3.7 *Assume Γ is a connected finite graph with multiple edges. Then*

D_5

Figure 3: The Dynkin diagram D_5 as a bipartite graph

1. *if $\|\Gamma\| \leq 2$ and if Γ has a multiple edge, $\|\Gamma\| = 2$ and $\Gamma = \tilde{A}_1$.*

2. *$\|\Gamma\| < 2$ if and only if Γ is one of the Dynkin diagrams A, D, E. In this case $\|\Gamma\| = 2\cos(\pi/h)$, where h is the Coxeter number of Γ.*

3. *$\|\Gamma\| = 2$ if and only if Γ is one of the extended Dynkin diagrams $\tilde{A}, \tilde{D}, \tilde{E}$.*

Lemma 3.7 is easy to prove. For instance, if there is a row or column vector of Γ with norm a, then $\|\Gamma\| \geq a$. See also [GHJ89], p. 19.

3.8   The tower of semisimple algebras

Why is Theorem 3.3, (2) true? The interested reader is invited to see [GHJ89]. Here we explain it in a much simpler situation.

Recall that a matrix algebra of finite rank is a finite factor by definition. This is an elementary analogue of a finite II_1 factor with a finite dimensional Hilbert space. So let us see what happens if we consider the fundamental construction for a pair $N \subset M$ of (sums of) matrix algebras. We call N and M (a pair of) semisimple algebras (over $\mathbb{C}$).

Let Γ be a connected bipartite graph with multiple edges, $v(\Gamma)$ and $e(\Gamma)$ its set of vertices and edges. Let $W(w)$ be a $\mathbb{C}$-vector space for a white vertex w. Let $W(b, w)$ be a $\mathbb{C}$-vector space for an edge e with $\partial e = b - w$ and $V(b) = \bigoplus_{\partial e = b - w} W(b, w) \otimes W(w)$ for a black vertex b, where the sum runs over all edges of Γ ending at b. Set

$$N := \bigoplus_{w:\text{white}} \text{End}_{\mathbb{C}}(W(w)),$$

$$M := \bigoplus_{b:\text{black}} \text{End}_{\mathbb{C}}(V(b))$$

$$= \bigoplus_{b:\text{black}} \bigoplus_{\partial e = b - w} \text{End}_{\mathbb{C}}(W(b, w)) \otimes \text{End}_{\mathbb{C}}(W(w)).$$

Now let $\varphi_0 \colon N \to M$ be the homomorphism defined by

$$\varphi_0 = \bigoplus_b \varphi_{0,b}, \quad \varphi_{0,b} = \bigoplus_{\partial e = b - w} \text{id}_{W(b,w)} \otimes \text{id}_{\text{End}(W(w))},$$

where $\mathrm{id}_{W(b,w)}$ is the identity homomorphism of $W(b,w)$. This is a representation of the oriented graph Γ in the sense of Definition 1.11 if $m(e) = \dim W(b,w) \leq 1$ for any edge e.

We set $\Lambda(M,N) := \Lambda(\Gamma)$ and call it the *inclusion matrix* of M in N.

Let us consider a tower of semisimple algebras arising from the fundamental construction for the pair $N \subset M$. We define $M_0 = N$, $M_1 = M$ and $M_{k+1} := \mathrm{End}_{M_{k-1}}(M_k)$ inductively.

Let $M_2 = \mathrm{End}_N M$, φ_1 the monomorphism of M_1 into M_2 by right multiplication. Let $V(b,w) = \mathrm{End}_{\mathbb{C}}(W(b,w))$. Then we see that

$$\mathrm{End}_N M = \bigoplus_{w:\text{white}} U(w),$$

$$U(w) := \bigoplus_{\partial e = b-w} \mathrm{End}_{W(w)}(V(b))$$

$$= \bigoplus_{\partial e = b-w} \mathrm{End}_{\mathbb{C}}(V(b,w)) \otimes \mathrm{End}_{\mathbb{C}}(W(w)),$$

$$\varphi_1 = \bigoplus_w \varphi_{1,w}, \quad \varphi_{1,w} = \bigoplus_{\partial e = b-w} \text{right mult.}_{V(b,w)} \otimes \mathrm{id}_{\mathrm{End}(W(w))}.$$

The construction shows that the graph Γ describes the inclusion of M_{k-1} into M_k by interchanging the roles of white and black vertices, and reversing the orientation of edges at each step. We see $\Lambda(M_{2k+1}, M_{2k}) = \Lambda(M,N)^t$, $\Lambda(M_{2k}, M_{2k-1}) = \Lambda(M,N)$.

We set $[M : N] := \lim_{k\to\infty}(\dim M_k / \dim M_0)^{1/k}$. (This is one of the equivalent definitions of the Jones index $[M : N]$.) We compute this in the simplest case when Γ is a connected graph with two vertices and a single edge e. Let $m(e)$ be the multiplicity of e, and $\partial e = b - w$. Then we see that

$$M_0 = N = \mathrm{End}_{\mathbb{C}}(W(w)),$$
$$M_1 = M = \mathrm{End}_{\mathbb{C}}(V(b)) \simeq \mathrm{End}_{\mathbb{C}}(W(b,w)) \otimes M_0,$$
$$M_2 = \mathrm{End}_{\mathbb{C}}(\mathrm{End}_{\mathbb{C}}(W(b,w))) \otimes \mathrm{End}_{\mathbb{C}}(W(w)),$$
$$\simeq \mathrm{End}_{\mathbb{C}}(W(b,w)) \otimes \mathrm{End}_{\mathbb{C}}(V(b)) \simeq \mathrm{End}_{\mathbb{C}}(W(b,w)) \otimes M_1.$$

Hence we see that $\dim_{\mathbb{C}} M_k / M_{k-1} = \dim_{\mathbb{C}} \mathrm{End}_{\mathbb{C}}(W(b,w)) = \dim_{\mathbb{C}}(M/N)$. It follows readily that $[M : N] = \dim_{\mathbb{C}}(M/N)$, as was remarked in 3.1.

In this situation, the following result is proved.

Theorem 3.9 ([GHJ89], pp. 32–33)   *1. The following are equivalent:*

(a) *there exists a row $b(\Gamma)$-vector s and $\beta \in \mathbb{C}^*$ with $s\Lambda\Lambda^t = \beta s$ such that every coordinate of s and $s\Lambda$ is nonzero,*

(b) *there exist $\mathbb{C}$-linear maps $e_k \colon M_k \to M_{k-1}$ such that $e_k^2 = e_k$ and*

(i) M_k is generated by M_{k-1} and e_k,

(ii) e_k satisfies $e_i = \beta e_i e_j e_i$ if $|i - j| = 1$ and $e_i e_j = e_j e_i$ if $|i - j| \geq 2$.

2. *If one of the equivalent conditions in (1) holds, then*

$$\beta = \|\Lambda(\Gamma)\Lambda(\Gamma)^t\| = \|\Lambda(\Gamma)\|^2 = [M : N].$$

This is nontrivial, but is just linear algebra. By Theorem 3.9, we have a situation similar to a pair of II_1 factors $N \subset M$ as well as a Temperly–Lieb algebra A_β.

From Lemma 3.7, we infer the following result.

Corollary 3.10 *Let* $M_0 = N \subset M_1 = M \subset \cdots \subset M_k \subset \cdots$ *be a tower of semisimple algebras. We have a Temperly–Lieb algebra A_β from the tower if and only if $\beta = [M : N]$ and $\beta \geq 4$ or $\beta = 4\cos^2(\pi/h)$ for $h = 3, 4, 5, \ldots$. Moreover*

1. *if $\beta = 4\cos^2(\pi/h)$, then the graph Γ is one of A, D, E;*

2. *if $\beta = 4$, then the graph Γ is one of $\widetilde{A}$, $\widetilde{D}$, $\widetilde{E}$.*

For a pair of II_1 factors $N \subset M$, we can always carry out the same construction as for a pair of semisimple algebras, and we find the same graphs (principal graphs), because the pair in fact satisfies the stronger restrictions of (infinite dimensional) II_1 factors. As a consequence, the cases D_{odd} and E_7 are excluded.

4 Two dimensional McKay correspondence

4.1 Finite subgroups of $\text{SL}(2, \mathbb{C})$

Up to conjugacy, any finite subgroup of $\text{SL}(2, \mathbb{C})$ is one of the subgroups listed in Table 4; see [Klein]. The triple (d_1, d_2, d_3) specifies the degrees of the generators of the G-invariant polynomial ring (compare Section 11).

4.2 McKay's observation

As we mentioned in Section 1, any simple singularity is a quotient singularity by a finite subgroup G of $\text{SL}(2, \mathbb{C})$, and so has a corresponding Dynkin diagram. McKay [McKay80] showed how one can recover the same graph purely in terms of the representation theory of G, without passing through the geometry of $\mathbb{A}^2/G$.

Type	G	name	order	h	(d_1, d_2, d_3)
A_n	$\mathbb{Z}_{n+1}$	cyclic	$n+1$	$n+1$	$(2, n+1, n+1)$
D_n	$\mathbb{D}_{n-2}$	binary dihedral	$4(n-2)$	$2n-2$	$(4, 2n-4, 2n-2)$
E_6	$\mathbb{T}$	binary tetrahedral	24	12	$(6, 8, 12)$
E_7	$\mathbb{O}$	binary octahedral	48	18	$(8, 12, 18)$
E_8	$\mathbb{I}$	binary icosahedral	120	30	$(12, 20, 30)$

Table 4: Finite subgroups of $SL(2, \mathbb{C})$

To be more precise, let G be a finite subgroup of $SL(2, \mathbb{C})$. Clearly, G has a two dimensional representation, which maps G injectively into $SL(2, \mathbb{C})$; we call this the *natural representation* ρ_{nat}. Let $\text{Irr}_* G$, respectively $\text{Irr}\, G$, be the set of all equivalence classes of irreducible representations, respectively nontrivial ones. (Caution: note that this goes against the familiar notation of group theory.) Thus by definition, $\text{Irr}_* G = \text{Irr}\, G \cup \{\rho_0\}$, where ρ_0 is the one dimensional trivial representation. Any representation of G over $\mathbb{C}$ is completely reducible, that is, it is a direct sum of irreducible representations up to equivalence. Therefore for any $\rho \in \text{Irr}_* G$, we have

$$\rho \otimes \rho_{\text{nat}} = \sum_{\rho' \in \text{Irr}_* G} a_{\rho, \rho'} \rho'$$

where $a_{\rho, \rho'}$ are certain nonnegative integers. In our situation, we see that $a_{\rho, \rho'} = 0$ or 1 (except for the case A_1, when $a_{\rho, \rho'} = 0$ or 2).

Let us look at the example D_5, the case of a binary dihedral group $G := \mathbb{D}_3$ of order 12. The group G is generated by σ and τ:

$$\sigma = \begin{pmatrix} \varepsilon & 0 \\ 0 & \varepsilon^{-1} \end{pmatrix}, \quad \tau = \begin{pmatrix} 0 & 1 \\ -1 & 0 \end{pmatrix} \quad \text{where } \varepsilon = e^{2\pi\sqrt{-1}/6}.$$

We note that $\text{Tr}(\sigma) = 1$, $\text{Tr}(\tau) = 0$, hence in this case, the natural representation is ρ_2 in Table 5.

Definition 4.3 The graph $\tilde{\Gamma}_{\text{GROUP}}(G)$ is defined to be the graph consisting of vertices $v(\rho)$ for $\rho \in \text{Irr}_* G$, and simple edges connecting any pair of vertices $v(\rho)$ and $v(\rho')$ with $a_{\rho, \rho'} = 1$. We denote by $\Gamma_{\text{GROUP}}(G)$ the full subgraph of $\tilde{\Gamma}_{\text{GROUP}}(G)$ consisting of the vertices $v(\rho)$ for $\rho \in \text{Irr}\, G$ and all the edges between them.

For example, let us look at the D_5 case. Let $\chi_j := \text{Tr}(\rho_j)$ be the character of ρ_j. Then from Table 5 we see that

$$\chi_2(g)\chi_2(g) = \chi_0(g) + \chi_1(g) + \chi_3(g), \quad \text{for } g = 1, \sigma \text{ or } \tau.$$

ρ	$\mathrm{Tr}\,\rho$	1	σ	τ
ρ_0	χ_0	1	1	1
ρ_1	χ_1	1	1	-1
ρ_2	χ_2	2	1	0
ρ_3	χ_3	2	-1	0
ρ_4	χ_4	1	-1	$\sqrt{-1}$
ρ_5	χ_5	1	-1	$-\sqrt{-1}$

Table 5: Character table of D_5

Hence $\chi_2\chi_2 = \chi_0 + \chi_1 + \chi_3$. General representation theory says that an irreducible representation of G is uniquely determined up to equivalence by its character. Therefore $\rho_2 \otimes \rho_2 = \rho_0 + \rho_1 + \rho_3$. Hence $a_{\rho_2,\rho_j} = 1$ for $j = 0, 1, 3$ and $a_{\rho_2,\rho_j} = 0$ for $j = 2, 4, 5$. Similarly, we see that

$$\chi_0\chi_2 = \chi_2, \quad \chi_1\chi_2 = \chi_2,$$
$$\chi_3\chi_2 = \chi_0 + \chi_1 + \chi_4,$$
$$\chi_4\chi_2 = \chi_3 \quad \text{and} \quad \chi_5\chi_2 = \chi_3.$$

In this way we obtain a graph – the extended Dynkin diagram $\widetilde{D}_5$ of Figure 4. It is also interesting to note that the degrees of the characters $\deg \rho_j = \chi_j(1)$ are equal to the multiplicities of the fundamental cycle we computed in Section 1. This is true in the other cases. Namely the graph $\Gamma_{\mathrm{GROUP}}(G)$ turns out to be one of the Dynkin diagrams ADE, while $\widetilde{\Gamma}_{\mathrm{GROUP}}(G)$ is the corresponding extended Dynkin diagram (see Figure 5). This is the observation of [McKay80].

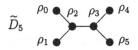

$$\widetilde{D}_5$$

Figure 4: McKay correspondence for $\widetilde{D}_5$

4.4   The Gonzalez-Sprinberg–Verdier construction

Let G be a finite subgroup of $\mathrm{SL}(2, \mathbb{C})$, X the minimal resolution of $S := \mathbb{A}^2/G$, and E the exceptional set. Gonzalez-Sprinberg and Verdier [GSV83] constructed a locally free sheaf V_ρ on X for any $\rho \in \mathrm{Irr}\,G$ such that there exists a unique $E_\rho \in \mathrm{Irr}\,E$ satisfying

$$\deg(c_1(V_\rho)_{|E_\rho}) = 1 \quad \text{and} \quad \deg(c_1(V_\rho)_{|E'}) = 0 \text{ for } E' \neq E_\rho, E' \in \mathrm{Irr}\,E.$$

Thus the map $\rho \mapsto E_\rho$ turns out to be a bijection from $\mathrm{Irr}\,G$ onto $\mathrm{Irr}\,E$.

Their construction of V_ρ is essentially as follows [Knörrer85], p. 178. Let $\rho\colon G \to \mathrm{GL}(V(\rho))$ be a nontrivial irreducible representation of G. Then the associated free $\mathcal{O}_{\mathbb{A}^2}$-module $\mathcal{V}(\rho) := \mathcal{O}_{\mathbb{A}^2} \otimes_{\mathbb{C}} V(\rho)$ admits a canonical G-action defined by $g \cdot (x, v) = (gx, gv)$. Let $\mathcal{V}(\rho)^G$ be the $\mathcal{O}_S$-module consisting of G-invariant sections in $\mathcal{V}(\rho)$. The (locally free) $\mathcal{O}_X$-module V_ρ is defined as

$$V_\rho := \mathcal{O}_X \otimes_{\mathcal{O}_S} \mathcal{V}(\rho)^G / \mathcal{O}_X\text{-torsion}.$$

Theorem 4.5 *Let G be a finite subgroup of $\mathrm{SL}(2, \mathbb{C})$, $S = \mathbb{A}^2/G$, X the minimal resolution of S and E the exceptional set. Then there is a bijection j of $\mathrm{Irr}_* G$ to $\mathrm{Irr}_* E$ such that*

    1. $j(\rho_0) = E_0 =: E_{\rho_0}$ and $j(\rho) = E_\rho$ for $\rho \in \mathrm{Irr}\,G$;

    2. $\deg(\rho) = m_{E_\rho}^{\mathrm{SING}}$ for all $\rho \in \mathrm{Irr}_* G$;

    3. $a_{\rho,\rho'} = (E_\rho, E_{\rho'})_{\mathrm{SING}}$ for $\rho \neq \rho' \in \mathrm{Irr}_* G$.

In particular:

Corollary 4.6 $\Gamma_{\mathrm{GROUP}}(G) = \Gamma_{\mathrm{SING}}(\mathbb{A}^2/G)$ and $\widetilde{\Gamma}_{\mathrm{GROUP}}(G) = \widetilde{\Gamma}_{\mathrm{SING}}(\mathbb{A}^2/G)$.

See [McKay80] and [GSV83]. Using invariant theory, [Knörrer85] gave a different proof of Theorem 4.5 based on the construction in [GSV83]. We discuss again the construction of [GSV83] from the viewpoint of Hilbert schemes in Sections 8–16, and give there our own proof of Theorem 4.5.

5   Missing links and problems

5.1   Known links

We review briefly what is known about links between any pair of the objects (a)–(f) – namely,

    (a) simple singularities, (b) finite subgroups of $\mathrm{SL}(2, \mathbb{C})$,

    (c) simple Lie algebras, (d) quivers, (e) CFT, (f) subfactors.

A very deep understanding of the link from (c) to (a) is provided by work of Grothendieck, Brieskorn, Slodowy and Springer. See [Slodowy80]. However, no intrinsic converse construction of simple Lie algebras starting from (a) is known.

The link from (b) to (a) is on the one hand the obvious quotient singularity construction, and on the other the very nontrivial McKay correspondence.

The construction of [GSV83] gives an explanation for the McKay correspondence. See also [Knörrer85] and Section 4. We will show a new way of understanding the link (the McKay correspondence) in Sections 8–16. Quivers of finite type appear in the course of this, which provides a link from (b) to (d) alongside the link from (b) to (a). This path has already been found in [Kronheimer89] in a slightly different manner.

For a given pair of II_1 factors one can construct a tower of II_1 factors by a certain procedure which specialists call mirror image transformations. In order to have an ADE classification we had better look at the same tower construction for a pair of semisimple algebras (semisimple algebras over $\mathbb{C}$ are sums of matrix algebras). In the tower of semisimple algebras the initial pair $N \subset M$ is described as a representation of an ADE quiver, while the rest of the tower is generated automatically from this. Therefore the link between (d) and (f) is firmly established, though the subfactors are only possible with the exception of D_{odd} and E_7. The link between (e) and (f) does not seem to be perfectly understood. See [EK98].

Infinite dimensional Heisenberg/Clifford algebras and their representations on Fock space enter the theory of Hilbert schemes. See [Nakajima96b], [Grojnowski96] and Section 6. This strongly suggests as yet unrevealed relations between the theory of Hilbert schemes with modular invariant partitions and II_1 (sub)factors.

The most desirable outcome would be a theory in which all six kinds of objects (a)–(f) arise naturally in various forms from one and the same object, for instance, from a finite subgroup of $SL(2, \mathbb{C})$.

5.2   Problems

The following problems are worth further investigation.

1. What are the Coxeter exponents and the Coxeter number for a finite subgroup of $SL(2, \mathbb{C})$, and why? (It is known that the Coxeter number equals the largest degree of the three homogeneous generators of the G-invariant polynomial ring. But why?)

2. What are the multiplicities of the highest weight for (e) and (f)?

3. Why do indices other than Coxeter exponents appear in Table 3 of Theorem 2.7?

4. The link from (b) to (c)? Can we recover the Lie algebras?

5. The link from (a) to (c)? Can we recover the Lie algebras?

6. The links from (b) to (e) and (f)?

7. Theorem 2.9 and Theorem 3.3 hint at an ADE classification of $c < 1$ minimal unitary series. If so, what do they look like? What is the link from (e) to (f) via this route?

6   Hilbert schemes of n points

6.1   Existence and projectivity

Let X be a projective scheme over $\mathbb{C}$. The *n-point Hilbert scheme* Hilb_X^n is by definition the universal scheme parametrizing all zero dimensional subschemes $Z \subset X$ such that $h^0(Z, \mathcal{O}_Z) = \dim(\mathcal{O}_Z) = n$. A zero dimensional subscheme $Z \in \operatorname{Hilb}_X^n$ has a defining ideal $I \subset \mathcal{O}_X$ that fits in an exact sequence

$$0 \to I \to \mathcal{O}_X \to \mathcal{O}_Z \to 0.$$

Thus, set theoretically,

$$\operatorname{Hilb}_X^n = \{Z \subset X; \dim(\mathcal{O}_Z) = n\}$$
$$\simeq \{I \subset \mathcal{O}_X; I \text{ an ideal of } \mathcal{O}_X, \dim(\mathcal{O}_X/I) = n\}.$$

See [Mumford], Lectures 3–4 or Grothendieck [FGA], Exposé 221 for an explanation of Hilbert schemes and a general treatment of their universal properties. A theorem of Grothendieck [FGA], Exposé 221 guarantees the existence of Hilbert schemes in a fairly general context; we give an elementary proof that Hilb_X^n exists and is a projective scheme, following suggestions of Y. Miyaoka and M. Reid.

Let $\mathcal{O}_X(1)$ be a very ample invertible sheaf on X defining an embedding $X \hookrightarrow \mathbb{P}^N$, and set $\mathcal{O}_X(m) := \mathcal{O}_X(1)^{\otimes m}$. We prove first that Hilb_X^n for fixed n can be viewed as a subscheme of the Grassmann variety of codimension n vector subspaces of $H^0(X, \mathcal{O}_X(n))$.

Lemma 6.2 *Let $Z \subset X \subset \mathbb{P}^N$ be a zero dimensional subschemes of degree n. Then*

(i) *The restriction map $r_Z \colon H^0(\mathcal{O}_X(m)) \to \mathcal{O}_Z(m) \simeq \mathcal{O}_Z$ is surjective for any $m \geq n-1$;*

(ii) *$I\mathcal{O}_X(m)$ is generated by its H^0 for any $m \geq n$.*

Proof   Write $\operatorname{Supp} Z = \{P_1, \ldots, P_s\}$, and $\deg_{P_i} Z = n_i$, so that $\sum n_i = n$. Now for each P_i, the map

$$r_i \colon H^0(\mathbb{P}^N, \mathcal{O}(m)) \to \mathcal{O}_{\mathbb{P}^N}/m_{P_i}^{n_i}$$

is surjective for any $m \geq n_i - 1$. Moreover, for $k \geq n_i$, the kernel of r_i contains forms not vanishing at any given point $Q \neq P_i$. This is obvious, because, if P_i is taken as the centre of inhomogeneous coordinates, then $\mathcal{O}_{\mathbb{P}^N}/m_{P_i}^{n_i}$ is just the vector space of polynomials of degree $\leq n_i - 1$. Clearly $\mathcal{O}_{\mathbb{P}^N}/m_{P_i}^{n_i} \to \mathcal{O}_{Z,P_i}$ is also surjective.

The lemma now follows on taking the product of forms of degree $\geq n_i$. □

Corollary 6.3 *Let X be a projective scheme and $\mathcal{O}_X(1)$ a very ample line bundle on X. Then $\operatorname{Hilb}^n X$ is a closed subscheme of the Grassmann variety of codimension n subspaces of $H^0(\mathcal{O}_X(n))$.*

Proof It is not hard to see that a subspace $V \subset H^0(\mathcal{O}_X(n))$ of codimension n generates a subsheaf $\mathcal{O}_X \cdot V = I(n) \subset \mathcal{O}_X(n)$ with $\dim(\mathcal{O}_X/I) = n$ if and only if the map $V \otimes H^0(\mathcal{O}_X(1)) \to H^0(\mathcal{O}_X(n+1))$ also has corank n. (This is the condition that V is closed under multiplication by linear forms.) This condition clearly defines a Zariski closed subscheme of the Grassmann variety. The alternative proof of the corollary uses the standard flattening stratifications of [Mumford], Lecture 8. □

The construction of $\operatorname{Hilb}^n X$ in Corollary 6.3 makes clear that $X \times \operatorname{Hilb}^n X$ has a sheaf of ideals I defining a 0-dimensional subscheme $Z^n \subset X \times \operatorname{Hilb}^n X$ satisfying the following universality property, a special case of a theorem of Grothendieck [FGA], Exposé 221. We will use this theorem to determine the precise structure of Hilb_X^G defined in Section 8.

Theorem 6.4 (existence and universality of Hilb_X^n) *Let X be a projective scheme and n any positive integer. Then there exists a projective scheme Hilb_X^n (possibly with finitely many irreducible components) and a universal proper flat family $\pi_{\mathrm{univ}} \colon Z^n \to \operatorname{Hilb}_X^n$ of zero dimensional subschemes of X such that:*

1. *any fibre of π_{univ} belongs to Hilb_X^n;*

2. *$Z_t^n = Z_s^n$ if and only if $t = s$, where $Z_t^n := \pi_{\mathrm{univ}}^{-1}(t)$ for $t \in \operatorname{Hilb}_X^n$;*

3. *given any flat family $\pi \colon Y \to S$ of zero dimensional subschemes of X with length n, there exists a unique morphism $\varphi \colon S \to \operatorname{Hilb}_X^n$ such that $(Y, \pi) \simeq \varphi^*(Z^n, \pi_{\mathrm{univ}})$.*

Let U be an open subscheme of X. Then Hilb_U^n is an open subscheme of Hilb_X^n consisting of the subschemes of X with support contained in U. We call Hilb_U^n the *n-point Hilbert scheme of U*.

6.5   Hilbert–Chow morphism

Write $S^n(\mathbb{A}^2)$ for the nth symmetric product of the affine plane $\mathbb{A}^2$. This is by definition the quotient of the products of n copies of $\mathbb{A}^2$ by the natural permutation action of the symmetric group S_n on n letters. It is the set of formal sums of n points, in other words, the set of unordered n-tuples of points.

We call $\mathrm{Hilb}^n(\mathbb{A}^2)$ the *Hilbert scheme of n points in* $\mathbb{A}^2$. It is a quasiprojective scheme of dimension $2n$. Any $Z \in \mathrm{Hilb}^n(\mathbb{A}^2)$ is a zero dimensional subscheme with $h^0(Z, \mathcal{O}_Z) = \dim(\mathcal{O}_Z) = n$. Suppose that Z is reduced. Then Z is a union of n distinct points. Since being reduced is an open and generic condition, $\mathrm{Hilb}^n(\mathbb{A}^2)$ contains a Zariski open subset consisting of formal sums of n distinct points. This is why we call $\mathrm{Hilb}^n(\mathbb{A}^2)$ the Hilbert scheme of n points on $\mathbb{A}^2$.

We have a natural morphism π from $\mathrm{Hilb}^n(\mathbb{A}^2)$ onto $S^n(\mathbb{A}^2)$ defined by

$$\pi \colon Z \mapsto \sum_{p \in \mathrm{Supp}(Z)} \dim(\mathcal{O}_{Z,p}) p.$$

We call π the *Hilbert–Chow morphism* (of $\mathbb{A}^2$). Let D be the subset of $S^n(\mathbb{A}^2)$ consisting of formal sums of n points with at least two coincident points. It is clear that π is the identity over $S^n(\mathbb{A}^2) \setminus D$, hence is birational. If $n = 2$ and if Z is nonreduced with $\mathrm{Supp}(Z)$ the origin, then Z is a subscheme defined by the ideal

$$I = (ax + by, x^2, xy, y^2), \quad \text{where} \quad (a, b) \neq (0, 0).$$

Thus the set of these subschemes is $\mathbb{P}^1$ parametrizing the ratios $a : b$. It follows that $\mathrm{Hilb}^2(\mathbb{A}^2)$ is the quotient by the symmetric group S_2 of the blowup of the nonsingular fourfold $\mathbb{A}^2 \times \mathbb{A}^2$ along the diagonal $\mathbb{A}^2$. For all n there is a relatively simple description, due to Barth, of $\mathrm{Hilb}^n_{\mathbb{A}^2}$ as a scheme, in terms of monads. See [OSS80] and [Nakajima96b], Chapter 2. We write some of these down explicitly in Sections 12–16.

One of the most remarkable features of $\mathrm{Hilb}^n(\mathbb{A}^2)$ is the following result.

Theorem 6.6 ([Fogarty68]) $\mathrm{Hilb}^n(\mathbb{A}^2)$ *is a smooth quasiprojective scheme, and* $\pi \colon \mathrm{Hilb}^n(\mathbb{A}^2) \to S^n(\mathbb{A}^2)$ *is a resolution of singularities of the symmetric product.*

A simpler proof of Theorem 6.6 is given in [Nakajima96b]. We note that smoothness of $\mathrm{Hilb}^n(\mathbb{A}^2)$ is peculiar to $\dim \mathbb{A}^2 = 2$. If $k \geq 3$, then a subscheme $Z \subset \mathbb{A}^k$ can be very complicated in general [Göttsche91]. See [Iarrobino77], [Briançon77]. [Göttsche91], p. 60 writes that $\mathrm{Hilb}^n(\mathbb{A}^k)$ is known to be singular for $k \geq 3$ and $n \geq 4$, while it is smooth for any k if $n = 3$. $\mathrm{Hilb}^n(\mathbb{A}^k)$ is connected for any n and k by [Fogarty68], while it is reducible, hence singular for any k and any large $n \gg k$ by [Iarrobino72].

Besides smoothness, $\text{Hilb}^n(\mathbb{A}^2)$ has various mysterious nice properties. Among others, the following is relevant to our subsequent study of $\text{Hilb}^G(\mathbb{A}^2)$.

Theorem 6.7 ([Beauville83]) $\text{Hilb}^n(\mathbb{A}^2)$ *admits a holomorphic symplectic structure.*

Proof See also [Fujiki83] for $n = 2$, and [Mukai84] for a more general case. The sketch proof below, mostly taken from [Beauville83], shows that the theorem also holds for $\text{Hilb}^n(S)$ if S is a smooth complex surface with a nowhere vanishing holomorphic two form. Let ω be a nowhere vanishing closed holomorphic 2-form on $S := \mathbb{A}^2$, say $dx \wedge dy$ in terms of the linear coordinates on S. The product S^n of n copies of S has the holomorphic 2-form $\psi := \sum_{i=1}^n p_i^*(\omega)$, where p_i is the ith projection. We show that ψ induces a symplectic form on $S^{[n]} := \text{Hilb}^n(S)$.

We write $S^{(n)} = \text{S}^n(S)$ for the nth symmetric product of S, that is, by definition, the quotient of the products of n copies of S by the natural permutation action of the symmetric group S_n on n letters. Let $\varepsilon \colon S^n \to S^{(n)}$ be the natural morphism. Let D_* be the open subset of D consisting of all 0-cycles of the form $2x_1 + x_2 + \cdots + x_{n-1}$ with all the x_i distinct. We set $S_*^{(n)} := S^{(n)} \setminus (D \setminus D_*)$, $S_*^{[n]} = \pi^{-1}(S_*^{(n)})$, $S_*^n := \varepsilon^{-1}(S_*^{(n)})$ and $\Delta_* = \varepsilon^{-1}(D_*)$. Then Δ_* is smooth and of codimension 2 in S_*^n. Then by [Beauville83], p. 766, $S_*^{[n]}$ is isomorphic to the quotient of the blowup of $\text{Bl}_{\Delta_*}(S_*^{(n)})$ of $S_*^{(n)}$ along Δ_* by the symmetric group S_n. Hence we have a natural morphism $\rho \colon \text{Bl}_D(S_*^{(n)}) \to S_*^{[n]}$. We see easily that ψ induces a holomorphic 2-form φ on $S_*^{[n]}$, which extends to $S^{[n]}$ because the codimension of the inverse image of $S^{[n]} \setminus S_*^{[n]}$ in $S^{[n]}$ is greater than one.

Let E_* be the inverse image of Δ_* in $\text{Bl}_{\Delta_*}(S_*^{(n)})$. Then the canonical bundle of $\text{Bl}_{\Delta_*}(S_*^{(n)})$ is E_*, because that of S^n is trivial. On the other hand, it is the sum of the divisor $\rho^*(\varphi^n)$ and the ramification divisor R of ρ. Since $R = E_*$ on $\text{Bl}_{\Delta_*}(S_*^{(n)})$, we see that $(\varphi)^n$ is everywhere nonvanishing on $S_*^{[n]}$, hence also on $S^{[n]}$ [Beauville83]. Thus φ is a nowhere degenerate 2-form, that is, a holomorphic symplectic form on $S^{[n]}$. $\quad\square$

Definition 6.8 The *infinite dimensional Heisenberg algebra* $\mathfrak{s}$ is by definition the Lie algebra generated by p_i, q_i for $i \geq 1$ and c, subject to the relations

$$[p_i, q_j] = c\delta_{ij}, \quad [p_i, p_j] = [q_i, q_j] = [p_i, c] = [q_i, c] = 0.$$

It is known that for any $a \in \mathbb{C}^*$, the Lie algebra $\mathfrak{s}$ has the *canonical commutation relations representation* σ_a on the *Fock space* $R := \mathbb{C}[x_1, x_2, \ldots]$, that is, the ring of polynomials in infinitely many indeterminates x_i; the representation is defined by

$$\sigma_a(p_i) = a\frac{\partial}{\partial x_i}, \quad \sigma_a(q_i) = x_i, \quad \sigma_a(c) = a \cdot \mathrm{id}_R.$$

We denote this $\mathfrak{s}$-module by R_a. We also define a derivation d_0 of $\mathfrak{s}$ by

$$[d_0, q_i] = iq_i, \quad [d_0, p_i] = -ip_i, \quad [d_0, c] = 0.$$

The following fact is important (see [Kac90], pp. 162–163):

Theorem 6.9 *An irreducible $\mathfrak{s}$-module with generator v_0 is isomorphic to R_a if $p_i(v_0) = 0$ for all i and $c(v_0) = av_0$ for some $a \neq 0$. The character of R_a is given by*

$$\mathrm{Tr}_{R_a}(q^{d_0}) = \prod_{i=1}^{\infty}(1 - q^i)^{-1}.$$

The vector v_0 in the above theorem is called a *vacuum vector* of V. We quote one of the surprising results of [Nakajima96b].

Theorem 6.10 *Let $\mathfrak{s}$ be the infinite dimensional Heisenberg algebra. Then the direct sum of all the cohomology groups $\bigoplus_{n \geq 0} H^*(\mathrm{Hilb}^n(\mathbb{A}^2), \mathbb{C})$ is an irreducible $\mathfrak{s}$-module with $a = 1$ whose vacuum vector v_0 is a generator of $H^0(\mathrm{Hilb}^0(\mathbb{A}^2), \mathbb{C})$.*

By Theorem 6.9, the above theorem gives in a sense the complete structure of the $\mathfrak{s}$-module. However we should mention that its irreducibility follows from comparison with the following Theorem 6.11.

[Nakajima96b] derives a similar conclusion when $\mathbb{A}^2$ is replaced by a smooth quasiprojective complex surface X. Then $\bigoplus_{n \geq 0} H^*(\mathrm{Hilb}^n(X), \mathbb{C})$ is an infinite dimensional Heisenberg/Clifford algebra module. Its irreducibility again follows from Theorem 6.11.

Cell decompositions of $\mathrm{Hilb}^n(\mathbb{P}^2)$ and $\mathrm{Hilb}^n(\mathbb{A}^2)$, and hence complete formulas for the Betti numbers of $\mathrm{Hilb}^n(\mathbb{P}^2)$ and $\mathrm{Hilb}^n(\mathbb{A}^2)$, are known by Ellingsrud and Strømme [ES87]. The formulas for the Betti numbers of $\mathrm{Hilb}^n(\mathbb{P}^2)$ and $\mathrm{Hilb}^n(\mathbb{A}^2)$ are written by [Göttsche91] more generally in the following beautiful manner.

To state the theorem, we define the Poincaré polynomial $p(X, z)$ of a smooth complex variety X by $p(X, z) := \sum_{i=0}^{\infty} \dim H^i(X, \mathbb{Q})z^i$. Moreover we define $p(X, z, t) := \sum_{n=0}^{\infty} p(\mathrm{Hilb}^n(X), z)t^n$ for a smooth complex surface X.

Theorem 6.11 ([Göttsche91]) *Let X be a smooth projective complex surface. Then*

$$p(X, z, t) = \prod_{m=1}^{\infty} \frac{(1 + z^{2m-1}t^m)^{b_1(X)}(1 + z^{2m+1}t^m)^{b_3(X)}}{(1 - z^{2m-2}t^m)^{b_0(X)}(1 - z^{2m}t^m)^{b_2(X)}(1 - z^{2m+2}t^m)^{b_4(X)}},$$

where $b_i(S)$ is the ith Betti number of S.

7 Three dimensional quotient singularities

7.1 Classification of finite subgroups of $\mathrm{SL}(3, \mathbb{C})$

Threefold Gorenstein quotient singularities have attracted the attention of both mathematicians and physicists in connection with Calabi–Yau threefolds, mirror symmetry and superstring theory. For a finite subgroup G of $\mathrm{GL}(n, \mathbb{C})$, the quotient $\mathbb{A}^n/G$ is Gorenstein if and only if $G \subset \mathrm{SL}(n, \mathbb{C})$; see [Khinich76] and [Watanabe74].

Now we review the classification of finite subgroups of $\mathrm{SL}(3, \mathbb{C})$ from the very classical works of [Blichfeldt17], and Miller, Blichfeldt and Dickson [MBD16]. In these works they nearly completed the classification of finite subgroups of $\mathrm{SL}(3, \mathbb{C})$ up to conjugacy. Unfortunately, however, there were two missing classes, which were supplemented later by Stephen S.-T. Yau and Y. Yu [YY93], p. 2.

There is an obvious series of finite subgroups coming from subgroups of $\mathrm{GL}(2, \mathbb{C})$. In fact, associating $(\det g)^{-1} \oplus g$ to each $g \in \mathrm{GL}(2, \mathbb{C})$, we have a finite subgroup of $\mathrm{SL}(3, \mathbb{C})$ for any subgroup of $\mathrm{GL}(2, \mathbb{C})$. Including this series, there are exactly four infinite series of finite subgroups of $\mathrm{SL}(3, \mathbb{C})$:

1. diagonal Abelian groups;

2. groups coming from finite subgroups in $\mathrm{GL}(2, \mathbb{C})$;

3. groups generated by (1) and T;

4. groups generated by (3) and Q.

Here

$$T = \begin{pmatrix} 0 & 1 & 0 \\ 0 & 0 & 1 \\ 1 & 0 & 0 \end{pmatrix}, \quad Q = \frac{1}{\sqrt{-3}} \begin{pmatrix} 1 & 1 & 1 \\ 1 & \omega & \omega^2 \\ 1 & \omega^2 & \omega \end{pmatrix}, \quad \text{where } \omega := e^{2\pi\sqrt{-1}/3}.$$

There are exactly eight sporadic classes, each of which contains a unique finite subgroup up to conjugacy, of order 108, 216, 648, 60, 168, 180, 504 and

1080 respectively. Only two finite simple groups appear: A_5 ($\simeq$ PSL$(2, \mathbb{F}_5)$) of order 60, and PSL$(2, \mathbb{F}_7)$ of order 168.

The subgroup PSL$(2, \mathbb{F}_7)$ of SL$(3, \mathbb{C})$ is the automorphism group of the Klein quartic curve $x_0^3 x_1 + x_1^3 x_2 + x_2^3 x_0 = 0$. On the other hand, A_5 is realized as a subgroup of SL$(3, \mathbb{C})$ as follows. Let G be the binary icosahedral subgroup of SL$(2, \mathbb{C})$ of order 120 (compare Section 16). This acts on the space of polynomials of homogeneous degree two on $\mathbb{A}^2$, with $\pm 1 \in G$ acting trivially. Therefore this is an irreducible representation of $G/\{\pm 1\}$ ($\simeq A_5$) of rank three. This realizes A_5 as a finite subgroup of SL$(3, \mathbb{C})$. Or, more simply, $A_5 \subset$ SO(3) is the group of automorphisms of the icosahedron.

In the case of order 108, the quotient $\mathbb{A}^3/G$ is a complete intersection defined by two equations, while it is a hypersurface in the remaining seven cases. The defining equations are completely known; in contrast with the two dimensional case, they are not all weighted homogeneous. The weighted homogeneous ones are the cases of order 108, 648, 60, 180 and 1080 [YY93].

All finite subgroups of GL$(2, \mathbb{C})$ are known by Behnke and Riemenschneider [BR95]. We note that in the easiest series (1) the quotients are torus embeddings. Therefore their smooth resolutions are constructed through torus embeddings. See [Roan89].

Outstanding in this area is the following theorem, which generalizes the two dimensional McKay correspondence to some extent.

Theorem 7.2 *For any finite subgroup G of* SL$(3, \mathbb{C})$*, there exists a smooth resolution X of the quotient $\mathbb{A}^3/G$ such that the canonical bundle of X is trivial (X is then called a* crepant *resolution of $\mathbb{A}^3/G$). For any such resolution X, $H^*(X, \mathbb{Z})$ is a free $\mathbb{Z}$-module of rank equal to the number of the conjugacy classes of G.*

[Ito95a], [Ito95b], [Markushevich92], [Roan94] and [Roan96] contributed to the proof of this theorem. It seems desirable to simplify the proofs for the complicated sporadic classes. Ito and Reid [IR96] generalized the theorem and sharpened it especially in dimension three by finding a bijective correspondence between irreducible exceptional divisors of the resolution and conjugacy classes of G (called *junior*) with certain type of eigenvalues: they defined the notion of *age* of a conjugacy class; the junior conjugacy classes are those of age equal to one. The junior ones play a more important role in the study of crepant resolutions.

8 Hilbert schemes and simple singularities: Introduction

The second half of the article starts here. In it, we study the link from (b) to (a).

8.1 Abstract

For any finite subgroup G of $SL(2,\mathbb{C})$ of order n, we consider the *G-orbit Hilbert scheme*, namely, a certain subscheme $\mathrm{Hilb}^G(\mathbb{A}^2)$ of $\mathrm{Hilb}^n(\mathbb{A}^2)$ that parametrizes G-invariant subschemes. We first give a direct proof, independent of the classification of finite subgroups of $SL(2,\mathbb{C})$, that $\mathrm{Hilb}^G(\mathbb{A}^2)$ is a minimal resolution of a simple singularity $\mathbb{A}^2/G$. Any point of the exceptional set E is a G-invariant 0-dimensional subscheme Z of $\mathbb{A}^2$ with support the origin. Let I be the ideal sheaf defining Z. Then I is an infinite dimensional G-module. Dividing it by a natural G-submodule of I gives a finite G-module $V(I)$, which turns out to be either an irreducible G-module or the sum of two inequivalent irreducible G-modules. This gives the McKay correspondence as described in Section 4.

8.2 Summary of main results

We explain in a little more detail. Let $S^n(\mathbb{A}^2)$ be the nth symmetric product of $\mathbb{A}^2$ (that is, the Chow variety $\mathrm{Chow}^n(\mathbb{A}^2)$), and $\mathrm{Hilb}^n(\mathbb{A}^2)$ the Hilbert scheme of n points of $\mathbb{A}^2$. By Theorems 6.6 and 6.7, $\mathrm{Hilb}^n(\mathbb{A}^2)$ is a crepant resolution of $S^n(\mathbb{A}^2)$ with a holomorphic symplectic structure.

Let G be an arbitrary finite subgroup of $SL(2,\mathbb{C})$; it acts on $\mathbb{A}^2$, and therefore has a canonical action on both $\mathrm{Hilb}^n(\mathbb{A}^2)$ and $S^n(\mathbb{A}^2)$. Now we consider the particular case where n equals the order of G. Then it is easy to see that the G-fixed point set $S^n(\mathbb{A}^2)^G$ in $S^n(\mathbb{A}^2)$ is isomorphic to the quotient $\mathbb{A}^2/G$. The G-fixed point set $\mathrm{Hilb}^n(\mathbb{A}^2)^G$ in $\mathrm{Hilb}^n(\mathbb{A}^2)$ is always nonsingular, but could a priori be disconnected. There is however a unique irreducible component of $\mathrm{Hilb}^n(\mathbb{A}^2)^G$ dominating $S^n(\mathbb{A}^2)^G$, which we denote by $\mathrm{Hilb}^G(\mathbb{A}^2)$. Since $\mathrm{Hilb}^G(\mathbb{A}^2)$ inherits a holomorphic symplectic structure from $\mathrm{Hilb}^n(\mathbb{A}^2)$, $\mathrm{Hilb}^G(\mathbb{A}^2)$ is a crepant (that is, minimal) resolution of $\mathbb{A}^2/G$ (see Theorem 9.3).

Our aim in this part is to study in detail the structure of $\mathrm{Hilb}^G(\mathbb{A}^2)$ using representations of G defined in terms of spaces of homogeneous polynomials or symmetric tensors.

Let $\mathfrak{m}$ (respectively $\mathfrak{m}_S$) be the maximal ideal of the origin of $\mathbb{A}^2$ (respectively $S := \mathbb{A}^2/G$) and set $\mathfrak{n} = \mathfrak{m}_S \mathcal{O}_{\mathbb{A}^2}$. A point $\mathfrak{p}$ of $\mathrm{Hilb}^G(\mathbb{A}^2)$ is a G-invariant

0-dimensional subscheme Z of $\mathbb{A}^2$, and to it we associate the G-invariant ideal subsheaf I defining Z, and the exact sequence

$$0 \to I \to \mathcal{O}_{\mathbb{A}^2} \to \mathcal{O}_Z \to 0.$$

We assume that $\mathfrak{p}$ is in the exceptional set E of $\mathrm{Hilb}^G(\mathbb{A}^2)$; since G acts freely outside the origin, Z is then supported at the origin, and $I \subset \mathfrak{m}$. As is easily shown, I contains $\mathfrak{n}$ (Corollary 9.6). Let $V(I) := I/(\mathfrak{m}I + \mathfrak{n})$. The finite G-module $V(I)$ is isomorphic to a minimal G-submodule of $I/\mathfrak{n}$ generating the $\mathcal{O}_{\mathbb{A}^2}$-module $I/\mathfrak{n}$.

If $\mathfrak{p}$ is a smooth point of E, we prove that $V(I)$ is a nontrivial irreducible G-module; while if $\mathfrak{p} \in E$ is a singular point, $V(I)$ is the direct sum of two inequivalent nontrivial irreducible G-modules. For any equivalence class of a nontrivial irreducible G-module ρ we define the subset $E(\rho)$ of E consisting of all $I \in \mathrm{Hilb}^G(\mathbb{A}^2)$ such that $V(I)$ contains ρ as a G-submodule. We will see that $E(\rho)$ is naturally identified with the set of all nontrivial proper G-submodules of $\rho^{\oplus 2}$, which is isomorphic to a smooth rational curve by Schur's lemma (Theorem 10.7). The map $\rho \mapsto E(\rho)$ gives a bijective correspondence (Theorem 10.4) between the set $\mathrm{Irr}\,G$ of all the equivalence classes of irreducible G-modules and the set $\mathrm{Irr}\,E$ of all the irreducible components of E, which turns out to be the classical McKay correspondence [McKay80].

We also give an explanation of why it is that tensoring by the natural representation appears as the key ingredient in the McKay correspondence. An outline of the story is given in Section 13.5. The most remarkable point, in addition to the McKay correspondence itself, is that there are two kinds of dualities (Theorems 10.6 and 12.4) in the G-module decomposition of the algebra $\mathfrak{m}/\mathfrak{n}$. (After completing the present work, we were informed by Shinoda that the dualities also follow from [Steinberg64].) It is the second duality (for instance, Theorem 10.6) that explains why tensoring by the natural representation appears in the McKay correspondence.

Our results hold also in characteristic p provided that the ground field k is algebraically closed and the order of G is coprime to p.

The research part of the article is organized as follows. In Section 9 we prove that $\mathrm{Hilb}^G(\mathbb{A}^2)$ is a crepant (or minimal) resolution of $\mathbb{A}^2/G$. We also give some elementary lemmas on representations of finite groups. In Section 10 we formulate our main theorem and relevant theorems. We give a complete description of the ideals corresponding to the points of the exceptional set E. In Section 11 we prove the dualities independently of the classification of finite subgroups of $\mathrm{SL}(2, \mathbb{C})$. In Sections 12–16 we study $\mathrm{Hilb}^G(\mathbb{A}^2)$ and prove the main theorem separately in the cases A_n, D_n, E_6, E_7 and E_8 respectively.

In Section 17, we raise some unsolved questions.

9   The crepant (minimal) resolution

Lemma 9.1 *Let G be a finite subgroup of $\mathrm{GL}(2,\mathbb{C})$, and $\mathrm{Hilb}^n(\mathbb{A}^2)^G$ the subset of $\mathrm{Hilb}^n(\mathbb{A}^2)$ consisting of all points fixed by G. Then $\mathrm{Hilb}^n(\mathbb{A}^2)^G$ is nonsingular.*

Proof By Theorem 6.6, $\mathrm{Hilb}^n(\mathbb{A}^2)$ is nonsingular. Let $\mathfrak{p}$ be a point of $\mathrm{Hilb}^n(\mathbb{A}^2)^G$. The action of G on $\mathrm{Hilb}^n(\mathbb{A}^2)$ at $\mathfrak{p}$ is linearized; in other words we see that there exist local parameters x_i of $\mathrm{Hilb}^n(\mathbb{A}^2)$ at $\mathfrak{p}$ and some constants $a_{ij}(g) \in \mathbb{C}$ such that $g^*x_i = \sum a_{ij}(g)x_j$ for any $g \in G$. The fixed locus $\mathrm{Hilb}^n(\mathbb{A}^2)^G$ at $\mathfrak{p}$ is by definition the reduced subscheme of $\mathrm{Hilb}^n(\mathbb{A}^2)^G$ defined by $x_i - \sum a_{ij}(g)x_j = 0$ for all $g \in G$. Hence it is nonsingular. $\square$

Lemma 9.2 *Let G be a finite subgroup of $\mathrm{SL}(2,\mathbb{C})$ of order n, and $S^n(\mathbb{A}^2)^G$ the subset of $S^n(\mathbb{A}^2)$ consisting of all points of $S^n(\mathbb{A}^2)$ fixed by G. Then $S^n(\mathbb{A}^2)^G \simeq \mathbb{A}^2/G$.*

Proof Let $0 \neq \mathfrak{q} \in \mathbb{A}^2$ be a point. Then since $\mathfrak{q}$ is not fixed by any element of G other than the identity, the set $G \cdot \mathfrak{q} := \{g(\mathfrak{q}); g \in G\}$ determines a point in $S^n(\mathbb{A}^2)^G$. Conversely, any point of $S^n(\mathbb{A}^2)^G$ is an unordered G-invariant set Σ in $\mathbb{A}^2$. If Σ contains a point $\mathfrak{q} \neq 0$, it must contain the set $G \cdot \mathfrak{q}$. Since $|\Sigma| = n = |G|$, we have $\Sigma = G \cdot \mathfrak{q}$. Note $G \cdot \mathfrak{q} = G \cdot \mathfrak{q}'$ for a pair of points $\mathfrak{q}, \mathfrak{q}' \neq 0$ if and only if $\mathfrak{q}' \in G \cdot \mathfrak{q}$. Therefore we have the isomorphism $S^n(\mathbb{A}^2 \setminus \{0\})^G \simeq (\mathbb{A}^2 \setminus \{0\})/G$, which extends naturally to a bijective morphism of $S^n(\mathbb{A}^2)^G$ onto $\mathbb{A}^2/G$. It follows that $S^n(\mathbb{A}^2)^G \simeq \mathbb{A}^2/G$ because $\mathbb{A}^2/G$ is normal. $\square$

Theorem 9.3 *Let $G \subset \mathrm{SL}(2,\mathbb{C})$ be a finite subgroup of order n. Then there is a unique irreducible component $\mathrm{Hilb}^G(\mathbb{A}^2)$ of $\mathrm{Hilb}^n(\mathbb{A}^2)^G$ dominating $\mathbb{A}^2/G$, which is a crepant (or equivalently a minimal) resolution of $\mathbb{A}^2/G$.*

Proof The Hilbert–Chow morphism of $\mathrm{Hilb}^n(\mathbb{A}^2)$ onto $S^n(\mathbb{A}^2)$ is defined by $\pi(Z) = \mathrm{Supp}(Z)$ (counted with the appropriate multiplicities) for a zero dimensional subscheme Z of $\mathbb{A}^2$. Since $\mathrm{Hilb}^n(\mathbb{P}^2)$ is a projective scheme by Theorem 6.4, the Hilbert–Chow morphism of $\mathrm{Hilb}^n(\mathbb{P}^2)$ is proper. Hence the Hilbert–Chow morphism of $\mathrm{Hilb}^n(\mathbb{A}^2)$ is proper, because it is obtained by restricting the image variety $S^n(\mathbb{P}^2)$ to $S^n(\mathbb{A}^2)$. This induces a natural morphism of $\mathrm{Hilb}^G(\mathbb{A}^2)$ onto $S^n(\mathbb{A}^2)^G \simeq \mathbb{A}^2/G$. Any point of $S^n(\mathbb{A}^2)^G \setminus \{0\}$ is a G-orbit of a point $0 \neq \mathfrak{p} \in \mathbb{A}^2$, which is a reduced zero dimensional subscheme invariant under G. It follows that $\mathrm{Hilb}^G(\mathbb{A}^2)$ is birationally equivalent to $S^n(\mathbb{A}^2)^G$, so that it is a resolution of $S^n(\mathbb{A}^2)^G \simeq \mathbb{A}^2/G$.

By [Fujiki83], Proposition 2.6, $\text{Hilb}^G(\mathbb{A}^2)$ inherits a canonical holomorphic symplectic structure from $\text{Hilb}(\mathbb{A}^2)$. Since $\dim \text{Hilb}^G(\mathbb{A}^2) = \dim \mathbb{A}^2/G = 2$, this implies that the dualizing sheaf of $\text{Hilb}^G(\mathbb{A}^2)$ is trivial. This completes the proof. $\square$

Lemma 9.4 *Let G be a finite subgroup of $\text{GL}(n, \mathbb{C})$. Let S be a connected reduced scheme, and $\mathcal{I}$ an ideal of $\mathcal{O}_{\mathbb{A}^n \times S}$ such that $\mathcal{O}_{\mathbb{A}^n \times S}/\mathcal{I}$ is flat over S. Let $\mathcal{I}_s := \mathcal{I} \otimes \mathcal{O}_{\mathbb{A}^n \times \{s\}}$. Suppose that we are given a regular action of G on $\mathbb{A}^n \times S$ possibly depending nontrivially on S. If $\dim \text{Supp}(\mathcal{O}_{\mathbb{A}^n \times \{s\}}/\mathcal{I}_s) = 0$ for any $s \in S$, then the equivalence class of the G-module $\mathcal{O}_{\mathbb{A}^n \times \{s\}}/\mathcal{I}_s$ is independent of s.*

Proof By the assumption $h^1(\mathcal{O}_{\mathbb{A}^n \times \{s\}}/\mathcal{I}_s) = 0$. Thus $h^0(\mathcal{O}_{\mathbb{A}^n \times \{s\}}/\mathcal{I}_s)$ is constant on S, because $\chi(\mathcal{O}_{\mathbb{A}^n \times \{s\}}/\mathcal{I}_s)$ is constant by [Hartshorne77], Chap. III. Hence again by [ibid.] $\mathcal{O}_{\mathbb{A}^n \times S}/\mathcal{I}$ is a locally free sheaf of $\mathcal{O}_S$-modules of finite rank. Let $E := \mathcal{O}_{\mathbb{A}^n \times S}/\mathcal{I}$ and $\Delta(g, x) := \det(x \cdot \text{id} - T(g))$ be the characteristic polynomial of the action $T(g)$ of $g \in G$ on E. Clearly $\Delta(g, x)$ is independent of a local trivialization of the sheaf E. It follows that $\Delta(g, x) \in \text{Hom}(\det E, \det E)[x] \simeq \Gamma(\mathcal{O}_S)[x]$, the polynomial ring of x over $\Gamma(\mathcal{O}_S)$. Moreover coefficients of the polynomial $\Delta(g, x)$ in x are elementary symmetric polynomials of eigenvalues of $T(g)$. Since all the eigenvalues of $T(g)$ are nth roots of unity where $n = |G|$, coefficients of $\Delta(g, x)$ take values in a finite subset of $\mathbb{C}$ over S. Since S is connected and reduced, they are constant. It follows that $\Delta(g, x) \in \mathbb{C}[x]$. In particular the character $\text{Tr}\, T(g)$, the coefficient of x in $\Delta(g, x)$ is independent of $s \in S$. Since any finite G-module is uniquely determined up to equivalence by its character, the equivalence class of the G-module $\mathcal{O}_{\mathbb{A}^n \times \{s\}}/\mathcal{I}_s$ is independent of $s \in S$. $\square$

Corollary 9.5 *Let G be a finite subgroup of $\text{SL}(2, \mathbb{C})$, and I an ideal of $\mathcal{O}_{\mathbb{A}^2}$ with $I \in \text{Hilb}^G(\mathbb{A}^2)$. Then as G-modules $\mathcal{O}_{\mathbb{A}^2}/I \simeq \mathbb{C}[G]$, the regular representation of G.*

Corollary 9.6 *Let I be an ideal of $\mathcal{O}_{\mathbb{A}^2}$ with $I \in \text{Hilb}^G(\mathbb{A}^2)$. Any G-invariant function vanishing at the origin is contained in I.*

Proof $\mathcal{O}_{\mathbb{A}^2}/I \simeq \mathbb{C}[G]$ by Corollary 9.5. This implies that $\mathcal{O}_{\mathbb{A}^2}/I$ has a unique trivial G-submodule spanned by constant functions of $\mathbb{A}^2$. It follows that any G-invariant function vanishing at the origin is contained in I. $\square$

Remark 9.7 By [Nakajima96b], Theorem 4.4, for $I \in \text{Hilb}^n(\mathbb{A}^2)$, the following conditions are equivalent,

1. $I \in \mathrm{Hilb}^G(\mathbb{A}^2)$;

2. $\mathcal{O}_{\mathbb{A}^2}/I \simeq \mathbb{C}[G]$;

3. $\mathrm{Hom}_{\mathcal{O}_{\mathbb{A}^2}}(I, \mathcal{O}_{\mathbb{A}^2}/I)^G \neq 0$.

10 The Main Theorem

10.1 Stratification of $\mathrm{Hilb}^G(\mathbb{A}^2)$ by $\mathrm{Irr}\, G$

Let G be a finite subgroup of $\mathrm{SL}(2, \mathbb{C})$. As in 4.2, we write $\mathrm{Irr}\, G$ for the set of all the equivalence classes of nontrivial irreducible G-modules, and $\mathrm{Irr}_*\, G$ for the union of $\mathrm{Irr}\, G$ and the trivial one dimensional G-module. Let $V(\rho) \in \mathrm{Irr}\, G$ be a G-module, and $\rho\colon G \to \mathrm{GL}(V(\rho))$ the corresponding homomorphism.

Let $X = X_G := \mathrm{Hilb}^G(\mathbb{A}^2)$ and $S = S_G := \mathbb{A}^2/G$. Write $\mathfrak{m}$ (respectively $\mathfrak{m}_S$) for the maximal ideal of $\mathbb{A}^2$ (respectively S) at the origin 0, and set $\mathfrak{n} := \mathfrak{m}_S \mathcal{O}_{\mathbb{A}^2}$. Let $\pi\colon X \to S$ be the natural morphism and E the exceptional set of π. Let $\mathrm{Irr}\, E$ be the set of irreducible components of E. Any $I \in X$ contained in E (to be exact, the subscheme defined by I belongs to X) is a G-invariant ideal of $\mathcal{O}_{\mathbb{A}^2}$ which contains $\mathfrak{n}$ by Corollary 9.6. For any ρ, ρ', and $\rho'' \in \mathrm{Irr}\, G$, we define

$$V(I) := I/(\mathfrak{m}I + \mathfrak{n}),$$
$$E(\rho) := \left\{ I \in \mathrm{Hilb}^G(\mathbb{A}^2); V(I) \supset V(\rho) \right\},$$
$$P(\rho, \rho') := \left\{ I \in \mathrm{Hilb}^G(\mathbb{A}^2); V(I) \supset V(\rho) \oplus V(\rho') \right\},$$
$$Q(\rho, \rho', \rho'') := \left\{ I \in \mathrm{Hilb}^G(\mathbb{A}^2); V(I) \supset V(\rho) \oplus V(\rho') \oplus V(\rho'') \right\}.$$

Remark 10.2 Note that we allow $\rho = \rho'$ in the definition of $P(\rho, \rho')$. Of course if $\rho \neq \rho'$, then $P(\rho, \rho') = E(\rho) \cap E(\rho')$.

Definition 10.3 Two irreducible G-modules ρ and ρ' are said to be *adjacent* if $\rho \otimes \rho_{\mathrm{nat}}$ contains ρ', which happens if and only if $\rho' \otimes \rho_{\mathrm{nat}}$ contains ρ.

In fact, since $G \subset \mathrm{SL}(2, \mathbb{C})$, we have $\chi_{\mathrm{nat}}(x^{-1}) = \chi_{\mathrm{nat}}(x)$ for all $x \in G$ where $\chi_{\mathrm{nat}} := \mathrm{Tr}(\rho_{\mathrm{nat}})$. Hence for any characters χ and χ' of G

$$(\chi\chi_{\mathrm{nat}}, \chi') = (1/|G|) \sum_{x \in G} \chi(x)\chi_{\mathrm{nat}}(x)\chi'(x^{-1}) = (\chi, \chi'\chi_{\mathrm{nat}}).$$

Thus the multiplicity of ρ' in $\rho \otimes \rho_{\mathrm{nat}}$ equals that of ρ in $\rho' \otimes \rho_{\mathrm{nat}}$.

The *Dynkin diagram* $\Gamma(\mathrm{Irr}\, G)$ or the *extended Dynkin diagram* $\Gamma(\mathrm{Irr}_*\, G)$ of G is the graph whose vertices are $\mathrm{Irr}\, G$ or $\mathrm{Irr}_*\, G$ respectively, with ρ and ρ' joined by a simple edge if and only if ρ and ρ' are adjacent.

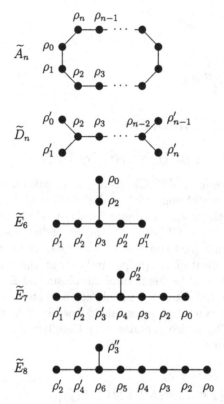

Figure 5: The extended Dynkin diagrams and representations

Then our main theorem is stated as follows.

Theorem 10.4 *Let G be a finite subgroup of* $\mathrm{SL}(2, \mathbb{C})$. *Then*

1. *the map $\rho \mapsto E(\rho)$ is a bijective correspondence between* $\mathrm{Irr}\, G$ *and* $\mathrm{Irr}\, E$;

2. *$E(\rho)$ is a smooth rational curve with $E(\rho)^2 = -2$ for any $\rho \in \mathrm{Irr}\, G$;*

3. *$P(\rho, \rho') \neq \emptyset$ if and only if ρ and ρ' are adjacent. In this case $P(\rho, \rho')$ is a single (reduced) point, at which $E(\rho)$ and $E(\rho')$ intersect transversally;*

4. *$P(\rho, \rho) = Q(\rho, \rho', \rho'') = \emptyset$ for any $\rho, \rho', \rho'' \in \mathrm{Irr}\, G$.*

In the A_n case, Theorem 10.4 follows from Theorem 9.3 and the theorems in Section 12; in the other cases, it follows from Theorem 9.3, Theorem 10.7 and Remark 10.8.

By Theorem 10.4, (3), $\Gamma(\mathrm{Irr}\, G)$ is the same thing as the dual graph $\Gamma(\mathrm{Irr}\, E)$ of E, in other words, the Dynkin diagram of the singularity S_G. Let h be the

Coxeter number of $\Gamma(\text{Irr } E)$. We also call h the Coxeter number of G. See Table 2 and Section 11.1.

We define nonnegative integers $d(\rho)$ for any $\rho \in \text{Irr } G$ as follows. If G is cyclic, choose a character χ of G such that $\rho_{\text{nat}} = \chi \oplus \chi^{-1}$, and define $e(\chi^k) = k$, $d(\chi^k) = |\frac{n+1}{2} - k|$. Although there are two choices of the generator χ, the definition of the pair $(\frac{h}{2} - d(\rho), \frac{h}{2} + d(\rho)) = (e(\rho), n + 1 - e(\rho))$ is independent of the choice. If G is not cyclic, then $\Gamma(\text{Irr } G)$ is star-shaped with a unique centre. For any $\rho \in \text{Irr } G$, we define $d(\rho)$ to be the distance from the vertex ρ to the centre. It is obvious that $d(\rho) = d(\rho') \pm 1$ if ρ and $\rho' \in \text{Irr } G$ are adjacent. Also in the cyclic case if we define the centre to be the midpoint of the graph, then $d(\rho)$ is the distance from the centre.

For any positive integer m let $S_m := S_m(\rho_{\text{nat}})$ be the symmetric m-tensors of ρ_{nat}, that is, the space of homogeneous polynomials of degree m. We say that a G-submodule W of $\mathfrak{m}/\mathfrak{n}$ is *homogeneous* of degree m if it is generated over $\mathbb{C}$ by homogeneous polynomials of degree m.

The G-module $\mathfrak{m}/\mathfrak{n}$ splits as a direct sum of irreducible homogeneous G-modules. If W is a direct sum of homogeneous G-submodules, then we denote the homogeneous part of W of degree m by $S_m(W)$. For any G-module W in some $S_m(\mathfrak{m}/\mathfrak{n})$, we write $S_j \cdot W$ for the G-submodule of $S_{m+j}(\mathfrak{m}/\mathfrak{n})$ generated over $\mathbb{C}$ by the products of $S_j(\mathfrak{m}/\mathfrak{n})$ and W. We denote by $W[\rho]$ the ρ factor of W, that is, the sum of all the copies of ρ in W; and similarly, we denote by $[W : \rho]$ the multiplicity of $\rho \in \text{Irr } G$ in a G-module W.

We define

$$S_{\text{McKay}}(\mathfrak{m}/\mathfrak{n}) = \sum_{\rho \in \text{Irr } G} S_{\frac{h}{2} \pm d(\rho)}(\mathfrak{m}/\mathfrak{n})[\rho].$$

Theorem 10.5 (First duality theorem) *Let G be any finite subgroup of $SL(2, \mathbb{C})$ and h its Coxeter number. Then as G-modules, we have*

1. $\mathfrak{m}/\mathfrak{n} = \sum_{\rho \in \text{Irr } G} 2(\deg \rho)\rho$;

2. $S_{\text{McKay}}(\mathfrak{m}/\mathfrak{n}) \simeq \sum_{\rho \in \text{Irr } G} 2\rho$;

3. $S_{\frac{h}{2} - k}(\mathfrak{m}/\mathfrak{n}) \simeq S_{\frac{h}{2} + k}(\mathfrak{m}/\mathfrak{n})$ *for any k;*

4. $S_k(\mathfrak{m}/\mathfrak{n}) = 0$ *for $k \geq h$.*

Theorem 10.6 (Second duality theorem) *Assume that G is not cyclic. Let h be the Coxeter number of G and $V_{\frac{h}{2} \pm d(\rho)}(\rho) := S_{\frac{h}{2} \pm d(\rho)}(\mathfrak{m}/\mathfrak{n})[\rho]$ for any $\rho \in \text{Irr } G$. Then*

1. $V_{\frac{h}{2} - d(\rho)}(\rho) \simeq V_{\frac{h}{2} + d(\rho)}(\rho) \simeq \rho^{\oplus 2}$ *or ρ if $d(\rho) = 0$, respectively $d(\rho) \geq 1$.*

2. If ρ and ρ' are adjacent with $d(\rho') = d(\rho) + 1 \geq 2$, then

$$V_{\frac{h}{2}-d(\rho)}(\rho) = \{S_1 \cdot V_{\frac{h}{2}-d(\rho')}(\rho')\}[\rho],$$
$$\text{and} \quad V_{\frac{h}{2}+d(\rho')}(\rho') = \{S_1 \cdot V_{\frac{h}{2}+d(\rho)}(\rho)\}[\rho'].$$

3. If $d(\rho) = 0$, we write $\rho_i \in \text{Irr}\, G$ for $i = 1, 2, 3$ for the three irreducible representations adjacent to ρ; then

$$\{S_1 \cdot V_{\frac{h}{2}-1}(\rho_i)\}[\rho] \simeq \rho,$$
$$V_{\frac{h}{2}+1}(\rho_i) = \{S_1 \cdot V_{\frac{h}{2}}(\rho)\}[\rho_i] \simeq \rho_i \quad \text{for } i = 1, 2, 3; \text{ and}$$
$$V_{\frac{h}{2}}(\rho) = \{S_1 \cdot V_{\frac{h}{2}-1}(\rho_i)\}[\rho] + \{S_1 \cdot V_{\frac{h}{2}-1}(\rho_j)\}[\rho] \simeq \rho^{\oplus 2} \quad \text{for } i \neq j.$$

See Section 11 for the proof of Theorems 10.5–10.6. It is the detailed form of the duality in Theorems 10.6 and 12.4 that we need for the explanation of the McKay observation in Section 13.5.

The exceptional sets of $\text{Hilb}^G(\mathbb{A}^2)$ are described in Theorems 10.7 and 12.3.

Theorem 10.7 *Assume that G is not cyclic.*

1. *Assume that ρ is one of the endpoints of the Dynkin diagram. Then $I \in E(\rho) \setminus \left(\bigcup_{\rho'} P(\rho, \rho') \right)$ if and only if $V(I)$ is a nonzero irreducible G-submodule ($\simeq \rho$) of $V_{\frac{h}{2}-d(\rho)}(\rho) \oplus V_{\frac{h}{2}+d(\rho)}(\rho)$ different from $V_{\frac{h}{2}+d(\rho)}(\rho)$.*

2. *Assume $d(\rho) \geq 1$ and that ρ is not one of the endpoints of the Dynkin diagram. Then $I \in E(\rho) \setminus \left(\bigcup_{\rho'} P(\rho, \rho') \right)$ if and only if $V(I)$ is a nonzero irreducible G-submodule ($\simeq \rho$) of $V_{\frac{h}{2}-d(\rho)}(\rho) \oplus V_{\frac{h}{2}+d(\rho)}(\rho)$ different from $V_{\frac{h}{2}-d(\rho)}(\rho)$ and $V_{\frac{h}{2}+d(\rho)}(\rho)$.*

3. *Let ρ and ρ' be an adjacent pair with $d(\rho') = d(\rho) + 1 \geq 2$. Then $I \in P(\rho, \rho')$ if and only if*

$$V(I) = V_{\frac{h}{2}-d(\rho)}(\rho) \oplus V_{\frac{h}{2}+d(\rho')}(\rho').$$

We define the latter to be $W(\rho, \rho')$.

4. *Assume $d(\rho) = 0$.*

    (a) *$I \in E(\rho) \setminus \left(\bigcup_{\rho'} P(\rho, \rho') \right)$ if and only if $V(I)$ is a nonzero irreducible G-module of $V_{\frac{h}{2}}(\rho)$ different from $\{S_1 \cdot V_{\frac{h}{2}-1}(\rho')\}[\rho]$ for any ρ' adjacent to ρ where we note that $V_{\frac{h}{2}}(\rho) \simeq \rho^{\oplus 2}$.*

(b) $I \in P(\rho, \rho') \neq \emptyset$ if and only if

$$V(I) = \{S_1 \cdot V_{\frac{h}{2}-1}(\rho')\}[\rho] \oplus V_{\frac{h}{2}+1}(\rho').$$

We define the latter to be $W(\rho, \rho')$.

The proofs of Theorems 10.4–10.7 are given in Sections 12–16 in the respective cases.

Remark 10.8 One can recover I from $V(I)$ by defining $I = V(I)\mathcal{O}_{A^2} + \mathfrak{n}$. By Theorem 10.7, the curve $E(\rho)$ is identified with $\mathbb{P}(\rho \oplus \rho) \simeq \mathbb{P}^1$, the projective space of nontrivial proper G-submodules ρ in $\rho \oplus \rho$.

Remark 10.9 The relations in Theorem 10.6, (2)–(3) as well as the following observation explain why tensoring by ρ_{nat} enters the McKay correspondence. We observe

$$\begin{aligned}
W(\rho, \rho') &= V_{\frac{h}{2}-d(\rho)}(\rho) \oplus V_{\frac{h}{2}+d(\rho')}(\rho') \qquad \text{for } d(\rho) \geq 1, d(\rho') = d(\rho) + 1 \\
&= \{S_1 \cdot V_{\frac{h}{2}-d(\rho')}(\rho')\}[\rho] \oplus V_{\frac{h}{2}+d(\rho')}(\rho') \\
&= V_{\frac{h}{2}-d(\rho)}(\rho) \oplus \{S_1 \cdot V_{\frac{h}{2}+d(\rho)}(\rho)\}[\rho'], \\
W(\rho, \rho') &= \{S_1 \cdot V_{\frac{h}{2}-1}(\rho')\}[\rho] \oplus V_{\frac{h}{2}+1}(\rho') \qquad \text{for } d(\rho) = 0, d(\rho') = 1 \\
&= \{S_1 \cdot V_{\frac{h}{2}-1}(\rho')\}[\rho] \oplus \{S_1 \cdot V_{\frac{h}{2}}(\rho)\}[\rho'].
\end{aligned}$$

11 Duality

11.1 Degrees of homogeneous generators

Let G be a noncyclic finite subgroup of $SL(2, \mathbb{C})$. In this section we prove Theorem 10.5, (3) and (4). Also assuming Theorem 10.6, (1) we prove Theorem 10.6, (2) and the first half of (3). Theorem 10.5, (2) follows readily from Theorem 10.6, (1). It remains to prove Theorem 10.5, (1), Theorem 10.6, (1) and the second half of (3), which we prove by case by case examinations in Sections 13–16. The cyclic case is treated in Section 12.

There are three G-invariant homogeneous polynomials φ_i for $i = 1, 2, 3$ which generate the ring of all G-invariant polynomials. Let $d_i := \deg \varphi_i$. We may assume that $d_1 \leq d_2 \leq \deg d_3 = h$, where h is the Coxeter number of G. We know that $d_1 + d_2 = d_3 + 2$. We note that the triple d_i can computed without using the classification of G, using instead the method of [Pinkham80]. See Section 4, Table 4 for the values of the d_i. We set $\overline{S}_m := S_m(\mathfrak{m}/\mathfrak{n})$.

Lemma 11.2 $\overline{S}_m \neq 0$ for $1 \leq m \leq h-1$ and $\overline{S}_m = 0$ for $m \geq h$.

Proof Choosing suitable φ_i, we may assume that the quotient space $\mathbb{A}^2/G$ is defined by one of the equations $\varphi_3^2 = F(\varphi_1, \varphi_2)$ given in 1.1. See [Klein] and [Pinkham80]. We also see $h = \deg \varphi_3 = \deg \varphi_1 + \deg \varphi_2 - 2$ by [Pinkham80]. Now we prove that φ_1 and φ_2 have no common factors as polynomials in x and y. For otherwise, there is $\varphi \in \mathbb{C}[x, y]$ such that $\deg \varphi < d_1$, and φ divides φ_i. Therefore φ also divides φ_3, because of the relation $\varphi_3^2 = F(\varphi_1, \varphi_2)$. This implies that the one dimensional subscheme of $\mathbb{A}^2$ defined by $\varphi = 0$ is mapped to the origin of $\mathbb{A}^2/G$. This contradicts that $\mathbb{A}^2$ is finite over $\mathbb{A}^2/G$.

Thus φ_1 and φ_2 have no common factors. Hence $\varphi_1 S_{m-d_1} \cap \varphi_2 S_{m-d_2} = \varphi_1 \varphi_2 S_{m-d_1-d_2} = 0$ for $m \leq h$. It follows that $\dim \overline{S}_m = \dim S_m - \dim S_{m-d_1} - \dim S_{m-d_2}$ for $m < h$, and thus

$$\dim \overline{S}_m = \begin{cases} m + 1 & \text{for } 1 \leq m \leq d_1 - 1, \\ d_1 & \text{for } d_1 \leq m \leq d_2 - 1, \\ d_1 + d_2 - m - 1 & \text{for } d_2 \leq m \leq d_3 - 1. \end{cases}$$

Similarly we have

$$\dim \overline{S}_h = \dim S_h/\mathbb{C}\varphi_3 - \dim S_{h-d_1} - \dim S_{h-d_2}$$
$$= h - (h + 1 - d_1) - (h + 1 - d_2) = d_1 + d_2 - h - 2 = 0.$$

$\square$

Corollary 11.3 $\dim \mathfrak{m}/\mathfrak{n} = d_1 d_2 - 2 = 2|G| - 2.$

Proof The first equality is clear from the proof of Lemma 11.2. The second $d_1 d_2 = 2|G|$ follows from the classification of G. $\square$

This corollary is not used elsewhere.

11.4   The bilinear form (f, g) on $\mathfrak{m}/\mathfrak{n}$

Let $f, g \in \mathfrak{m}$ be homogeneous. Then we define a bilinear form (f, g) as follows. First we define $(f, g) = 0$ if $\deg(f) + \deg(g) \neq h$. If $\deg(f) + \deg(g) = h$, then in view of Lemma 11.2 we can express fg as a linear combination of φ_i with coefficients in $\mathcal{O}_{\mathbb{A}^2}$, say $fg = a_1\varphi_1 + a_2\varphi_2 + a_3\varphi_3$ where a_i is homogeneous and a_3 is a constant. We define

$$(f, g) := a_3.$$

This is well defined. In fact, assume that $fg = b_1\varphi_1 + b_2\varphi_2 + b_3\varphi_3$. Then we have $(a_3 - b_3)\varphi_3 = (b_1 - a_1)\varphi_1 + (b_2 - a_2)\varphi_2$. By the proof of Lemma 11.2, φ_3 is not a linear combination of φ_1 and φ_2 with coefficients in $\mathcal{O}_{\mathbb{A}^2}$. It follows that $a_3 = b_3$. Moreover if either $f \in \mathfrak{n}$ or $g \in \mathfrak{n}$, then $(f, g) = 0$. Therefore the bilinear form is well defined on $\mathfrak{m}/\mathfrak{n}$.

Lemma 11.5   *1.* $(fg, h) = (f, gh)$ *for all* $f, g, h \in \mathfrak{m}$;

2. $(f, g) = (\sigma^*(f), \sigma^*(g))$ *and* $(\sigma^*(f), g) = (f, (\sigma^{-1})^*(g))$ *for all* $f, g \in \mathfrak{m}$, *and all* $\sigma \in G$;

3. $(\,,\,) : f \times g \mapsto (f, g)$ *is a nondegenerate bilinear form on* $\mathfrak{m}/\mathfrak{n}$.

Proof   (1) and (2) are clear. We prove (3). For it, we prove the following claim.

Claim 11.6 *Let* $f(x, y)$ *be a homogeneous polynomial of degree* $p < h$. *If* $xf(x, y) = yf(x, y) = 0$ *in* $\mathfrak{m}/\mathfrak{n}$, *then* $f(x, y) = 0$ *in* $\mathfrak{m}/\mathfrak{n}$.

In fact, by the assumption, there exist homogeneous a_i and $b_i \in \mathfrak{m}$ such that $xf = a_1\varphi_1 + a_2\varphi_2$ and $yf = b_1\varphi_1 + b_2\varphi_2$. Hence we have

$$(ya_1 - xb_1)\varphi_1 + (ya_2 - xb_2)\varphi_2 = 0.$$

We see that $\deg(ya_i - xb_i) = p + 2 - d_i < h + 2 - d_i \le d_1 + d_2 - d_i$ for $i = 1, 2$, because $h + 2 = d_1 + d_2$. Meanwhile φ_1 and φ_2 have no nontrivial common factors. It follows that $ya_i - xb_i = 0$. This implies that $x \mid a_i$ and $y \mid b_i$. Hence $f = 0$ in $\mathfrak{m}/\mathfrak{n}$. $\square$

We now proceed with the proof of Lemma 11.5, (3). Let $f \in \mathfrak{m}$ be homogeneous. Assume that $(f, g) = 0$ for any $g \in \mathfrak{m}/\mathfrak{n}$. We prove that $f = 0$ in $\mathfrak{m}/\mathfrak{n}$ by descending induction on $p := \deg f$. If $p = h - 1$, then $f = 0$ by Claim 11.6. Assume $p < h - 1$. By the assumption, we get $(xf, g) = (f, xg) = 0$ and $(yf, g) = (f, yg) = 0$ for any $g \in \mathfrak{m}/\mathfrak{n}$. By the induction hypothesis, $xf = 0$ and $yf = 0$ in $\mathfrak{m}/\mathfrak{n}$. Then by Claim 11.6 we have $f = 0$ in $\mathfrak{m}/\mathfrak{n}$. $\square$

Lemma 11.7 *Let* V *be a* G-*submodule of* $\overline{S}_{(h/2)-k}$, *and* V^* *a* G-*submodule of* $\overline{S}_{(h/2)+k}$ *dual to* V *with respect to the bilinear form* $(\,,\,)$, *in the sense that* $(\,,\,)$ *defines a perfect pairing between* V *and* V^*. *Then* V *is isomorphic to the complex conjugate of* V^* *as* G-*modules.*

Proof   Let V^c be an arbitrary G-module of $\overline{S}_{(h/2)-k}$ complementary to V. Then we define V^* to be the orthogonal complement in $\overline{S}_{(h/2)+k}$ to V^c. By Lemma 11.5, (2), $\sigma^*(V^*) \subset V^*$ for any $\sigma \in G$. Moreover by Lemma 11.5, (2) $\mathrm{Tr}(\sigma^*_{|V}) = \mathrm{Tr}((\sigma^{-1})^*_{|V^*})$, which is equal to the complex conjugate of $\mathrm{Tr}(\sigma^*_{|V^*})$ because any eigenvalue of $\mathrm{Tr}(\sigma^*_{|V^*})$ is a root of unity. Although the definition of V^* depends on the choice of V^c, we always have $V \simeq$ the complex conjugate of V^*. $\square$

Corollary 11.8 *Let V, V' be G-submodules of $\mathfrak{m}/\mathfrak{n}$. If V and the complex conjugate of V' are not isomorphic as G-modules, then V and V' are orthogonal.*

Lemma 11.9 *Let ρ and ρ' be equivalence classes of irreducible G-modules with $\rho \neq \rho'$. Let $V \simeq \rho$ and $W \simeq \rho'$ be G-submodules in $\overline{S}_{(h/2)-k}$ and $\overline{S}_{(h/2)-k+1}$ respectively, and $W^* \simeq (\rho')^*$ a dual to W in $\overline{S}_{(h/2)+k-1}$. If $W \subset S_1 \cdot V$, there is a G-submodule V^* of $S_1 \cdot W^*$ dual to V. If $[\rho_{\mathrm{nat}} \otimes (\rho')^* : \rho] = 1$, then V^* is uniquely determined.*

Proof Let V^c and W^c be (homogeneous) complementary G-submodules to V and W respectively. Thus by definition,

$$V \oplus V^c = \overline{S}_{(h/2)-k} \quad \text{and} \quad W \oplus W^c = \overline{S}_{(h/2)-k+1}.$$

Let W^* be the orthogonal complement to W^c in $\overline{S}_{(h/2)+k-1}$ with respect to $(\ ,\ )$. If $W \subset S_1 V$, then there exists $g, h \in V$ such that $xg + yh \in W$. By Lemma 11.5, (3), there exists $f^* \in W^*$ such that $(f^*, xg+yh) \neq 0$ so that we first assume that $(xf^*, g) = (f^*, xg) \neq 0$. Let U be a minimal G-submodule of $\mathfrak{m}/\mathfrak{n}$ containing xf^*. Then U contains V^* dual to V by Lemma 11.5, (3) and $(xf^*, g) \neq 0$. Obviously $V^* \subset S_1 W^*$ and $V^* \simeq$ the complex conjugate of V by Lemma 11.7. If $[S_1 \cdot W^* : \rho'] \leq [\rho_{\mathrm{nat}} \otimes (\rho')^* : \rho] = 1$, then the uniqueness of V^* is clear. If $(yf^*, g) = (f^*, yg) \neq 0$, then we see the same by the same argument. $\square$

Remark 11.10 For any $\rho'' \in \mathrm{Irr}\, G$, $\rho_{\mathrm{nat}} \otimes \rho''$ is a sum of G-submodules *with multiplicity one* [McKay80] (recall that $G \subset \mathrm{SL}(2, \mathbb{C})$), so that ρ has multiplicity at most one in $S_1 \cdot W^*$. Therefore the dual V^* is uniquely determined and it is the orthogonal complement of V^c in $(S_1 \cdot W^*) \cap \overline{S}_{(h/2)+k-1}$.

Lemma 11.9 implies the following. In the case of E_6, since

$$S_1 \cdot \overline{S}_3[\rho_2'] = \overline{S}_4[\rho_1'] + \overline{S}_4[\rho_3] \quad \text{and} \quad S_1 \cdot \overline{S}_3[\rho_2''] = \overline{S}_4[\rho_1''] + \overline{S}_4[\rho_3],$$

we have $S_1 \cdot \overline{S}_8[\rho_1'] = \overline{S}_9[\rho_2']$, $S_1 \cdot \overline{S}_8[\rho_1''] = \overline{S}_9[\rho_2'']$ and $S_1 \cdot \overline{S}_8[\rho_3] = \overline{S}_9[\rho_2'] + \overline{S}_9[\rho_2'']$, and vice versa. See Section 14.

11.11   Partial proofs of Theorems 10.5 and 10.6.

Since $\mathrm{Tr}_{\overline{S}_k}$ is real for any k, $\overline{S}_k$ contains any G-module and its complex conjugate with equal multiplicities. Theorem 10.5, (3) is clear from Lemma 11.5, (3) and Lemma 11.7. Theorem 10.5, (4) follows from Lemma 11.2. Theorem 10.6, (2) as well as the first half of (3) are clear from Lemma 11.9.

12 The cyclic groups A_n

12.1 Characters

Let x, y be coordinates on $\mathbb{A}^2$ and $\mathfrak{m} = (x, y)$ be the maximal ideal of $\mathbb{A}^2$ at the origin. Let G be the cyclic group of order $n + 1$ with generator σ. Let ε be a primitive $(n + 1)$st root of unity. We define the action of the generator σ on $\mathbb{C}^2$ by $(x, y) \mapsto (x, y)\sigma = (\varepsilon x, \varepsilon^{-1} y)$. The simple singularity of type A_n is the quotient $S_G = \mathbb{A}^2/G$. Let $\mathfrak{m}_S$ be the maximal ideal of S_G at the origin and $\mathfrak{n} := \mathfrak{m}_S \mathcal{O}_{\mathbb{A}^2}$.

The Coxeter number h of A_n is equal to $n + 1$. Let ρ_0 be the trivial character, and ρ_i for $1 \le i \le n$ the character with $\rho_i(\sigma) = \varepsilon^i$. Then $e(\rho_i) = i$ and $h - e(\rho_i) = n + 1 - i$.

Lemma 12.2 *Any $I \in \mathrm{Hilb}^G(\mathbb{A}^2)$ is one of the following ideals of colength $n + 1$:*

$$I(\Sigma) := \prod_{p \in \Sigma} \mathfrak{m}_p = (x^{n+1} - a^{n+1}, xy - ab, y^{n+1} - b^{n+1}),$$

where $\Sigma = G \cdot (a, b)$ is a G-orbit of $\mathbb{A}^2$ disjoint from the origin; or

$$I_i(p_i : q_i) := (p_i x^i - q_i y^{n+1-i}, xy, x^{i+1}, y^{n+2-i}),$$

for some $1 \le i \le n$ and some $[p_i, q_i] \in \mathbb{P}^1$.

Proof Let $I \in \mathrm{Hilb}^G(\mathbb{A}^2)$ with $I \subset \mathfrak{m}$. Then by Corollary 9.5, $\mathcal{O}_{\mathbb{A}^2}/I \simeq \mathbb{C}[G] \simeq \bigoplus_{i=0}^n \rho_i$ as G-modules. Thanks to Corollary 9.6, we define $N := \mathfrak{m}/\mathfrak{n}$ and $M := I/\mathfrak{n}$, and for each $i \ne 0$, let $M[\rho_i]$ and $N[\rho_i]$ be the ρ_i-part of M, respectively N. Then $N[\rho_i] \simeq \rho_i^{\oplus 2}$, spanned by x^i and y^{n+1-i}, while $M[\rho_i] \simeq \rho_i$ for all $i \ne 0$. It follows that for each i, there exists $[p_i, q_i] \in \mathbb{P}^1$ such that $p_i x^i - q_i y^{n+1-i} \in M$. If $p_i q_i \ne 0$ for some i, then setting $u := p_i x^i - q_i y^{n+1-i}$, we have $M = (u) + \mathfrak{n}/\mathfrak{n}$ and $I = (u, xy)$ where i is obviously uniquely determined by I. If M contains no $p_i x^i - q_i y^{n+1-i}$ with $p_i q_i \ne 0$ for any i, then $I = (x^j, y^{n+2-j}, xy)$ for some j. □

Theorem 12.3 *Let a and b be the parameters of $\mathbb{A}^2$ on which the group G acts by $g(a, b) = (\varepsilon a, \varepsilon^{-1} b)$.*

Let $S = \mathbb{A}^2/G := \mathrm{Spec}\,\mathbb{C}[a^{n+1}, ab, b^{n+1}]$ and $\widetilde{S} \to S$ its toric minimal resolution, with affine charts U_i defined by

$$U_i := \mathrm{Spec}\,\mathbb{C}[s_i, t_i] \quad \text{for } 1 \le i \le n+1,$$

where $s_i := a^i/b^{n+1-i}$ and $t_i := b^{n+2-i}/a^{i-1}$. Then the isomorphism of $\widetilde{S}$ with $\mathrm{Hilb}^G(\mathbb{A}^2)$ is given by (the morphism defined by the universal property of

Hilb^n(A^2) *from) two dimensional flat families of subschemes defined by the G-invariant ideals of* $\mathcal{O}_{A^2}$

$$\mathcal{I}_i(s_i, t_i) := (x^i - s_i y^{n+1-i}, xy - s_i t_i, y^{n+2-i} - t_i x^{i-1})$$

for $1 \le i \le n+1$.

Proof   Note first that $\mathcal{I}_i(s_i, 0) = I_i(1 : s_i)$ and $\mathcal{I}_i(0, t_i) = I_{i-1}(t_i : 1)$ for $i \ge 2$.

If $ab = s_i t_i \ne 0$, we see $\mathcal{I}_i(s_i, t_i) = (x^{n+1} - a^{n+1}, xy - ab, y^{n+1} - b^{n+1})$. In fact, let $p = (a, b) \ne (0, 0) \in A^2$ and $\Sigma := \{p \cdot g; g \in G\}$. It is clear that $\mathcal{I}_i(s_i, t_i) \subset \mathfrak{m}_p$ so that $\mathcal{I}_i(s_i, t_i) \subset I_\Sigma$ by the G-invariance of $\mathcal{I}_i(s_i, t_i)$. Since the colengths of $\mathcal{I}_i(s_i, t_i)$ and I_Σ in $\mathcal{O}_{A^2}$ are equal to $n + 1$, $\mathcal{I}_i(s_i, t_i) = I_\Sigma = (x^{n+1} - a^{n+1}, xy - ab, y^{n+1} - b^{n+1})$.

By the universality of Hilb^n(A^2) and by Lemma 12.2, we have a finite birational morphism of $\widetilde{S}$ onto a smooth surface Hilb^G(A^2). It follows that $\widetilde{S} \simeq$ Hilb^G(A^2). $\square$

Theorem 12.4 (Duality for A_n)   *Assume that G is cyclic. Then for any $\rho \in \operatorname{Irr} G$ there exists a unique pair* $V^+_{e(\rho)}(\rho)$ *and* $V^-_{n+1-e(\rho)}(\rho)$ *of homogeneous G-submodules of* $S_{e(\rho)}(\mathfrak{m}/\mathfrak{n})[\rho]$ *and* $S_{n+1-e(\rho)}(\mathfrak{m}/\mathfrak{n})[\rho]$ *such that*

1. $V^+_{e(\rho)}(\rho) \simeq V^-_{n+1-e(\rho)}(\rho) \simeq \rho$, *and*

2. if ρ and ρ' are adjacent with $e(\rho) = e(\rho') + 1$, then

$$V^+_{e(\rho)}(\rho) = \{S_1 \cdot V^+_{e(\rho')}(\rho')\}[\rho], \quad V^-_{n+1-e(\rho')}(\rho') = \{S_1 \cdot V^-_{n+1-e(\rho)}(\rho)\}[\rho'].$$

Proof   First we prove uniqueness of $V^\pm_j(\rho)$. Since $S_1 = \rho_1 \oplus \rho_n$, we have unique choices $V^+_1(\rho_1) = S_1[\rho_1] = \{x\}$ and $V^-_1(\rho_n) = S_1[\rho_n] = \{y\}$. Then we have

$$V^+_{i+1}(\rho_{i+1}) = \{S_1 \cdot V^+_i(\rho_i)\}[\rho_{i+1}] = \{x^{i+1}\},$$
$$V^-_{n+1-i}(\rho_i) = \{S_1 \cdot V^-_{n-i}(\rho_{i+1})\}[\rho_i] = \{y^{n+1-i}\}.$$

In fact, this follows from (2) by induction. This proves Theorem 12.4. $\square$

Theorem 10.4 for G cyclic follows from setting $E(\rho_i) = E_i$. There is a way of understanding $I_i(p_i, q_i)$ similar to that of Theorem 10.7.

13   The binary dihedral groups D_n

13.1   Binary dihedral group

Let G be the subgroup of $\mathrm{SL}(2,\mathbb{C})$ of order $4n-8$ generated by two elements σ and τ:

$$\sigma = \begin{pmatrix} \varepsilon & 0 \\ 0 & \varepsilon^{-1} \end{pmatrix}, \quad \tau = \begin{pmatrix} 0 & 1 \\ -1 & 0 \end{pmatrix},$$

where ε is a primitive $\ell := (2n-4)$th root of unity. Then we have

$$\sigma^{2n-4} = 1, \quad \tau^4 = 1, \quad \sigma^{n-2} = \tau^2, \quad \tau\sigma\tau^{-1} = \sigma^{-1}.$$

The group G is called the *binary dihedral group* $\mathbb{D}_{n-2}$. The Coxeter number h of D_n is equal to $2n-2$. See Table 6 for the characters of D_n.

G acts on $\mathbb{A}^2$ from the right by $(x,y) \mapsto (x,y)g$ for $g \in G$. The ring of all G-invariant polynomials is generated by $x^\ell + y^\ell$, $xy(x^\ell - y^\ell)$ and x^2y^2. By Theorem 9.3, $X_G := \mathrm{Hilb}^G(\mathbb{A}^2)$ is a minimal resolution of $S_G := \mathbb{A}^2/G$ with a simple singularity of type D_n.

Remark 13.2 We note that if we let H be the (normal) subgroup of G generated by σ and $N := G/H$, N acts on $\mathrm{Hilb}^H(\mathbb{A}^2)$ so that we have a minimal resolution $\mathrm{Hilb}^N(\mathrm{Hilb}^H(\mathbb{A}^2))(\simeq X_G)$ of S_G.

13.3   Symmetric tensors modulo $\mathfrak{n}$

Recall $\ell := 2n-4$. Let S_m be the space of symmetric m-tensors of $\rho_{\mathrm{nat}} := \rho_2$, that is, the space of homogeneous polynomials of degree m and $\overline{S}_m$ the image of S_m in $\mathfrak{m}/\mathfrak{n}$. The spaces $\overline{S}_m$ decompose into irreducible G-modules as follows. Let $\rho_1 := \rho_0' + \rho_1'$, $\rho_{n-1} := \rho_{n-1}' + \rho_n'$ and $\rho_k := \rho_j$ if $k \equiv j \mod 2n-4$. Then we have

$$S_m = \begin{cases} \rho_0' + \rho_3 + \rho_5 + \cdots + \rho_{m-1} + \rho_{m+1} & \text{for } m \equiv 0 \mod 4, \\ \rho_1' + \rho_3 + \rho_5 + \cdots + \rho_{m-1} + \rho_{m+1} & \text{for } m \equiv 2 \mod 4, \\ \rho_2 + \rho_4 + \rho_6 + \cdots + \rho_{m-1} + \rho_{m+1} & \text{for } m \equiv 1,3 \mod 4. \end{cases}$$

13.4   The submodules $V_i(\rho_j)$

By Table 7 we see that $\mathfrak{m}/\mathfrak{n} \simeq (\mathbb{C}[G] \ominus \rho_0)^{\oplus 2}$. This isomorphism is realized by giving G-submodules $2\rho_i'$ for $i = 1, n-1, n$ and $4\rho_i$ for $2 \le i \le n-2$ explicitly as follows. We define a G-submodule of $\mathfrak{m}/\mathfrak{n}$ by $\overline{V}_i(\rho_j) := S_i(\mathfrak{m}/\mathfrak{n})[\rho_j]$, and define $V_i(\rho_j)$ to be a G-submodule of S_i such that $V_i(\rho_j) \simeq \overline{V}_i(\rho_j)$ and $V_i(\rho_j) \equiv$

ρ	1	σ	τ	d	$(\frac{h}{2}\pm d)$
ρ'_0	1	1	1	$(n-3)$	–
ρ'_1	1	1	-1	$n-3$	$(2,\ell)$
ρ_2	2	$\varepsilon+\varepsilon^{-1}$	0	$n-4$	$(3,\ell-1)$
ρ_k	2	$\varepsilon^{k-1}+\varepsilon^{-(k-1)}$	0	$n-2-k$	$(k+1,\ell+1-k)$
ρ_{n-2}	2	$\varepsilon^{n-3}+\varepsilon^{-(n-3)}$	0	0	$(n-1,n-1)$
ρ'_{n-1}	1	-1	i^n	1	$(n-2,n)$
ρ'_n	1	-1	$-i^n$	1	$(n-2,n)$

Table 6: Character table of D_n

m	$\overline{S}_m$		m	$\overline{S}_m$
0	0		$\ell+2$	0
1	ρ_2		$\ell+1$	ρ_2
2	$\rho'_1+\rho_3$		ℓ	$\rho'_1+\rho_3$
3	$\rho_2+\rho_4$		$\ell-1$	$\rho_2+\rho_4$
$\cdots$	$\cdots$		$\cdots$	$\cdots$
k	$\rho_{k-1}+\rho_{k+1}$		$\ell-k+2$	$\rho_{k-1}+\rho_{k+1}$
$n-2$	$\rho_{n-3}+\rho'_{n-1}+\rho'_n$		n	$\rho_{n-3}+\rho'_{n-1}+\rho'_n$
$n-1$	$2\rho_{n-2}$			

Table 7: Irreducible decompositions of $\overline{S}_m(D_n)$

$V_2(\rho'_1)$	xy		$V_\ell(\rho'_1)$	$x^\ell-y^\ell$
$\cdots$	$\cdots$		$\cdots$	$\cdots$
$V_{k-1}(\rho_k)$	x^{k-1},y^{k-1}		$V_{k+1}(\rho_k)$	$x^k y, xy^k$
$V_{\ell-k+1}(\rho_k)$	$x^{\ell-k+1},y^{\ell-k+1}$		$V_{\ell-k+3}(\rho_k)$	$x^{\ell-k+2}y, xy^{\ell-k+2}$
$\cdots$	$\cdots$		$\cdots$	$\cdots$
$V_{n-3}(\rho_{n-2})$	x^{n-3},y^{n-3}		$V_{n+1}(\rho_{n-2})$	$x^n y, xy^n$
$V_{n-1}(\rho_{n-2})$	$x^{n-1},y^{n-1},x^{n-2}y,xy^{n-2}$			
$V'_{n-1}(\rho_{n-2})$	x^{n-1},y^{n-1}		$V''_{n-1}(\rho_{n-2})$	$x^{n-2}y, xy^{n-2}$
$V_{n-2}(\rho'_{n-1})$	$x^{n-2}-i^n y^{n-2}$		$V_n(\rho'_{n-1})$	$xy(x^{n-2}+i^n y^{n-2})$
$V_{n-2}(\rho'_n)$	$x^{n-2}+i^n y^{n-2}$		$V_n(\rho'_n)$	$xy(x^{n-2}-i^n y^{n-2})$

Table 8: $V_m(\rho)(D_n)$

$\bar{V}_i(\rho_j)$ mod $\mathfrak{n}$. We use $V_i(\rho_j)$ and $\bar{V}_i(\rho_j)$ interchangeably whenever this is harmless. We see easily that $V_i(\rho_j) \simeq \rho_j$ or 0 except for $(i,j) = (n-1, n-2)$, while $V_{n-1}(\rho_{n-2}) \simeq \rho_{n-2}^{\oplus 2}$. We list the nonzero G-submodules of $\mathfrak{m}/\mathfrak{n}$: it is easy to see that $\mathfrak{n}$ is generated by $x^\ell + y^\ell$, $(x^\ell - y^\ell)xy$ and $x^2 y^2$. We also note that $x^{\ell+2}, y^{\ell+2} \in \mathfrak{n}$ and that $\mathfrak{m}/\mathfrak{n}$ is spanned by x^i, y^i, $x^i y$ and xy^i for $1 \le i \le \ell$ with the single relation $x^\ell + y^\ell \equiv 0 \mod \mathfrak{n}$. Hence we see easily that $\mathfrak{m}/\mathfrak{n}$ is the sum of the above $V_i(\rho_j)$. It follows that $\mathfrak{m}/\mathfrak{n} \simeq \sum_{\rho \in \mathrm{Irr}\, G} 2 \deg(\rho)\rho \simeq (\mathbb{C}[G] \ominus \rho_0)^{\oplus 2}$.

13.5   A sketch for D_5

Before starting on the general case, we sketch the case of D_5 without rigorous proofs. First we recall

$$V_2(\rho'_1) = \{xy\}, \quad V_6(\rho'_1) = \{x^6 - y^6\},$$
$$V_3(\rho_2) = \{x^2 y, xy^2\}, \quad V_5(\rho_2) = \{x^5, y^5\}.$$

We consider the case $\mathcal{I}(W) \in E(\rho'_1) \setminus P(\rho'_1, \rho_2)$. Let $\mathcal{I}(W) := W\mathcal{O}_{\mathbb{A}^2} + \mathfrak{n}$ for any nonzero G-module $W \in \mathbb{P}(V_2(\rho'_1) + V_6(\rho'_1)) = \mathbb{P}(\{xy, x^6 - y^6\})$ such that $W \ne V_6(\rho'_1)$, that is, $W \ne \{x^6 - y^6\}$. Then we see that

$$\mathcal{I}(W)/\mathfrak{n} = W + \sum_{k=1}^5 S_k W + \mathfrak{n}/\mathfrak{n} = W + \sum_{k=1}^5 S_k V_2(\rho'_1) + \mathfrak{n}/\mathfrak{n}$$

$$\simeq W + \rho_2 + \rho_3 + (\rho'_4 + \rho'_5) + \rho_3 + \rho_2 \simeq \sum_{\rho \in \mathrm{Irr}\, G} \deg(\rho)\rho.$$

Thus $\mathcal{I}(W) \in \mathrm{Hilb}^G(\mathbb{A}^2)$. It is clear that $V(\mathcal{I}(W)) := \mathcal{I}(W)/\mathfrak{m}\mathcal{I}(W) + \mathfrak{n} \simeq W \simeq \rho'_1$. It follows that $\mathcal{I}(W) \in E(\rho'_1) \setminus P(\rho'_1, \rho_2)$. Hence we have

$$\lim_{W \to V_6(\rho'_1)} \mathcal{I}(W) = V_6(\rho'_1) + \sum_{k \ge 1} S_k V_2(\rho'_1)$$

$$= \mathcal{I}(V_6(\rho'_1) \oplus S_1 V_2(\rho'_1)) = \mathcal{I}(V_6(\rho'_1) \oplus V_3(\rho_2)) \in P(\rho'_1, \rho_2),$$

where $S_1 \otimes V_2(\rho'_1) \simeq S_1 V_2(\rho'_1) \simeq V_3(\rho_2) \simeq \rho_2$. The factor $S_1 \otimes V_2(\rho'_1) \simeq \rho_2$ among generators of $P(\rho'_1, \rho_2)$ explains the relation between tensoring by $S_1 \simeq \rho_2$ and the intersection of $E(\rho'_1)$ with $E(\rho_2)$ in McKay's observation.

Next we consider $W \in \mathbb{P}(V_3(\rho_2) \oplus V_5(\rho_2))$ with $W \ne V_3(\rho_2), V_5(\rho_2)$. We have

$$\mathcal{I}(W)/\mathfrak{n} := W + \sum_{k \ge 1} S_k W + \mathfrak{n}/\mathfrak{n}$$

$$= W + \sum_{k \ge 1}^2 S_k V_3(\rho_2) + \overline{S}_6 + \overline{S}_7 + \mathfrak{n}/\mathfrak{n}$$

$$\simeq W + \rho_3 + (\rho'_4 + \rho'_5) + (\rho'_1 + \rho_3) + \rho_2 \simeq \sum_{\rho \in \mathrm{Irr}\, G} \deg(\rho)\rho.$$

Since $\overline{S}_6 = V_6(\rho_1') + S_3 V_3(\rho_2) \neq S_3 V_3(\rho_2)$, we have

$$\lim_{W \to V_3(\rho_2)} \mathcal{I}(W) = V_6(\rho_1') + V_3(\rho_2) + \sum_{k \geq 1} S_k V_3(\rho_2)$$

$$= \mathcal{I}(V_6(\rho_1') \oplus V_3(\rho_2)) \in P(\rho_1', \rho_2)$$

$$= \mathcal{I}(\{S_1 V_5(\rho_2)\}[\rho_1'] \oplus V_3(\rho_2)),$$

where $V_6(\rho_1') = \{S_1 V_5(\rho_2)\}[\rho_1'] \simeq \rho_1'$, and $\{S_1 V_5(\rho_2)\}[\rho_1'] = V_6(\rho_1') \simeq \rho_1'$ is by definition the sum of all the ρ_1' factors of $S_1 V_5(\rho_2) \simeq S_1 \otimes V_5(\rho_2)$. Hence

$$\lim_{\substack{W \to V_6(\rho_1') \\ W \simeq \rho_1'}} \mathcal{I}(W) = \lim_{\substack{W \to V_3(\rho_2) \\ W \simeq \rho_2}} \mathcal{I}(W) \in P(\rho_1', \rho_2).$$

The above argument explains the relation between tensoring by $\rho_2 = \rho_{\text{nat}}$ and the intersection of two rational curves. The argument also shows that $E(\rho)$ is naturally identified with $\mathbb{P}(V_{4-d(\rho)}(\rho) + V_{4+d(\rho)}(\rho))$, the set of all nontrivial proper G-submodules of $V_{4-d(\rho)}(\rho) + V_{4+d(\rho)}(\rho) \simeq \rho^{\oplus 2}$, which is isomorphic to $\mathbb{P}^1$ by Schur's lemma.

Now we consider the general case. We restate Theorem 10.7 as follows.

Theorem 13.6 *Let E be the exceptional set of the morphism $\pi \colon X_G \to S_G$, and $\mathrm{Sing}(E)$ the singular points of E. Let $E(\rho)$ be an irreducible component of E for $\rho \in \mathrm{Irr}\, G$ and $E^0(\rho) := E(\rho) \setminus \mathrm{Sing}(E)$. Then $E^0(\rho)$ and $\mathrm{Sing}(E)$ are as follows:*

$$E^0(\rho_1') = \left\{ \mathcal{I}(W); \begin{array}{l} W \subset V_2(\rho_1') \oplus V_\ell(\rho_1') \\ W \neq 0, V_\ell(\rho_1') \end{array} \right\},$$

$$E^0(\rho_k) = \left\{ \mathcal{I}(W); \begin{array}{l} W \subset V_{k+1}(\rho_k) \oplus V_{\ell-k+1}(\rho_k) \\ W \neq 0, V_{k+1}(\rho_k), V_{\ell-k+1}(\rho_k) \end{array} \right\} \quad \text{for } 2 \leq k \leq n-3,$$

$$E^0(\rho_{n-2}) = \left\{ \mathcal{I}(W); \begin{array}{l} W \subset V_{n-1}(\rho_{n-2}), W \neq 0, V_{n-1}''(\rho_{n-2}) \\ W \neq S_1 \cdot V_{n-2}(\rho_j') \quad \text{for } j = n-1, n \end{array} \right\},$$

$$E^0(\rho_j) = \left\{ \mathcal{I}(W); \begin{array}{l} W \subset V_{n-2}(\rho_j') \oplus V_n(\rho_j') \\ W \neq 0, V_n(\rho_j') \end{array} \right\} \quad \text{for } j = n-1, n;$$

and

$$\mathrm{Sing}(E) = \left\{ \begin{array}{ll} P(\rho_1', \rho_2), & P(\rho_k, \rho_{k+1}) \quad \text{for } 2 \leq k \leq n-3 \\ P(\rho_{n-2}, \rho_{n-1}'), & P(\rho_{n-2}, \rho_n') \end{array} \right\},$$

where

$$P(\rho_1', \rho_2) = \mathcal{I}(V_\ell(\rho_1') \oplus V_3(\rho_2)),$$
$$P(\rho_k, \rho_{k+1}) = \mathcal{I}(V_{\ell-k+1}(\rho_k) \oplus V_{k+2}(\rho_{k+1})) \quad for\ 2 \leq k \leq n-4,$$
$$P(\rho_{n-3}, \rho_{n-2}) = \mathcal{I}(V_n(\rho_{n-3}) \oplus V_{n-1}''(\rho_{n-2})),$$
$$P(\rho_{n-2}, \rho_j') = \mathcal{I}(S_1 V_{n-2}(\rho_j') \oplus V_n(\rho_j')).$$

13.7 Proof of Theorem 13.6 – Start

For $2 \leq k \leq n-2$, write $C(\rho_k)$ for the set of all proper G-submodules of $V_{k+1}(\rho_k) \oplus V_{\ell-k+1}(\rho_k)$; similarly, let $C(\rho_1')$ be the set of all proper G-submodules of $V_2(\rho_1') \oplus V_\ell(\rho_1')$ and for $i = n-1, n$, let $C(\rho_i')$ be the set of all proper G-submodules of $V_{n-2}(\rho_i') \oplus V_n(\rho_i')$. It is clear that the $C(\rho_k)$ and $C(\rho_i')$ are rational curves. As we will see in the sequel, they are embedded naturally into $\mathrm{Grass}(\mathfrak{m}/\mathfrak{n}, 2\ell - 2)$.

Case $\mathcal{I}(W) \in E(\rho_1') \setminus P(\rho_1', \rho_2)$ Let $\mathcal{I}(W) := W\mathcal{O}_{\mathbb{A}^2} + \mathfrak{n}$ for any nonzero G-module $W \in C(\rho_1')$ with $W \neq V_\ell(\rho_1')$. First assume $W = V_2(\rho_1')$. Then it is easy to see that $\mathcal{I}(W)/\mathfrak{n}$ contains $V_{k+1}(\rho_k)$, $V_{\ell-k+3}(\rho_k)$, $V_{n-1}''(\rho_{n-2})$ and $V_{n+1}(\rho_{n-2})$ for any $2 \leq k \leq n-3$. Similarly $\mathcal{I}(W)/\mathfrak{n}$ contains $V_n(\rho_{n-1}')$ and $V_n(\rho_n')$ as well as $W = V_2(\rho_1')$. It follows that

$$\mathcal{I}(W)/\mathfrak{n} = W + \sum_{k=1}^{\ell-1} S_k \bar{V}_2(\rho_1') = W + \sum_{k=1}^{\ell-2} S_k \bar{V}_2(\rho_1') + \overline{S}_{\ell+1}.$$

In particular, $\mathcal{I}(W)/\mathfrak{n} \simeq \sum_{\rho \in \mathrm{Irr}\,G} \deg(\rho)\rho$. Hence $\mathcal{I}(W) \in \mathrm{Hilb}^G(\mathbb{A}^2)$. We see that

$$V(\mathcal{I}(W)) := \mathcal{I}(W)/\{\mathfrak{m}\mathcal{I}(W) + \mathfrak{n}\} \simeq W \simeq \rho_1'.$$

It follows that $\mathcal{I}(W) \in E(\rho_1')$.

Next we assume $W \neq V_2(\rho_1'), V_\ell(\rho_1')$. Then we first see that $x^3 y \in \mathcal{I}(W)$ because $x^3 y - (x^3 y - 2tx^{\ell+2}) = 2tx^{\ell+2} \in \mathfrak{n}$. It follows that $\mathcal{I}(W)/\mathfrak{n}$ contains $V_{\ell+1}(\rho_2)$, $V_{k+1}(\rho_k)$, $V_{\ell-k+3}(\rho_k)$, $V_{n-1}''(\rho_{n-2})$, $V_{n+1}(\rho_{n-2})$, $V_n(\rho_{n-1}')$ and $V_n(\rho_n')$ where $3 \leq k \leq n-3$. Since $S_1 \cdot W + V_{\ell+1}(\rho_2) = V_3(\rho_2) + V_{\ell+1}(\rho_2) \simeq \rho_2^{\oplus 2}$, $\mathcal{I}(W)/\mathfrak{n}$ also contains $2\rho_2$. It follows that

$$\mathcal{I}(W)/\mathfrak{n} = W + \sum_{m \geq 0}^{\ell-2} S_m \bar{V}_2(\rho_1') = W + \sum_{m=0}^{\ell-3} S_m \bar{V}_2(\rho_1') + \overline{S}_{\ell+1}.$$

Hence we have $\mathcal{I}(W)/\mathfrak{n} \simeq \sum_{\rho \in \mathrm{Irr}\,G} \deg(\rho)\rho$. Therefore $\mathcal{I}(W) \in \mathrm{Hilb}^G(\mathbb{A}^2)$. By the above structure of $\mathcal{I}(W)/\mathfrak{n}$, $V(\mathcal{I}(W)) \simeq W \simeq \rho_1'$. It follows that $\mathcal{I}(W) \in E(\rho_1') \setminus P(\rho_1', \rho_2)$.

Case $\mathcal{I}(W) \in P(\rho_1', \rho_2)$ Let $W = W(\rho_1', \rho_2) := V_\ell(\rho_1') \oplus V_3(\rho_2)$. Now $\mathcal{I}(W)/\mathfrak{n}$ contains $x^2 y$ and xy^2, hence also $V_{i+1}(\rho_i)$, $V_{\ell-i+3}(\rho_i)$ for $3 \leq i \leq n-3$, $V_{\ell+1}(\rho_2)$, $V_{n+1}(\rho_{n-2})$, $V_n(\rho_{n-1}')$ and $V_n(\rho_n')$. Similarly, $\mathcal{I}(W)/\mathfrak{n}$ contains $V_{n-1}''(\rho_{n-2})$. We note that $\{\mathcal{I}(W)/\mathfrak{n}\}\,[\rho_1'] = W = V_\ell(\rho_1') = \{S_1 \cdot V_{\ell-1}(\rho_2)\}\,[\rho_1']$ and $\{\mathcal{I}(W)/\mathfrak{n}\}\,[\rho_2] = V_3(\rho_2) \oplus V_{\ell+1}(\rho_2) = S_1 \cdot V_2(\rho_1') \oplus V_{\ell+1}(\rho_2)$. It follows that

$$\mathcal{I}(W)/\mathfrak{n} = W + \sum_{m=0}^{\ell-2} S_m \bar{V}_3(\rho_2) = W + \sum_{m=0}^{\ell-3} S_m \bar{V}_3(\rho_2) + \overline{S}_{\ell+1}.$$

Hence we have $\mathcal{I}(W)/\mathfrak{n} \simeq \sum_{\rho \in \mathrm{Irr}\, G} \deg(\rho)\rho$. Therefore $\mathcal{I}(W) \in \mathrm{Hilb}^G(\mathbb{A}^2)$. We also see that $\mathcal{I}(W) \in P(\rho_1', \rho_2)$, because

$$V(\mathcal{I}(W)) = V_\ell(\rho_1') \oplus \{S_1 \cdot V_2(\rho_1')\}\,[\rho_2]$$
$$= \{S_1 \cdot V_{\ell-1}(\rho_2)\}\,[\rho_1'] \oplus V_3(\rho_2) \simeq \rho_1' \oplus \rho_2.$$

Case $\mathcal{I}(W) \in E(\rho_k) \setminus P(\rho_{k\pm 1}, \rho_k)$ **for** $2 \leq k \leq n-3$ We consider now $W \in C(\rho_k) = \mathbb{P}(\rho_k \subset V_{k+1}(\rho_k) \oplus V_{\ell-k+1}(\rho_k))$ with $W \neq V_{k+1}(\rho_k)$, $V_{\ell-k+1}(\rho_k)$. Let $\mathcal{I}(W) = W\mathcal{O}_{\mathbb{A}^2} + \mathfrak{n}$.

Hence we may assume that $x^{k+1}y - ty^{\ell-k+1} \in W$ for a nonzero constant t. Since $x^{k+3}y = x^2(x^{k+1}y - ty^{\ell-k+1}) + tx^2 y^{\ell-k+1}$, and $x^2 y^2 \in \mathfrak{n}$, $\mathcal{I}(W)$ contains $x^{k+3}y$. Similarly, $ty^{\ell-k+2} = -y(x^{k+1}y - ty^{\ell-k+1}) + x^{k+1}y^2$ gives $y^{\ell-k+2} \in \mathcal{I}(W)$. Hence we see that $\mathcal{I}(W)/\mathfrak{n}$ contains $V_{\ell-i+1}(\rho_i)$ for $2 \leq i \leq k-1$, $V_{i+1}(\rho_i)$ for $k+2 \leq i \leq n-3$, $V_{\ell-i+3}(\rho_i)$ for $2 \leq i \leq n-3$, $V_{n-1}'(\rho_{n-2})$, $V_{n-1}''(\rho_{n-2})$, $V_\ell(\rho_1')$, $V_n(\rho_{n-1}')$ and $V_n(\rho_n')$. Since $xy^{\ell-k+1} \in V_{\ell-k+2}(\rho_{k+1})$, we have $V_{\ell-k+3}(\rho_k) \subset \mathcal{I}(W)/\mathfrak{n}$ and $x^{k+2}y = x(x^{k+1}y - ty^{\ell-k+1}) + txy^{\ell-k+1} \in \mathcal{I}(W)/\mathfrak{n}$. Hence $V_{k+2}(\rho_{k+1}) \subset \mathcal{I}(W)/\mathfrak{n}$ if $k \leq n-4$. It follows that

$$\mathcal{I}(W)/\mathfrak{n} = W + \sum_{m=1}^{\ell-k} S_m \bar{V}_{k+1}(\rho_k) + \sum_{m=0}^{k-1} S_m \bar{V}_{\ell-k+2}(\rho_{k-1})$$
$$= W + \sum_{m=1}^{\ell-2k} S_m \bar{V}_{k+1}(\rho_k) + \sum_{m=\ell-k+2}^{\ell+1} \overline{S}_m.$$

It follows from $W \simeq \rho_k$ that $\mathcal{I}(W)/\mathfrak{n} \simeq \sum_{\rho \in \mathrm{Irr}\, G} \deg(\rho)\rho$. Therefore $\mathcal{I}(W) \in \mathrm{Hilb}^G(\mathbb{A}^2)$. It is easy to see that $V(\mathcal{I}(W)) \simeq W \simeq \rho_k$ so that $\mathcal{I}(W) \in E(\rho_k)$.

Case $\mathcal{I}(W) \in P(\rho_k, \rho_{k+1})$ Let $W = W(\rho_k, \rho_{k+1}) := V_{\ell-k+1}(\rho_k) \oplus V_{k+2}(\rho_{k+1})$ for $2 \leq k \leq n-4$. For $k = n-3$, set

$$W = W(\rho_{n-3}, \rho_{n-2}) := V_n(\rho_{n-3}) \oplus V_{n-1}''(\rho_{n-2}).$$

Now $\mathcal{I}(W)/\mathfrak{n}$ contains $V_{\ell-i+1}(\rho_i)$ for $2 \leq i \leq k$, $V_{i+1}(\rho_i)$ for $k+1 \leq i \leq n-3$, $V_{\ell-i+3}(\rho_i)$ for $2 \leq i \leq n-2$, $V_{n-1}''(\rho_{n-2})$ and $V_n(\rho_i')$ for $i = n-1, n$. Similarly

$V_\ell(\rho'_1) \subset \mathcal{I}(W)/\mathfrak{n}$. Hence $\mathcal{I}(W) \in P(\rho_k, \rho_{k+1}) \subset \mathrm{Hilb}^G(\mathbb{A}^2)$. We also see that

$$V(\mathcal{I}(W)) = \begin{cases} V_{\ell-k+1}(\rho_k) \oplus \{S_1 \cdot V_{k+1}(\rho_k)\}\,[\rho_{k+1}] & \text{for } 2 \le k \le n-4, \\ V_n(\rho_{n-3}) \oplus \{S_1 \cdot V_{n-2}(\rho_{n-3})\}\,[\rho_{n-2}] & \text{for } k = n-3 \end{cases}$$

$$= \begin{cases} \{S_1 \cdot V_{\ell-k}(\rho_{k+1})\}\,[\rho_k] \oplus V_{k+1}(\rho_k) \simeq \rho_k \oplus \rho_{k+1}, \\ \{S_1 \cdot V'_{n-1}(\rho_{n-2})\}\,[\rho_{n-3}] \oplus V''_{n-1}(\rho_{n-2}) \simeq \rho_{n-3} \oplus \rho_{n-2}. \end{cases}$$

Case $\mathcal{I}(W) \in E(\rho_{n-2}) \setminus \big(P(\rho_{n-2}, \rho_{n-3}) \cup P(\rho_{n-2}, \rho'_{n-1}) \cup P(\rho_{n-2}, \rho'_n)\big)$ Let $W \in C(\rho_{n-2}) = \mathbb{P}(V_{n-1}(\rho_{n-2}))$, and define $\mathcal{I}(W) := W\mathcal{O}_{\mathbb{A}^2} + \mathfrak{n}$. Set

$$W_0 = S_1 \cdot V_{n-2}(\rho'_{n-1}), \quad W_\infty = S_1 \cdot V_{n-2}(\rho'_n) \quad \text{and} \quad W_1 = V''_{n-1}(\rho_{n-2}).$$

Let $H = x^{n-2} - i^{n/2}y^{n-2}$ and $G = x^{n-2} + i^{n/2}y^{n-2}$. Then for some t, we have $W = \langle xH - txG, yH + tyG \rangle$. Assume $t \ne 0, 1, \infty$, or equivalently, $W \ne W_\lambda$ for $\lambda = 0, 1, \infty$. Then $x^n \in \mathcal{I}(W)/\mathfrak{n}$, so that $V_\ell(\rho'_1)$, $V_{\ell-i+1}(\rho_i)$ for $2 \le i \le n-3$ and $V_{\ell-i+3}(\rho_i)$ for $2 \le i \le n-2$ are contained in $\mathcal{I}(W)/\mathfrak{n}$. We also see that $xyH \in V_n(\rho'_{n-1}) \subset \mathcal{I}(W)/\mathfrak{n}$ and $xyG \in V_n(\rho'_n) \subset \mathcal{I}(W)/\mathfrak{n}$. It follows that

$$\mathcal{I}(W)/\mathfrak{n} = W + \sum_{m=n}^{\ell+1} \overline{S}_m.$$

Since $W \simeq \rho_{n-2}$, we have $\mathcal{I}(W)/\mathfrak{n} \simeq \sum_{\rho \in \mathrm{Irr}\,G} \deg(\rho)\rho$ with $V(\mathcal{I}(W)) \simeq W$. It follows that $\mathcal{I}(W) \in \mathrm{Hilb}^G(\mathbb{A}^2)$.

Case $\mathcal{I}(W) \in E(\rho'_{n-1}) \setminus P(\rho_{n-2}, \rho'_{n-1})$ Let $W \in C(\rho'_{n-1}) := \mathbb{P}(V_{n-2}(\rho'_{n-1}) \oplus V_n(\rho'_{n-1}))$. Assume $W \ne V_n(\rho'_{n-1})$. Then $\mathcal{I}(W)/\mathfrak{n}$ contains $x^n y$ and hence x^n. It follows that $\mathcal{I}(W)/\mathfrak{n}$ contains $V_{\ell-i+1}(\rho_i)$, $V_{\ell-i+3}(\rho_i)$ for $2 \le i \le n-3$, and $V_{n+1}(\rho_{n-2})$. We also see that $\mathcal{I}(W)/\mathfrak{n}$ contains $x^{n-1} - i^{n/2}xy^{n-2}$ so that $\{\mathcal{I}(W)/\mathfrak{n}\} \cap V_{n-1}(\rho_{n-2}) \simeq \rho_{n-2}$. Similarly we see easily that $V_\ell(\rho'_1), V_n(\rho'_n) \subset \mathcal{I}(W)/\mathfrak{n}$. It follows that

$$\mathcal{I}(W)/\mathfrak{n} = W + \sum_{m=1}^{2} S_m \overline{V}_{n-2}(\rho'_{n-1}) + \sum_{m=n+1}^{\ell+1} \overline{S}_m.$$

Since $W \simeq \rho'_{n-1}$, $\mathcal{I}(W)/\mathfrak{n} \simeq \sum_{\rho \in \mathrm{Irr}\,G} \deg(\rho)\rho$. Therefore $\mathcal{I}(W) \in E(\rho'_{n-1}) \subset \mathrm{Hilb}^G(\mathbb{A}^2)$ with $V(\mathcal{I}(W)) \simeq W$.

Case $\mathcal{I}(W) \in P(\rho_{n-2}, \rho'_{n-1})$ We consider

$$W = W(\rho_{n-2}, \rho'_{n-1}) := S_1 \cdot V_{n-2}(\rho'_{n-1})[\rho_{n-2}] \oplus V_n(\rho'_{n-1}) = W_0 \oplus V_n(\rho'_{n-1}).$$

Then $\mathcal{I}(W)/\mathfrak{n}$ contains x^n, therefore $\mathcal{I}(W)/\mathfrak{n}$ contains $V_\ell(\rho'_1)$, $V_{\ell-i+1}(\rho_i)$, $V_{\ell-i+3}(\rho_i)$ for $2 \leq i \leq n-3$, $V_{n+1}(\rho_{n-2})$ and $V_n(\rho'_n)$. Since $W \subset \mathcal{I}(W)/\mathfrak{n}$, we see that $\mathcal{I}(W)/\mathfrak{n} \simeq \sum_{\rho \in \mathrm{Irr}\,G} \deg(\rho)\rho$. Hence $\mathcal{I}(W) \in P(\rho_{n-2}, \rho'_{n-1}) \subset \mathrm{Hilb}^G(\mathbb{A}^2)$ with $V(\mathcal{I}(W)) \simeq W$.

Case $\mathcal{I}(W) \in E(\rho'_n) \setminus P(\rho_{n-2}, \rho'_n)$ or $\mathcal{I}(W) \in P(\rho_{n-2}, \rho'_n)$ This is similar to the above, and we omit the details. $\square$

Lemma 13.8 *For ρ' adjacent to ρ, the limit of $\mathcal{I}(W)$ as $\mathcal{I}(W) \in E(\rho)$ approaches $P(\rho, \rho')$ is $\mathcal{I}(W(\rho, \rho'))$.*

Proof We first consider $W \in C(\rho'_1)$ with $W \neq V_\ell(\rho'_1)$. Then by 13.7 we see that $\mathcal{I}(W) = W + V_3(\rho_2) + \sum_{m \geq 1} S_m V_3(\rho_2)$. Hence we have

$$\lim_{\substack{W \to V_\ell(\rho'_1) \\ W \in C(\rho'_1)}} \mathcal{I}(W) = V_\ell(\rho'_1) + V_3(\rho_2) + \sum_{m \geq 1} S_m V_3(\rho_2)$$

$$= \mathcal{I}(V_\ell(\rho'_1) \oplus V_3(\rho_2)) = \mathcal{I}(W(\rho'_1, \rho_2)).$$

Next we consider $W \in C(\rho_2)$ with $W \neq V_3(\rho_2), V_{\ell-1}(\rho_2)$. Then by 13.7 we have $\mathcal{I}(W) = W + V_\ell(\rho'_1) + \sum_{m \geq 0} S_m V_4(\rho_3)$. Since $V_4(\rho_3) \subset S_1 V_3(\rho_2)$, we have

$$\lim_{\substack{W \to V_3(\rho_2) \\ W \in C(\rho_2)}} \mathcal{I}(W) = V_\ell(\rho'_1) + V_3(\rho_2) + \sum_{m \geq 1} S_m V_3(\rho_2)$$

$$= \mathcal{I}(W(\rho'_1, \rho_2)) = \lim_{\substack{W \to V_\ell(\rho'_1) \\ W \in C(\rho'_1)}} \mathcal{I}(W).$$

Suppose that $W \in C(\rho_k) = \mathbb{P}(V_{\ell-k+1}(\rho_k) \oplus V_{k+1}(\rho_k))$ with $W \neq V_{k+1}(\rho_k)$, $V_{\ell-k+1}(\rho_k)$. By 13.7 we see

$$\mathcal{I}(W) = W + \sum_{m \geq 0} S_m V_{k+2}(\rho_{k+1}) + \sum_{m \geq 0} S_m V_{\ell-k+2}(\rho_{k-1}).$$

Thus for $2 \leq k \leq n-4$ we see that

$$\lim_{W \to V_{\ell-k+1}(\rho_k)} \mathcal{I}(W) = \mathcal{I}(W(\rho_k, \rho_{k+1})) = \lim_{W \to V_{k+2}(\rho_{k+1})} \mathcal{I}(W).$$

Similarly for $W \in C(\rho_{n-2})$ with $W \neq W_\lambda$ for $\lambda = 0, 1, \infty$ we have

$$\mathcal{I}(W) = W + \sum_{m \geq 0} S_m V_n(\rho_{n-3}) + \sum_{\substack{m \geq 0 \\ j=n-1,n}} S_m V_n(\rho'_j) = W + \sum_{m \geq n} S_m,$$

$$\lim_{W \to W_1} \mathcal{I}(W) = \sum_{m \geq 0} S_m V_n(\rho_{n-3}) + \sum_{m \geq 0} S_m W_1 = \mathcal{I}(W_1 \oplus V_n(\rho_{n-3})),$$

because $V_n(\rho_{n-3}) \subset S_1 W_0 + \mathfrak{n}$. Consequently

$$\lim_{W' \to V_n(\rho_{n-3})} \mathcal{I}(W') = V_n(\rho_{n-3}) + \sum_{m \geq 0} S_m V_{n+1}(\rho_{n-4}) + \sum_{m \geq 0} S_m V''_{n-1}(\rho_{n-2})$$

$$= \sum_{m \geq 0} S_m V_n(\rho_{n-3}) + \sum_{m \geq 0} S_m W_1 = \lim_{W'' \to W_1} \mathcal{I}(W''),$$

where $W' \in C(\rho_{n-3})$, $W'' \in C(\rho_{n-2})$. The limit when W approaches W_0 or W_∞ is similar. $\square$

To complete the proofs of Theorem 13.6, we also need to prove:

Lemma 13.9 $E(\rho)$ and $E(\rho')$ intersects at $P(\rho, \rho')$ transversally if ρ and ρ' are adjacent.

Proof By the proof of Theorem 9.3, $X_G = \text{Hilb}^G(\mathbb{A}^2)$ is smooth, with tangent space $T_{[I]}(X_G)$ at $[I]$ the G-invariant subspace $\text{Hom}_{\mathcal{O}_{\mathbb{A}^2}}(I, \mathcal{O}_{\mathbb{A}^2}/I)^G$ of $T_{[I]}(\text{Hilb}^n(\mathbb{A}^2))$, which is isomorphic to $\text{Hom}_{\mathcal{O}_{\mathbb{A}^2}}(I, \mathcal{O}_{\mathbb{A}^2}/I)$, where $n = |G|$. Assume that ρ and ρ' are adjacent with $d(\rho') = d(\rho) + 1$. Let $W(\rho, \rho') = V_{\frac{h}{2}-d(\rho)}(\rho) \oplus V_{\frac{h}{2}-d(\rho')}(\rho')$. Then $\mathcal{I}(W(\rho, \rho')) \in P(\rho, \rho')$. We prove the following formula

$$T_{[I]}(X_G) \simeq \text{Hom}_{\mathcal{O}_{\mathbb{A}^2}}(I, \mathcal{O}_{\mathbb{A}^2}/I)^G \simeq$$
$$\text{Hom}_G(V_{\frac{h}{2}-d(\rho)}(\rho), V_{\frac{h}{2}+d(\rho)}(\rho)) \oplus \text{Hom}_G(V_{\frac{h}{2}+d(\rho')}(\rho'), V_{\frac{h}{2}-d(\rho')}(\rho')),$$

where $I = \mathcal{I}(W(\rho, \rho'))$. First assume $\rho = \rho_2$ and $\rho' = \rho'_1$. Then

$$\text{Hom}_{\mathcal{O}_{\mathbb{A}^2}}(I, \mathcal{O}_{\mathbb{A}^2}/I)^G \subset$$
$$\text{Hom}_G(V_\ell(\rho'_1), V_2(\rho'_1)) \oplus \text{Hom}_G(V_3(\rho_2), V_1(\rho_2) \oplus V_{\ell-1}(\rho_2)).$$

Let φ be any element of $\text{Hom}_{\mathcal{O}_{\mathbb{A}^2}}(I, \mathcal{O}_{\mathbb{A}^2}/I)^G$. A nontrivial G-isomorphism φ_0 of $V_3(\rho_2)$ onto $V_1(\rho_2)$ is given by $\varphi_0(x^2 y) = x$, $\varphi_0(xy^2) = -y$. Therefore we may assume $\varphi = c\varphi_0 \mod V_{\ell-1}(\rho_2)$ for some constant c. Since φ defines an $\mathcal{O}_{\mathbb{A}^2}$-homomorphism, we have $y\varphi(x^2 y) = x\varphi(xy^2)$, so that $2cxy = 0$ in $\mathcal{O}_{\mathbb{A}^2}/I$. It follows that $c = 0$, and $\varphi(V_3(\rho_2)) \subset V_{\ell-1}(\rho_2)$. Thus the formula for $I = \mathcal{I}(W(\rho'_1, \rho_2))$ is proved.

Now we consider the general case. By 13.7 we see that $\{\mathfrak{m}/I\}[\rho]$ contains $V_{\frac{h}{2}+d(\rho)}(\rho)$ as a nontrivial factor, while $\{\mathfrak{m}/I\}[\rho']$ contains $V_{\frac{h}{2}-d(\rho')}(\rho')$ similarly. Moreover by the proof in 13.7 we see that either of the linear subspaces $\text{Hom}_G(V_{\frac{h}{2}-d(\rho)}(\rho), V_{\frac{h}{2}+d(\rho)}(\rho))$ and $\text{Hom}_G(V_{\frac{h}{2}+d(\rho')}(\rho'), V_{\frac{h}{2}-d(\rho')}(\rho'))$ yield nontrivial deformations of the ideal I inside the exceptional set E.

Since $\dim T_{[I]}(X_G) = 2$ by Theorem 9.3, these linear subspaces span $T_{[I]}(X_G)$. Hence we have

$$T_{[I]}(X_G) \simeq$$
$$\mathrm{Hom}_G(V_{\frac{h}{2}-d(\rho)}(\rho), V_{\frac{h}{2}+d(\rho)}(\rho)) \oplus \mathrm{Hom}_G(V_{\frac{h}{2}+d(\rho')}(\rho'), V_{\frac{h}{2}-d(\rho')}(\rho')),$$

with

$$T_{[I]}(E(\rho)) \simeq \mathrm{Hom}_G(V_{\frac{h}{2}-d(\rho)}(\rho), V_{\frac{h}{2}+d(\rho)}(\rho)),$$
$$T_{[I]}(E(\rho')) \simeq \mathrm{Hom}_G(V_{\frac{h}{2}+d(\rho')}(\rho'), V_{\frac{h}{2}-d(\rho')}(\rho')).$$

This completes the proof of Lemma 13.9 for $\rho, \rho' \neq \rho_{n-2}$. The cases $\rho = \rho_{n-2}$ are proved similarly. $\quad\square$

Lemma 13.10 *Let $E^*(\rho)$ be the closure in E of the set*

$$\{\mathcal{I}(W); W \in C(\rho), W \neq V_{\frac{h}{2}\pm d(\rho)}\}.$$

Then $E^*(\rho)$ is a smooth rational curve.

Proof By Lemma 13.9, $E^*(\rho)$ is smooth at $\mathcal{I}(W(\rho, \rho'))$ for ρ' adjacent to ρ. It remains to prove the assertion elsewhere on $E^*(\rho)$.

Let $C^0(\rho) := \{W \in C(\rho); W \neq V_{\frac{h}{2}\pm d(\rho)}\}$ and $I := \mathcal{I}(W)$ for $W \in C^0(\rho)$. Since we have a flat family of ideals $\mathcal{I}(W)$ for $W \in C^0(\rho)$, we have a natural morphism $\iota: C^0(\rho) \to \mathrm{Hilb}^G(\mathbf{A}^2)$, and a natural homomorphism $(d\iota)_*: T_{[W]}(C(\rho)) \to T_{[I]}(\mathrm{Hilb}^G(\mathbf{A}^2))$. Equivalently there is a homomorphism

$$(d\iota)_*: \mathrm{Hom}(W, V_{\frac{h}{2}-d(\rho)}(\rho) + V_{\frac{h}{2}+d(\rho)}(\rho)/W) \to \mathrm{Hom}_{\mathcal{O}_{\mathbf{A}^2}}(I, \mathcal{O}_{\mathbf{A}^2}/I)^G.$$

Let $\varphi \in T_{[W]}(C(\rho))$. Then $(d\iota)_*(\varphi)(I) \subset \mathfrak{m}/I$ because $C(\rho) \subset E$. Recall that $\{\mathfrak{m}/I\}[\rho_0] = 0$ by Corollary 9.6. Hence $(d\iota)_*(\varphi)(\mathfrak{n}) = 0$. Since $I/\mathfrak{n}$ is generated by W by 13.7, $(d\iota)_*(\varphi)$ is induced from φ by extending it to $\bigoplus S_k W$ as an $\mathcal{O}_{\mathbf{A}^2}$-homomorphism. Note that we have

$$V_{\frac{h}{2}-d(\rho)}(\rho) + V_{\frac{h}{2}+d(\rho)}(\rho)/W \subset \mathfrak{m}/I.$$

It follows that $(d\iota)_*$ is injective and that $C^0(\rho)$ is immersed at $\mathcal{I}(W)$. The same argument applies as well when $W = V_{\frac{h}{2}+d(\rho)}$ if there is no adjacent ρ' with $d(\rho') > d(\rho)$. Hence $E^*(\rho)$ is a smooth rational curve. $\quad\square$

We will see $E(\rho) = E^*(\rho)$ soon in 13.11.

13.11   Proof of Theorem 13.6 – Conclusion

Let E be the exceptional set of π, and E^* the union of all $E^*(\rho)$ for $\rho \in \operatorname{Irr} G$. Since $E^*(\rho) \subset E(\rho)$ by 13.7, E^* is a subset of E. Since π is a birational morphism, E is connected and set theoretically, it is the total fiber $\pi^{-1}(0)$ over the singular point $0 \in S_G$. Hence in particular $P(\rho, \rho') \subset E$ for any ρ, ρ'. By Lemma 13.9, the dual graph of E^* is the same as the Dynkin diagram $\Gamma(\operatorname{Irr} G)$ of $\operatorname{Irr} G$. Hence E^* is connected because $\Gamma(\operatorname{Irr} G)$ is connected. By Lemma 13.10 E^* is smooth except at $\mathcal{I}(W(\rho, \rho'))$, while E^* has two smooth irreducible components $E^*(\rho)$ and $E^*(\rho')$ meeting transversally at $\mathcal{I}(W(\rho, \rho'))$ by Lemma 13.9. It follows that E^* is a connected component of E. Hence $E^* = E$. It follows that $E(\rho) = E^*(\rho)$ for all $\rho \in \operatorname{Irr} G$, $P(\rho, \rho') = \{\mathcal{I}(W(\rho, \rho'))\}$ for ρ, ρ' adjacent, and $P(\rho, \rho') = \emptyset$ otherwise. Similarly $Q(\rho, \rho', \rho'') = \emptyset$. Thus Theorem 13.6 is proved.

13.12   Conclusion

The proof of Theorem 13.6 also proves Theorems 10.4 and 10.7 automatically. Theorems 10.5–10.6 are clear from Tables 7–8. Since any subscheme in $\operatorname{Hilb}^G(\mathbb{A}^2)$ with support outside the exceptional set E is a G-orbit of $|G|$ distinct points in $\mathbb{A}^2 \setminus \{0\}$, the defining ideal I of it is given by using G-invariant functions as follows

$$I = \big(F(x, y) - F(a, b), G(x, y) - G(a, b), H(x, y) - H(a, b)\big),$$

where $F(x, y) = x^\ell + y^\ell$, $G(x, y) = xy(x^\ell - y^\ell)$, $H(x, y) = x^2 y^2$ and $(a, b) \neq (0, 0)$. Thus we obtain a complete description of the ideals in $\operatorname{Hilb}^G(\mathbb{A}^2)$.

14   The binary tetrahedral group E_6

14.1   Character table

The binary tetrahedral group $G = \mathbb{T}$ is defined as the subgroup of $\operatorname{SL}(2, \mathbb{C})$ of order 24 generated by $\mathbb{D}_2 = \langle \sigma, \tau \rangle$ and μ:

$$\sigma = \begin{pmatrix} i & 0 \\ 0 & -i \end{pmatrix}, \quad \tau = \begin{pmatrix} 0 & 1 \\ -1 & 0 \end{pmatrix}, \quad \mu = \frac{1}{\sqrt{2}} \begin{pmatrix} \varepsilon^7, & \varepsilon^7 \\ \varepsilon^5, & \varepsilon \end{pmatrix},$$

where $\varepsilon = e^{2\pi i/8}$ [Slodowy80], p. 74. G acts on $\mathbb{A}^2$ from the right by $(x, y) \mapsto (x, y)g$ for $g \in G$. $\mathbb{D}_2$ is a normal subgroup of G and the following is exact:

$$1 \to \mathbb{D}_2 \to G \to \mathbb{Z}/3\mathbb{Z} \to 1.$$

See Table 9 for the character table of G [Schur07] and the other relevant invariants. The Coxeter number h of E_6 is equal to 12; here $\omega = (-1 + \sqrt{3}i)/2$.

ρ	1	2	3	4	5	6	7	d	$(\frac{h}{2} \pm d)$
	1	-1	τ	μ	μ^2	μ^4	μ^5		
(♯)	1	1	6	4	4	4	4		
ρ_0	1	1	1	1	1	1	1	(2)	–
ρ_2	2	-2	0	1	-1	-1	1	1	(5,7)
ρ_3	3	3	-1	0	0	0	0	0	(6,6)
ρ_2'	2	-2	0	ω^2	$-\omega$	$-\omega^2$	ω	1	(5,7)
ρ_1'	1	1	1	ω^2	ω	ω^2	ω	2	(4,8)
ρ_2''	2	-2	0	ω	$-\omega^2$	$-\omega$	ω^2	1	(5,7)
ρ_1''	1	1	1	ω	ω^2	ω	ω^2	2	(4,8)

Table 9: Character table of E_6

14.2 Symmetric tensors modulo $\mathfrak{n}$

Let S_m be the space of homogeneous polynomials in x and y of degree m. The G-modules S_m and $\overline{S}_m := S_m(\mathfrak{m}/\mathfrak{n})$ by ρ_2 decompose into irreducible G-modules. We define a G-submodule of $\mathfrak{m}/\mathfrak{n}$ by $\overline{V}_i(\rho_j) := S_i(\mathfrak{m}/\mathfrak{n})[\rho_j]$ the sum of all copies of ρ in $S_i(\mathfrak{m}/\mathfrak{n})$, and define $V_i(\rho_j)$ to be a G-submodule of S_i such that $V_i(\rho_j) \simeq \overline{V}_i(\rho_j)$, $V_i(\rho_j) \equiv \overline{V}_i(\rho_j) \mod \mathfrak{n}$. We use $V_i(\rho_j)$ and $\overline{V}_i(\rho_j)$ interchangeably whenever this is harmless. For a G-module W we define $W[\rho]$ to be the sum of all the copies of ρ in W.

It is known by [Klein], p. 51 that there are G-invariant polynomials A_6, A_8, A_6^2 and A_{12} respectively of homogeneous degrees 6, 8, 12 and 12. In his notation, we may assume that $A_6 = T$, $A_8 = W$ and $A_{12} = \varphi^3$. See Section 14.3.

The decomposition of S_m and $\overline{S}_m$ for small values of m are given in Table 10. The factors of $\overline{S}_m$ in brackets are those in S_{McKay}. We see by Table 10 that $V_{6 \pm d(\rho)}(\rho) \simeq \rho^{\oplus 2}$ if $d(\rho) = 0$, or ρ if $d(\rho) \geq 1$. We also see that $\overline{S}_{6-k} \simeq \overline{S}_{6+k}$ for any k. Thus Theorems 10.5–10.6 for E_6 follows from Table 10 immediately.

m	S_m	$\overline{S}_m$
0	ρ_0	0
1	ρ_2	ρ_2
2	ρ_3	ρ_3
3	$\rho_2' + \rho_2''$	$\rho_2' + \rho_2''$
4	$\rho_1' + \rho_1'' + \rho_3$	$(\rho_1' + \rho_1'') + \rho_3$
5	$\rho_2 + \rho_2' + \rho_2''$	$(\rho_2 + \rho_2' + \rho_2'')$
6	$\rho_0 + 2\rho_3$	$(2\rho_3)$
7	$2\rho_2 + \rho_2' + \rho_2''$	$(\rho_2 + \rho_2' + \rho_2'')$
8	$\rho_0 + \rho_1' + \rho_1'' + 2\rho_3$	$(\rho_1' + \rho_1'') + \rho_3$
9	$\rho_2 + 2\rho_2' + 2\rho_2''$	$\rho_2' + \rho_2''$
10	$\rho_1' + \rho_1'' + 3\rho_3$	ρ_3
11	$2\rho_2 + 2\rho_2' + 2\rho_2''$	ρ_2
12	$2\rho_0 + \rho_1' + \rho_1'' + 3\rho_3$	0

Table 10: Irreducible decompositions of $\overline{S}_m(E_6)$

m	ρ	$V_m(\rho)$	m	ρ	$V_m(\rho)$
1	ρ_2	x, y	7	ρ_2	$s_1\varphi, s_2\varphi$
2	ρ_3	x^2, xy, y^2	7	ρ_2'	$s_1\psi, s_2\psi$
3	ρ_2'	q_1, q_2	7	ρ_2''	$q_1\varphi, q_2\varphi$
3	ρ_2''	s_1, s_2	8	ρ_1'	ψ^2
4	ρ_1'	φ	8	ρ_1''	φ^2
4	ρ_1''	ψ	8	ρ_3	$p_1p_2\varphi, p_2p_3\varphi, p_3p_1\varphi$
4	ρ_3	p_1p_2, p_2p_3, p_3p_1	9	ρ_2'	$x\psi^2, y\psi^2$
5	ρ_2	γ_1, γ_2	9	ρ_2''	$x\varphi^2, y\varphi^2$
5	ρ_2'	$x\varphi, y\varphi$	10	ρ_3	$x^2\varphi^2, xy\varphi^2, y^2\varphi^2$
5	ρ_2''	$x\psi, y\psi$	11	ρ_2	$q_1\varphi^2, q_2\varphi^2$
6	ρ_3	$V_2(\rho_3)\varphi \oplus V_2(\rho_3)\psi$			

Table 11: $V_m(\rho)(E_6)$

14.3   Generators of $V_j(\rho)$

We prepare some notation for Table 11. Let

$$p_1 = x^2 - y^2, \quad p_2 = x^2 + y^2, \quad p_3 = xy,$$
$$q_1 = x^3 + (2\omega + 1)xy^2, \quad q_2 = y^3 + (2\omega + 1)x^2y,$$
$$s_1 = x^3 + (2\omega^2 + 1)xy^2, \quad s_2 = y^3 + (2\omega^2 + 1)x^2y,$$
$$\gamma_1 = x^5 - 5xy^4, \quad \gamma_2 = y^5 - 5x^4y, \quad T = p_1p_2p_3,$$
$$\varphi = p_2^2 + 4\omega p_3^2, \quad \psi = p_2^2 + 4\omega^2 p_3^2, \quad W = \varphi\psi.$$

We note that $\mathfrak{n}$ is generated by T, W and φ^3 (or ψ^3) by [Klein], p. 51. Computations give Table 11. We note the relations

$$\rho_2' = \rho_1' \cdot \rho_2 = \rho_1'' \cdot \rho_2'', \qquad \rho_2'' = \rho_1' \cdot \rho_2' = \rho_1'' \cdot \rho_2,$$
$$\rho_2 = \rho_1' \cdot \rho_2'' = \rho_1'' \cdot \rho_2', \qquad \rho_3 = \rho_1' \cdot \rho_3 = \rho_1'' \cdot \rho_3.$$

In view of Table 10, each irreducible G factor appears in $\overline{S}_m$ with multiplicity at most one except when $m = 6$, $\rho = \rho_3$. Therefore the following congruences of G-modules modulo $\mathfrak{n}$ are clear from the fact that these G-modules are nontrivial modulo $\mathfrak{n}$:

$$V_3(\rho_2'')\varphi \equiv V_3(\rho_2')\psi, \qquad V_4(\rho_3)\varphi \equiv V_4(\rho_3)\psi,$$
$$V_1(\rho_2)\varphi^2 \equiv V_5(\rho_2)\psi, \qquad V_5(\rho_2)\varphi \equiv V_1(\rho_2)\psi^2,$$
$$V_2(\rho_3)\varphi^2 \equiv V_2(\rho_3)\psi^2, \qquad V_3(\rho_2')\varphi^2 \equiv V_3(\rho_2'')\psi^2.$$

For instance, $s_i\varphi - q_i\psi \equiv 0 \mod T$, so that $V_3(\rho_2'')\varphi \equiv V_3(\rho_2')\psi$. Since $p_1p_2(\varphi - \psi) \equiv 0 \mod T$, $p_2p_3(\varphi - \omega\psi) \equiv 0 \mod T$ and $p_3p_1(\varphi - \omega^2\psi) \equiv 0 \mod T$ so that $V_4(\rho_3)\varphi \equiv V_4(\rho_3)\psi$.

Lemma 14.4   *1.*

$$S_m\bar{V}_4(\rho_1') = \begin{cases} \rho_2' & \text{for } m = 1, \\ \rho_3 & \text{for } m = 2, \\ \rho_2 + \rho_2'' & \text{for } m = 3, \\ \rho_1'' + \rho_3 & \text{for } m = 4. \end{cases}$$

2. $S_m\bar{V}_4(\rho_1') = \overline{S}_{m+4}$ for $m \geq 5$, and $S_m\bar{V}_5(\rho_2') = S_{m+1}\bar{V}_4(\rho_1')$ for $m \geq 1$.

3. $S_m\bar{V}_5(\rho_2) = \rho_3$ for $m = 1$, $\rho_2' + \rho_2''$ for $m = 2$, and $\overline{S}_{k+5}$ for $m \geq 3$.

4. $S_1\bar{V}_7(\rho_2') = \rho_1' + \rho_3$.

Proof   (1) is clear for $k = 1, 2$. Next we consider $S_3 V_4(\rho_1')$. By Table 10 $S_3 V_4(\rho_1') \simeq S_3 \otimes V_4(\rho_1') \simeq \rho_2'' + \rho_2$. We prove $S_1 \cdot A_6 \neq \{S_3 V_4(\rho_1')\}[\rho_2] = V_3(\rho_2'') V_4(\rho_1')$. For otherwise, A_6 is divisible by $\varphi \in V_4(\rho_1')$, whence $A_6/\varphi \in V_2(\rho_1') = \{0\}$, a contradiction. Hence we have $S_3 \bar{V}_4(\rho_1') = \rho_2 + \rho_2''$. Similarly $S_4 V_4(\rho_1') = \rho_0 + \rho_1'' + \rho_3$ where $\{S_4 V_4(\rho_1')\}[\rho_0] = S_0 \cdot A_8$. The factors ρ_1'' and ρ_3 in $S_4 V_4(\rho_1')$ are not divisible by A_6. In fact, otherwise $\{S_4 V_4(\rho_1')\}[\rho_3] = S_2 \cdot A_6$ because $S_2 \simeq \rho_3$. It follows that A_6 is divisible by φ, which is a contradiction. Therefore $S_4 \bar{V}_4(\rho_1') = \rho_1'' + \rho_3$. Finally we see $S_5 V_4(\rho_1') = \rho_2 + \rho_2' + \rho_2''$ where $\{S_5 V_4(\rho_1')\}[\rho_2] = S_1 \cdot A_8$. The factors ρ_2' and ρ_2'' in $S_5 V_4(\rho_1')$ are not divisible by A_6. For instance if $\{S_5 V_4(\rho_1')\}[\rho_2'] = V_3(\rho_2') \cdot A_6$, then since the generators of $V_3(\rho_2')$ are coprime, A_6 is divisible by φ, a contradiction. It follows that $S_5 \bar{V}_4(\rho_1') = \rho_2' + \rho_2'' = \bar{S}_9$. The rest of (1) is clear. (2) is clear from (1).

Next, we prove that $S_1 \bar{V}_5(\rho_2) = \rho_3$. First, Table 11 gives $\dim S_1 V_5(\rho_2) = 4$. Thus $S_1 V_5(\rho_2) \simeq \rho_2 \otimes \rho_2 \simeq \rho_0 + \rho_3$. Hence $\{S_1 V_5(\rho_2)\}[\rho_0] = S_0 \cdot A_6$. It follows that $S_1 \bar{V}_5(\rho_2) = \rho_3$. Now consider $S_2 V_5(\rho_2)$. Since $\dim S_1 \otimes V_5(\rho_2) = 4$, we have $\dim S_2 \otimes V_5(\rho_2) \geq 5$. We see that $S_2 V_5(\rho_2) = S_2 \otimes V_5(\rho_2) = \rho_2 + \rho_2' + \rho_2''$, and that $\rho_2 \simeq S_1 \cdot A_6 \subset S_2 V_5(\rho_2)$, $V_3(\rho_2'') V_4(\rho_1') = V_7(\rho_2') \simeq \rho_2'$ and $V_3(\rho_2') V_4(\rho_1') = V_7(\rho_2'') \simeq \rho_2''$. Hence $S_2 \bar{V}_5(\rho_2) = \rho_2' + \rho_2''$.

On the other hand, $S_1 V_3(\rho_2'') = S_1 \otimes V_3(\rho_2'') = \rho_1' + \rho_3$, so that $S_1 V_7(\rho_2') = S_1 V_3(\rho_2'') V_4(\rho_1'') = \rho_1' + \rho_3$. We prove that $S_1 \bar{V}_7(\rho_2') = \rho_1' + \rho_3$. For otherwise, by Table 10, we have $\{S_1 \bar{V}_7(\rho_2')\}[\rho_3] = 0$ so that $\{S_1 V_7(\rho_2')\}[\rho_3] = S_2 A_6$. $V_7(\rho_2')$ is divisible by ψ, so that A_6 is divisible by ψ. Hence $A_6/\psi \in V_2(\rho_1')$, which contradicts $S_2 = \rho_3$. Therefore $\{S_1 V_7(\rho_2')\}[\rho_3] = \rho_3$ and $S_1 \bar{V}_7(\rho_2') = \rho_1' + \rho_3$. Similarly $S_1 \bar{V}_7(\rho_2'') = \rho_1'' + \rho_3$. This proves (4). Moreover $S_3 \bar{V}_5(\rho_2) = S_1 S_2 \bar{V}_5(\rho_2) = S_1(\bar{V}_7(\rho_2') + \bar{V}_7(\rho_2''))$ so that $S_3 \bar{V}_5(\rho_2) \supset \rho_1' + \rho_1'' + \rho_3 = \bar{S}_8$. This proves (3).   □

Lemma 14.5   Let $W_k = S_1 \cdot \bar{V}_5(\rho_2^{(k)})$ $(\simeq \rho_3)$ for any $k = 0, 1, 2$, where $\rho_2^{(k)} = \rho_2, \rho_2', \rho_2''$. Let $W \in \mathbb{P}(V_6(\rho_3))$. Then $S_1 W = \rho_2 + \rho_2' + \rho_2''$ if and only if $W \neq W_k$ for $k = 1, 2, 3$.

Proof   We see $S_1 \cdot W_1 = S_2 \cdot \bar{V}_5(\rho_2') = S_3 \cdot \bar{V}_4(\rho_1') = \rho_2 + \rho_2''$ by Lemma 14.4. Similarly $S_1 \cdot W_2 = S_3 \cdot \bar{V}_4(\rho_1'') = \rho_2 + \rho_2'$. Also by Lemma 14.4, (3) we have $S_1 \cdot W_0 = \rho_2' + \rho_2''$.

Conversely assume $W \neq W_k$ for any k. Choose and fix a G-module isomorphism $h: W_1 \to W_2$. For instance, $h(p_k \varphi) = \omega^{-k} p_k \psi$. Then h induces a natural isomorphism $\{S_1 \otimes h\}[\rho_2] : \{S_1 \otimes W_1\}[\rho_2] \to \{S_1 \otimes W_2\}[\rho_2]$, which induces an isomorphism $\{S_1 \cdot h\}[\rho_2] : \{S_1 \cdot W_1\}[\rho_2] \to \{S_1 \cdot W_2\}[\rho_2]$. Since $\bar{S}_7$ contains a single ρ_2, we have $\{S_1 \cdot W_1\}[\rho_2] \simeq \{S_1 \cdot W_2\}[\rho_2]$ $(\simeq \rho_2)$ by $\{S_1 \cdot h\}[\rho_2]$. It follows that $\{S_1 \cdot h\}[\rho_2]$ is a nonzero constant multiple of the identity. Since $V_6(\rho_3) = W_1 \oplus W_2$, this proves uniqueness of the G-submodule $W \simeq \rho_3$ of $V_6(\rho_3)$ such that $\{S_1 \cdot W\}[\rho_2] = 0$. Since $\{S_1 \cdot W_0\}[\rho_2] = 0$, we have $\{S_1 \cdot W\}[\rho_2] \neq 0$ by the assumption $W \neq W_0$. Similarly there exists a unique proper G-submodule $W \in V_6(\rho_3)$ such that $\{S_1 \cdot W\}[\rho_2'] = 0$ or $\{S_1 \cdot W\}[\rho_2''] = 0$. As we saw above, $\{S_1 \cdot W_1\}[\rho_2'] = 0$ and $\{S_1 \cdot W_2\}[\rho_2''] = 0$. Therefore $S_1 \cdot W = \rho_2 + \rho_2' + \rho_2''$ if $W \neq W_k$ for $k = 0, 1, 2$. □

14.6   Proof of Theorem 10.7 in the E_6 case

Consider $I \in X_G$ in the exceptional set E, or equivalently, $I \in X_G$ with $I \subset \mathfrak{m}$. For a finite submodule W of $\mathfrak{m}$ we define $\mathcal{I}(W) = W\mathcal{O}_{\mathbb{A}^2} + \mathfrak{n}$ and $V(\mathcal{I}(W)) := \mathcal{I}(W)/\mathfrak{m}\mathcal{I}(W) + \mathfrak{n}$. We write $\equiv$ for congruence modulo $\mathfrak{n}$.

Case $\mathcal{I}(W) \in E^0(\rho_1')$ Let $W \in \mathbb{P}(V_4(\rho_1')\oplus V_8(\rho_1'))$, so that $W \simeq \rho_1'$. Suppose that $W \neq V_8(\rho_1')$ and set $\mathcal{I}(W) = W\mathcal{O}_{\mathbb{A}^2} + \mathfrak{n}$. Since $\overline{S}_{12} = 0$, by Lemma 14.4 we have $S_k \cdot W \equiv S_k \cdot \bar{V}_4(\rho_1')$ for $k \geq 4$. Also by Lemma 14.4 $S_k \cdot \bar{V}_4(\rho_1') = \overline{S}_{k+4}$ for $k \geq 5$. Hence $\overline{S}_k \subset \mathcal{I}(W)/\mathfrak{n}$ for $k \geq 9$. Since $S_k \cdot W = S_k \cdot \bar{V}_4(\rho_1') \mod \overline{S}_9$ for $k \geq 1$, we deduce that

$$\mathcal{I}(W)/\mathfrak{n} = W + \sum_{k \geq 1} S_k \cdot \bar{V}_4(\rho_1') = W + \sum_{k=1}^{4} S_k \cdot \bar{V}_4(\rho_1') + \sum_{k=9}^{11} \overline{S}_k.$$

We see by Lemma 14.4

$$W + S_4\bar{V}_4(\rho_1') = \rho_1' + \rho_1'' + \rho_3 = \frac{1}{2}(\overline{S}_4 + \overline{S}_8),$$

$$S_1\bar{V}_4(\rho_1') + S_3\bar{V}_4(\rho_1') = \rho_2 + \rho_2' + \rho_2'' = \frac{1}{2}(\overline{S}_5 + \overline{S}_7),$$

$$S_2\bar{V}_4(\rho_1') = \rho_3 = \frac{1}{2}\overline{S}_6.$$

By duality, we have $\mathcal{I}(W)/\mathfrak{n} = \sum_{\rho \in \mathrm{Irr}\, G} \deg(\rho)\rho$. Thus $\mathcal{I}(W) \in X_G$ and $V(\mathcal{I}(W)) \simeq W$.

Case $\mathcal{I}(W) \in E^0(\rho_2')$ Let $W \in \mathbb{P}(V_5(\rho_2') \oplus V_7(\rho_2'))$ with $W \simeq \rho_2'$. Suppose $W \neq V_5(\rho_2'), V_7(\rho_2')$. Since $\overline{S}_{12} = 0$, we have $S_k \cdot W \equiv S_k \cdot \bar{V}_5(\rho_2') = \overline{S}_{k+5}$ for $k \geq 5$ by the condition $W \neq V_7((\rho_2'))$. We also see that $S_4 \cdot W = S_4 \cdot \bar{V}_5(\rho_2') \mod \overline{S}_{11} = \overline{S}_9$. Therefore $\overline{S}_9 \subset \mathcal{I}(W)/\mathfrak{n}$. Hence $S_k \cdot W = S_k \cdot \bar{V}_5(\rho_2') \mod \overline{S}_9$ for $k \geq 2$. Since $S_1 \cdot \bar{V}_5(\rho_2') = \rho_3$ and $S_1 \cdot \bar{V}_7(\rho_2') = \rho_1' + \rho_3$, we have $S_1 \cdot W \equiv \rho_1' + \rho_3$ and $\{S_1 \cdot W\}[\rho_1'] \equiv \bar{V}_8(\rho_1') \subset \mathcal{I}(W)/\mathfrak{n}$ by the assumption $W \neq V_5(\rho_2')$. Since $S_3\bar{V}_5(\rho_2') = \rho_1' + \rho_3$, we have $\overline{S}_8 = V_8(\rho_1') \oplus S_3\bar{V}_5(\rho_2') \subset \mathcal{I}(W)/\mathfrak{n}$. It follows that

$$\mathcal{I}(W)/\mathfrak{n} = W + \sum_{k\geq 1} S_k \cdot \bar{V}_5(\rho_2') = W + \sum_{k=1}^{2} S_k \cdot \bar{V}_5(\rho_2') + \sum_{k=8}^{11} \overline{S}_k \quad \text{and}$$

$$W + S_1\bar{V}_5(\rho_2') + S_2\bar{V}_5(\rho_2') = \rho_2 + \rho_2' + \rho_2'' + \rho_3 = \frac{1}{2}(\overline{S}_5 + \overline{S}_6 + \overline{S}_7).$$

Hence $\mathcal{I}(W)/\mathfrak{n} = \sum_{\rho\in\mathrm{Irr}\,G} \deg(\rho)\rho$. Thus $\mathcal{I}(W) \in X_G$ with $V(\mathcal{I}(W)) \simeq W$.

Case $\mathcal{I}(W) \in E^0(\rho_1'')$ or $\mathcal{I}(W) \in E^0(\rho_2'')$ These cases are similar.

Case $\mathcal{I}(W) \in E^0(\rho_2)$ Let $W \in \mathbb{P}(V_5(\rho_2)\oplus V_7(\rho_2))$, so that $W \simeq \rho_2$. Suppose that $W \neq V_7(\rho_2)$. As above, we see that $\overline{S}_k \subset \mathcal{I}(W)/\mathfrak{n}$ for $k \geq 10$. It follows that $S_3 \cdot W = S_3 \cdot \bar{V}_5(\rho_2) \mod \overline{S}_{10} = \overline{S}_8$. Therefore $\overline{S}_k \subset \mathcal{I}(W)/\mathfrak{n}$ for $k \geq 8$. Similarly $S_2 \cdot W \equiv S_2 \cdot \bar{V}_5(\rho_2) = \rho_2' + \rho_2'' \mod \overline{S}_8$ and $S_1 \cdot W \equiv S_1 \cdot \bar{V}_5(\rho_2) = \rho_3$ $\mod \overline{S}_8$. It follows that

$$\mathcal{I}(W)/\mathfrak{n} = W + \sum_{k\geq 1} S_k \cdot \bar{V}_5(\rho_2) = W + \sum_{k=1}^{2} S_k \cdot \bar{V}_5(\rho_2) + \sum_{k=8}^{11} \overline{S}_k, \quad \text{and}$$

$$W + S_1\bar{V}_5(\rho_2) + S_2\bar{V}_5(\rho_2) = \rho_2 + \rho_2' + \rho_2'' + \rho_3 = \frac{1}{2}(\overline{S}_5 + \overline{S}_6 + \overline{S}_7).$$

Hence $\mathcal{I}(W)/\mathfrak{n} = \sum_{\rho\in\mathrm{Irr}\,G} \deg(\rho)\rho$. Thus $\mathcal{I}(W) \in X_G$ with $V(\mathcal{I}(W)) \simeq W$.

Case $\mathcal{I}(W) \in E^0(\rho_3)$ Let $W \in \mathbb{P}(V_6(\rho_3))$. Let $W_k = S_1 \cdot V_5(\rho_2^{(k)})$ for any $k = 0, 1, 2$ where $\rho_2^{(k)} = \rho_2, \rho_2', \rho_2''$. Now we suppose that $W \neq W_k$. Then $S_1 \cdot W \equiv \overline{S}_7$ by Lemma 14.5 so that $\mathcal{I}(W)$ contains $\overline{S}_k$ for any $k \geq 7$. It follows that

$$\mathcal{I}(W)/\mathfrak{n} = W + \sum_{k\geq 1} S_k W = W + \sum_{k=7}^{11} \overline{S}_k.$$

Hence $\mathcal{I}(W)/\mathfrak{n} = \sum_{\rho\in\mathrm{Irr}\,G} \deg(\rho)\rho$, and so $\mathcal{I}(W) \in X_G$ with $V(\mathcal{I}(W)) \simeq W$.

Case $\mathcal{I}(W) \in P(\rho_1', \rho_2')$ Let $W = W(\rho_1', \rho_2') := V_8(\rho_1') \oplus V_5(\rho_2')$. Recall that $W = \{S_1 \cdot V_7(\rho_2')\}[\rho_1'] \oplus V_5(\rho_2') = V_8(\rho_1') \oplus S_1 \cdot V_4(\rho_1')$. By Lemma 14.4, we see that $S_1 \cdot \bar{V}_5(\rho_2') = \rho_3$, $S_2 \cdot \bar{V}_5(\rho_2') = \rho_2 + \rho_2''$, $S_3 \cdot \bar{V}_5(\rho_2') = \rho_1'' + \rho_3$ and $\overline{S}_k \subset \mathcal{I}(W)/\mathfrak{n}$ for $k \geq 8$. It follows that $\mathcal{I}(W)/\mathfrak{n} = \sum_{\rho\in\mathrm{Irr}\,G} \deg(\rho)\rho$ by Table 10. Therefore $\mathcal{I}(W) \in X_G$ with $V(\mathcal{I}(W)) \simeq W$.

Case $\mathcal{I}(W) \in P(\rho_2', \rho_3)$ Let $W = W(\rho_2', \rho_3) := V_7(\rho_2') \oplus S_1 V_5(\rho_2') = V_7(\rho_2') \oplus W_1$. We recall that $S_1 \cdot W_1 = \rho_2 + \rho_2''$, so that $\overline{S}_k \subset \mathcal{I}(W)/\mathfrak{n}$ for $k \geq 7$. Since $W_1 = \rho_3$ we have $\mathcal{I}(W)/\mathfrak{n} = \sum_{\rho \in \mathrm{Irr}\, G} \deg(\rho)\rho$ by Table 10. Therefore $\mathcal{I}(W) \in X_G$ with $V(\mathcal{I}(W)) \simeq W$.

Cases $\mathcal{I}(W) \in P(\rho_2, \rho_3)$ **or** $\mathcal{I}(W) \in P(\rho_2'', \rho_3)$ Similar.

The following Lemma is proved in the same manner as before. It allows us to complete the proof of Theorem 10.7 by the same argument as in Section 13.

Lemma 14.7 *Each $E(\rho)$ is a smooth rational curve. Moreover, if ρ and ρ' are adjacent then*

1. *as $\mathcal{I}(W) \in E(\rho)$ approaches the point $P(\rho, \rho')$, the limit of $\mathcal{I}(W)$ is $\mathcal{I}(W(\rho, \rho'))$;*

2. *$E(\rho)$ and $E(\rho')$ intersect transversally at $P(\rho, \rho')$.*

14.8   Conclusion

Theorem 10.4 also follows from Lemma 14.7. Theorem 10.7, (3) follows from Tables 10–11 and Lemma 14.5.

Let $I \in X_G$. If $\mathrm{Supp}(\mathcal{O}_{\mathbf{A}^2}/I)$ is not the origin, then

$$I = (T(x, y) - T(a, b), \varphi^3(x, y) - \varphi^3(a, b), W(x, y) - W(a, b))$$

where $(a, b) \neq (0, 0)$.

By the same argument as in Section 13 we thus obtain a complete description of the G-invariant ideals in X_G.

15   The binary octahedral group E_7

15.1   Character table

The binary octahedral group $\mathbb{O}$ is defined as the subgroup of $\mathrm{SL}(2, \mathbb{C})$ of order 48 generated by $\mathbb{T} = \langle \sigma, \tau, \mu \rangle$ and κ:

$$\sigma = \begin{pmatrix} i & 0 \\ 0 & -i \end{pmatrix}, \quad \tau = \begin{pmatrix} 0 & 1 \\ -1 & 0 \end{pmatrix}, \quad \mu = \frac{1}{\sqrt{2}} \begin{pmatrix} \varepsilon^7 & \varepsilon^7 \\ \varepsilon^5 & \varepsilon \end{pmatrix}, \quad \kappa = \begin{pmatrix} \varepsilon & 0 \\ 0 & \varepsilon^7 \end{pmatrix},$$

where $\varepsilon = e^{2\pi i/8}$ [Slodowy80], p. 73. G acts on $\mathbf{A}^2$ from the right by $(x, y) \mapsto (x, y)g$ for $g \in G$. $\mathbb{D}_2$ and $\mathbb{T}$ are normal subgroups of G and the following sequences are exact:

$$1 \to \mathbb{T} \to G \to \mathbb{Z}/2\mathbb{Z} \to 1$$

and

$$1 \to \mathbb{D}_2 \to G \to S_3 \to 1,$$

where S_3 is the symmetric group on 3 letters.

See Table 12 for the character table of G and other relevant invariants. E_7 has Coxeter number $h = 18$.

15.2   Symmetric tensors modulo $\mathfrak{n}$

The G-modules S_m and $\overline{S}_m := S_m(\mathfrak{m}/\mathfrak{n})$ by $\rho_{\text{nat}} := \rho_2$ for small values of m split into irreducible G-modules as in Table 13. The factors of $\overline{S}_m$ in brackets are those in S_{McKay}. We use the same notation $\overline{V}_m(\rho)$ and $V_m(\rho)$ for $\rho \in \text{Irr}\, G$ as before. Let $\varphi = p_2^2 + 4\omega p_3^2$, $\psi = p_2^2 + 4\omega^2 p_3^2$, $T(x,y) = (x^4 - y^4)xy$. In Table 14 we denote by $W_j^{(i)} \simeq \rho_4$ the G-submodules of $V_9(\rho_4) \simeq \rho_4^{\oplus 2}$; $W_2'' := S_1 \cdot V_8(\rho_2'')$, $W_3 := S_1 \cdot V_8(\rho_3)$, $W_3' := S_1 \cdot V_8(\rho_3')$.

Lemma 15.3 *The G-module $S_m \overline{V}_k(\rho)$ splits into irreducible G-submodules as in Table 15. We read the table as $S_2 \overline{V}_6(\rho_1') = \rho_3'$, $S_2 \overline{V}_8(\rho_2'') = \rho_3 + \rho_3'$ and so on.*

Proof   The assertions for $(m,k) = (1,6),(2,6),(3,6)$ are clear. There are three generators A_8, A_{12} and A_{18} of respective degrees 8, 12 and 18 for the ring of G-invariant polynomials. We know that $A_8 = \varphi\psi$, $A_{12} = T^2$ by [Klein], p. 54.

Note first that $S_m = S_{m-8} \cdot A_8 \oplus \overline{S}_m$ for $m = 10, 11$ and

$$S_4 V_6(\rho_1') = (\rho_2'' + \rho_3') \otimes \rho_1' = \rho_2'' + \rho_3, \quad S_5 V_6(\rho_1') = (\rho_2'' + \rho_4) \otimes \rho_1' = \rho_2 + \rho_4.$$

If $\{S_4 \overline{V}_6(\rho_1')\}[\rho_3] = 0$ in $\overline{S}_{10}$, then $\{S_4 \overline{V}_6(\rho_1')\}[\rho_3] = S_2 \cdot A_8$. A_8 would be divisible by T, a generator of $V_6(\rho_1')$. However, this is impossible. Hence $\{S_4 \overline{V}_6(\rho_1')\}[\rho_3] = \rho_3$ so that $S_4 \overline{V}_6(\rho_1') = \rho_2'' + \rho_3$. $S_5 \overline{V}_6(\rho_1') = \rho_2 + \rho_4$ is proved similarly.

Since $S_6 V_6(\rho_1') = (\rho_1')^2 + \rho_3 + \rho_3' = \rho_0 + \rho_3 + \rho_3'$, $S_6 \overline{V}_6(\rho_1') = \rho_3 + \rho_3'$ or ρ_3. If $S_6 \overline{V}_6(\rho_1') = \rho_3$, then $S_6[\rho_3] \cdot V_6(\rho_1')$ is divisible by T^2, so that $S_6[\rho_3]$ is divisible by T. Since $\deg T = 6$, this is impossible. Hence $S_6 \overline{V}_6(\rho_1') = \rho_3 + \rho_3'$.

Next we have $S_7 V_6(\rho_1') = \rho_2' + \rho_2 + \rho_4$ and $\{S_7 V_6(\rho_1')\}[\rho_2] = \rho_2 \cdot A_{12}$. If $\{S_7 \overline{V}_6(\rho_1')\}[\rho_4] = 0$, then $\{S_7 V_6(\rho_1')\}[\rho_4] = V_7[\rho_4]V_6(\rho_1') = \rho_4 \cdot A_{12}$ or $\rho_4 \cdot A_8$. In the first case, $V_7[\rho_4]$ is divisible by T, which is impossible because $\deg T = 6$ and $\dim S_1 = 2 < \deg \rho_4 = 4$. In the second case, $V_7[\rho_4]$ is divisible by A_8, which is impossible. It follows that $\{S_7 \overline{V}_6(\rho_1')\}[\rho_4] = \rho_4$. If $\{S_7 V_6(\rho_1')\}[\rho_2'] = 0$, then $V_7[\rho_2']V_6(\rho_1') = \rho_2' \cdot A_{12}$ or $\rho_2' \cdot A_8$. In the first case $V_7[\rho_2']$ is divisible by

ρ	1	2	3	4	5	6	7	8	d	$(\frac{h}{2} \pm d)$
	1	-1	μ	μ^2	τ	κ	$\tau\kappa$	κ^3		
$\sharp$	1	1	8	8	6	6	12	6		
ρ_0	1	1	1	1	1	1	1	1	(3)	$-$
ρ_2	2	-2	1	-1	0	$\sqrt{2}$	0	$-\sqrt{2}$	2	$(7,11)$
ρ_3	3	3	0	0	-1	1	-1	1	1	$(8,10)$
ρ_4	4	-4	-1	1	0	0	0	0	0	$(9,9)$
ρ_3'	3	3	0	0	-1	-1	1	-1	1	$(8,10)$
ρ_2'	2	-2	1	-1	0	$-\sqrt{2}$	0	$\sqrt{2}$	2	$(7,11)$
ρ_1'	1	1	1	1	1	-1	-1	-1	3	$(6,12)$
ρ_2''	2	2	-1	-1	2	0	0	0	1	$(8,10)$

Table 12: Character table of E_7

m	S_m	$\overline{S}_m$
1	ρ_2	ρ_2
2	ρ_3	ρ_3
3	ρ_4	ρ_4
4	$\rho_2'' + \rho_3'$	$\rho_2'' + \rho_3'$
5	$\rho_2' + \rho_4$	$\rho_2' + \rho_4$
6	$\rho_1' + \rho_3 + \rho_3'$	$(\rho_1') + \rho_3 + \rho_3'$
7	$\rho_2 + \rho_2' + \rho_4$	$(\rho_2 + \rho_2') + \rho_4$
8	$\rho_0 + \rho_2'' + \rho_3 + \rho_3'$	$(\rho_2'' + \rho_3 + \rho_3')$
9	$\rho_2 + 2\rho_4$	$(2\rho_4)$
10	$\rho_2'' + 2\rho_3 + \rho_3'$	$(\rho_2'' + \rho_3 + \rho_3')$
11	$\rho_2 + \rho_2' + 2\rho_4$	$(\rho_2 + \rho_2') + \rho_4$
12	$\rho_0 + \rho_1' + \rho_2'' + \rho_3 + 2\rho_3'$	$(\rho_1') + \rho_3 + \rho_3'$
13	$\rho_2 + 2\rho_2' + 2\rho_4$	$\rho_2' + \rho_4$
14	$\rho_1' + \rho_2'' + 2\rho_3 + 2\rho_3'$	$\rho_2'' + \rho_3'$
15	$\rho_2 + \rho_2' + 3\rho_4$	ρ_4
16	$\rho_0 + 2\rho_2'' + 2\rho_3 + 2\rho_3'$	ρ_3
17	$2\rho_2 + \rho_2' + 3\rho_4$	ρ_2
18	$\rho_0 + \rho_1' + \rho_2'' + 3\rho_3 + 2\rho_3'$	0

Table 13: Irreducible decompositions of $S_m(E_7)$ and $\overline{S}_m(E_7)$

m	ρ	$V_m(\rho)$
7	ρ_2	$7x^4y^3 + y^7, -x^7 - 7x^3y^4$
11	ρ_2	$x^{10}y - 6x^6y^5 + 5x^2y^9, -xy^{10} + 6x^5y^6 - 5x^9y^2$
8	ρ_3	$-2xy^7 - 14x^5y^3, x^8 - y^8, 2x^7y + 14x^3y^5$
10	ρ_3	$4x^{10} + 60x^6y^4, 5x^9y + 54x^5y^5 + 5xy^9$
		$60x^4y^6 + 4y^{10}$
9	ρ_4	$W_2'' + W_3 = W_3 + W_3' = W_3' + W_2'' \simeq \rho_4^{\oplus 2}$
9	W_2''	$12x^6y^3 + 12x^2y^7, x^9 - 10x^5y^4 + xy^8$
		$-x^8y + 10x^4y^5 - y^9, 12x^7y^2 + 12x^3y^6$
9	W_3	$21x^6y^3 + 3x^2y^7, -x^9 + 7x^5y^4 + 2xy^8$
		$-2x^8y - 7x^4y^5 + y^9, -3x^7y^2 - 21x^3y^6$
9	W_3'	x^3T, x^2yT, xy^2T, y^3T
8	ρ_3'	x^2T, xyT, y^2T
10	ρ_3'	$-3x^8y^2 - 14x^4y^6 + y^{10}, 8x^7y^3 + 8x^3y^7$
		$x^{10} - 14x^6y^4 - 3x^2y^8$
7	ρ_2'	xT, yT
11	ρ_2'	$-11x^8y^3 - 22x^4y^7 + y^{11}, 11x^3y^8 + 22x^7y^4 - x^{11}$
6	ρ_1'	T
12	ρ_1'	$x^{12} - 33x^8y^4 - 33x^4y^8 + y^{12}$
8	ρ_2''	$\psi^2, -\varphi^2$
10	ρ_2''	$x^5y\psi - xy^5\varphi, -x^5y\varphi + xy^5\psi$

Table 14: $V_m(\rho)(E_7)$

T, which contradicts Table 14. In the second case $V_7[\rho_2]$ is divisible by A_8, absurd. Hence $\{S_7\bar{V}_6(\rho_1')\}[\rho_2'] = \rho_2'$. It follows that $S_7\bar{V}_6(\rho_1') = \rho_2' + \rho_4 = \bar{S}_{13}$.

We note next that $\dim S_1V_{11}(\rho_2') \geq 3$. If $\dim S_1V_{11}(\rho_2') = 3$, then there exists an $f \in S_{10}$ such that $V_{11}(\rho_2') = S_1 \cdot f$. Hence $f \in S_{10}[\rho_1'] = \{0\}$, a contradiction. Hence $\dim S_1V_{11}(\rho_2') = 4$, so that $S_1V_{11}(\rho_2') = \rho_1' + \rho_3'$. If $\{S_1\bar{V}_{11}(\rho_2')\}[\rho_3'] = 0$, we have $\{S_1V_{11}(\rho_2')\}[\rho_3'] = V_4[\rho_3'] \cdot A_8$ by Table 13. Since $\dim S_1 < \deg\rho_3' = 3$, there exists a nontrivial element of $\{S_1V_{11}(\rho_2')\}[\rho_3']$ divisible by both x and A_8. Hence $V_{11}(\rho_2')$ contains a nontrivial element divisible by A_8. This implies that $V_{11}(\rho_2')$ is divisible by A_8. Then $V_3(\rho_2') = V_{11}(\rho_2')A_8^{-1} = \rho_2'$, which contradicts $S_3 = \rho_4$. Hence $S_1\bar{V}_{11}(\rho_2') = \rho_1' + \rho_3'$.

It is clear from $\rho_2 \otimes \rho_2'' = \rho_4$ and Table 13 that $S_1\bar{V}_8(\rho_2'') = \rho_4$.

Next $S_2 \otimes V_8(\rho_2'') = \rho_3 + \rho_3'$ by Table 12. Since $\dim S_2V_8(\rho_2'') \geq 4$, we have $S_2V_8(\rho_2'') = \rho_3 + \rho_3'$. If $\{S_2\bar{V}_8(\rho_2'')\}[\rho_3'] = 0$, then $\{S_2V_8(\rho_2'')\}[\rho_3'] = S_2 \cdot A_8$. Since

m	k	ρ	$S_m \bar{V}_k(\rho)$	m	k	ρ	$S_m \bar{V}_k(\rho)$
1	6	ρ'_1	ρ'_2	2	8	ρ''_2	$\rho_3 + \rho'_3$
2	6		ρ'_3	3	8		$\rho_2 + \rho'_2 + \rho_4$
3	6		ρ_4	1	7	ρ_2	ρ_3
4	6		$\rho''_2 + \rho_3$	2	7		ρ_4
5	6		$\rho_2 + \rho_4$	3	7		$\rho''_2 + \rho'_3$
6	6		$\rho_3 + \rho'_3$	4	7		$\rho'_2 + \rho_4$
7	6		$\rho'_2 + \rho_4$	5	7		$\rho'_1 + \rho_3 + \rho'_3$
1	11	ρ'_2	$\rho'_1 + \rho'_3$	1	10	ρ_3	$\rho_2 + \rho_4$
1	8	ρ''_2	ρ_4	1	10	ρ'_3	$\rho'_2 + \rho_4$

Table 15: Decomposition of $S_m \bar{V}_k(\rho)$

$\deg \rho''_2 < \deg \rho_3$ and $V_8(\rho''_2)$ is generated by φ^2 and ψ^2, there exists a nontrivial element of $\{S_2 V_8(\rho''_2)\}[\rho_3]$ divisible by both φ^2 and A_8. Since φ and ψ are coprime, S_{10} contains a nontrivial element divisible by $\varphi^2 \psi$, a contradiction. If $\{S_2 \bar{V}_8(\rho''_2)\}[\rho_3] = 0$, then $\{S_2 V_8(\rho''_2)\}[\rho_3] = S_2 \cdot A_8 = \rho_3$, a contradiction. Hence $S_2 \bar{V}_8(\rho''_2) = \rho_3 + \rho'_3$.

Next we consider $S_3 \bar{V}_8(\rho''_2)$. Since $\dim S_2 V_8(\rho''_2) = 6$ by the above proof, we have $\dim S_3 V_8(\rho''_2) \geq 7$. By Table 12 $S_3 \otimes V_8(\rho''_2) = \rho_2 + \rho'_2 + \rho_4$ so that $S_3 V_8(\rho''_2) = \rho_2 + \rho'_2 + \rho_4$. Assume $S_3 \bar{V}_8(\rho''_2) \neq \rho_2 + \rho'_2 + \rho_4$. Then by Table 13 the only possibility is that $\{S_3 \bar{V}_8(\rho''_2)\}[\rho_4] = 0$. Assume $\{S_3 V_8(\rho''_2)\}[\rho_4] = S_3 \cdot A_8$ so that there exists an element of $\{S_3 V_8(\rho''_2)\}[\rho_4]$ divisible by both φ^2 and A_8. Therefore there exists a nontrivial element of S_3 divisible by ψ, which is a contradiction. Hence $S_3 \bar{V}_8(\rho''_2) = \rho_2 + \rho'_2 + \rho_4$.

Clearly $S_1 V_7(\rho_2) = \rho_0 + \rho_3$, $S_2 V_7(\rho_2) = \rho_2 + \rho_4$. Hence $S_1 \bar{V}_7(\rho_2) = \rho_3$ and $S_2 \bar{V}_7(\rho_2) = \rho_4$.

Next $S_3 \otimes V_7(\rho_2) = \rho_4 \otimes \rho_2 = \rho''_2 + \rho_3 + \rho'_3$ by Table 12. Since $\dim S_2 V_7(\rho_2) = 6$, we have $\dim S_3 V_7(\rho_2) \geq 7$ so that $S_3 \otimes V_7(\rho_2) = \rho''_2 + \rho_3 + \rho'_3$. It is clear that $\{S_1 V_7(\rho_2)\}[\rho_0] = S_0 \cdot A_8$, $\{S_2 V_7(\rho_2)\}[\rho_2] = S_1 \cdot A_8$. Hence $\{S_3 V_7(\rho_2)\}[\rho_3] = S_2 \cdot A_8$. It is clear that $\{S_3 V_7(\rho_2)\}[\rho'_3] \neq S_2 \cdot A_8$ and $\{S_3 V_7(\rho_2)\}[\rho''_2] \neq S_2 \cdot A_8$. Hence $S_3 \bar{V}_7(\rho_2) = \rho''_2 + \rho'_3$.

Next we see $\dim S_4 V_7(\rho_2) = 10$, $S_4 V_7(\rho_2) \simeq S_4 \otimes V_7(\rho_2) = \rho'_2 + 2\rho_4$. Hence $S_4 \bar{V}_7(\rho_2) = \rho'_2 + \rho_4$ by Table 13. It is easy to see that $\dim S_5 V_7(\rho_2) = 12$. Hence $S_5 V_7(\rho_2) = S_5 \otimes V_7(\rho_2) = \rho'_1 + \rho''_2 + \rho_3 + 2\rho'_3$ so that $S_5 \bar{V}_7(\rho_2) = \rho'_1 + \rho_3 + \rho'_3 = \bar{S}_{12}$ by Table 13.

Similarly we see easily that $\dim S_1 V_{10}(\rho_3) = \dim S_1 V_{10}(\rho'_3) = 6$. Hence $S_1 V_{10}(\rho_3) = \rho_2 + \rho_4$, $S_1 V_{10}(\rho'_3) = \rho'_2 + \rho_4$. If $\{S_1 \bar{V}_{10}(\rho_3)\}[\rho_4] = 0$, then $\{S_1 V_{10}(\rho_3)\}[\rho_4] = S_3 \cdot A_8$. Therefore there exists a nontrivial element of

$V_{10}(\rho_3)$ divisible by A_8 so that $V_{10}(\rho_3)$ is divisible by A_8. This implies that $\bar{V}_{10}(\rho_3) = 0$. But by the choice of it, $V_{10}(\rho_3) \simeq \bar{V}_{10}(\rho_3)$, a contradiction. This completes the proof. $\square$

Corollary 15.4   *1.* $S_1\bar{V}_6(\rho_1') = \bar{V}_7(\rho_2')$, $S_2\bar{V}_6(\rho_1') = \bar{V}_8(\rho_3')$, $S_1\bar{V}_7(\rho_2) = \bar{V}_8(\rho_3)$.

   2. $S_3\bar{V}_8(\rho_2'') = \bar{S}_{11}$, $S_5\bar{V}_7(\rho_2) = \bar{S}_{12}$, $S_7\bar{V}_6(\rho_1') = \bar{S}_{13}$.

   3. $S_2\bar{V}_8(\rho_3') = \rho_2'' + \rho_3$, $S_2\bar{V}_8(\rho_2'') = \rho_3 + \rho_3'$, $S_2\bar{V}_8(\rho_3) = \rho_2'' + \rho_3'$.

Proof Clear. $\square$

   We omit the proof of Theorem 10.7 because we need only to follow the proof in the E_6 case verbatim.

15.5   Conclusion

We also can give a complete description of G-invariant ideals in X_G. Let

$$\chi = x^{12} - 33x^8y^4 - 33x^4y^8 + y^{12}, \quad F(x,y) = \chi T, \quad W(x,y) = \varphi\psi.$$

Let $I \in X_G$. If $\mathrm{Supp}(\mathcal{O}_{\mathbb{A}^2}/I)$ is not the origin, then we know that

$$I = \big(W(x,y) - W(a,b), T^2(x,y) - T^2(a,b), F(x,y) - F(a,b)\big),$$

where $(a,b) \neq (0,0)$.

16   The binary icosahedral group E_8

16.1   Character table

The binary icosahedral group $\mathbb{I}$ is defined as the subgroup of $\mathrm{SL}(2,\mathbb{C})$ of order 120 generated by σ and τ:

$$\sigma = -\begin{pmatrix} \varepsilon^3 & 0 \\ 0 & \varepsilon^2 \end{pmatrix}, \quad \tau = \frac{1}{\sqrt{5}}\begin{pmatrix} -(\varepsilon - \varepsilon^4) & \varepsilon^2 - \varepsilon^3 \\ \varepsilon^2 - \varepsilon^3 & \varepsilon - \varepsilon^4 \end{pmatrix}$$

where $\varepsilon = e^{2\pi i/5}$. We note $\sigma^5 = \tau^2 = -1$. G acts on $\mathbb{A}^2$ from the right by $(x,y) \mapsto (x,y)g$ for $g \in G$. G is isomorphic to $\mathrm{SL}(2,\mathbb{F}_5)$. An isomorphism of G with $\mathrm{SL}(2,\mathbb{F}_5)$ is given by $\sigma \mapsto \left(\begin{smallmatrix} 3 & 3 \\ 3 & 0 \end{smallmatrix}\right), \tau \mapsto \left(\begin{smallmatrix} 2 & 0 \\ 0 & 3 \end{smallmatrix}\right)$. Let $\eta = \varepsilon^2 = e^{4\pi i/5}$. In Slodowy's notation [Slodowy80], p. 74

$$\tau = \frac{1}{\eta^2 - \eta^3}\begin{pmatrix} \eta + \eta^4 & 1 \\ -1 & -\eta - \eta^4 \end{pmatrix}.$$

See Table 16 for the character table of G [Schur07] and the other relevant invariants. The Coxeter number h of E_8 is equal to 30; here $\mu^{\pm} = \frac{1\pm\sqrt{5}}{2}$.

ρ	1	2	3	4	5	6	7	8	9	d	$(\frac{h}{2}\pm d)$
	1	-1	σ	σ^2	σ^3	σ^4	τ	$\sigma^2\tau$	$\sigma^7\tau$		
$\sharp$	1	1	12	12	12	12	30	20	20		
ρ_0	1	1	1	1	1	1	1	1	1	(5)	–
ρ_2	2	-2	μ^+	$-\mu^-$	μ^-	$-\mu^+$	0	-1	1	4	$(11,19)$
ρ_3	3	3	μ^+	μ^-	μ^-	μ^+	-1	0	0	3	$(12,18)$
ρ_4	4	-4	1	-1	1	-1	0	1	-1	2	$(13,17)$
ρ_5	5	5	0	0	0	0	1	-1	-1	1	$(14,16)$
ρ_6	6	-6	-1	1	-1	1	0	0	0	0	$(15,15)$
ρ_4'	4	4	-1	-1	-1	-1	0	1	1	1	$(14,16)$
ρ_2'	2	-2	μ^-	$-\mu^+$	μ^+	$-\mu^-$	0	-1	1	2	$(13,17)$
ρ_3''	3	3	μ^-	μ^+	μ^+	μ^-	-1	0	0	1	$(14,16)$

Table 16: Character table of E_8

m	$\overline{S}_m$	m	$\overline{S}_m$
0	0	30	0
1	ρ_2	29	ρ_2
2	ρ_3	28	ρ_3
3	ρ_4	27	ρ_4
4	ρ_5	26	ρ_5
5	ρ_6	25	ρ_6
6	$\rho_3''+\rho_4'$	24	$\rho_3''+\rho_4'$
7	$\rho_2'+\rho_6$	23	$\rho_2'+\rho_6$
8	$\rho_4'+\rho_5$	22	$\rho_4'+\rho_5$
9	$\rho_4+\rho_6$	21	$\rho_4+\rho_6$
10	$\rho_3+\rho_3''+\rho_5$	20	$\rho_3+\rho_3''+\rho_5$
11	$(\rho_2)+\rho_4+\rho_6$	19	$(\rho_2)+\rho_4+\rho_6$
12	$(\rho_3)+\rho_4'+\rho_5$	18	$(\rho_3)+\rho_4'+\rho_5$
13	$(\rho_2'+\rho_4)+\rho_6$	17	$(\rho_2'+\rho_4)+\rho_6$
14	$(\rho_3''+\rho_4'+\rho_5)$	16	$(\rho_3''+\rho_4'+\rho_5)$
15	$(2\rho_6)$		

Table 17: Irreducible decompositions of $\overline{S}_m(E_8)$

m	k	ρ	$S_m \bar{V}_k(\rho)$	m	k	ρ	$S_m \bar{V}_k(\rho)$
1	11	ρ_2	ρ_3	1	16	ρ_5	$\rho_4 + \rho_6$
2	11		ρ_4	1	13	ρ_2'	ρ_4'
3	11		ρ_5	2	13		ρ_6
4	11		ρ_6	3	13		$\rho_3'' + \rho_5$
5	11		$\rho_3'' + \rho_4'$	4	13		$\rho_4 + \rho_6$
6	11		$\rho_2' + \rho_6$	5	13		$\rho_3 + \rho_4' + \rho_5$
7	11		$\rho_4' + \rho_5$	1	16	ρ_4'	$\rho_2' + \rho_6$
8	11		$\rho_4 + \rho_6$	1	14	ρ_3''	ρ_6
9	11		$\rho_3 + \rho_3'' + \rho_5$	2	14		$\rho_4' + \rho_5$
1	18	ρ_3	$\rho_2 + \rho_4$	3	14		$\rho_2' + \rho_4 + \rho_6$
1	17	ρ_4	$\rho_3 + \rho_5$				

Table 18: Irreducible decompositions of $S_m \bar{V}_k(\rho)$

16.2   Symmetric tensors modulo $\mathfrak{n}$

The G-modules $\overline{S}_m := S_m(\mathfrak{m}/\mathfrak{n})$ by $\rho_{\mathrm{nat}} := \rho_2$ for small values of m split into irreducible G-modules as in Table 17. The factors of $\overline{S}_m$ in brackets are those in S_{McKay}. We use the same notation $\bar{V}_m(\rho)$ and $V_m(\rho)$ for $\rho \in \mathrm{Irr}\, G$ as before. We define irreducible G-submodules of $V_{15}(\rho_6)$ ($\simeq \rho_6^{\oplus 2}$) and σ_i, τ_j by

$$W_3'' := S_1 V_{14}(\rho_3''), \quad W_4' := S_1 V_{14}(\rho_4'), \quad W_5 := S_1 V_{14}(\rho_5),$$
$$\sigma_1 := x^{10} + 66x^5 y^5 - 11y^{10}, \quad \sigma_2 := -11x^{10} - 66x^5 y^5 + y^{10},$$
$$\tau_1 := x^{10} - 39x^5 y^5 - 26y^{10}, \quad \tau_2 := -26x^{10} + 39x^5 y^5 + y^{10}.$$

Lemma 16.3 *The G-modules $S_m \bar{V}_k(\rho)$ split into irreducible G-submodules as in Table 18.*

Proof We give a brief proof of the lemma. Recall that the ring of G-invariant polynomials is generated by three elements A_{12}, A_{20} and A_{30} of degree 12, 20, 30 respectively. See [Klein], p. 55 or Table 4. Note that $S_1 \otimes V_{11}(\rho_2) = \rho_2 \otimes \rho_2 = \rho_0 + \rho_3$. Hence $S_1 \otimes V_{11}(\rho_2) = \rho_0 A_{12} + \rho_3$. In fact $A_{12} = xy(x^{10} + 11x^5 y^5 - y^{10})$ by [Klein], p. 56. It follows that $S_1 \bar{V}_{11}(\rho_2) = \rho_3$. Similarly $S_k \otimes V_{11}(\rho_2) \supset S_{k-1} A_{12}$. Therefore $S_2 \otimes V_{11}(\rho_2) = \rho_2 + \rho_4$, $S_2 \bar{V}_{11}(\rho_2) = \rho_4$, $S_3 \otimes V_{11}(\rho_2) = \rho_3 + \rho_5$, $S_3 \bar{V}_{11}(\rho_2) = \rho_5$, $S_4 \otimes V_{11}(\rho_2) = \rho_4 + \rho_6$, $S_4 \bar{V}_{11}(\rho_2) = \rho_6$, $S_5 \otimes V_{11}(\rho_2) = \rho_3'' + \rho_4' + \rho_5$, $S_5 \bar{V}_{11}(\rho_2) = \rho_3'' + \rho_4'$. All of these are proved as in Lemma 15.3. In fact, for instance $\dim S_5 V_{11}(\rho_2) = 7$ by Table 19, and $\rho_6 \otimes \rho_2 = \rho_3'' + \rho_4' + \rho_5$ so that $S_5 \bar{V}_{11}(\rho_2) = \rho_3'' + \rho_4'$.

We see $S_6 \bar{V}_{11}(\rho_2) = \rho_2' + \rho_6$ because $\bar{S}_{17} = \rho_2' + \rho_4 + \rho_6$ and $\rho_2 \otimes S_5 \bar{V}_{11}(\rho_2) = \rho_2 \otimes (\rho_3'' + \rho_4') = \rho_2' + 2\rho_6$ contains no ρ_4. $\bar{S}_{18} = \rho_3 + \rho_4' + \rho_5$ and $\rho_2 \otimes S_6 \bar{V}_{11}(\rho_2) = \rho_2 \otimes (\rho_2' + \rho_6)$ contains no ρ_3, whence $S_7 \bar{V}_{11}(\rho_2) = \rho_4' + \rho_5$. Similarly $S_8 \bar{V}_{11}(\rho_2) = \rho_4 + \rho_6$ because $\bar{S}_{19} = \rho_2 + \rho_4 + \rho_6$, $\rho_2 \otimes S_7 \bar{V}_{11}(\rho_2) = \rho_2' + \rho_4 + 2\rho_6$. By Table 17 $\bar{S}_{20} = \rho_3 + \rho_3'' + \rho_5$. $\rho_2 \otimes S_8 \bar{V}_{11}(\rho_2) = \rho_3 + 2\rho_5 + \rho_3'' + \rho_4'$. Hence $S_9 \bar{V}_{11}(\rho_2) = \rho_3 + \rho_3'' + \rho_5 = \bar{S}_{20}$.

$S_1 \bar{V}_{18}(\rho_3) = \rho_2 + \rho_4$ follows from comparison of $S_1 \otimes V_{18}(\rho_3)$ and S_{19} and the fact that any polynomial in $V_{18}(\rho_3)$ is not divisible by A_{12}.

Similarly $S_1 \bar{V}_{17}(\rho_4) = \rho_3 + \rho_5$, $S_1 \bar{V}_{16}(\rho_5) = \rho_4 + \rho_6$ and $S_1 \bar{V}_{13}(\rho_2') = \rho_4'$. Since $\rho_3 \otimes \rho_2' = \rho_6$, we see $S_2 \bar{V}_{13}(\rho_2') = \rho_6$. One checks $\dim S_3 V_{13}(\rho_2') = \dim S_1 W_4' = 8$ by using Table 19. It follows from this that $S_3 \bar{V}_{13}(\rho_2') = \rho_3'' + \rho_5$. Similarly it is clear that $S_4 V_{13}(\rho_2') = S_4 \otimes V_{13}(\rho_2') = \rho_4 + \rho_6$ and $S_5 \bar{V}_{13}(\rho_2') = S_5 \otimes \bar{V}_{13}(\rho_2') = \bar{S}_{18}$. Note $\dim S_k V_{14}(\rho_3'') = 3(k+1)$ for $k = 1, 2, 3$ so that $S_k \bar{V}_{14}(\rho_3'') = S_k \otimes V_{14}(\rho_3'')$. It follows from it that $S_k \bar{V}_{14}(\rho_3'') = S_k \otimes \rho_3''$ for $k = 1, 2, 3$. In particular, $S_2 \bar{V}_{14}(\rho_3'') = \rho_3 \otimes \rho_3' = \rho_4' + \rho_5$, $S_3 \bar{V}_{14}(\rho_3'') = \rho_2' + \rho_4 + \rho_6 = \bar{S}_{17}$. $\square$

Corollary 16.4   *1. $S_k \bar{V}_{11}(\rho_2) = \bar{V}_{11+k}(\rho_{k+2})$ for $1 \le k \le 3$; $S_1 \bar{V}_{13}(\rho_2') = \bar{V}_{14}(\rho_4')$.*

2. $S_9 \bar{V}_{11}(\rho_2) = \bar{S}_{20}$, $S_5 \bar{V}_{13}(\rho_2') = \bar{S}_{18}$, $S_3 \bar{V}_{14}(\rho_3'') = \bar{S}_{17}$.

3. $S_2 \bar{V}_{14}(\rho_5) = \rho_3'' + \rho_4'$, $S_2 \bar{V}_{14}(\rho_4') = \rho_3'' + \rho_5$, $S_2 \bar{V}_{14}(\rho_3'') = \rho_4' + \rho_5$.

Proof   By Table 19, $\dim S_1 W_3'' = 9$, $\dim S_1 W_4' = 8$, $\dim S_1 W_5 = 7$. Hence $S_2 V_{14}(\rho_3'') = S_1 W_3'' = \rho_4 + \rho_5$, $S_2 \bar{V}_{14}(\rho_4') = S_2 V_{14}(\rho_4') = S_1 W_4' = \rho_3'' + \rho_5$, $S_5 \bar{V}_{11}(\rho_2) = S_2 \bar{V}_{14}(\rho_5) = S_1 W_5 = \rho_3'' + \rho_4'$. $\square$

In order to prove Theorem 10.7 in the E_8 case we have only to follow the proof of Theorem 10.7 in the D_n or E_6 case verbatim. We omit the details.

17   *Fine*

We would like to mention some related problems that are unsolved or are the subject of current research.

Conjecture 17.1 *Let G be any finite subgroup of $SL(3, \mathbb{C})$. Then $\mathrm{Hilb}^G(\mathbb{A}^3)$ is a crepant smooth resolution of $\mathbb{A}^3/G$.*

The conjecture is solved affirmatively in the Abelian case [Nakamura98], where for any finite Abelian subgroup G of $GL(n, \mathbb{C})$ the Hilbert scheme $\mathrm{Hilb}^G(\mathbb{A}^n)$ is described as a (possibly nonnormal) toric variety. There is a McKay correspondence [Reid97], [INkjm98] similar to [GSV83]. See also

m	ρ	$V_m(\rho)$
11	ρ_2	$x\sigma_1, -y\sigma_2$
19	ρ_2	$-57x^{15}y^4 + 247x^{10}y^9 + 171x^5y^{14} + y^{19}$
		$-x^{19} + 171x^{14}y^5 - 247x^9y^{10} - 57x^4y^{15}$
12	ρ_3	$x^2\sigma_1, -5x^{11}y - 5xy^{11}, y^2\sigma_2$
18	ρ_3	$-12x^{15}y^3 + 117x^{10}y^8 + 126x^5y^{13} + y^{18}$
		$45x^{14}y^4 - 130x^9y^9 - 45x^4y^{14}$
		$x^{18} - 126x^{13}y^5 + 117x^8y^{10} + 12x^3y^{15}$
13	ρ_4	$x^3\sigma_1, -3x^{12}y + 22x^7y^6 - 7x^2y^{11}$
		$-7x^{11}y^2 - 22x^6y^7 - 3xy^{12}, y^3\sigma_2$
17	ρ_4	$-2x^{15}y^2 + 52x^{10}y^7 + 91x^5y^{12} + y^{17}$
		$10x^{14}y^3 - 65x^9y^8 - 35x^4y^{13}$
		$-35x^{13}y^4 + 65x^8y^9 + 10x^3y^{14}$
		$-x^{17} + 91x^{12}y^5 - 52x^7y^{10} - 2x^2y^{15}$
14	ρ_5	$x^4\sigma_1, -2x^{13}y + 33x^8y^6 - 8x^3y^{11}$
		$-5x^{12}y^2 - 5x^2y^{12}$
		$-8x^{11}y^3 - 33x^6y^8 - 2xy^{13}, -y^4\sigma_2$
16	ρ_5	$64x^{15}y + 728x^{10}y^6 + y^{16}$
		$66x^{14}y^2 + 676x^9y^7 - 91x^4y^{12}$
		$56x^{13}y^3 + 741x^8y^8 - 56x^3y^{13}$
		$91x^{12}y^4 + 676x^7y^9 - 66x^2y^{14}$
		$x^{16} + 728x^6y^{10} - 64xy^{15}$
13	ρ_2'	$y^3\tau_2, -x^3\tau_1$
17	ρ_2'	$x^{17} + 119x^{12}y^5 + 187x^7y^{10} + 17x^2y^{15}$
		$-17x^{15}y^2 + 187x^{10}y^7 - 119x^5y^{12} + y^{17}$
14	ρ_3''	$x^{14} - 14x^9y^5 + 49x^4y^{10}$
		$7x^{12}y^2 - 48x^7y^7 - 7x^2y^{12}$
		$49x^{10}y^4 + 14x^5y^9 + y^{14}$
16	ρ_3''	$3x^{15}y - 143x^{10}y^6 - 39x^5y^{11} + y^{16}$
		$-25x^{13}y^3 - 25x^3y^{13}$
		$x^{16} + 39x^{11}y^5 - 143x^6y^{10} - 3xy^{15}$

Table 19: $V_m(\rho)(E_8)$

m	ρ	$V_m(\rho)$
14	ρ_4'	$xy^3\tau_2, -x^4\tau_1, y^4\tau_2, -x^3y\tau_1$
16	ρ_4'	$-2x^{15}y + 77x^{10}y^6 - 84x^5y^{11} + y^{16}$
		$35x^{12}y^4 + 110x^7y^9 + 15x^2y^{14}$
		$15x^{14}y^2 - 110x^9y^7 + 35x^4y^{12}$
		$-x^{16} - 84x^{11}y^5 - 77x^6y^{10} - 2xy^{15}$
15	ρ_6	$W_3'' + W_4' = W_4' + W_5 = W_5 + W_3'' \simeq \rho_6^{\oplus 2}$
15	W_3''	$:= S_1V_{14}(\rho_3'')\ (\simeq \rho_6)$
		$x^{15} + 84x^{10}y^5 + 77x^5y^{10} + 2y^{15}$
		$-x^{14}y + 14x^9y^6 - 49x^4y^{11}$
		$-7x^{13}y^2 + 48x^8y^7 + 7x^3y^{12}$
		$7x^{12}y^3 - 48x^7y^8 - 7x^2y^{13}$
		$-49x^{11}y^4 - 14x^6y^9 - xy^{14}$
		$-2x^{15} + 77x^{10}y^5 - 84x^5y^{10} + y^{15}$
15	W_4'	$:= S_1V_{14}(\rho_4')\ (\simeq \rho_6)$
		$x^{15} + 39x^{10}y^5 - 143x^5y^{10} - 3y^{15}$
		$-2x^{14}y + 78x^9y^6 + 52x^4y^{11}$
		$x^{13}y^2 - 39x^8y^7 - 26x^3y^{12}$
		$-26x^{12}y^3 + 39x^7y^8 + x^2y^{13}$
		$52x^{11}y^4 - 78x^6y^9 - 2xy^{14}$
		$3x^{15} - 143x^{10}y^5 - 39x^5y^{10} + y^{15}$
15	W_5	$:= S_1V_{14}(\rho_5)\ (\simeq \rho_6)$
		$5x^{15} + 330x^{10}y^5 - 55x^5y^{10}$
		$-7x^{14}y + 198x^9y^6 - 43x^4y^{11}$
		$-19x^{13}y^2 + 66x^8y^7 - 31x^3y^{12}$
		$-31x^{12}y^3 - 66x^7y^8 - 19x^2y^{13}$
		$-43x^{11}y^4 - 198x^6y^9 - 7xy^{14}$
		$-55x^{10}y^5 - 330x^5y^{10} + 5y^{15}$

Table 19: $V_m(\rho)(E_8)$, continued

[Nakamura98]. In general the normalization of $\text{Hilb}^G(\mathbb{A}^n)$ is a torus embedding associated with a certain fan $\text{Fan}(G)$ given explicitly by using some combinatorial data arising from the given group G. However, in general it is not known whether $\text{Hilb}^G(\mathbb{A}^n)$ is normal. There are various examples of $\text{Hilb}^G(\mathbb{A}^n)$. Reid gave some examples of singular Hilb^G for finite Abelian subgroups G in $\text{GL}(3, \mathbb{C})$ in private correspondence.

If G is the cyclic subgroup of $\text{SL}(4, \mathbb{C})$ of order two generated by minus the identity then $\text{Hilb}^G(\mathbb{A}^4)$ is nonsingular; however, it is not a crepant resolution of $\mathbb{A}^4/G$. There are also some examples of Abelian subgroups of $\text{SL}(4, \mathbb{C})$ for which $\text{Hilb}^G(\mathbb{A}^4)$ is singular, although a crepant resolution does exist. The simplest example is the Abelian subgroup of order eight consisting of diagonal 4×4 matrices with diagonal coefficients ± 1. [Kidoh98] gave a concrete description of $\text{Hilb}^G(\mathbb{A}^2)$ for a finite Abelian subgroup G of $\text{GL}(2, \mathbb{C})$ by using two kinds of continued fractions.

We will treat the non-Abelian cases of Conjecture 17.1 elsewhere [GNS98]; in almost all the non-Abelian cases, a certain beautiful duality in $\mathfrak{m}/\mathfrak{n}$ is observed [GNS98]. See also Section 7.

The following question would be important for future applications:

Problem 17.2 *Let G be a finite subgroup of $\text{SL}(n, \mathbb{C})$, N a normal subgroup of G. When is $\text{Hilb}^G(\mathbb{A}^n) \simeq \text{Hilb}^{G/N}(\text{Hilb}^N(\mathbb{A}^n))$?*

Unfortunately the answer is negative in general in dimension three. This will appear in [GNS98].

References

[Arnold74]   V.I. Arnold, *Critical points of smooth functions*, in Proc. Internat. Congress of Math., Vancouver 1974, pp. 19–39.

[Beauville83] A. Beauville, *Variétés Kähleriennes dont la première classe de Chern est nulle*, J. Diff. Geom. **18** (1983) 755–782.

[Blichfeldt05] H.F. Blichfeldt, *The finite discontinuous primitive groups of collineations in three variables*, Math. Ann. **63** (1905) 552–572.

[Blichfeldt17] H.F. Blichfeldt, *Finite collineation groups*, The Univ. Chicago Press, Chicago, 1917.

[BR95]   K. Behnke and O. Riemenschneider, *Quotient surface singularities and their deformations*, in Singularity Theory, World Scientific, 1995, pp. 1–54.

[Briançon77]   J. Briançon, *Description de* Hilb^n $\mathbb{C}\{x,y\}$, Invent. Math. **41** (1977) 45–89.

[Bourbaki]   N. Bourbaki, *Groupes et algèbres de Lie*, Chap. 4,5 et 6 Masson, Paris–New York, 1981.

[BV92]   B. Blok, A. Varchenko, *Topological conformal field theories and the flat coordinates*, Internat. J. Mod. Phys. **7** (1992) 1467–1490.

[Cardy86]   J.L. Cardy, *Operator content of two-dimensional conformally invariant theories*, Nucl. Phys. B **270** (1986) 186–204.

[Cardy88]   J.L. Cardy, *Conformal invariance and statistical mechanics*, in Champs, cordes et phénomènes (Les Houches, 1988), pp. 169–245.

[Chevalley55]   C. Chevalley, *Invariants of finite groups generated by reflections*, Amer. J. Math. **77** (1955) 778–782.

[CIZ87]   A. Capelli, C. Itzykson, J.B. Zuber, *Modular invariant partition functions in two dimensions*, Nucl. Phys. B **280** (1987) 445–465.

[Durfee79]   A. Durfee, *Fifteen characterizations of rational double points and simple singularities*, L'Enseign. Math. **25** (1979) 131–163.

[EK98]   D. E. Evans, Y. Kawahigashi, *Quantum symmetries on operator algebras*, Oxford University Press, 1998.

[ES87]   G. Ellingsrud and S.A. Strømme, *On the homology of the Hilbert scheme of points in the plane*, Invent. Math. **87** (1987) 343–352.

[EY89]   T. Eguchi, S. K. Yang, *Virasoro algebras and critical phenomena* (in Japanese), J. Japan Phys. Soc. **44** (1988) 894–901.

[FGA]   A. Grothendieck, *Fondements de la géométrie algébrique*, Sém. Bourbaki, 1957–62, Secrétariat Math., Paris (1962).

[Fujiki83]   A. Fujiki, *On primitively symplectic compact Kähler V-manifolds*, Progress in Mathematics, Birkhäuser 39 (1983) 71–250.

[Fogarty68]   J. Fogarty, *Algebraic families on an algebraic surface*, Amer. J. Math. **90** (1968) 511–521.

[Gabriel72]   P. Gabriel, *Unzerlegbare Darstellungen I*, Manuscr. Math. **6** (1972) 71–103.

[Gawedzki89] K. Gawedzki, *Conformal field theory*, Sém. Bourbaki 1988/89, Astérisque (1989) No. 177–178, Exp. No. 704, 95–126.

[GHJ89]   F. M. Goodman, P. de la Harpe, V. F. R. Jones, *Coxeter graphs and towers of algebras*, Math. Sci. Research Inst. publications 14, Springer, 1989.

[GNS98]   Y. Gomi, I. Nakamura, K. Shinoda, (in preparation)

[Göttsche91] L. Göttsche, *Hilbert schemes of zero-dimensional subschemes of smooth varieities*, Lecture Notes in Math., 1572, Springer, 1994

[Grojnowski96] I. Grojnowski, *Instantons and affine algebras I: the Hilbert scheme and vertex operators*, Math. Res. Letters 3 (1996), 275–291.

[GSV83]   G. Gonzalez-Sprinberg, J.-L. Verdier, *Construction géométrique de la correspondence de McKay*, Ann. Sci. École. Norm. Sup. 16 (1983) 409–449.

[GW86]   D. Gepner, E. Witten, *String theory on group manifolds*, Nucl. Phys. B 278 (1986) 493–549.

[Hartshorne77] R. Hartshorne, *Algebraic geometry*, Graduate texts in math. 52, Springer, 1977.

[HHSV77]   M. Hazewinkel, W. Hesselink, D. Siersma, F. D. Veldkamp, *The ubiquity of Coxeter–Dynkin diagrams (an introduction to the A-D-E classification)*, Nieuw Archief voor Wiskunde 25 (1977) 257–307.

[Humphreys90] J. E. Humphreys, *Reflection groups and Coxeter groups*, Cambridge studies in advanced mathematics 29, Cambridge University Press, 1990.

[Iarrobino72] A. Iarrobino, *Reducibility of the families of 0-dimensional schemes on a variety*, Invent. Math. 15 (1972) 72-77.

[Iarrobino77] A. Iarrobino, *Punctual Hilbert schemes*, Memoirs Amer. Math. Soc. 188, Amer. Math. Soc. Providence, 1977.

[Ito95a]   Y. Ito, *Crepant resolution of trihedral singularities and the orbifold Euler characteristic*, Internat. J. Math., 6 (1995) 33–43.

[Ito95b]   Y. Ito, *Gorenstein quotient singularities of monomial type in dimension three*, J. Math. Sci. Univ. of Tokyo 2 (1995) 419–440.

[INkjm98]   Y. Ito and I. Nakajima, *McKay correspondence and Hilbert schemes in dimension three*, preprint math/9803120, to appear in Topology.

[IN96]   Y. Ito and I. Nakamura, *McKay correspondence and Hilbert schemes*, Proc. Japan Acad. **72** (1996) 135–138.

[IR96]   Y. Ito, M. Reid, *The McKay correspondence for finite groups of* SL(3, ℂ), in Higher dimensional complex varieties (Trento, 1996) 221–240.

[Izumi91]   M. Izumi, *Application of fusion rules to classification of subfactors*, Publ. RIMS, Kyoto Univ. **27** (1991) 953–994.

[Jones91]   V. F. R. Jones, *Subfactors and knots*, Regional conference series in mathematics **80**, Amer. Math. Soc., Providence, 1991.

[Kato87]   A. Kato, *Classification of modular invariant partition functions in two dimensions*, Modern Physics Letters A **2** (1987) 585–600.

[Kac90]   V. G. Kac, *Infinite dimensional Lie algebras*, Third edition, Cambridge University Press, Cambridge–New York, 1990.

[Khinich76]   V. A. Khinich, *On the Gorenstein property of the ring of invariants of a Gorenstein ring*, Izv. Akad. Nauk SSSR Ser. Mat. **40** (1976) 50–56, English transl. Math. USSR-Izv. **10** (1976) 47–53.

[Kidoh98]   R. Kidoh, *Hilbert schemes and cyclic quotient surface singularities*, preprint (1998).

[Klein]   F. Klein, *Vorlesungen über das Ikosaeder und die Auflösung der Gleichungen vom fünften Grade* (P. Slodowy, Ed.), Birkhäuser and B.G. Teubner, 1993. English transl. Lectures on the icosahedron and the solution of equations of the fifth degree, Dover, New York, 1956.

[Knop87]   F. Knop, *Ein neuer Zusammenhang zwischen einfachen Gruppen und einfachen Singularitäten*, Invent. Math. **90** (1987) 579–604.

[Knörrer85]   H. Knörrer, *Group representations and the resolution of rational double points*, in Finite groups – coming of age, Contemp. Math. **45**, 1985, pp. 175–222.

[Kronheimer89]   P. B. Kronheimer, *The construction of ALE spaces as hyperkähler quotients*, J. Diff. Geometry **29** (1989) 665–683.

[KW88]   V. G. Kac, M. Wakimoto, *Modular invariant representations of infinite dimensional Lie algebras and superalgebras*, Proc. Nat. Acad. Sci. U.S.A. **85** (1988) 4956–4960.

[Markushevich92] D. Markushevich, *Resolution of* $\mathbb{C}^3/H_{168}$, Math. Ann. **308** (1997) 279–289

[McKay80]   J. McKay, *Graphs, singularities, and finite group*, in Santa Cruz conference on finite groups (Santa Cruz, 1979), Proc. Symp. Pure Math. **37**, AMS, 1980, pp. 183–186.

[MBD16]   G. A. Miller, H. F. Blichfeldt, L. E. Dickson, *Theory and applications of finite groups*, Dover, New York, 1916.

[Mukai84]   S. Mukai, *Symplectic structure of the moduli of bundles on abelian or K3 surface*, Inv. Math. **77** (1984) 101–116.

[Mumford]   D. Mumford, *Lectures on curves on an algebraic surface*, Ann. of Math. Studies **66**, Princeton 1966.

[Nakajima96a] H. Nakajima, *Heisenberg algebra and Hilbert schemes of points on projective surfaces*, Ann. of Math. (2) **145** (1997) 379–388

[Nakajima96b] H. Nakajima, *Lectures on Hilbert schemes of points on surfaces*, 1996, available from http://www.kusm.kyoto-u.ac.jp

[Nakamura98] I. Nakamura, *Hilbert schemes of Abelian group orbits*, preprint (1998).

[Ocneanu88] A. Ocneanu, *Quantized group string algebras and Galois theory for algebras*, in Operator algebras and applications (Warwick, 1987), Vol. 2, London Math. Soc. Lecture Note Series **136**, Cambridge University Press, 1988, pp. 119–172.

[OSS80]   C. Okonek, M. Schneider and H. Spindler, *Vector bundles on complex projective spaces*, Progress in Math. **3**, Birkhäuser, 1980.

[OT92]   P. Orlik, H. Terao, *Arrangements of hyperplanes*, Grundlehren der mathematischen Wissenschaften, Springer, 1992.

[Pasquier87a] V. Pasquier, *Two dimensional critical systems labelled by Dynkin diagrams*, Nucl. Phys. B **285** (1987) 162–172.

[Pasquier87b] V. Pasquier, *Operator contents of the ADE lattice models*, J. Phys. A: Math. Gen **20** (1987) 5707–5717.

[Pinkham80] H. Pinkham, *Singularités de Klein, I*, in Séminaire sur les singularités de surfaces, Lecture Note in Math. **777**, Springer, 1980, pp. 1–9.

[PZ96] V. B. Petkova, J.-B. Zuber, *From CFT to graphs*, Nucl. Phys. B **463** (1996) 161–193.

[Reid97] M. Reid, *McKay correspondence*, in Proc. of algebraic geometry symposium (Kinosaki, Nov 1996), T. Katsura (Ed.), 14–41, Duke file server alg-geom 9702016, 30 pp.

[Roan89] S-S. Roan, *On the generalization of Kummer surfaces*, J. Diff. Geom. **30** (1989) 523–537.

[Roan94] S-S. Roan, *On $c_1 = 0$ resolution of quotient singularities*, Internat. J. Math. **5** (1994) 523–536.

[Roan96] S-S. Roan, *Minimal resolution of Gorenstein orbifolds in dimension three*, Topology **35** (1996) 489–508.

[Schur07] I. Schur, *Untersuchungen über die Darstellung der endlichen Gruppen durch gebrochene lineare Substitutionen*, J. reine angew. Math. **132** (1907) 85–137.

[Slodowy80] P. Slodowy, *Simple singularities and simple algebraic groups*, Lecture Notes in Math. **815**, Springer, 1980.

[Slodowy90] P. Slodowy, *A new A-D-E classification*, Bayreuth Math. Schr. **33** (1990) 197–213.

[Slodowy95] P. Slodowy, *Groups and special singularities*, in Singularity theory, World Scientific, 1995, pp. 731–799.

[Steinberg64] R. Steinberg, *Differential equations invariant under finite reflection groups*, Trans. Amer. Math. Soc. **112** (1964) 392–400.

[YY93] Stephen S.-T. Yau and Y. Yu, *Gorenstein quotient singularities in dimension three*, Memoirs Amer. Math. Soc., **105**, Amer. Math. Soc. 1993.

[Watanabe74] K. Watanabe, *Certain invariant subrings are Gorenstein I, II*, Osaka J. Math. **11** (1974) 1–8, 379–388.

[Wenzl87] H. Wenzl, *On sequences of projections*, C. R. Math. Rep. Acad. Sci. Canada. **9** (1987) 5–9.

[Zuber90]   J.-B. Zuber, *Graphs, algebras, conformal field theories and integrable lattice models*, Nucl. Phys. B (Proc. Suppl.) **18B** (1990) 313–326.

Yukari Ito,
Department of Mathematics, Tokyo Metropolitan University, Hachioji,
Tokyo 192–03, Japan
yukari@math.metro-u.ac.jp

Iku Nakamura,
Department of Mathematics, Hokkaido University,
Sapporo, 060, Japan
nakamura@math.hokudai.ac.jp

Bounds for Seshadri constants

Oliver Küchle*   Andreas Steffens*

Introduction

This paper develops a new approach to bounding Seshadri constants of nef and big line bundles at a general point of a complex projective variety. A modification of this approach even allows us to give bounds valid at arbitrary points.

The Seshadri constant $\varepsilon(L, x)$, introduced by Demailly [De92], measures the local positivity of a nef line bundle L at a point $x \in X$ of a complex projective variety X; it can be defined as

$$\varepsilon(L, x) := \inf_{C \ni x} \left\{ \frac{L \cdot C}{\mathrm{mult}_x(C)} \right\},$$

where the infimum is taken over all reduced irreducible curves $C \subset X$ passing through x. The interest in Seshadri constants comes in part from the fact that they govern an elementary method for producing sections in adjoint line bundles $\mathcal{O}_X(K_X + rL)$ with certain properties. This connects the theory of Seshadri constants to the famous conjectures of Fujita on global generation and very ampleness of such bundles (cf. §1 below for further characterizations and properties of Seshadri constants; see also [De92] and [EKL]).

There has been considerable progress in the study of Seshadri constants in recent years, starting with Ein and Lazarsfeld's result (cf. [EL]) that the Seshadri constant of an ample line bundle on a smooth surface is ≥ 1 for all except possibly countably many points. On the other hand, examples by Miranda (cf. [EKL, 1.5]) show that for any integer $n \geq 2$ and any real number $\delta > 0$ there is a smooth n-fold X, an ample line bundle L on X and a point $x \in X$ with $\varepsilon(L, x) < \delta$; in other words, there does not exist a universal lower bound for Seshadri constants valid for all X and ample L at every point $x \in X$.

Then it was proved by Ein, Küchle and Lazarsfeld [EKL] that, for a nef and big line bundle L on a projective n-fold, the Seshadri constant at very

---

*Supported by the Deutsche Forschungsgemeinschaft

general points (that is, outside a countable union of proper subvarieties) is bounded below by $1/n$, which implies the existence of a lower bound at general points depending only on n.

Finally, we want to mention the recent papers by Lazarsfeld [La] and Nakamaye [Na] dealing with Seshadri constants on Abelian varieties, as well as variants due to Küchle [Kü], Xu [Xu95] and Paoletti [Pa] concerning Seshadri constants along several points and higher dimensional subvarieties respectively.

Our approach to bounding Seshadri constants is based on considering families of divisors with high multiplicity at assigned points. The method itself relies upon ideas of Demailly's paper [De93] as explained and translated into the language of algebraic geometry by Ein, Lazarsfeld and Nakamaye [ELN]. Very roughly, the idea is to start with an effective divisor E in a linear system $|kL|$ (for $k \gg 0$) with large multiplicity at x, and to consider the locus V of points where the singularities of E are "concentrated" in a certain way. The possibility that V is zero dimensional imposes constraints on the local positivity of L at x in a sense. Otherwise one uses variational techniques to give a lower bound on the degree of V.

In contrast to [De93] and [ELN], instead of making a positivity assumption on the tangent bundle of the manifold in question to be able to "differentiate", we apply the strategy of differentiation in parameter directions in the spirit of [EKL]. The result of this method is the following theorem on linear system, which might also be of interest in other contexts:

Theorem 1 *Let X be a smooth projective n-fold, L a nef and big line bundle on X and $\alpha > 0$ a rational number such that $L^n > \alpha^n$. Let*

$$0 = \beta_1 < \beta_2 < \cdots < \beta_n < \beta_{n+1} = \alpha$$

be any sequence of rational numbers and $x \in X$ a very general point. Then one of the following holds:

(a) *there exist $k \gg 0$ and a divisor $E \in |kL|$ having an isolated singularity with multiplicity at least $k(\beta_{n+1} - \beta_n)$ at x, or*

(b) *there exists a proper subvariety $V \subset X$ through x of codimension $c < n$ such that*

$$\deg_L V = L^{n-c} \cdot V \le \frac{1}{(\beta_{c+1} - \beta_c)^c} \left(1 - \sqrt[n]{\left(1 - \frac{\alpha^n}{L^n}\right)^c}\right) L^n$$

$$< \frac{\alpha^n}{(\beta_{c+1} - \beta_c)^c}.$$

To pass from Theorem 1 to actually bounding the Seshadri constant, we use a rescaling trick (cf. Remark 3.2) in combination with the well-known characterization of Seshadri constants via the generation of s-jets by certain adjoint linear systems (cf. Theorem 1.5). The bound we get does not improve that of [EKL] in general, although the method at hand may give better results in certain cases, since our bound can be expressed more flexibly in terms of the degrees of subvarieties with respect to the line bundle in question:

Theorem 2 *Let L be a nef line bundle on an irreducible projective n-fold X, and $x \in X$ a very general point. Suppose given positive rational numbers $\alpha_1, \ldots, \alpha_n$, and set $\gamma = 1 + \sum_{i=1}^{n-1} \alpha_i$. For some real number $\varepsilon > 0$, suppose that every d-dimensional subvariety $V \subset X$ containing x (for any d with $1 \leq d \leq n$) satisfies*

$$\deg_L V = L^d \cdot V \geq \varepsilon^d \gamma^n \alpha_d^{d-n}.$$

Then $\varepsilon(L, x) \geq \varepsilon$.

After writing a first draft of this paper, we realized that the rescaling argument mentioned above can also be applied in the context of [ELN], leading, somewhat surprisingly, to bounds for Seshadri constants valid at *arbitrary* points; however, these bounds depend on the line bundle L and the manifold X, or rather, its tangent bundle T_X:

Corollary 3 *Let X be a smooth projective n-fold, $x \in X$ any point, A an ample line bundle and $\delta \geq 0$ a real number such that $T_X(\delta A)$ is nef. Then*

$$\varepsilon(A, x) \geq \min \left\{ \frac{1}{(n-1)^{n-1}(2n-1)^n}, \frac{1}{\delta} \right\}.$$

This gives, in particular, bounds valid at arbitrary points for the Seshadri constants of the canonical line bundle $K_X = (\bigwedge^n T_X)^*$ or its inverse if these are ample, or for any ample line bundle in case K_X is trivial (cf. Remark 4.4).

Corollary 3 accords with, and should be compared to, bounds following from recent work of Angehrn and Siu [AS] on the basepoint freeness of adjoint linear systems (cf. 4.5), and Demailly's original very ampleness criteria [De93].

The paper is organized as follows. After fixing notation and establishing the general setup, we recall some basic facts about Seshadri constants and collect some auxiliary statements in §1. Then, in §2, we prove the main technical result, Theorem 1. Finally we give the applications to bounding Seshadri constants at general points in §3, and at arbitrary points in §4.

Acknowledgments

We are grateful to Rob Lazarsfeld, both for having drawn our attention to Seshadri constants, and for sharing his insights in this topic with us. The reader will notice that our treatment relies on ideas and prior works due to Demailly [De93], Ein, Küchle and Lazarsfeld [EKL], Fujita [Fu94], and especially Ein, Lazarsfeld and Nakamaye [ELN], to whom we are indebted.

Notation and the general setup

0.1 Throughout this paper we work over the field $\mathbb{C}$ of complex numbers. Given a variety Y, a statement valid for a *very general* point $y \in Y$ is a statement which holds for all points in the complement of some countable union of proper subvarieties of Y.

0.2 Given a smooth variety Y, an integer $m \geq 0$ and a subvariety $W \subset Y$ we denote by $\mathcal{I}_W^{(m)}$ the symbolic power sheaf of all functions vanishing to order at least m along W. Then $\mathcal{I}_W^{(1)} = \mathcal{I}_W$ is the ideal sheaf of W, and $\mathcal{I}_Z^{(m)} = \mathcal{I}_Z^m$ for a smooth subvariety $Z \subset Y$.

0.3 Let M be a line bundle on a smooth variety Y. Given a divisor $E \in |kM|$ we call the normalized multiplicity

$$\operatorname{ind}_y(E) = \frac{\operatorname{mult}_y(E)}{k}$$

the *index* of E at a point $y \in Y$.

0.4 For a line bundle L and a coherent sheaf of ideals $\mathcal{J}$ on Y, we denote by $|L \otimes \mathcal{J}|$ the linear subsystem of the complete linear system $|L|$ corresponding to sections in $L \otimes \mathcal{J}$. Given such a system $|L \otimes \mathcal{J}| \neq \emptyset$ on Y, the *base locus* $\operatorname{Bs}|L \otimes \mathcal{J}|$ is the support of the intersection of all members of $|L \otimes \mathcal{J}|$.

0.5 We will be concerned with the following setup: let X be a smooth irreducible n-dimensional projective variety, T a smooth irreducible affine variety and $g \colon T \to X$ a quasi-finite dominant morphism with graph Γ; examples of this situation are provided by Zariski open subsets $T \subset X$. Let pr_X and pr_T denote the projections from $Y = X \times T$ to its factors. Note that these restrict to dominant maps from Γ to X respectively T. Given a Zariski closed subset (or subscheme) $Z \subset X \times T$, we consider the fibre Z_t of pr_T over $t \in T$ as a subset (or subscheme) of X. Similarly, $Z_x \subset T$ is the fibre of Z over $x \in X$. Given a sheaf $\mathcal{F}$ on $X \times T$ we write $\mathcal{F}_t$ for the induced sheaf on X.

1 Preliminaries

In this section we collect some preliminary results needed in the sequel. We start with some remarks concerning multiplicity loci in a family. Let E be an effective divisor on a smooth variety Y. Then the function $y \mapsto \text{mult}_y(E)$ is Zariski upper-semicontinuous on Y. For any given irreducible subvariety $Z \subset Y$ we write $\text{mult}_Z(E)$ for the value of $\text{mult}_z(E)$ at a general point $z \in Z$. The following lemma allows us to calculate multiplicities fibrewise.

Lemma 1.1 (cf. [EKL, 2.1]) *Let X and T be smooth irreducible varieties, and suppose that $Z \subset X \times T$ is an irreducible subvariety which dominates T. Let $E \subset X \times T$ be an effective divisor. Then for a general point $t \in T$, and any irreducible component $W_t \subset Z_t$ of the fibre Z_t, we have*

$$\text{mult}_{W_t}(E_t) = \text{mult}_Z(E). \quad \square$$

The next two elementary lemmas give a way of detecting irreducible components of base loci, and show that being an irreducible component is well-behaved in families.

Lemma 1.2 *Let m be a positive integer, M a line bundle on a smooth variety Y and $V \subset W \subset Y$ subvarieties such that V is an irreducible component of W. Suppose that*

(1) $\mathcal{I}_W \otimes M$ is generated by global sections, and

(2) $\mathcal{I}_W \subset \mathcal{I}_V^{(m)}$.

Then V is an irreducible component of $\text{Bs}\,|\mathcal{I}_V^{(m)} \otimes M|$.

Proof Consider the inclusions

$$V \subset \text{Bs}\,|\mathcal{I}_V^{(m)} \otimes M| \subset \text{Bs}\,|\mathcal{I}_W \otimes M| \subset W, \qquad (*)$$

where the first inclusion is obvious, the second follows from (2), and the third from the fact that $\text{Bs}\,|\mathcal{I}_W \otimes M| = W$ because $\mathcal{I}_W \otimes M$ is globally generated according to (1).

Now let $Z \subset \text{Bs}\,|\mathcal{I}_V^{(m)} \otimes M|$ be any irreducible component containing V. Then $(*)$ and the assumption that $V \subset W$ is an irreducible component imply $V = Z$, hence the claim follows. $\square$

Lemma 1.3 *Let $f\colon Y \to Z$ be a morphism between irreducible varieties, and $V \subset W \subset Y$ subvarieties such that V is an irreducible component of W, and f restricts to a dominant map from V to Z.*

Then over a general point $z \in Z$, every irreducible component U_z of V_z has dimension $\dim V - \dim Z$ and is an irreducible component of W_z.

Proof For the first part we refer to [Ha, II, Ex 3.22]. The second assertion comes down to an easy dimension count as follows: write

$$W = V \cup V'$$

with $V' \subset Y$ a subvariety not containing V, so that $\dim V > \dim(V \cap V')$. If the map from $V \cap V'$ to Z is not dominant, then V_z and V'_z do not meet in a general fibre and the claim follows.

Otherwise, over general $z \in Z$, we obtain:

$$\dim V_z = \dim V - \dim Y$$
$$> \dim(V \cap V') - \dim Y = \dim(V \cap V')_z,$$

and therefore $\dim U_z > \dim(V \cap V')_z \geq \dim(U_z \cap V'_z)$, where $U_z \subset V_z$ is any irreducible component. Considering the decomposition $V_z = U_z \cup U'_z$, where $U_z \not\subset U'_z$, we conclude $\dim U_z > \dim(U_z \cap (U'_z \cup V'_z))$, which shows that U_z is an irreducible component of $W_z = U_z \cup (V'_z \cup U'_z)$. $\square$

For the reader's convenience we recall some well-known facts concerning Seshadri constants. The next lemma deals with the Seshadri constant at general points versus that at very general points (compare [EKL, 1.4]).

Lemma 1.4 *Let X be a smooth projective variety and L a nef and big line bundle on X. Suppose $\varepsilon(L, y) = \varepsilon$ for a point $y \in X$. Then for any real $\delta > 0$ there exists a Zariski open neighbourhood $U(\delta) \subset X$ of y such that*

$$\varepsilon(L, x) \geq \varepsilon - \delta \quad \text{for all } x \in U(\delta). \quad \square$$

Finally we recall the relations between isolated singularities, generation of higher jets, and Seshadri constants (cf. [ELN, 1.1], [EKL, 1.3]).

Theorem 1.5 *Let X be a smooth projective n-fold and L a nef and big line bundle on X.*

(1) *Suppose there exists a divisor $E \in |kL|$ having an isolated singularity of index $\geq n + s$ at $x \in X$. Then $H^1(X, \mathcal{O}_X(K_X + L) \otimes \mathcal{I}_x^{s+1}) = 0$. In particular, $|K_X + L|$ generates s-jets at x, i.e., we have a surjective evaluation map*

$$H^0(X, \mathcal{O}_X(K_X + L)) \to H^0(X, \mathcal{O}_X(K_X + L) \otimes \mathcal{O}_X/\mathcal{I}_x^{s+1}).$$

(2) *Let $\varepsilon(L, x)$ be the Seshadri constant of L at x. If*

$$r > \frac{s}{\varepsilon(L, x)} + \frac{n}{\varepsilon(L, x)},$$

then $|K_X + rL|$ generates s-jets at $x \in X$.

(3) Conversely, if $\varepsilon(L,x) < \alpha$, then for all $s_0 > 0$ and all real c there exists an $s \geq s_0$ and an $r > \dfrac{s}{\alpha} + c$ such that $|K_X + rL|$ does not generate s-jets at x. $\square$

2 Proof of Theorem 1

2.1 Partitions of the interval $[0, \alpha]$

Let $\alpha \in \mathbb{Q}^+$ be such that $L^n > \alpha$, and $0 = \beta_1 < \beta_2 < \cdots < \beta_{n+1} = \alpha$ a partition of the interval $[0, \alpha]$ with rational β_i. In the applications in §3 we will use a clever choice of the β_i, which we call "rescaling" of the interval $[0, \alpha]$.

2.2 Families of divisors

Pick an arbitrary point $y \in X$ and a smooth affine neighbourhood $T \subset X$ of y in X. Then the embedding $g \colon T \hookrightarrow X$ satisfies the properties of 0.5. Note that (very) general points of T correspond to (very) general points of X. We will use the notations introduced in 0.5 henceforth.

Arguing as in [EKL, 3.8], for $k \gg 0$ with $\alpha k \in \mathbb{Z}$ we obtain divisors $\mathcal{E}_k \in |\operatorname{pr}_X^*(kL)|$ in $X \times T$ satisfying

$$\operatorname{ind}_\Gamma(\mathcal{E}_k) > \alpha.$$

The argument is, in brief, that for any $x \in X$ one finds a divisor $E \in |kL|$ with $\operatorname{mult}_x(E) > \alpha k$ by using Riemann-Roch and a parameter count. Hence the torsion free $\mathcal{O}_T$-module

$$R = \operatorname{pr}_{T*}\left(\operatorname{pr}_X^*(kL) \otimes \mathcal{I}_\Gamma^{(\alpha k)}\right)$$

has positive rank, and is globally generated since T is affine. Therefore via the evaluation map $\operatorname{pr}_T^* R \to \operatorname{pr}_X^*(kL) \otimes \mathcal{I}_\Gamma^{(\alpha k)}$, a nonzero section Φ of R gives the desired divisor $\mathcal{E}_k$.

2.3 The multiplicity schemes $\mathcal{Z}_\sigma(\mathcal{E})$

For $k \geq 0$ with $\alpha k \in \mathbb{Z}$, we set

$$A_k = \left|\mathcal{I}_\Gamma^{(\alpha k)} \otimes \operatorname{pr}_X^*(kL)\right|.$$

Then by 2.2, A_k is nonempty for sufficiently large k. For nonzero $\mathcal{E}_k \in A_k$ and any rational number $\sigma > 0$, we define

$$\mathcal{Z}_\sigma(\mathcal{E}_k) = \{y = (x,t) \in \mathcal{E}_k \mid \operatorname{ind}_y(\mathcal{E}_k) \geq \sigma\}.$$

Note that $\mathcal{Z}_\sigma(\mathcal{E}_k)$ is a Zariski closed subset of $X \times T$. Its natural scheme structure is given locally by the vanishing of all partial derivatives of order $< k\sigma$ of a local equation of $\mathcal{E}_k$. We are only interested in $\mathcal{Z}_\sigma(\mathcal{E}_k)$ as an algebraic set, and for a general choice of $\mathcal{E}_k$.

The following lemma, which is an analog of [ELN, 3.8], says that the multiplicity loci $\mathcal{Z}_\sigma(\mathcal{E}_k)$ are independent of k and the choice of a general $\mathcal{E}_k \in A_k$ as soon as k is sufficiently large.

Lemma 2.3.1 *For fixed σ there is a positive integer k_0 such that $\mathcal{Z}_\sigma(\mathcal{E}_{k_1}) = \mathcal{Z}_\sigma(\mathcal{E}_{k_2})$ for all $k_1, k_2 \geq k_0$.*

Proof (compare [ELN])   For simplicity, as σ and α are fixed, we write $\mathcal{Z}(k)$ for $\mathcal{Z}_\sigma(\mathcal{E}_k)$. Choose an integer $m \geq 2$ such that $A_k \neq \emptyset$ for $k \geq m$. Fixing an integer $a \geq m$, we claim that there exists a positive integer $k(a)$ such that

$$\mathcal{Z}(r) \subset \mathcal{Z}(a) \quad \text{whenever} \quad r \geq k(a). \tag{$*$}$$

To prove the claim, suppose that $y \notin \mathcal{Z}(a)$, so that there exists $\eta > 0$ satisfying

$$\text{mult}_y(\mathcal{E}_a) \leq a\sigma - \eta.$$

Since the index is a discrete invariant, η is bounded below independently of y; in fact if $m\sigma \in \mathbb{Z}$ then $\eta \geq 1/m$. Suppose $b \geq m$ is an integer coprime to a. Then any integer $r \geq ab$ can be expressed as $r = \alpha a + \beta b$ with $\alpha, \beta \in \mathbb{Z}$ and $0 \leq \beta \leq a$. Consider the divisor $\mathcal{E}'_r = \alpha \mathcal{E}_a + \beta \mathcal{E}_b \in A_r$. Then

$$
\begin{aligned}
\text{mult}_y(\mathcal{E}_r) &\leq 1 + \text{mult}_y(\mathcal{E}'_r) = 1 + \alpha \cdot \text{mult}_y(\mathcal{E}_a) + \beta \cdot \text{mult}_y(\mathcal{E}_b) \\
&\leq 1 + \alpha a \sigma - \alpha \eta + \beta \cdot \text{mult}_y(\mathcal{E}_b) \\
&= r \left(\sigma - \frac{\eta \left(1 - \frac{\beta b}{r} \right)}{a} + \frac{1 + \beta \left(\text{mult}_y(\mathcal{E}_b) - \sigma b \right)}{r} \right),
\end{aligned}
$$

where the first inequality is a consequence of a general version of Bertini's Theorem for the general $\mathcal{E}_r \in A_r$ (see, for example, [Xu96]). Since η, β and b are bounded independently of r, it follows that $\text{mult}_y(\mathcal{E}_r) < r\sigma$ for $r \gg 0$. Hence $y \notin \mathcal{Z}(r)$ for all sufficiently large r, as claimed.

If $\mathcal{Z}(r) = \mathcal{Z}(a)$ for all $r \gg 0$ we are finished. Otherwise by $(*)$, there exists $a' > 0$ such that $\mathcal{Z}(a') \subsetneq \mathcal{Z}(a)$. The argument can then be repeated with a' instead of a. Since this process must stop after a finite number of steps, the lemma follows. $\square$

The next lemma is an adaptation of [ELN, 1.5, 1.6] to our situation. We present a sketch proof for the reader's convenience.

Lemma 2.3.2 (Gap Lemma) *Let $\mathcal{E} \subset X \times T$ be a family of effective divisors on X with $\operatorname{ind}_\Gamma(\mathcal{E}) > \alpha$ along the graph $\Gamma \subset X \times T$ of $g\colon T \to X$. Define*

$$\mathcal{Z}_0 = X \times T \quad and$$
$$\mathcal{Z}_j = \mathcal{Z}_{\beta_j}(\mathcal{E}) = \{y \in \mathcal{E} \mid \operatorname{ind}_y(\mathcal{E}) \geq \beta_j\} \quad for\ 1 \leq j \leq n+1$$

(with β_i as above). Then there exists an index c with $1 \leq c \leq n$, and an irreducible subvariety $\mathcal{V} \subset X \times T$ such that:

(1) $\operatorname{codim}(\mathcal{V}) = c$,

(2) $\Gamma \subset \mathcal{V}$, and

(3) $\mathcal{V}$ is an irreducible component of both $\mathcal{Z}_c$ and $\mathcal{Z}_{c+1}$.

This means that the index of $\mathcal{E}$ "jumps" by at least $\beta_{c+1} - \beta_c$ along $\mathcal{V}$, that is, $\operatorname{ind}_y(\mathcal{E}) \geq \beta_{c+1}$ for every $y \in \mathcal{V}$ and there is an open set $U \subset X \times T$ meeting $\mathcal{V}$ such that $\operatorname{ind}_v(\mathcal{E}) < \beta_c$ for every $v \in U \setminus \mathcal{V}$.

Sketch of Proof The sets $\mathcal{Z}_i$ lie in a chain

$$\Gamma \subset \mathcal{Z}_{n+1} \subset \cdots \subset \mathcal{Z}_1 = \mathcal{E} \subsetneq \mathcal{Z}_0 = X \times T.$$

Starting with $\mathcal{Z}_{n+1}$ and working up in dimension, we can choose irreducible components $\mathcal{V}_j$ of $\mathcal{Z}_j$ containing Γ such that $\mathcal{V}_{j+1} \subset \mathcal{V}_j$. So we arrive at a chain of irreducible varieties

$$\Gamma \subset \mathcal{V}_{n+1} \subset \mathcal{V}_n \subset \ldots \subset \mathcal{V}_1 \subsetneq \mathcal{V}_0 = X \times T,$$

and since $X \times T$ is irreducible of dimension $2n = \dim(\Gamma) + n$, at least two consecutive links in the chain must coincide, say $\mathcal{V}_c = \mathcal{V}_{c+1}$, and we take $\mathcal{V} = \mathcal{V}_c$. Using elementary combinatorial arguments one shows that also the condition $\operatorname{codim}(\mathcal{V}) = c$ can be achieved. For details we refer to [ELN], proof of Lemma 1.6. $\square$

2.4 The "jumping" locus as base locus of a certain linear system

By Lemma 2.3.1, there exists an integer k_0 such that the multiplicity loci $\mathcal{Z}_{\beta_i}(\mathcal{E}_k)$, for $1 \leq i \leq n+1$, are independent of k as soon as $k \geq k_0$ and $\mathcal{E}_k \in A_k$ is general. Therefore also the multiplicity "jumping" loci $\mathcal{V}$ obtained by the Gap Lemma 2.3.2 can be chosen independently of $\mathcal{E} = \mathcal{E}_k$ and k up to the above restrictions. Fix such a $\mathcal{V}$ and put $\beta = \beta_{c+1} - \beta_c$.

Proposition 2.4.1 *For all sufficiently divisible $k \gg 0$ the jumping locus $\mathcal{V}$ is an irreducible component of the base locus of the linear system*

$$\left| \mathcal{I}_{\mathcal{V}}^{(\beta k)} \otimes \mathrm{pr}_X^*(kL) \right|.$$

Here sufficiently divisible means that $\beta_i k \in \mathbb{Z}$ for all $i = 1, \ldots, n+1$.

We start by recalling some general facts concerning differentiating sections of line bundles $\mathrm{pr}_X^*(kL)$ in parameter directions and its connection with certain multiplicity loci (cf. also [EKL, §2] and [ELN, §2]).

Let $\mathcal{D}_{X \times T}^{\ell}(\mathrm{pr}_X^*(kL))$ be the sheaf of differential operators of order $\leq \ell$ on $\mathrm{pr}_X^*(kL)$ and $\mathcal{D}_T^{\ell}$ the sheaf of differential operators of order $\leq \ell$ on T. Since there is a canonical inclusion of vector bundles

$$\mathrm{pr}_T^*(\mathcal{D}_T^{\ell}) \hookrightarrow \mathcal{D}_{X \times T}^{\ell}(\mathrm{pr}_X^*(kL)),$$

the sections of $\mathcal{D}_T^{\ell}$ act naturally on the space of sections of $\mathrm{pr}_X^*(kL)$. A section $\psi \in \Gamma(X \times T, \mathrm{pr}_X^*(kL))$ determines a sheaf homomorphism

$$\eth_{\ell}(\psi) \colon\ \mathrm{pr}_T^*(\mathcal{D}_T^{\ell}) \to \mathrm{pr}_X^*(kL).$$

If we represent ψ locally by a function f, then $\eth_{\ell}(\psi)$ just takes a differential operator D to the function $D(f)$. Since $\mathrm{pr}_X^*(kL)$ is a line bundle, there exists a sheaf of ideals $\mathcal{I}_{\Sigma_{\ell}(\psi)}$ such that

$$\mathrm{Im}\Big(\eth_{\ell}(\psi) \colon\ \mathrm{pr}_T^*(\mathcal{D}_T^{\ell}) \to \mathrm{pr}_X^*(kL) \Big) = \mathcal{I}_{\Sigma_{\ell}(\psi)} \otimes \mathrm{pr}_X^*(kL).$$

Let ψ be a defining section for a divisor $\mathcal{E} \in |\,\mathrm{pr}_X^*(kL)|$. Then we claim that

$$\Sigma_{\ell}(\psi) = \big\{ (x, t) \in X \times T \mid \mathrm{mult}_t(\mathcal{E}_x) > \ell \big\}. \tag{2.4.2}$$

Indeed, the scheme structure on the right hand side is given locally by the vanishing of all partial derivatives of order $\leq \ell$ in the T direction of a local equation for $\mathcal{E}$.

Note also that the sheaves $\mathcal{I}_{\Sigma_{\ell}(\psi)} \otimes \mathrm{pr}_X^*(kL)$ are generated by global sections, because they are quotients of the globally generated sheaf $\mathrm{pr}_T^*\big(\mathcal{D}_T^{\ell} \big)$.

Proof of Proposition 2.4.1   The plan is to apply Lemma 1.2. By assumption,

$$\alpha k, \quad p = \beta_c k \quad \text{and} \quad q = \beta k \quad \text{are integers.}$$

Let $\mathcal{E} \in A_k$ be a divisor determining $\mathcal{V}$ and ψ a section defining $\mathcal{E}$. For integers ℓ put $\Sigma_{\ell} = \Sigma_{\ell}(\psi)$. We claim that

$$\mathcal{I}_{\Sigma_{p-1}} \subset \mathcal{I}_{\mathcal{V}}^{(q)}. \tag{2.4.3}$$

To prove this, let f be a local equation for $\mathcal{E}$ over some open set U. Then $\mathcal{I}_{\Sigma_{p-1}}$ is locally generated by all functions

$$\{D(f) \mid D \in (\mathrm{pr}_T^* \mathcal{D}_T^{p-1})(U)\}.$$

On the other hand we have $\widetilde{D}(f) \in \mathcal{I}_\mathcal{V}$ for every $\widetilde{D} \in \mathcal{D}_{X \times T}^{p+q-1}(\mathrm{pr}_X^*(kL))(U)$, since $\mathcal{V}$ is an irreducible component of $\mathcal{Z}_{\frac{p+q}{k}}(\mathcal{E})$. And in particular $\widetilde{D}(f) \in \mathcal{I}_\mathcal{V}$ for every $\widetilde{D} \in (\mathrm{pr}_T^* \mathcal{D}_T^{p+q-1})(U)$. Hence for all $D \in \mathrm{pr}_T^* \mathcal{D}_T^{p-1}$ the function $D(f)$ vanishes to order $\geq q$ on $\mathcal{V}$, which proves (2.4.3).

Since $\mathcal{I}_{\Sigma_{p-1}} \otimes \mathrm{pr}_X^*(kL)$ is globally generated, applying Lemma 1.2 to our situation will give the desired result once we show that $\mathcal{V}$ is also an irreducible component of Σ_{p-1}. This is the content of Lemma 2.4.4 below. $\square$

Lemma 2.4.4 *For σk a positive integer, let $\mathcal{E} \in |\mathrm{pr}_X^*(kL)|$ be an effective divisor on $X \times T$ and $V \subset \mathcal{Z}_\sigma(\mathcal{E})$ an irreducible component dominating X. Then V is also an irreducible component of $\Sigma_{k\sigma-1}(\mathcal{E})$.*

Proof By definition $\mathcal{Z}_\sigma(\mathcal{E}) \subset \Sigma_{k\sigma-1}(\mathcal{E})$. Let $W \subset \Sigma_{k\sigma-1}(\mathcal{E})$ be an irreducible component containing V. If we can show that $W \subset \mathcal{Z}_\sigma(\mathcal{E})$, then we are done, because that implies $V = W$. Lemma 1.1 shows that $\mathrm{ind}_W(\mathcal{E}) = \mathrm{ind}_{W'_x}(\mathcal{E})$ for general $x \in X$ and any irreducible component W'_x of W_x. Hence the assertion follows from $\mathrm{ind}_{W_x}(\mathcal{E}_x) \geq \sigma$, where we used (2.4.2). $\square$

Corollary 2.4.5 *For all sufficiently divisible $k \gg 0$, we have*

$$\left| \mathcal{I}_\Gamma^{(\alpha k)} \otimes \mathrm{pr}_X^*(kL) \right| \subset \left| \mathcal{I}_\mathcal{V}^{(\beta k)} \otimes \mathrm{pr}_X^*(kL) \right|.$$

Proof By the above, any sufficiently general $\mathcal{E} \in A_k$ determines the same $\mathcal{V}$, in particular, such an $\mathcal{E}$ satisfies $\mathrm{ind}_\mathcal{V}(\mathcal{E}) \geq \beta_c$, and this implies that $\mathcal{E} \in |\mathcal{I}_{\Sigma_{p-1}} \otimes \mathrm{pr}_X^*(kL)|$ by (2.4.2), where again we assume that αk, $p = \beta_c k$ and $q = \beta k$ are integers. The claim then follows from (2.4.3). $\square$

Proposition 2.4.6 *For all sufficiently divisible $k \gg 0$ and for very general $t \in T$, the following hold:*

(1) There exists an irreducible subvariety $V \subset X$ of codimension c containing $g(t)$ which is an irreducible component of the base locus of the linear system

$$|\mathcal{J}_k| := \left| \left(\mathcal{I}_\mathcal{V}^{(\beta k)} \right)_t \otimes kL \right|$$

on X such that $\mathrm{mult}_V D \geq k\beta$ for all $D \in |\mathcal{J}_k|$. In particular, if $c = n$, i.e., $\mathcal{V} = \Gamma$, then by Bertini's Theorem there exists a divisor $D \in |kL|$ having an isolated singularity of index $\geq \beta$ at the point $x = g(t)$.

(2) $\dim H^0(X, \mathcal{I}_{g(t)}^{\alpha k} \otimes kL) \leq \dim H^0(X, \mathcal{J}_k)$.

Proof   To begin with, we study the situation for a fixed k. First we note that after possibly shrinking T, we can assume that the coherent sheaf $\mathcal{F} = \mathcal{I}_{\mathcal{V}}^{(\beta k)} \otimes \mathrm{pr}_X^*(kL)$ is flat over T. In fact $\mathrm{pr}_T \colon X \times T \to T$ is projective and T is affine and integral, hence the assertion follows by considering the Hilbert polynomials of the $\mathcal{F}_t$ (cf. also [Ha, III.9.9]): these do not depend on t for t in an open dense subset of T.

After possibly shrinking T more, it follows from semicontinuity that there is a natural isomorphism

$$H^0(X \times T, \mathcal{F}) \otimes k(t) \simeq H^0(X, \mathcal{F}_t) \tag{$*$}$$

(cf. [Ha, III.12.9]). In other words, taking global sections of $\mathcal{F}$ commutes with restricting to fibres over general $t \in T$, and therefore $(\mathrm{Bs}\,|\mathcal{F}|)_t = \mathrm{Bs}\,|\mathcal{J}_k|$ for such t.

Now we can prove (1). By Proposition 2.4.1 we have $\Gamma \subset \mathcal{V} \subset \mathrm{Bs}\,|\mathcal{F}|$ with $\mathcal{V}$ an irreducible component of $\mathrm{Bs}\,|\mathcal{F}|$, hence Lemma 1.3 shows that any irreducible component V of $\mathcal{V}_t$ is an $(n-c)$-dimensional irreducible component of $(\mathrm{Bs}\,|\mathcal{F}|)_t = \mathrm{Bs}\,|\mathcal{J}_k|$. It remains to show that $\mathrm{mult}_V D \geq k\beta$ for all $D \in |\mathcal{J}_k| = |\mathcal{F}_t|$. But this follows from $(*)$ and Lemma 1.2.

Assertion (2) follows in the same way from Corollary 2.4.5 and the fact that

$$\left(\mathcal{I}_\Gamma^{(\alpha k)}\right)_t = \left(\mathcal{I}_\Gamma^{\alpha k}\right)_t = \mathcal{I}_{g(t)}^{\alpha k}.$$

To complete the proof of the proposition we only have to remark that, since $\mathcal{V}$ does not depend on k, the above arguments work simultaneously for all divisible $k \gg 0$ if we replace the general $t \in T$ by a very general $t \in T$. $\square$

2.5   Bounding the degree of irreducible components of base loci

In this section we complete the proof of Theorem 1 by bounding the degree of the irreducible component V of Proposition 2.4.6, using a strategy essentially due to Fujita (see [Fu82], [Fu94]). Alternatively one could carry out an approach via graded linear systems as in [ELN], leading to slightly weaker bounds.

Let $k \gg 0$ be sufficiently large and divisible, and fix a very general $t \in T$. Let $\mathcal{J}_k = (\mathcal{I}_{\mathcal{V}}^{(\beta k)})_t \otimes kL$ and $V \subset X$ be as in Proposition 2.4.6; recall that V depends on t but *not* on k, and that V is an $(n - c)$-dimensional irreducible component of $\mathrm{Bs}\,|\mathcal{J}_k|$. We may assume that $\dim V > 0$, since otherwise the assertion of Theorem 1 follows from Proposition 2.4.6, (1).

2.5.1 Resolving the base locus of $\Lambda := |\mathcal{J}_k|$, we can find a sequence

$$X' = X_s \to \cdots \to X_r \to X_{r-1} \to \cdots \to X_1 \to X_0 = X$$

of birational morphisms $\tau_i \colon X_i \to X_{i-1}$ together with linear systems Λ_i on X_i such that

(1) $\Lambda_0 = \Lambda$.

(2) $\tau_i \colon X_i \to X_{i-1}$ is the blowup of a smooth subvariety C_i of X_{i-1}.

(3) $\tau_i^* \Lambda_{i-1} = \Lambda_i + m_i E_i$ for some nonnegative integers m_i, where E_i is the exceptional divisor on X_i lying over C_i and $E_i \not\subset \mathrm{Bs}\,\Lambda_i$.

(4) $\mathrm{Bs}\,\Lambda_s = \emptyset$.

Let $\tau = \tau_s \circ \cdots \circ \tau_1$ be the composite, E_i^* the pullback of E_i to X', and $Y_i = \tau(E_i^*)$, so that Y_i coincides with the image of C_i in X.

Let F_i denote the pullback to X' of the general member of the linear system Λ_i on X_i. Finally, let H be a general member of Λ_s, $F = \tau^*(kL)$, and E the fixed part of $\tau^*\Lambda$, so that $H = F - E$, where $E = \sum_{i=1}^s m_i E_i^*$.

By assumption there exists an index r with $Y_r = V$, and

$$m_r \geq k\beta \qquad (2.5.2)$$

since $\mathrm{mult}_V(D) \geq k\beta$ for all $D \in |\mathcal{J}_k|$. We also may and will assume that the resolution τ is chosen in such a way that $\dim Y_i < \dim V = n - c$ for all $i < r$, and in particular $V = Y_r$ is birational to C_r.

Lemma 2.5.3 $F^{n-c} \cdot E \cdot H^{c-1} \geq k^{n-c} m_r^c \deg_L V$.

Proof First of all note that, since F and H are nef and $E - m_r E_r^*$ is an effective divisor, we have

$$F^{n-c} \cdot E \cdot H^{c-1} \geq m_r F^{n-c} \cdot E_r^* \cdot H^{c-1}. \qquad (*)$$

Then one proceeds exactly as in the proof of [Fu82, Lemma 3.2] by first showing

$$F^{n-c} \cdot F_r^c \geq F^{n-c} \cdot F_r \cdot H^{c-1}. \qquad (**)$$

This inequality is proved inductively, and reduces to checking the inequality $H^{c-b-1} F_r^b F^{n-c}(F_r - H) \geq 0$ for $b \geq 0$, which follows from the semipositivity of F_r^b on the class representing $F^{n-c} \cdot E'$ for all components $E' \subset E$.

Finally one computes both sides of the inequality (∗∗). Since $\dim Y_i < n{-}c$ for all $i < r$, we have

$$F^{n-c} \cdot F_r^c = F^{n-c} \cdot (F - \sum_{i \le r} m_i E_i^*)^c = F^{n-c} \cdot (F - m_r E_r^*)^c,$$

hence $F^{n-c} \cdot F_r^c = F^n - m_r^c \deg_{kL}(V)$ by the birationality of the morphism $C_r \to Y_r = V$. A similar argument shows $F^{n-c} \cdot F_r \cdot H^{c-1} = F^n - m_r E_r^* \cdot H^{c-1} \cdot F^{n-c}$. Combining this with (∗) and (∗∗) proves the lemma. □

Lemma 2.5.4 *For any $\varepsilon > 0$ there exist a sufficiently large and divisible k and a resolution $\tau\colon X' \to X$ of the rational map given by $|\mathcal{J}_k|$ satisfying the properties in 2.5.1 and such that $H^n = (F - E)^n \ge k^n(L^n - \alpha^n - \varepsilon)$.*

Proof The proof follows closely the proof of [Fu94, Theorem]. We therefore only give an outline, indicating the necessary modifications. For varying k, consider $|\mathcal{J}_k|$ and denote by (X_k', H_k) the pair (X', H) obtained as in 2.5.1. We derive a contradiction assuming that $H_k^n < k^n(L^n - \alpha^n - \varepsilon)$ on X_k' for all large and divisible k.

Letting ε grow if necessary, we may assume that

$$H_\ell^n \ge \ell^n \left(L^n - \alpha^n - \varepsilon - \frac{\varepsilon}{(2n)!} \right)$$

on X_ℓ' for one fixed large and divisible ℓ. Now, for any integer $s > 0$, we claim that

$$h^0(X, \mathcal{J}_{s\ell}) \le h^0(X_\ell', sH_\ell) + \frac{n(s\ell)^n \varepsilon}{(2n)!},$$

which is proved exactly as in [Fu94] by using the lower bound on H_ℓ^n and considering an appropriate resolution of $\Lambda = |\mathcal{J}_{s\ell}|$. From Proposition 2.4.6, (2) and asymptotic Riemann–Roch we then obtain

$$\frac{(s\ell)^n}{n!}(L^n - \alpha^n) + \varphi(s) \le h^0(X, I_x^{\alpha s\ell} \otimes s\ell L) \le h^0(X, \mathcal{J}_{s\ell})$$
$$\le h^0(X_\ell', sH_\ell) + \frac{n(s\ell)^n \varepsilon}{(2n)!}$$
$$\le \frac{s^n}{n!}\ell^n (L^n - \alpha^n - \varepsilon) + \frac{n(s\ell)^n \varepsilon}{(2n)!} + \psi(s),$$

where φ and ψ are functions with $\lim_{s\to\infty} \frac{\varphi(s)}{s^n} = \lim_{s\to\infty} \frac{\psi(s)}{s^n} = 0$. This gives the desired contradiction. □

Before stating the main result of this section which will complete the proof of Theorem 1, we need to recall some well-known facts (cf. [De93, 5.2], [Fu82, 1.2]).

Lemma 2.5.5 *Let F, H be nef divisors on a smooth projective n-fold Y. Then:*

(1) $F^d \cdot H^{n-d} \geq \sqrt[n]{(F^n)^d} \sqrt[n]{(H^n)^{n-d}}$ holds for all $0 \leq d \leq n$.

(2) If $E = F - H$ is effective, then $F^a \cdot H^{n-a} \geq F^b \cdot H^{n-b}$ for any $a \geq b$. $\square$

Proposition 2.5.6 *In the above notation the degree of V satisfies*

$$\deg_L V \leq \frac{1}{\beta^c} \left(1 - \sqrt[n]{\left(1 - \frac{\alpha^n}{L^n}\right)^c}\right) L^n.$$

Proof By Lemma 2.5.3 we have

$$\deg_L V \leq \frac{1}{k^{n-c}m_r^c} F^{n-c} \cdot E \cdot H^{c-1}.$$

Note that $F^{n-c} \cdot E \cdot H^{c-1} = (F^{n-c+1} \cdot H^{c-1} - F^{n-c} \cdot H^c)$. So if we bound the first term using Lemma 2.5.5, (2) and the second using (1), we find

$$\deg_L V \leq \frac{1}{k^{n-c}m_r^c} \left(F^{n-c+1} \cdot H^{c-1} - F^{n-c} \cdot H^c\right)$$

$$\leq \frac{1}{k^{n-c}m_r^c} \left(F^n - (F^n)^{\frac{n-c}{n}}(H^n)^{\frac{c}{n}}\right)$$

$$\leq \frac{k^n}{k^{n-c}m_r^c} \left(L^n - (L^n)^{\frac{n-c}{n}}(L^n - \alpha^n)^{\frac{c}{n}}\right)$$

$$\leq \frac{1}{\beta^c} \left(1 - \sqrt[n]{\left(1 - \frac{\alpha^n}{L^n}\right)^c}\right) L^n,$$

where the last steps are Lemma 2.5.4 plus the fact that $\deg_L V$ is integral, and (2.5.2). $\square$

3   Applications

Theorem 3.1 *Let X be a smooth projective n-fold, $x \in X$ a very general point and L a nef and big line bundle. Let r and s be positive integers, $\gamma > 1$ a rational number satisfying $(rL)^n > (\gamma(n+s))^n$, and $\alpha_1, \ldots, \alpha_{n-1}$ positive rational numbers with $\sum_{i=1}^{n-1} \alpha_i = \gamma - 1$. Suppose that $|K_X + rL|$ does not generate s-jets at x.*

Then there is a proper subvariety $V \subset X$ of positive dimension d containing x with

$$\deg_L V = L^d \cdot V \leq \frac{(rL)^n}{\alpha_d^{n-d}(n+s)^{n-d}r^d} \left(1 - \sqrt[n]{\left(1 - \frac{\gamma^n(n+s)^n}{(rL)^n}\right)^{n-d}}\right).$$

Proof   Put $L' = rL$, $\alpha = \beta_{n+1} = \gamma(n+s)$, $\beta_n = (\gamma-1)(n+s)$, and recursively downwards $\beta_i = \beta_{i+1} - \alpha_{n-i}(n+s)$ for $i = n-1, \ldots, 1$. Let $x \in X$ be a very general point and apply Theorem 1. In case (a) there exists $k \gg 0$ and a divisor $E \in |kL'|$ having an isolated singularity of index $\geq n+s$ at x, hence by Theorem 1.5, (1) the linear system $|K_X + L'| = |K_X + rL|$ generates s-jets at x. Therefore there is a subvariety $V \subset X$ with the claimed properties.   □

Remark 3.2   Up to the assumption on the positivity of T_X and the genericity of $x \in X$, Theorem 3.1 looks similar to [ELN, Theorem 4.1]. Note however that in the estimate of the degree $\deg_L V$ we have $(n+s)^d$ compared to $(n+s)^n$ in [ELN], which turns out to be crucial when bounding the Seshadri constant. This improvement is achieved by "rescaling" the interval $[0, \alpha]$ as in 2.1.

3.3   Proof of Theorem 2

First we note that, since we are only considering general points, there is no loss of generality in supposing that X is smooth (see [EKL, 3.2] for the precise argument).

Suppose that $\varepsilon(L, x) < \varepsilon$. Then by Theorem 1.5, (3) there exist positive integers s and r with $r > \frac{s+n}{\varepsilon}$ such that $|K_X + rL|$ does not separate s-jets at x. By assumption we have

$$(rL)^n \geq r^n(\varepsilon\gamma)^n > (\gamma(s+n))^n.$$

So we apply Theorem 3.1 to obtain a subvariety $V \ni x$ of positive dimension d satisfying a degree bound which, due to the trivial estimate $1 - \sqrt[n]{(1-a)^{n-d}} < 1$ for $0 < a < 1$ can be replaced by

$$L^d \cdot V < \varepsilon^d \gamma^n \alpha_d^{d-n},$$

leading to a contradiction.   □

Remark 3.4   From Theorem 1 one can deduce easily various boundedness statements by specifying the α_i:

(a) Setting $\alpha_1 = \cdots = \alpha_n = 1$, we get the universal bound $\varepsilon(L, x) \geq n^{-n}$ for the Seshadri constant.

(b) With the following choice, one comes closer to the bound of [EKL]: put $\alpha_i = (n-1)(2^{-i}(1 - 2^{1-n})^{-1}$, and define $\mu(d) := \min_V \{L^d \cdot V\}$, where the minimum runs over all d-dimensional subvarieties $V \subset X$ containing very general points. Then

$$\varepsilon(L, x) \geq \min_{1 \leq d \leq n} \left\{ \sqrt[d]{\mu(d) \frac{(n-1)^{n-d}}{n^n[2^d(1 - 2^{1-n})]^{n-d}}} \right\}.$$

4   Bounds at an arbitrary point

In this section, we show how to apply the strategy of §3 to obtain certain bounds for Seshadri constants at arbitrary points using the following result of [ELN]:

Theorem 4.1 (Ein–Lazarsfeld–Nakamaye) *Let $x \in X$ be any point of a smooth n-fold X. Let A be an ample line bundle on X and $\delta \geq 0$ a real number such that $T_X(\delta A)$ is nef. Suppose that*

$$0 = \beta_1 < \cdots < \beta_{n+1} < \sqrt[n]{L^n}$$

are rational numbers. Then either

(a) *there exists $E \in |kA|$ (for $k \gg 0$) with an isolated singularity at x of index*

$$\frac{\beta_{n+1} - \beta_n}{1 + \delta\beta_n}; \quad or$$

(b) *there exists an irreducible subvariety $V \ni x$ of codimension $c \neq n$ with*

$$A^{n-c}V = \deg_A V \leq \left(\frac{1 + \delta\beta_c}{\beta_{c+1} - \beta_c}\right)^c \cdot \beta_{n+1}^n.$$

Theorem 4.1 follows from [ELN, Theorem 3.9] together with [ELN, 1.5, 1.6] and the remark (from the proof of Theorem 3.9) that V is in any event an irreducible component of the base locus of $|\mathcal{I}_V^{(k\varepsilon)} \otimes \mathcal{O}_X(k(1+\delta\sigma)A))|$ on X; when V is 0-dimensional, this gives the divisor in (a) by Bertini's Theorem.

Then the argument proceeds as in §3:

Corollary 4.2 *Let X, A and δ be as in (4.1), and moreover, let r, s be positive integers, and $\gamma > 1$ a rational number such that $(rA)^n > (\gamma(n+s))^n$. Let $\alpha_1, \ldots, \alpha_{n-1}$ be positive rational numbers with*

$$1 + \left(1 + \frac{\delta}{r}(n+s)\right) \cdot \sum_{i=1}^{n-1} \alpha_i = \gamma.$$

Then for any $x \in X$ either $K_X + rA$ separates s-jets at x, or there exists a subvariety $V \subset X$ through x of dimension $d \geq 1$ and degree

$$A^{n-c} \cdot V \leq \frac{\gamma^n(n+s)^d}{r^d}\left(\frac{1 + \frac{\delta}{r}(n+s)\sum_{i=d+1}^{n-1}\alpha_i}{\alpha_d}\right)^{n-d}$$

Proof Put $A' = rA, \beta_{n+1} = \gamma(n+s), \beta_n = (n+s) \cdot \sum_{i=1}^{n-1} \alpha_i$, and $\beta_i = \beta_{i+1} - \alpha_{n-i}(n+s)$. Then apply Theorem 4.1 to A' and $\delta' = \frac{\delta}{r}$, and use Theorem 1.5, (1). $\square$

Theorem 4.3 *Let X, A and δ be as in 4.1, moreover let $\varepsilon > 0$ be real and $\alpha_1, \ldots, \alpha_n$ positive rational numbers. Let $x \in X$ be any point and suppose that, for any $1 \le d \le n$, any d-dimensional subvariety $V \subset X$ containing x satisfies*

$$\deg_A V = A^d \cdot V \ge \varepsilon^d \left(\frac{1 + \delta\varepsilon \sum_{i=d+1}^{n-1} \alpha_i}{\alpha_d} \right)^{n-d} \cdot \left(1 + (1+\delta\varepsilon) \sum_{i=1}^{n-1} \alpha_i \right)^n .$$

Then $\varepsilon(A, x) \ge \varepsilon$.

Proof Fix $x \in X$ and suppose $\varepsilon(A, x) < \varepsilon$. Then there exist positive integers r, s with $r > (s+n)/\varepsilon$ such that $|K_X + rA|$ does not generate s-jets at x. Put $\gamma = 1 + \left(1 + \frac{\delta}{r}(n+s) \right) \sum_{i=1}^{n-1} \alpha_i$. Then by assumption

$$(rA)^n \ge r^n \varepsilon^n \left(1 + (1+\delta\varepsilon) \sum_{i=1}^{n-1} \alpha_i \right)^n$$

$$> (s+n)^n \left(1 + \left(1 + \frac{\delta}{r}(s+n) \right) \sum_{i=1}^{n-1} \alpha_i \right)^n = ((s+n)\gamma)^n .$$

Then we arrive at a contradiction because Corollary 4.2 gives the existence of $V \ni x$ of dimension $d \ge 1$ with

$$A^d \cdot V \le \frac{\gamma^n (n+s)^d}{r^d} \left(\frac{1 + \frac{\delta}{r}(n+s) \sum_{i=d+1}^{n-1} \alpha_i}{\alpha_d} \right)^{n-d}$$

$$< \varepsilon^d \gamma^n \left(\frac{1 + \delta\varepsilon \sum_{i=d+1}^{n-1} \alpha_i}{\alpha_d} \right)^{n-d}$$

$$< \varepsilon^d \left(\frac{1 + \delta\varepsilon \sum_{i=d+1}^{n-1} \alpha_i}{\alpha_d} \right)^{n-d} \cdot \left(1 + (1+\delta\varepsilon) \sum_{i=1}^{n-1} \alpha_i \right)^n . \quad \square$$

Setting $\alpha_1 = \cdots = \alpha_n = 1$ one obtains Corollary 3 of the introduction.

Remark 4.4 It is well known that, for a very ample line bundle H on X, the twisted tangent bundle $T_X(K_X + nH)$ is globally generated, and in particular nef (cf. [De93, 12.1]). In case $A = K_X$ is ample on X one therefore can use one of the available effectivity statements for very ampleness of ample line bundles (e.g. [De93]) to determine explicit values for δ, making Theorem 4.3 or Corollary 3 effective, in the sense that the bounds for the Seshadri constant of K_X at any $x \in X$ only depend on the dimension n. The same argument works in case $A = -K_X$ is ample, or for any ample A in case K_X is trivial.

Remark 4.5 Finally, let us compare Corollary 3 with the bounds that can be obtained using Angehrn and Siu's basepoint-free Theorem. Namely, Angehrn and Siu prove that, for an ample line bundle A on X, the adjoint line bundles $mA + K_X$ are free for $m \geq \frac{1}{2}n(n+1) + 1$. An elementary argument (see for example [Kü, 3.3]) shows that $\varepsilon(A, x) \geq 1$ for all ample basepoint-free line bundles A. Moreover, if $T_X(\delta A)$ is nef, then so is the $\mathbb{Q}$-line bundle $M := \det(T_X(\delta A)) = -K_X + n\delta A$. Seshadri constants by definition have the sublinearity property

$$\varepsilon(\lambda L + \mu M, x) \geq \lambda \cdot \varepsilon(L, x) + \mu \cdot \varepsilon(M, x)$$

for nef line bundles L, M and any rational $\lambda, \mu \geq 0$. This shows that

$$\varepsilon(A, x) \geq \frac{2}{n(n + 2\delta + 1) + 2}.$$

References

[AS] U. Angehrn and Y.-T. Siu, Effective freeness and point separation for adjoint bundles, Invent. Math. **122** (1995) 291–308

[De92] J.-P. Demailly, Singular Hermitian metrics on positive line bundles, in Hulek et al., eds., Complex algebraic varieties (Bayreuth 1990), Lect. Notes in Math **1507** (1992) 87–104

[De93] J.-P. Demailly, A numerical criterion for very ample line bundles, J. Diff. Geom. **37** (1993) 323–374

[EKL] L. Ein, O. Küchle and R. Lazarsfeld, Local positivity of ample line bundles, J. Diff. Geom. **43** (1995) 193–219

[EL] L. Ein and R. Lazarsfeld, Seshadri constants on smooth surfaces, Astérisque **218** (1993) 177–185

[ELN] L. Ein, R. Lazarsfeld and M. Nakamaye, Zero-estimates, intersection theory and a theorem of Demailly, in M. Andreatta and T. Peternell, eds., Higher dimensional algebraic varieties (Trento, 1994) de Gruyter (1996) 183–208

[Fu82] T. Fujita, Theorems of Bertini type for certain types of polarized manifolds, J. Math. Soc. Japan **34**, No. 4 (1982) 709–717

[Fu94] T. Fujita, Approximating Zariski decomposition of big line bundles, Kodai Math. J. **17** (1994) 1–3

[Ha]   R. Hartshorne, Algebraic geometry, Graduate Texts in Mathematics 52, Springer

[Kü]   O. Küchle, Multiple point Seshadri constants and the dimension of adjoint linear series, Ann. Inst. Four. **46** (1996) 63–71

[La]   R. Lazarsfeld, Lengths of periods and Seshadri constants of Abelian varieties, Math. Res. Lett. **3** (1996) 439–447

[Na]   M. Nakamaye, Seshadri constants on Abelian varieties, Amer. J. of Math. **118** (1996) 621–636

[Pa]   R. Paoletti, Seshadri constants, gonality of space curves, and restrictions of stable bundles, J. Diff. Geom. **40** (1994) 475–504

[Xu95] G. Xu, Ample line bundles on smooth surfaces, Crelle's J. **469** (1995) 199–209

[Xu96] G. Xu, On Bertini's theorem, Manuscripta Math **89** (1996) 237–244

Oliver Küchle and Andreas Steffens,
Mathematisches Institut, Universität Bayreuth,
D–95440 Bayreuth, Deutschland
Kuechle@btm8x4.mat.uni-bayreuth.de
Andreas.Steffens@freeway.de

Degenerate double covers
of the projective plane

Marco Manetti [*]

Abstract

We prove that the set of canonical models of surfaces of general type which are double covers of $\mathbb{P}^2$ branched over a plane curve of degree $2h$ is a connected component of the moduli space if and only if h is even. To get a connected component when h is odd, we must add some special surfaces called *degenerate double covers of $\mathbb{P}^2$*.

Moreover, we show that the theory of simple iterated double covers (cf. [Ma3]) "works" for every degenerate double cover of $\mathbb{P}^2$; this allows us to construct many examples of connected components of the moduli space having simple iterated double covers of $\mathbb{P}^2$ as generic members.

0   Introduction

Double covers of rational surfaces play an important role in the theory of minimal surfaces of general type, especially those with small c_1^2 (cf. [Hor1]). For example, if S is a smooth minimal surface with $K_S^2 = 2$ and $p_g(S) = 3$, then its canonical model is a double cover of $\mathbb{P}^2$, while if $K_S^2 = 8$ and $p_g(S) = 6$ then S is either a deformation of a double cover of $\mathbb{P}^2$, or a deformation of a double cover of $\mathbb{P}^1 \times \mathbb{P}^1$ ([B–P–V, p. 231]).

One of the main goals of this paper is to determine all smooth surfaces of general type that are (smooth) deformations in the large of a double cover of $\mathbb{P}^2$. In other words, if $\mathcal{M}$ denotes the coarse moduli space of surfaces of general type ([Gi]) and $S_0 \to \mathbb{P}^2$ is a double cover (branched over a nonsingular curve), we want to describe all surfaces S whose classes $[S] \in \mathcal{M}$ are in the same connected component as $[S_0]$. It is important to say that, as we consider only smooth deformations, the minimal resolution of double covers of $\mathbb{P}^2$ branched over very singular curves are not in the same connected component of $[S_0]$.

[*]Research carried out under the EU HCM project AGE (Algebraic Geometry in Europe), contract number ERBCHRXCT 940557.

For every $h \geq 4$, let $N(\mathbb{P}^2, \mathcal{O}(h)) \subset \mathcal{M}$ be the (irreducible) subset of surfaces whose canonical model is a double cover of $\mathbb{P}^2$ branched over a plane curve $D \subset \mathbb{P}^2$ of degree $2h$. Equivalently $N(\mathbb{P}^2, \mathcal{O}(h)) \subset \mathcal{M}$ is the set of double covers of $\mathbb{P}^2$ branched over a curve of degree $2h$ with at most simple singularities [B–P–V, II.8]. It is not difficult to prove that $N(\mathbb{P}^2, \mathcal{O}(h))$ is an open subset of the moduli space, but the above examples show that it is not closed in general. Our first result is the following:

Theorem A *Let $h \geq 4$ be a fixed integer and set $N = N(\mathbb{P}^2, \mathcal{O}(h))$.*

(1) N is open in the moduli space.

(2) N is closed in the moduli space if and only if h is even.

(3) If h is odd then the closure of N in the moduli space is open.

(4) The closure of $N(\mathbb{P}^2, \mathcal{O}(h))$ in the moduli space is a connected component.

We recall that the local analytic structure of $\mathcal{M}$ at a point $[S]$ is isomorphic to the quotient of the base space of the semiuniversal deformation of the canonical model S^{can} by the (finite) group of automorphisms of S (cf. [Gi], [Ma4]) and the subset $\mathcal{M}_{K^2,\chi} \subset \mathcal{M}$ of minimal surfaces with fixed K^2, χ is a quasiprojective variety. (1) is therefore an easy application of well-known theorems about deformations of double covers of smooth surfaces, while (4) is an immediate consequence of (1), (2) and (3).

The idea of the proof of (2) and (3) is the following: let $\{S_t\}$, $t \in \Delta$ be a flat family of minimal surfaces of general type such that $[S_t] \in N(\mathbb{P}^2, \mathcal{O}(h))$ for every $t \neq 0$. Denoting by Y_0 the canonical model of S_0, we use the results of [Ma1] to prove that either $[S_0] \in N(\mathbb{P}^2, \mathcal{O}(h))$, or Y_0 is a double cover of W_0, the projective cone over the rational normal curve of degree 4 in $\mathbb{P}^4$, nonflat over the vertex $w_0 \in W_0$; in the latter case we call Y_0 a *degenerate double cover* of $\mathbb{P}^2$. (As the referee points out, degenerate double cover of $\mathbb{P}^2$ and their deformations can also be described easily in terms of subvarieties of the weighted projective space $\mathbb{P}(1, 1, 1, 2, d)$.)

It is then clear that (3) is a consequence of the local irreducibility of the moduli space $\mathcal{M}$ at every degenerate double cover Y_0 of $\mathbb{P}^2$; this is proved in §4 by giving an explicit description of the Kuranishi family of Y_0.

We also prove the vanishing of some Ext groups on Y_0; these results allow us to apply the machinery of [Ma3] of simple iterated double covers to give a large number of examples of connected components of the moduli space. A finite map between normal algebraic surfaces $p: X \to Y$ is called a *simple iterated double cover* associated to a sequence of line bundles $L_1, \ldots, L_n \in$ Pic Y if the following conditions hold:

(1) There exist normal surfaces $X = X_0, \ldots, X_n = Y$ and flat double covers $\pi_i \colon X_{i-1} \to X_i$ such that $p = \pi_n \circ \cdots \circ \pi_1$.

(2) If $p_i = \pi_n \circ \cdots \circ \pi_{i+1} \colon X_i \to Y$ is the composite of the π_j for $j > i$, then we have the eigensheaf decomposition $\pi_{i*}\mathcal{O}_{X_{i-1}} = \mathcal{O}_{X_i} \oplus p_i^*(-L_i)$ for each $i = 1, \ldots, n$.

We say that a simple iterated double cover $\pi \colon X \to Y$ is *smooth* if each X_i is smooth. By [Ma3, 2.1], $\pi \colon X \to Y$ is smooth if and only if X is smooth.

For any sequence $L_1, \ldots, L_n \in \operatorname{Pic} \mathbb{P}^2$, define $N(\mathbb{P}^2, L_1, \ldots, L_n)$ to be the locus in the moduli space $\mathcal{M}$ of surfaces of general type whose canonical model is a simple iterated double cover of $\mathbb{P}^2$ associated to $L_1, \ldots, L_n$. The subset $N(\mathbb{P}^2, L_1, \ldots, L_n)$ is parametrized by a Zariski open subset of the space of sections of a decomposable vector bundle of rank $2^n - 1$ over $\mathbb{P}^2$ (cf. [Ma3]), and therefore it is irreducible and unirational.

Theorem B *Set $N = N(\mathbb{P}^2, L_1, \ldots, L_n)$ with $l_i = \deg L_i$, and write $\overline{N}$ for the closure of N in the moduli space.*

(1) If $l_n \geq 4$ and $l_i > 2l_{i+1}$ for $i = 1, \ldots, n-1$ then N is an open subset of the moduli space.

(2) Assume in addition to (1) that $l_1, \ldots, l_{n-1}$ are even integers, l_n is odd and $[S] \in \overline{N} \setminus N$; then the canonical model of S is a simple iterated double cover of a degenerate double cover of $\mathbb{P}^2$ and has unobstructed deformations.

(3) If $l_n \geq 5$ is odd, l_i is even and $l_i > 2l_{i+1}$ for each $i = 1, \ldots, n-1$ then $\overline{N}$ is a connected component of the moduli space $\mathcal{M}$.

In the last section of this paper we see, using Theorem B, (3) that simple iterated double covers of $\mathbb{P}^2$ give examples of distinct connected components of moduli space whose general members are smooth simple iterated double covers of $\mathbb{P}^2$ with fixed numerical invariants.

This paper is considered as the ideal continuation of [Ma1] and [Ma3]. To avoid the excessive number of pages necessary for a selfcontained proof of Theorems A and B, we often use the results of the above papers.

An earlier version of this paper formed part of the author's thesis [Ma4]. It is a pleasure to thank my advisor F. Catanese, and also L. Badescu and C. Ciliberto for their interest in this work and many fruitful discussions. Thanks also to the referee for several useful remarks.

Notation

We work exclusively over the complex field $\mathbb{C}$; for any algebraic variety X, we write Ω_X^1 for the sheaf of Kähler differentials and $\theta_X = (\Omega_X^1)^\vee$ for the sheaf of tangent vector fields. If X is normal we denote by K_X its Weil canonical divisor.

By a *deformation* we mean any flat family over a connected base; a *small* deformation means a deformation over a germ of a complex space. Thus to say that a class $\mathcal{C}$ of surfaces is *stable under small deformations* means that for every deformation $f\colon X \to B$ and every point $b \in B$ such that $f^{-1}(b) \in \mathcal{C}$, there exists an analytic neighbourhood $b \in U \subset B$ such that $f^{-1}(u) \in \mathcal{C}$ for every $u \in U$.

We denote by Def_X the functor of infinitesimal deformations of X from the category of local Artinian $\mathbb{C}$-algebras to pointed sets ([Sch]), by $T_X^1 = \mathrm{Def}_X(\mathbb{C}[t]/(t^2))$ its tangent space, and, if T_X^1 is finite dimensional, by $\mathrm{Def}\,X$ the base space of the semiuniversal deformation of X (also called Kuranishi family). We define in a similar way $\mathrm{Def}_{(X,0)}$, $T_{(X,0)}^1$ and $\mathrm{Def}(X,0)$ for any isolated singularity $(X,0)$.

According to [Ca3] a surface singularity $(X,0)$ is called a *half rational double point* (or half R.D.P.) if it is the quotient of a rational double point by an analytic involution. The half rational double points are completely classified in [Ca3].

For every $q \geq 0$ we denote by $\mathbb{F}_q = \mathbb{P}(\mathcal{O}_{\mathbb{P}^1} \oplus \mathcal{O}_{\mathbb{P}^1}(q))$ the Segre–Hirzebruch surface; if $p\colon \mathbb{F}_q \to \mathbb{P}^1$ is the natural fibration we denote by $\sigma_\infty, f, \sigma_0 \in \mathrm{Pic}\,\mathbb{F}_q$ the classes of the unique section of p with negative selfintersection, of the fibre of p and of a section disjoint from σ_∞. It is well known that σ_0, f are generators of the Picard group, $\sigma_0 \sim \sigma_\infty + qf$, $\sigma_0^2 = -\sigma_\infty^2 = q$, and the canonical bundle is $K = -\sigma_0 - \sigma_\infty - 2f$.

1 Degenerations of double covers of $\mathbb{P}^2$

Throughout this paper we denote by $W_0 \subset \mathbb{P}^5$ the projective cone over the rational normal curve of degree 4 in $\mathbb{P}^4$ and by $w_0 \in W_0$ its singular point. The minimal resolution of W_0 is the Segre–Hirzebruch surface $\mathbb{F}_4$. It is well known that W_0 is a degeneration of $\mathbb{P}^2$ and, according to [Ca3, §2] and [Ma1, Theorem 15 and Theorem 8], $\mathbb{P}^2$ and W_0 are the only degenerations of the projective plane with at worst half R.D.P.s

Lemma 1.1 *Let $\sigma \subset W_0 \subset \mathbb{P}^5$ be a generic hyperplane section. Then σ is a generator of $\mathrm{Pic}\,W_0 = \mathbb{Z}$.*

If $W \to \Delta$ is a deformation of W_0 such that $W_t = \mathbb{P}^2$ for every $t \neq 0$, then every line bundle on W_0 extends to a line bundle on W. Moreover, if L

is a line bundle on W such that $L_0 = a\sigma$ then $L_t = \mathcal{O}_{\mathbb{P}^2}(2a)$ for $t \neq 0$.

Proof Let $\gamma\colon X = \mathbb{F}_4 \to W_0$ be the minimal resolution. Since σ doesn't contain the vertex w_0 of the cone and $\sigma^2 = 4$, $\gamma^{-1}(\sigma)$ must be a section σ_0. The singularity at w_0 is rational, which identifies $\operatorname{Pic} W_0$ with the set of line bundle L_0 on X such that $L_0 \cdot \sigma_\infty = 0$. Since $q(W_0) = p_g(W_0) = 0$, the restriction $\operatorname{Pic} W \to \operatorname{Pic} W_0$ is an isomorphism by ([Ma1, Lemma 2]). After a possible restriction of the family $W \to \Delta$ to an open disk $0 \in \Delta' \subset \Delta$ of smaller radius we can assume W embedded in $\mathbb{P}^5 \times \Delta$ (cf. [Ma1, Prop. 3]) and the restriction of $\mathcal{O}_{\mathbb{P}^5}(a)$ to W_t for $t \neq 0$ is a very ample line bundle with selfintersection $4a^2$. The conclusion is now trivial. $\square$

Lemma 1.2 *Let $f\colon \mathcal{Y} \to \Delta$ be a proper flat family of normal surfaces such that Y_t is a smooth surface for every $t \neq 0$, and Y_0 has at worst R.D.P.s.*

Let $\tau\colon \mathcal{Y} \to \mathcal{Y}$ be an involution preserving f such that $Y_t/\tau = \mathbb{P}^2$ for every $t \neq 0$. Then either:

(i) $Y_0/\tau = \mathbb{P}^2$, or

(ii) $Y_0/\tau = W_0$. The double cover $\pi\colon Y_0 \to W_0$ is branched exactly over the vertex $w_0 \in W_0$ and over a divisor $D' \sim (2a-1)\sigma$ with $w_0 \notin D'$. For $t \neq 0$, $Y_t \to Y_t/\tau = \mathbb{P}^2$ is branched over $D'_t \sim \mathcal{O}(4a-2)$ and the divisibility of the canonical class $r(Y_t)$ is even.

Proof The fact that the quotient family $\mathcal{Y}/\tau \to \Delta$ has normal fibres follows from the general fact [L–W, 5.6] that smoothings of normal two dimensional singularities are preserved under finite group actions which are free in codimension 3. We also note that in our particular case, if $\pi\colon \mathcal{Y} \to \mathcal{Y}/\tau$ is the projection, then $\pi_*\mathcal{O}_{\mathcal{Y}} = \mathcal{O}_{\mathcal{Y}/\tau} \oplus M$, where M is the $\mathcal{O}_\Delta$ module of anti-invariant functions, and this decomposition commutes with base change $B \to \Delta$.

Y_0/τ is a normal degeneration of $\mathbb{P}^2$ with at worst half R.D.P.s, and therefore either $Y_0/\tau = \mathbb{P}^2$ or $Y_0/\tau = W_0$. Assume $Y_0/\tau = W_0$; then, since (W_0, w_0) is not a rational double point, $y_0 = \pi^{-1}(w_0)$ is a fixed point of the involution τ. By [Ma3, Prop. 3.2 and Table 2], the singularity (Y_0, y_0) is a simple node defined by the equation $x_0^2 + x_1^2 + x_2^2 = 0$ and the involution τ is conjugate to $x_i \mapsto -x_i$ for $i = 0, 1, 2$. In particular y_0 is an isolated fixed point of the involution.

Let $\delta\colon S \to Y_0$ be the resolution of the node (Y_0, y_0) and $E = \delta^{-1}(y_0) \subset S$ the corresponding nodal curve, i.e., a smooth rational curve with selfintersection $E^2 = -2$. The action of τ can be lifted to an action on S, and it is easy to see that $S/\tau = X = \mathbb{F}_4$. Moreover the flat double cover $\pi\colon S \to \mathbb{F}_4$ is branched over $D = \sigma_\infty \cup D'$, where $\sigma_\infty \cap D' = \emptyset$, and since this divisor must be 2-divisible in $\operatorname{Pic}\mathbb{F}_4$, $D' \sim (2a - 1)\sigma_0$ and $\frac{1}{2}(\sigma_\infty \cup D') = a\sigma_0 - 2f$ where f denotes the fibre of $\mathbb{F}_4$. $\square$

We now recall some results of [Ma3] about flat double covers of surfaces and their deformations. Let $\pi\colon X \to Y$ be a flat double cover of normal surfaces and $L \to Y$ the line bundle such that $\pi_*\mathcal{O}_X = \mathcal{O}_Y \oplus \mathcal{O}_Y(-L)$.

The surface X can be described as a hypersurface of L defined by the equation $z^2 = f$, where z is a coordinate in the fibres of L and $f \in H^0(\mathcal{O}_Y(2L))$. Clearly $R = \{z = 0\} \subset X$ and $D = \{f = 0\} \subset Y$ are exactly the ramification and the branch divisors of π. The line bundle L and the branch divisor D determine the double cover uniquely up to isomorphism. In this situation we say that X is a flat double cover of Y associated to the line bundle L. If $\varphi \in H^0(\mathcal{O}_Y(2L))$ the surface given by the equation $z^2 = f + \varphi$ is called a *natural deformation* of X.

Applying the functor $\operatorname{Hom}_{\mathcal{O}_X}(-, \mathcal{O}_X)$ to the exact sequence of sheaves

$$0 \to \pi^*\Omega_Y^1 \to \Omega_X^1 \to \mathcal{O}_R(-R) \to 0 \tag{1.3}$$

on X gives the exact sequence

$$\operatorname{Ext}^1_{\mathcal{O}_X}(\mathcal{O}_R(-R), \mathcal{O}_X) \xrightarrow{\varepsilon} \operatorname{Ext}^1_{\mathcal{O}_X}(\Omega_X^1, \mathcal{O}_X)$$
$$\to \operatorname{Ext}^1_{\mathcal{O}_Y}(\Omega_Y^1, \mathcal{O}_Y) \oplus \operatorname{Ext}^1_{\mathcal{O}_Y}(\Omega_Y^1, -L). \tag{1.3'}$$

Now there exists an isomorphism $\operatorname{Ext}^1_{\mathcal{O}_X}(\mathcal{O}_R(-R), \mathcal{O}_X) = H^0(\mathcal{O}_R(\pi^*D)) = H^0(\mathcal{O}_D(D))$, and if $q(Y) = 0$ then the image of ε is exactly the space of first order natural deformations of X. In particular if $Y = \mathbb{P}^2$ and the degree of L is ≥ 4 then ε is surjective, so that by the Kodaira–Spencer criterion the family of natural deformations is complete, and then double covers of $\mathbb{P}^2$ are stable under small deformations.

This proves that $N(\mathbb{P}^2, \mathcal{O}(h))$ is open in the moduli space for every $h \geq 4$ and that

$$N_0(\mathbb{P}^2, \mathcal{O}(h)) = \{[S] \in N(\mathbb{P}^2, \mathcal{O}(h)) \mid K_S \text{ is ample}\}$$

is an open dense subset of it. More generally, the following holds:

Proposition 1.4 ([Ma3], 2.3) *In the above notation, let $\widetilde{X} \to \widetilde{Y} \to H$ be a deformation of the map π parametrized by a smooth germ $(H, 0)$ and let $r_X\colon (H, 0) \to \operatorname{Def} X$, $r_Y\colon (H, 0) \to \operatorname{Def} Y$ be the induced maps. Assume:*

  1. r_Y is smooth.

  2. The image of r_X contains the natural deformations.

3. $\mathrm{Ext}^1_{\mathcal{O}_Y}(\Omega^1_Y, -L) = 0$ *and* $H^1(\mathcal{O}_Y) = 0$.

Then the morphism r_X *is smooth.*

Definition 1.5 Let $a \geq 3$ be an integer and $\pi \colon S \to \mathbb{F}_4$ the double cover associated to $L = a\sigma_0 - 2f$ branched over the disjoint union of σ_∞ and a divisor $D' \sim (2a - 1)\sigma_0$ with at worst simple singularities ([B–P–V, II.8]). $E = \pi^{-1}(\sigma_\infty)$ is a nodal curve; taking its contraction $\delta \colon S \to Y_0$ we get a surface with at worst rational double points which is a double cover of the cone W_0. We say that Y_0 is a *degenerate double cover of* $\mathbb{P}^2$. The number a determines $K^2_{Y_0} = 8(a - 2)^2$ and is called the *discrete building datum* of Y_0.

If $a = 2$ the above construction still makes sense. In this case, we obtain a well-studied class of K3 surfaces [Sai, §5], [Hor2].

Theorem 1.6 *For even* $h \geq 4$, *the set* $N = N(\mathbb{P}^2, \mathcal{O}(h))$ *is a connected component of the moduli space. If* h *is odd then the set* $\overline{N} \setminus N$ *is contained in the set of degenerate double covers of* $\mathbb{P}^2$ *with discrete building datum* $a = \frac{1}{2}(h + 1)$.

Proof Note first that N and N_0 have the same closure in the moduli space. If $[S_0] \in \overline{N}_0$, then by the valuative criterion, there exists a deformation $f \colon S \to \Delta$ of S_0 with $[S_t] \in N_0$ for every $t \neq 0$, and an involution τ acting on the punctured family $S^* \to \Delta^*$ such that $S_t/\tau = \mathbb{P}^2$ for every $t \neq 0$. Let $Y \to \Delta$ be the relative canonical model of $S \to \Delta$. Then Y_0 is a normal surface with at worst rational double points and ample canonical bundle and $Y_t = S_t$ for every $t \neq 0$. It is now an immediate consequence of [F–P, Prop. 4.4] (cf. also [Ma5]) that τ extends to a biregular involution of Y. As is well known, in general τ does not necessarily extend to a biregular involution of S (cf. [Ca3]).

The theorem now follows from Lemma 1.2. $\square$

Remark 1.7 In the notation of 1.4, if Def_X is prorepresentable (e.g., if $H^0(\theta_X) = 0$), then it can be proved that Proposition 1.4 holds without the assumption that $(H, 0)$ is smooth. Philosophically, this means that if $X \to Y$ is a flat double cover of normal surfaces ramified over a sufficiently ample divisor then $\mathrm{Def}\,X$ is isomorphic to the product of $\mathrm{Def}\,Y$ with a smooth germ $(\mathbb{C}^n, 0)$.

One of the main goals of the next sections is to prove that every degenerate double cover Y_0 has unobstructed deformations. The natural double cover $\pi \colon Y_0 \to W_0$ is not flat and so we cannot apply the above result (fortunately, since $\mathrm{Def}\,W_0$ is not irreducible, cf. [Rie]).

2   Vanishing theorems for degenerate double covers of $\mathbb{P}^2$

Before proving the main results of this section we need an explicit description of all cotangent vector fields with fixed poles on a Segre–Hirzebruch surface $\mathbb{F}_q$, $q > 1$. For this, we use the description

$$\mathbb{F}_q = (\mathbb{C}^2 - \{0\}) \times (\mathbb{C}^2 - \{0\})/ \sim$$

where $(l_0, l_1, t_0, t_1) \sim (\lambda l_0, \lambda l_1, \lambda^q \mu t_0, \mu t_1)$ for any $\lambda, \mu \in \mathbb{C}^*$.

From now on, by the standard torus action on $\mathbb{F}_q$ we mean the faithful $(\mathbb{C}^*)^2$ action given by

$$(\mathbb{C}^*)^2 \ni (\xi, \eta) \colon (l_0, l_1, t_0, t_1) \mapsto (l_0, \xi l_1, \eta t_0, t_1).$$

$\mathbb{F}_q$ is covered by four affine planes $\mathbb{C}^2 \simeq U_{i,j} = \{l_i t_j \neq 0\}$, which are invariant under the standard torus action. In this affine covering, we choose local coordinates as follows:

$$
\begin{aligned}
U_{0,1} : \quad & z = l_1/l_0 \quad \text{and} \quad s = t_0/t_1 l_0^q, \\
U_{0,0} : \quad & z = l_1/l_0 \quad \text{and} \quad s' = t_1 l_0^q/t_0, \\
U_{1,0} : \quad & z' = l_0/l_1 \quad \text{and} \quad y' = t_1 l_1^q/t_0, \\
U_{1,1} : \quad & z' = l_0/l_1 \quad \text{and} \quad y = t_0/t_1 l_1^q.
\end{aligned}
\tag{2.1}
$$

We call $U_{0,1}$ the principal affine subset and z, s principal affine coordinates. The other pairs of affine coordinates are related to s, z by

$$z' = z^{-1}, \quad s' = s^{-1} \quad y = sz^{-q} \quad \text{and} \quad y' = s^{-1}z^q = y^{-1}.$$

The map $\mathbb{F}_q \to \mathbb{P}^1$, $(l_0, l_1, t_0, t_1) \mapsto (l_0, l_1)$ represents the Segre–Hirzebruch surface as a rational geometrically ruled surface, where $\sigma_\infty : \{t_1 = 0\}$, $\sigma_0 : \{t_0 = 0\}$ and $f : \{l_1 = 0\}$. Note that the rational function y gives the rational equivalence $\sigma_\infty \sim \sigma_0 - qf$.

Lemma 2.2 $h^0(\mathbb{F}_q, \Omega^1(p\sigma_0 + r\sigma_\infty)) = qp^2 - 1$ *for every* $p, q > 0$ *and* $r \geq 0$.

Proof $H^0(\Omega^1(p\sigma_0 + r\sigma_\infty))$ is the vector space of rational cotangent vector fields having at most poles of order p and r along σ_0 and σ_∞ respectively. The standard torus action induces an eigenspace decomposition

$$H^0(\Omega^1(p\sigma_0 + r\sigma_\infty)) = \bigoplus_{a,b \in \mathbb{Z}} M_{a,b}.$$

where $\omega \in M_{a,b}$ if and only if

$$\omega = \alpha_{a,b} z^{a-1} s^b dz + \beta_{a,b} z^a s^{b-1} ds$$

on the open set $U_{0,1}$, for some complex numbers $\alpha_{a,b}, \beta_{a,b}$.
The same ω is written in $U_{0,0}$ as

$$\omega = \alpha_{a,b} z^{a-1} s'^{-b} dz - \beta_{a,b} z^a s'^{-b-1} ds'$$

and in $U_{1,1}$

$$\omega = -(\alpha_{a,b} + q\beta_{a,b}) z'^{-(a+1+qb)} y^b dz' + \beta_{a,b} z'^{-(a+qb)} y^{b-1} dy.$$

Note that $\sigma_0 \cap U_{0,1} : \{s = 0\}$, $\sigma_\infty \cap U_{0,1} = \emptyset$, $\sigma_0 \cap U_{0,0} = \emptyset$, $\sigma_\infty \cap U_{0,0} : \{s' = 0\}$, $\sigma_0 \cap U_{1,1} : \{y = 0\}$ and $\sigma_\infty \cap U_{1,1} = \emptyset$. From the above local description of ω it follows immediately that $\omega \neq 0$ implies $b < 0$, and then there exists an isomorphism $H^0(\Omega^1(p\sigma_0 + r\sigma_\infty)) = H^0(\Omega^1(p\sigma_0))$. By reflexivity, every section of $\Omega^1(p\sigma_0)$ on $U_{0,1} \cup U_{0,0} \cup U_{1,1}$ extends to a unique section on $\mathbb{F}_q$, so that the following set of rational cotangent vector fields

$$\begin{cases} z^{a-1} s^b dz & \text{for} \quad a \geq 1, \ 0 \geq b \geq -p, \ a+1+qb \leq 0, \\ z^a s^{b-1} ds & \text{for} \quad a \geq 0, \ -1 \geq b \geq 1-p, \ a+bq < 0, \\ -q z^{a-1} s^b dz + z^a s^{b-1} ds & \text{for} \quad -1 \geq b \geq 1-p, \ a+bq = 0 \end{cases}$$

are $qp^2 - 1$ bihomogeneous sections of $\Omega^1(p\sigma_0)$; to prove that these form a basis is an easy calculation that we omit. $\square$

Corollary 2.3 *For every $p, q, r > 0$, $h^1(\mathbb{F}_q, \theta) = q - 1$, $h^1(\mathbb{F}_q, \Omega^1(p\sigma_0)) = 1$ and $h^2(\mathbb{F}_q, \theta) = h^2(\mathbb{F}_q, \Omega^1(p\sigma_0)) = h^1(\mathbb{F}_q, \Omega^1(p\sigma_0 + rf)) = 0$.*

Proof The equality $h^1(\mathbb{F}_q, \theta) = q - 1$ is well known ([Ko]). By the Hodge decomposition and Serre duality we have $h^0(\mathbb{F}_q, \Omega^1) = h^2(\mathbb{F}_q, \theta(K)) = 0$ and $h^2(\mathbb{F}_q, \Omega^1) = 0$, and since both $-K$ and $p\sigma_0$ are effective divisors also $h^2(\mathbb{F}_q, \theta)$ and $h^2(\mathbb{F}_q, \Omega^1(p\sigma_0))$ vanish. By Riemann–Roch and Lemma 2.2 we then get $h^1(\mathbb{F}_q, \Omega^1(p\sigma_0)) = 1$.

For every $p, r > 0$ it follows from standard exact sequences that

$$h^1(\mathbb{F}_q, \Omega^1(p\sigma_0 + rf)) \leq h^1(\mathbb{F}_q, \Omega^1(\sigma_0 + f)) = h^0(\mathbb{F}_q, \Omega^1(\sigma_0 + f)) - q,$$

and using the method of proof of Lemma 2.2 we easily see that $z^{a-1} s^{-1} dz$, for $0 \leq a \leq q - 1$ is a basis of $H^0(\mathbb{F}_q, \Omega^1(\sigma_0 + f))$, and the right-hand side above is 0. $\square$

Proposition 2.4 *For the surface $\mathbb{F}_q$, $q > 0$ we have:*

(i) $H^0(a\sigma_0 + bf) \neq 0$ *if and only if* $a \geq 0$ *and* $aq + b \geq 0$.

(ii) *The linear system* $|a\sigma_0 + bf|$ *contains a reduced divisor if and only if either* $a > 0, b \geq -q$ *or* $a = 0, b > 0$.

(iii) $H^1(a\sigma_0 + bf) = 0$ *if and only if either* $a = -1$ *or* $a \geq 0, b \geq -1$ *or* $a \leq -2, b \leq q - 1$.

(iv) *For every pair of positive integers* p, r, *the natural map*

$$H^0(p\sigma_0) \otimes H^0(r\sigma_0) \to H^0((p+r)\sigma_0)$$

is surjective; in particular the image of $\mathbb{F}_q$ *by the complete linear system* $|\sigma_0|$ *is projectively normal.*

(v) $P_{-1}(\mathbb{F}_q) = h^0(-K_{\mathbb{F}_q}) = \max(9, q + 6)$.

Proof (i) and (ii) are clear since $|\sigma_0|$, $|f|$ are base point free and $\sigma_\infty \in |\sigma_0 - qf|$. By Serre duality it is sufficient to study the vanishing of h^1 only for $a \geq -1$. Using standard exact sequences and induction on $|b|$ we have

$$h^1(-\sigma_0 + bf) = h^1(-\sigma_0) = 0 \quad \text{for every } b \in Z,$$

and if $b \geq -1$, by induction on $a \geq 0$ we have

$$h^1(a\sigma_0 + bf) \leq h^1(-\sigma_0 + bf) = 0.$$

If $a \geq 0$ and $b \leq -2$ then we can write $a\sigma_0 + bf = \sigma_\infty + D$ where by (i) and Serre duality $h^2(D) = 0$, thus

$$h^1(a\sigma_0 + bf) \geq h^1(\mathcal{O}_{\sigma_\infty}(a\sigma_0 + bf)) = h^1(\mathcal{O}_{\mathbb{P}^1}(b)) > 0.$$

In the principal affine coordinates z, s a bihomogeneous basis of $H^0(p\sigma_0)$ is given by the monomials $s^{-a}z^b$ with $0 \leq a \leq p$ and $0 \leq b \leq aq$, so that (iv) follows immediately.

For every $q \geq 0$ we have $-K = \sigma_0 + \sigma_\infty + 2f$ and $K^2 = 8$. If $q \leq 3$ by (iii) and Serre duality $H^1(-K) = H^2(-K) = 0$ and $P_{-1} = 9$ by Riemann–Roch. If $q \geq 3$ then $-K \cdot \sigma_\infty < 0$ and $P_{-1} = h^0(\sigma_0 + 2f) = q + 6$. $\square$

Throughout the rest of this section a is a fixed integer ≥ 3. Let X be the Segre–Hirzebruch surface $\mathbb{F}_4$ and $\pi \colon S \to X$ the double cover ramified over $D = \sigma_\infty \cup D'$ with D' a reduced divisor linearly equivalent to $(2a - 1)\sigma_0$. We assume that S has at worst rational double points and write $R \subset S$ for the ramification divisor.

We have $\pi_*\mathcal{O}_S = \mathcal{O}_X \oplus \mathcal{O}_X(-L)$ where $L = a\sigma_0 - 2f$ and $E = \pi^{-1}(\sigma_\infty)$ is a nodal curve. Denote by $\delta \colon S \to Y_0$ the contraction of E; then Y_0 is a surface with at worst rational double points and ample canonical bundle. We call $\delta(E) = y_0$ the *vertex* of the degenerate double cover Y_0.

By abuse of notation we use the same letter σ to denote the line bundles $\sigma_0 \in \text{Pic} \, X$, $\pi^* \sigma_0 \in \text{Pic} \, S$ and $\delta_* \pi^* \sigma_0 \in \text{Pic} \, Y_0$. By the Hurwitz formula $K_S = \pi^*(K_X + L) = (a - 2)\sigma$.

Lemma 2.5 $H^1(Y_0, p\sigma) = 0$ *for every integer* p.

Proof By the Leray spectral sequence we have

$$H^1(Y_0, p\sigma) = H^1(S, p\sigma) = H^1(X, p\sigma) \oplus H^1(X, (p - a)\sigma + 2f)$$

and the lemma follows from Proposition 2.4, (iii). $\square$

Lemma 2.6 *For every smooth curve* C *contained in a smooth surface* S, $H^1_C(\Omega^1_S) \neq 0$.

Proof There exists an inclusion $H^0(\mathcal{F} \otimes \mathcal{O}_C(C)) \subset H^1_C(\mathcal{F})$ for any locally free sheaf $\mathcal{F}$ on S (this is proved in [B–W, 1.5] for the tangent sheaf, but the same proof works for any locally free sheaf) and according to the exact sequence of differentials $H^0(\mathcal{O}_C) \subset H^0(\Omega^1_S \otimes \mathcal{O}_C(C))$. $\square$

Lemma 2.7 *If* $p \geq 2a$ *then* $h^1(S, \Omega^1_S(K_S + p\sigma)) \leq 1$.

Proof We consider the exact sequence on S (cf. (1.3))

$$0 \to \pi^*(\Omega^1_X(K_X + L + p\sigma)) \to \Omega^1_S(K_S + p\sigma) \to \mathcal{O}_R(\pi^*(K_X + p\sigma)) \to 0,$$

where $R \subset S$ is the ramification divisor.

Using the previous results, we get

$$h^1(\mathcal{O}_D(K_X + p\sigma))$$
$$\leq h^1(X, (p - 2)\sigma + 2f) + h^2(X, K_X + (p - 2a)\sigma + 4f) = 0,$$
$$h^1(\pi^*\Omega^1_X(K_X + L + p\sigma)) =$$
$$h^1(\Omega^1_X(K_X + L + p\sigma)) + h^1(\Omega^1_X(K_X + p\sigma)) = 1$$

for $p \geq 2a$, and the proof follows from the equality $h^1(\mathcal{O}_R(\pi^*(K_X + p\sigma))) = h^1(\mathcal{O}_D(K_X + p\sigma))$. $\square$

Theorem 2.8 *In the above notation*, $\text{Ext}^1_{Y_0}(\Omega^1_{Y_0}, -p\sigma) = 0$ *for every* $p \geq 2a$.

Proof Y_0 is a Gorenstein surface, in particular $K_{Y_0} + p\sigma$ is a Cartier divisor. By Serre duality ([Ha1, p. 243])

$$\text{Ext}^1_{Y_0}(\Omega^1_{Y_0}, -p\sigma)^\vee = \text{Ext}^1_{Y_0}(\Omega^1_{Y_0}(K_{Y_0} + p\sigma), K_{Y_0})^\vee = H^1(\Omega^1_{Y_0}(K_{Y_0} + p\sigma)).$$

We use the following exact sequence of sheaves on Y_0 ([Kas], [Pi2])

$$0 \to \Omega^1_{Y_0} \to \delta_*\Omega^1_S \xrightarrow{\alpha} \mathbb{C}_{y_0} \to 0,$$

where for every open subset $E \subset U \subset S$ and every $\omega \in H^0(U, \Omega^1_S)$, $\alpha(\omega) = 0$ if and only if the holomorphic 2-form $d\omega$ vanishes on E. We observe immediately that $\Omega^1_{Y_0}$, being locally generated by closed 1-forms, is contained in the kernel of α; the converse inclusion requires some computation ([Kas, p. 55]). Note moreover that, according to ([Ste], [Pi1, App.]), the sheaf $\delta_*\Omega^1_S$ is reflexive, and then the exactness of the above sequence is equivalent to the equality $H^1_{\{y_0\}}(Y_0, \Omega^1_{Y_0}) = \mathbb{C}$.

Twisting the above exact sequence by $K_{Y_0} + p\sigma = \delta_*(K_S + p\sigma)$ we get

$$0 \to \Omega^1_Y(K_{Y_0} + p\sigma) \to \delta_*\Omega^1_S(K_S + p\sigma) \xrightarrow{\alpha} \mathbb{C}_{y_0} \to 0.$$

Our first step is to prove that $H^1(\Omega^1_{Y_0}(K_{Y_0} + p\sigma)) = H^1(\delta_*\Omega^1_S(K_S + p\sigma))$ for $p \geq 2a$, that is, that α is surjective on global sections. Actually the following stronger result holds:

Lemma 2.9 *In the above notation if $p \geq 2$ then the composite of $H^0(\alpha)$ with the pullback map $\pi^*: H^0(\Omega^1_X(K_X + p\sigma)) \to H^0(\Omega^1_S(K_S + p\sigma))$ is surjective.*

Proof Let s, z be the principal affine coordinates on $X = \mathbb{F}_4$ and consider $\omega = s^{-2}dz(dz \wedge ds) \in H^0(\Omega^1_X(K_X + p\sigma))$. In the open set $U_{0,0} \subset X$ with coordinates z, s', we have $\omega = dz(ds' \wedge dz)$, and $\sigma_\infty = \{s' = 0\}$; locally, S is the double cover of X defined by the equation $\xi^2 = s'$, so that $\pi^*\omega = 2\xi dz(d\xi \wedge dz)$.

Now $d\xi \wedge dz$ extends to a holomorphic invertible section of K_S in a neighbourhood of E and then, up to nonzero scalar multiplication, $\alpha(\pi^*\omega) = \alpha(\xi dz) \neq 0$ since $d(\xi dz) = d\xi \wedge dz$. $\square$

The Leray spectral sequence gives an exact sequence

$$0 \to H^1(\delta_*\Omega^1_S(K_S + p\sigma)) \to H^1(\Omega^1_S(K_S + p\sigma)) \xrightarrow{r} H^0(R^1\delta_*\Omega^1_S(K_S + p\sigma)),$$

and if $r \neq 0$ then by Lemma 2.7 the proof is complete.

For any open set $E \subset U \subset S$ there exists an exact sequence

$$0 \to H^0(U, \Omega^1_S(K_S + p\sigma)) \xrightarrow{\beta} H^0(U \setminus E, \Omega^1_S(K_S + p\sigma))$$
$$\xrightarrow{d} H^1_E(\Omega^1_S(K_S + p\sigma)) \xrightarrow{r_U} H^1(U, \Omega^1_S(K_S + p\sigma)).$$

On the open set $V = \delta(U) \subset Y$, the coherent sheaf $\delta_* \Omega_S^1(K_S + p\sigma)$ is reflexive, in particular the above map β is an isomorphism and the map r_U is injective. Since $H_E^1(\Omega_S^1(K_S + p\sigma)) = H_E^1(\Omega_S^1) \neq 0$ the above inclusion factors as

$$H_E^1(\Omega_S^1) \subset H^1(S, \Omega_S^1(K_S + p\sigma)) \xrightarrow{r_U} H^1(U, \Omega_S^1(K_S + p\sigma)),$$

and then $r = \varinjlim r_U \neq 0$. $\quad\square$

3 Deformations of degenerate double covers locally trivial at the vertex

Let a, $\pi \colon S \to X$ and $\delta \colon S \to Y_0$ be as in Definition 1.5. Then π is a flat double cover, and there exists a family of natural deformations of S obtained by deforming the branch divisor $D = \sigma_\infty \cup D' \sim 2a\sigma - 4f$. Since σ_∞ is a fixed part of the linear system $|D|$, the natural deformations are parametrized by $H^0(X, (2a - 1)\sigma)$.

The singularity (Y_0, y_0) is rational, so that, as in [B–W], we can define the blowdown morphism $\beta \colon \mathrm{Def}_S \to \mathrm{Def}_{Y_0}$. It is clear that every (infinitesimal) natural deformation of S is trivial in a neighbourhood of E and its blowdown is a deformation of Y_0, locally trivial at y_0.

Thus taking first order deformations gives a commutative diagram

$$
\begin{array}{ccc}
H^0(X, (2a - 1)\sigma) & \xrightarrow{\mathrm{Nat}} & T_S^1 \\
\rho \downarrow & & \downarrow \beta \\
T^1 LT(Y_0, y_0) & \longrightarrow & T_{Y_0}^1
\end{array}
$$

where β is the blowdown map and $T^1 LT(Y_0, y_0)$ the kernel of the natural restriction map $T_{Y_0}^1 \to T_{(Y_0, y_0)}^1$. Note that natural deformations never give a complete family of deformations of S, because the nodal curve E contributes to the space T_S^1 ([B–W]).

Theorem 3.1 *The above map ρ is surjective and the blowdown of the family of natural deformations of S is a complete family of deformations of Y_0, locally trivial at the vertex, with smooth base space.*

Proof The exact sequence (1.3′) in this particular case becomes

$$H^0(\mathcal{O}_R(\pi^* D)) \xrightarrow{\varepsilon} \mathrm{Ext}_S^1(\Omega_S^1, \mathcal{O}_S) \xrightarrow{\sigma} H^1(\theta_X) \oplus H^1(\theta_X(-L)),$$

and the image of ε is the set of first order natural deformations. Given an open subset $V \subset X$, the inclusion $\pi^* \Omega_X^1 \to \Omega_S^1$ induces a commutative diagram

$$
\begin{array}{ccc}
\mathrm{Ext}_S^1(\Omega_S^1, \mathcal{O}_S) & \longrightarrow & \mathrm{Ext}_{\pi^{-1}(V)}^1(\Omega_{\pi^{-1}(V)}^1, \mathcal{O}_{\pi^{-1}(V)}) \\
\sigma \downarrow & & \downarrow \\
H^1(\theta_X) \oplus H^1(\theta_X(-L)) & \xrightarrow{\gamma_V} & H^1(\theta_V) \oplus H^1(\theta_V(-L)).
\end{array}
$$

Lemma 3.2 *In the above set-up, if $\sigma_\infty \subset V$, then γ_V is injective.*

Proof of 3.2 It is clearly sufficient to prove that the two natural maps

$$
\gamma_1 \colon H^1(\theta_X) \to H^1(\theta_X \otimes \mathcal{O}_{\sigma_\infty}), \quad \gamma_2 \colon H^1(\theta_X(-L)) \to H^1(\theta_X(-L) \otimes \mathcal{O}_{\sigma_\infty})
$$

are isomorphisms.

Note first that $h^1(\theta_X \otimes \mathcal{O}_{\sigma_\infty}) = 3$, $h^1(\theta_X(-L) \otimes \mathcal{O}_{\sigma_\infty}) = 1$ and by Corollary 2.3, $h^1(\theta_X) = 3$, $h^1(\theta_X(-L)) = h^1(\Omega_X^1((a-2)\sigma_0)) = 1$, $h^2(\theta_X(-\sigma_\infty)) = h^0(\Omega_X^1(-\sigma_0 - 2f)) = 0$.

Hence γ_1 is surjective and then it is an isomorphism. To show that γ_2 is surjective, we prove that the natural map $H^2(\theta_X(-L-\sigma_\infty)) \to H^2(\theta_X(-L))$ or its Serre dual $H^0(\Omega_X^1((a-2)\sigma_0)) \to H^0(\Omega_X^1((a-2)\sigma_0 + \sigma_\infty))$ is an isomorphism, but this is exactly the result of Lemma 2.2. $\square$

Returning to the proof of Theorem 3.1, we note that the open sets $\pi^{-1}(V)$, $\sigma_\infty \subset V$ are a fundamental system of neighbourhoods of E. Thus from Lemma 3.2 it follows that for every open subset $U \subset S$ with $E \subset U$, the kernel of the natural map

$$
\alpha \colon \mathrm{Ext}_S^1(\Omega_S^1, \mathcal{O}_S) \to \mathrm{Ext}_U^1(\Omega_U^1, \mathcal{O}_U)
$$

is contained in the set of first order natural deformations $\ker \sigma = \mathrm{im}\, \varepsilon$.

We now apply this fact to a smooth open subset $E \subset U$ such that $\delta(U)$ is an affine open neighbourhood of y_0. According to the Cartesian diagram ([B–W])

$$
\begin{array}{ccc}
T_S^1 & \xrightarrow{\alpha} & H^1(U, \theta_U) \\
\beta \downarrow & & \downarrow \beta_U \\
T_{Y_0}^1 & \xrightarrow{r} & T_{Y_0, y_0}^1
\end{array}
$$

we have $\beta(\ker \alpha) = \ker r = T^1 LT(Y_0, y_0)$ and since $\rho = \beta \circ \varepsilon$, the first part of the theorem is proved.

For the second part, we introduce the functor on Artin rings $LT(Y_0, y_0)$ of deformations of Y_0 which are locally trivial at the point y_0. More generally, for every complex space Z with isolated singularities and for every finite subset $\{z_1, \ldots, z_n\} \subset Z$, we can define the functor D of deformations of Z which are locally trivial at the points $z_1, \ldots, z_n$. This functor has been studied by several authors (cf. [F–M]); for example in [G–K, §1], it is proved that:

1. D satisfies the Schlessinger conditions [Sch, H1–H3].

2. There exists a closed analytic subgerm (possibly nonreduced) V of Def Z such that the restriction of the semiuniversal deformation of Z to V is a complete family of deformations locally trivial at $z_1, \ldots, z_n$.

3. The Zariski tangent space of V is the kernel of the differential of the natural morphism Def $Z \to \prod_i \text{Def}(Z, z_i)$.

Applying these results to the functor $LT(Y_0, y_0)$ concludes the proof. $\square$

4   The Kuranishi family of a degenerate double cover

Let $\pi \colon Y_0 \to W_0$ be a degenerate double cover of $\mathbb{P}^2$ ramified over the union of the vertex w_0 and a divisor $D' \sim (2a - 1)\sigma$ with $a \geq 3$. Here we construct explicitly a smooth complete family of deformations of Y_0. This will imply in particular that the moduli space at Y_0 is locally irreducible and then the closure in the moduli space of the set $N(\mathbb{P}^2, \mathcal{O}(h))$ is a connected component for every $h \geq 4$.

The idea is to describe deformations of Y_0 as canonical coverings of suitable deformations of the cone W_0 and then prove that they give a complete family.

We first recall some well-known facts about cyclic coverings associated to $\mathbb{Q}$-Cartier divisors. For every normal complex space X we denote by $\mathcal{M}_X$ the sheaf of meromorphic functions on X and for every analytic Weil divisor $D \subset X$ by $\mathcal{O}_X(D)$ the reflexive subsheaf of $\mathcal{M}_X$ of meromorphic functions f such that $\text{div}(f) + D \geq 0$. We keep this explicit description of $\mathcal{O}_X(D)$ throughout this section.

Let L be a Weil divisor on a normal irreducible variety X such that nL is Cartier and let $s \in H^0(X, nL)$ be a meromorphic function such that the divisor $D = \text{div}(s) + nL$ is reduced and is contained in the set of points where L is Cartier.

Multiplication by s gives a morphism of $\mathcal{O}_X$-modules $\mathcal{O}_X(-nL) \to \mathcal{O}_X$ and we may define in a natural way a coherent analytic reflexive $\mathcal{O}_X$-algebra (cf. [Reid, 3.6], [E–V, 1.4]):

$$\mathcal{A}(L, s) = \bigoplus_{i=0}^{n-1} \mathcal{A}_i = \bigoplus_{i=0}^{n-1} \mathcal{O}_X(-iL).$$

If (X,x) is a normal analytic singularity, its local analytic class group is by definition the quotient of the free Abelian group generated by the germs of analytic Weil divisors modulo the subgroup of principal divisors. For a 2-dimensional rational singularity, it is a finite group naturally isomorphic to the first homology group of the link of X ([Bri]).

Lemma 4.1 *Let n, L, s, D be as above. If $x \notin D$ then up to isomorphism, the local analytic $\mathcal{O}_x$-algebra $\mathcal{A}_x(L, s)$ only depends on the class of L in the local analytic class group of the analytic singularity (X, x).*

Proof Let n, L', s', D' be another set of data with $x \notin D'$ and assume that $L - L'$ is principal at x. This means that there exists an analytic open neighbourhood U of x and a meromorphic function f on U such that $L = L' + \mathrm{div}(f)$ and $\mathrm{div}(s)|_U = -nL$, $\mathrm{div}(s')|_U = -nL'$.

Therefore $s^{-1}s'f^{-n}$ is an invertible holomorphic function on U and, possibly shrinking U, we may assume that it admits an nth root g. Thus $s = s'(fg)^n$ and the multiplication map $(fg)^i \colon \mathcal{O}_U(-iL') \to \mathcal{O}_U(-iL)$ gives the required isomorphism. $\square$

The cyclic group μ_n acts on the algebra $\mathcal{A}$ by

$$\mu_n \times \mathcal{A}_i \ni (\xi, h) \mapsto \xi^{-i}h \in \mathcal{A}_i,$$

and then the finite map

$$\pi \colon Z = \mathrm{Specan}_X(\mathcal{A}(L, s)) \to X$$

is a cyclic covering of normal varieties (here Specan ([Fi, 1.14]) is the analytic spectrum; if X is projective then by GAGA principles it is the same as the usual algebraic spectrum ([Ha1, II, Ex. 5.17)]).

According to Lemma 4.1, if $x \notin \mathrm{div}(s)+nL$, the germ of the covering over the point x is independent of s.

Corollary 4.2 *In the above set-up, assume X compact and let T be a sufficiently small analytic open neighbourhood of s in $H^0(X, nL)$. Let $\pi \colon Z_T \to X \times T$ be the cyclic covering of degree n associated to the Weil divisor $L \times T$ and multiplication given by $s(x,t) = t(x)$, $t \in T$.*

If $X \to S$ is a flat map such that the composite $Z \to X \to S$ is flat then also the composite $Z_T \to X \times T \to S \times T$ is flat.

Proof Let $U \subset X$ be the open subset where L is Cartier. If T is sufficiently small then $s_t(x) = 0$ for some $t \in T$ implies that $x \in U$. Hence if $x \notin U$ then by Lemma 3.1 the germ of Z_T over (x, s) is locally isomorphic to $Z \times T$. On the other hand the map $U \times T \to S \times T$ is flat and the restriction of the algebra $\mathcal{A}$ over $U \times T$ is locally free and then the restriction of π over $U \times T$ is a flat map. □

Therefore, in the case $S = \text{point}$, we have a morphism from deformations of s to deformations of Z. Consider for example the hypersurface $Z \subset \mathbb{P}^3 \times \mathbb{C}$ with equation $z_1 z_2 - z_3^2 = t z_0^2$ for $t \in \mathbb{C}$, and the involution $\tau \colon Z \to Z$ given by $\tau(t, z_0, z_1, z_2, z_3) = (t, z_0, -z_1, -z_2, -z_3)$.

Let $t \colon Z \to \mathbb{C}$ be the projection on the coordinate t and Z_t the projective subvariety of Z of points with fixed t. It is immediate to observe that Z_t is a smooth quadric for $t \neq 0$, whereas Z_0 is the cone over a nonsingular conic, and t gives the semiuniversal deformation of the isolated singularity $(Z_0, (1,0,0,0,0))$.

The quotient Z/τ is the variety $W \subset \mathbb{P}^5 \times \mathbb{C}$ defined by the equation

$$\text{rank} \begin{pmatrix} x_1 & x_2 & x_3 + t x_0 \\ x_2 & x_3 & x_4 \\ x_3 + t x_0 & x_4 & x_5 \end{pmatrix} \leq 1, \tag{4.3}$$

where $x_0 = z_0^2$, $x_1 = z_1^2$, $x_2 = z_1 z_3$, $x_3 = z_3^2$, $x_4 = z_2 z_3$, $x_5 = z_2^2$.

The quotient family $W \to \mathbb{C}$, $(x, t) \mapsto t$ is a deformation of W_0 and is exactly the degeneration of $\mathbb{P}^2$ obtained by sweeping out the cone over the Veronese surface $V \subset \mathbb{P}^5$. To see this, let $C(V, v) \subset \mathbb{P}^6$ be the projective cone over the image of the map $\mathbb{P}^2_u \to \mathbb{P}^5_x$, $x_1 = u_0^2$, $x_2 = u_0 u_1$, $x_3 = u_1^2$, $x_4 = u_1 u_2$, $x_5 = u_2^2$, $x_6 = u_0 u_2 - u_1^2$. It is defined by the equation

$$\text{rank} \begin{pmatrix} x_1 & x_2 & x_3 + x_6 \\ x_2 & x_3 & x_4 \\ x_3 + x_6 & x_4 & x_5 \end{pmatrix} \leq 1. \tag{4.4}$$

V is the intersection of $C(V, v)$ with the hyperplane $x_0 = 0$ and the vertex v is the point with homogeneous coordinates $(1, 0, 0, 0, 0, 0, 0)$.

Let $H_t \subset \mathbb{P}^6$, $t \in \mathbb{C}$ be the hyperplane given by the equation $x_6 - t x_0 = 0$. Then $H_t \cap V = V \cap \{x_6 = 0\}$ is a smooth hyperplane section and the surface $W_t = C(V, v) \cap H_t$ is exactly the surface defined in (4.3).

Let $H \subset W$ be the Weil divisor defined by the equation $x_2 = x_3 = x_4 = 0$. Then $\mathcal{O}_W(-H)$ is the ideal sheaf of H, and $2H$ is the hyperplane section $x_3 = 0$ of W. In fact the closed subset $\{x_1 = x_3 = x_5 = 0\}$ has codimension 3 in W and then it is sufficient to prove the equality $2H = \text{div}(x_3)$ on its complement. An easy computation shows that on every affine subset $W \cap \{x_i \neq 0\}$ $i = 1, 3, 5$ the equality of ideals $(x^2 x_i^{-1}, x^3 x_i^{-1}, x^4 x_i^{-1})^2 = (x^3 x_i^{-1})$ holds.

Note that $\pi_* \mathcal{O}_Z = \mathcal{O}_W \oplus (z_0/z_3) \mathcal{O}_W(-H)$ and then there exists an isomorphism of $\mathcal{O}_W$-algebras $\pi_* \mathcal{O}_Z = \mathcal{O}_W \oplus \mathcal{O}_W(-H)$, where the algebra structure on the right is induced by multiplication $x_0/x_3 \colon \mathcal{O}_W(-2H) \to \mathcal{O}_W$.

Now let $\pi_0 \colon Y_0 \to W_0 \subset \mathbb{P}^5$ be a fixed degenerate double cover. Then, according to Proposition 2.4, W_0 is projectively normal in $\mathbb{P}^5$ and then there exists a section $s_0 \in H^0(\mathbb{P}^5, \mathcal{O}(2a-1))$ such that π_0 is ramified over the union of $\{w_0\}$ with the divisor of the restriction of s_0 to W_0.

Let T be a small open neighbourhood of s_0 and consider the double cover

$$Y_T = \operatorname{Specan}_{W \times T}\left(\mathcal{O}_{W \times T} \oplus \mathcal{O}_{W \times T}(-(2a-1)H \times T)\right) \to W \times T,$$

where the algebra structure is induced by the section $s(x,t) = s_t(x)$ for $s_t \in T$ and $x \in W$. This makes sense since $2H \times T$ is a Cartier divisor linearly equivalent to $\{s(x,t) = 0\}$.

By our previous results (4.1, 4.2) it follows that:

(i) The map $Y_T \to T$ is a deformation of the space

$$Y = \operatorname{Specan}_W\left(\mathcal{O}_W \oplus \mathcal{O}_W(-(2a-1)H)\right)$$

with the algebra structure induced by s_0.

(ii) Over the vertex w_0 the space Y is isomorphic to the above space Z and then the composite $Y \to W \to \mathbb{C}$ gives a complete deformation of the node (Y_0, y_0).

It is now easy to prove the following

Theorem 4.5 *In the above notation the composite*

$$f \colon Y_T \to W \times T \to \mathbb{C} \times T$$

is a smooth complete family of deformations of Y_0.

Proof We need to prove that $f^{-1}(0, s_0) = Y_0$ and that the Kodaira–Spencer map of the family is surjective.

By definition $f^{-1}(0, s_0) = \operatorname{Spec}_{W_0}(\mathcal{O}_{W_0} \oplus (\mathcal{O}_W(-(2a-1)H) \otimes \mathcal{O}_{W_0}))$ while by definition and from the normality of Y_0 we have $Y_0 = \operatorname{Spec}_{W_0}(\mathcal{O}_{W_0} \oplus \mathcal{O}_{W_0}(-L))$ where $L = a\sigma - 2l$, and $l \subset W_0$ is a line through w_0.

Note that all lines through w_0 are linearly equivalent, L is linearly equivalent to $(4a-2)l$, the intersection $H_0 = H \cap W_0$ is the union of the two lines $l_1 = \{x_1 = x_2 = x_3 = x_4 = 0\}$, $l_2 = \{x_5 = x_2 = x_3 = x_4 = 0\}$ and then the natural map $j_n \colon \mathcal{O}_W(nH) \otimes \mathcal{O}_{W_0} \to \mathcal{O}_{W_0}(2nl)$ is an isomorphism over $W_0 \setminus \{w_0\}$ for every integer n.

In a neighbourhood of the vertex w_0, since the sheaf $\mathcal{O}_W(nH)$ is reflexive on W and invertible for even n, by [E–V, 2.1], the map j_n is injective for every n and an isomorphism for even n. Moreover, the ideal of $H_0 \subset W_0$ is generated by $x_2 x_0^{-1}, x_3 x_0^{-1}, x_4 x_0^{-1}$ and then j_{-1} is also surjective. Tensoring with the line bundle $\mathcal{O}_W(2pH)$, $p \in \mathbb{Z}$, we get the surjectivity of j_n for every integer n. In particular, since j_{1-2a} is an isomorphism, Y_0 is a fibre of f.

By (ii) the composite of the Kodaira–Spencer map of f with the natural restriction map $r \colon T^1(Y_0) \to T^1(Y_0, y_0)$ is surjective, therefore it is sufficient to prove that Y_T contains every deformation locally trivial at the vertex. But this is an immediate consequence of Theorem 3.1 and the surjectivity of the map $H^0(\mathbb{P}^5, \mathcal{O}(2a-1)) \to H^0(W_0, (2a-1)\sigma) = H^0(\mathbb{F}_4, (2a-1)\sigma)$. $\square$

Corollary 4.6 *Every degenerate double cover deforms to a smooth double cover of $\mathbb{P}^2$; in particular, for every odd integer $h \geq 5$, the subset $N(\mathbb{P}^2, \mathcal{O}(h))$ is not closed in the moduli space.*

Corollary 4.7 *The line bundle σ of Y_0 can be extended to every deformation of Y_0.*

Proof The pullback of the hyperplane section $2H$ to Y_T is an extension of σ to a complete family. $\square$

Proof of Theorem A (1) and (2) follow from Theorem 1.6. According to Theorem 4.5 the set of (possibly degenerate) double covers of $\mathbb{P}^2$ is stable under small deformations, therefore $\overline{N}$ is locally open in the moduli space, proving (3) and (4). $\square$

5 Simple iterated double covers of $\mathbb{P}^2$ and their deformations

This section is almost entirely devoted to the proof of Theorem B of the introduction. The first preliminary result we need is the following

Lemma 5.1 *Let $\pi \colon X \to Y$ be a simple iterated double cover associated to a sequence $L_1, \ldots, L_n \in \operatorname{Pic} Y$. Assume that Y and $L_1, \ldots, L_n$ satisfy the conditions:*

(a) $\operatorname{Def} Y$ *is smooth.*

(b) $L_1, \ldots, L_n$ *extends to a complete deformation of Y.*

(c) *For every $h \geq 0$ and $0 < i < j_1 < \cdots < j_h \leq n$,*

$$H^1\left(Y, 2L_i - \sum_{s=1}^{h} L_{j_s}\right) = 0.$$

(d) $H^1(\mathcal{O}_Y) = 0$.

(e) For every $h \geq 0$ and every sequence $1 \leq j_1 < j_2 < \cdots < j_h \leq n$,

$$\operatorname{Ext}^1_{\mathcal{O}_Y}(\Omega^1_Y, \sum_{s=1}^{h} -L_{j_s}) = 0 \quad \text{and} \quad H^1(Y, \sum_{s=1}^{h} -L_{j_s}) = 0.$$

(f) $H^0(L_i) \neq 0$ for every i, and $H^0(Y, 2L_i - L_j) = 0$ for every $j < i$.

Then $\operatorname{Def} X$ is smooth, every deformation of X is a simple iterated double cover of a deformation of Y and if $M \in \operatorname{Pic} Y$ extends to a complete deformation of Y then $\pi^* M$ extends to a complete deformation of X.

Proof Using the computation (involving (f)) about Ext groups in the proof of [Ma3, 2.7] and induction on n, we can reduce the proof of the lemma to the case $n = 1$. In this case the cohomological conditions (c), (d) and (e) become

$$H^1(\mathcal{O}_Y) = H^1(Y, -L) = H^1(Y, 2L) = \operatorname{Ext}^1_{\mathcal{O}_Y}(\Omega^1_Y, -L) = 0.$$

Then X is defined in the total space of the line bundle L by the equation $z^2 = v_0$ with $v_0 \in H^0(Y, 2L)$. Let $\widetilde{Y} \to \operatorname{Def} Y$ be the Kuranishi family of Y and $\widetilde{L} \in \operatorname{Pic} \widetilde{Y}$ the extension of L (which is unique by (d)). Since $H^1(Y, 2L) = 0$, by semicontinuity and base change there exists a subspace $V \subset H^0(\widetilde{Y}, 2\widetilde{L})$ such that the natural restriction $V \to H^0(Y, 2L)$ is an isomorphism.

Taking $H \subset V$ a small neighbourhood of v_0 we can consider the flat double cover $\widetilde{\pi} \colon \widetilde{X} \to \widetilde{Y} \times V$ defined by the equation

$$z^2 = v(y) \quad \text{with} \quad y \in \widetilde{Y} \text{ and } v \in V.$$

By construction the flat maps

$$\widetilde{X} \xrightarrow{\widetilde{\pi}} \widetilde{Y} \times V \to \operatorname{Def} Y \times V$$

are deformations of the double cover $\pi \colon X \to Y$ and satisfy the assumptions of Proposition 1.4. Therefore $\widetilde{X}$ is a complete deformation of X, $\operatorname{Def} X$ is smooth and it is clear that if $\widetilde{M} \in \operatorname{Pic} \widetilde{Y}$ extends M then $\widetilde{\pi}^* \widetilde{M}$ extends $\pi^* M$. Note that if the generic deformation of Y is smooth and the linear system $|2L|$ is base point free then a generic small deformation of X is also smooth. $\square$

We recall that $\operatorname{Ext}^1_{\mathcal{O}_{\mathbb{P}^2}}(\Omega^1_{\mathbb{P}^2}, \mathcal{O}_{\mathbb{P}^2}(-h)) = 0$ for every $h \neq 3$ and therefore we have the result:

Corollary 5.2 *Let $\pi \colon X \to \mathbb{P}^2$ be a simple iterated double cover associated to the sequence of line bundles $L_1, \ldots, L_n$ with $\deg L_i = l_i$. If $l_i > 2l_{i+1}$ and $l_n \geq 4$ then every deformation of X is a simple iterated double cover of $\mathbb{P}^2$ and the set $N(\mathbb{P}^2, L_1, \ldots, L_n)$ is open in the moduli space.*

Proof Immediate consequence of Lemma 5.1. □

Next we want to classify all the possible degenerations of simple iterated double covers of $\mathbb{P}^2$. For reasons that will be clear later, we consider only the case where the degrees l_i satisfy certain numerical conditions.

Definition 5.3 A sequence of line bundles $L_1, \ldots, L_n$, $L_i = \mathcal{O}_{\mathbb{P}^2}(l_i)$ is called a *good sequence* if it satisfies the following 3 conditions:

(1) $l_i \geq 4$ for every $i = 1, \ldots, n$.

(2) $l_i > 2l_{i+1}$ for every $i = 1, \ldots, n-1$.

(3) l_n is odd and l_i is even for $i = 1, \ldots, n-1$.

A *good* simple iterated double cover of $\mathbb{P}^2$ is, by definition, a simple iterated double cover associated to a good sequence.

Proposition 5.4 *Let* $L_1, \ldots, L_n \in \operatorname{Pic} \mathbb{P}^2$ *be a good sequence,* $L_i = \mathcal{O}(l_i)$ *and let* X_0 *be the canonical model of a surface belonging to the closure of* $N(\mathbb{P}^2, L_1, \ldots, L_n)$. *Then either* X_0 *is a simple iterated double cover of* $\mathbb{P}^2$ *associated to* $L_1, \ldots, L_n$, *or there exists a degenerate double cover* Y_0 *of* $\mathbb{P}^2$ *with discrete building datum* $a = \frac{1}{2}(l_n + 1)$ *such that* X_0 *is a simple iterated double cover of* Y_0 *associated to the sequence* $M_1, \ldots, M_{n-1}$, $M_i = (l_i/2)\sigma$.

Proof Note that if $\pi \colon X \to \mathbb{P}^2$ is a smooth simple iterated double cover associated to a good sequence $\mathcal{O}(l_i)$ then $K_X = \pi^*\mathcal{O}(\sum l_i - 3)$ is ample and then the subset $N_0 \subset N$ of surfaces with smooth canonical models is a Zariski open dense subset, in particular the closure of N_0 is the same as the closure of N in the moduli space.

Let $f \colon X \to \Delta$ be a deformation of X_0 such that X_t is a smooth simple iterated double cover of $\mathbb{P}^2$ associated to $L_1, \ldots, L_n$ for $t \neq 0$.

We now prove by induction on n that, up to base change, there exists a factorization

$$f \colon X \xrightarrow{\ p\ } Y \xrightarrow{\ g\ } \Delta,$$

where g is a deformation of a (possibly degenerate) double cover Y_0 of the projective plane with Y_t a smooth double cover associated to L_n for $t \neq 0$, and p is a simple iterated double cover of Y associated to $\widetilde{M}_1, \ldots, \widetilde{M}_{n-1}$ with $\widetilde{M}_i$ the unique extension of M_i to Y (if Y_0 is not degenerate we set $M_i = g_0^*L_i$).

This holds trivially if $n = 1$, so we assume $n > 1$. As in the proof of Theorem 1.6, after a possible base change, the action of the trivial involution

τ of the surfaces X_t extends to an action over X, and we have a quotient family

$$f' \colon Z = X/\tau \to \Delta.$$

Now Z_t is a simple iterated double cover of $\mathbb{P}^2$ associated to the sequence $L_2, \ldots, L_n$ for $t \neq 0$. By the Hurwitz formula, for $t \neq 0$ the canonical bundle of Z_t is the pullback of $\mathcal{O}_{\mathbb{P}^2}(l_2 + \cdots + l_n - 3)$ and then its divisibility in $H^2(Z_t, \mathbb{Z})$ is even. By [Ma3, 3.1], Z_0 has at worst rational double points and the double cover $\pi \colon X \to Z$ is flat. Now by the Brieskorn–Tyurina simultaneous resolution and by semicontinuity, the minimal resolution of Z_0 is a smooth minimal surface of general type and, taking the relative canonical model of Z we have, up to a base change, a factorization

$$X \xrightarrow{\pi} Z = X/\tau \xrightarrow{\delta} Z_{\mathrm{can}} \xrightarrow{p} Y \xrightarrow{g} \Delta,$$

with p, g and Y as in the induction hypothesis.

Thus, to conclude the proof we only need to show that δ is an isomorphism, i.e., that the canonical line bundle of Z_0 is ample. Since π is flat there exists a decomposition

$$\pi_* \mathcal{O}_X = \mathcal{O}_Z \oplus p^* \mathcal{O}_Z(-H),$$

where H is a line bundle over Z such that H_t is the pullback of L_1 for every $t \neq 0$. If $\widetilde{M}_1 \in \operatorname{Pic} Y$ is the extension of $\frac{1}{2}(l_n + 1)\sigma$ then $\widetilde{M}_{1,t}$ is the pullback of L_1 over Y_t; moreover, since $q(Z_0) = 0$, we have $H = \delta^* p^* \widetilde{M}_1$ by injectivity of the extension ([Ma3, 3.8]).

Assume that K_{Z_0} is not ample; then δ contracts some irreducible curve C with $K_{Z_0} \cdot C = 0$, $H_0 \cdot C = 0$. If $C' \subset X_0$ is the strict transform of C then by the Hurwitz formula $K_{X_0} \cdot C' = 0$, which is a contradiction. $\square$

Proof of Theorem B Theorem B, (1) is exactly Lemma 5.1 applied to the case $Y = \mathbb{P}^2$, $L_i = \mathcal{O}_{\mathbb{P}^2}(l_i)$. In B.2, $L_1, \ldots, L_n$ is a good sequence (Definition 5.3), so that by Proposition 5.4, if $[S] \in \overline{N} \setminus N$ then the canonical model of S is a simple iterated double cover of Y, a degenerate double cover of $\mathbb{P}^2$ with discrete building datum $a = \frac{1}{2}(l_n + 1)$, associated to the sequence $M_1, \ldots, M_{n-1} \in \operatorname{Pic} Y$, $M_i = (l_i/2)\sigma$ and the smoothness of $\operatorname{Def} X$ follows from Lemma 5.1 and the vanishing theorems of §2.

B.3 follows immediately from B.1, B.2 and the local properties of the moduli space of surfaces of general type. $\square$

6 Some examples

For every smooth surface of general type S we denote by $I_S = K_S^2 - 8\chi(\mathcal{O}_S)$ its index and by $r(S)$ the divisibility of the canonical class:

$$r(S) = \max\{r \in \mathbb{N} \mid k_S = rc \text{ for some } c \in H^2(S, \mathbb{Z})\}.$$

If S is a smooth simple iterated double cover of $\mathbb{P}^2$ associated to a good sequence $L_1, \ldots, L_n$ with $\deg(L_i) = l_i$, then using the formulas of [Ma3] we have $\pi_1(S) = 0$, and

$$K_S^2 = 2^n(\sum l_i - 3)^2, \quad I_S = 2^n(1 - \sum l_i^2), \quad r(S) = (\sum l_i - 3),$$

so that the invariants K^2, I and r only depend on the two positive integers $\sum l_i$, and $\sum l_i^2$.

Example 6.1 For $n = 3$ we can consider the two sequences

$$(l_1, l_2, l_3) = (3T - 24, T, 5) \quad \text{and} \quad (l_1', l_2', l_3') = (3T - 22, T - 6, 9).$$

Then $\sum l_i = \sum l_i'$, $\sum l_i^2 = \sum l_i'^2$ and for every even $T \geq 26$, both $L_i = \mathcal{O}(l_i)$ and $L_i' = \mathcal{O}(l_i')$ are good sequences.

For $T = 26$ the associated simple iterated double covers have $K^2 = 53792$, $I = -28928$, $c_2 = 70288$, $r = 82$. It is not difficult to prove that every two distinct good sequences $L_1, \ldots, L_n$, $M_1, \ldots, M_m$ give distinct connected components $N(\mathbb{P}^2, L_1, \ldots, L_n)$ and $N(\mathbb{P}^2, M_1, \ldots, M_m)$. To see this, it is enough to show that for the generic $[S] \in N(\mathbb{P}^2, L_1, \ldots, L_n)$ we have $\operatorname{Aut} S = \mathbb{Z}/2\mathbb{Z} = \{1, \tau\}$. In fact S/τ must necessarily belongs to $N(\mathbb{P}^2, L_2, \ldots, L_n)$, and then we use induction on n.

We don't prove here the above statement about automorphisms. However, this is a straightforward generalization of the analogous result for simple iterated double covers of $\mathbb{P}^1 \times \mathbb{P}^1$ proved in [Ma3], as well as an immediate application of a general result about automorphisms of generic simple cyclic covers proved in [Ma5].

Example 6.2 Let $X \to \mathbb{P}^2$, $Y \to \mathbb{P}^1 \times \mathbb{P}^1$ be simple iterated double covers associated to $L_1 = \mathcal{O}(26), L_2 = \mathcal{O}(12), L_3 = \mathcal{O}(5)$ and $L_1' = \mathcal{O}(20, 40), L_2' = \mathcal{O}(22, 2)$. A calculation shows that X and Y have the same invariants K^2, I, r, and it is not difficult to see that X, Y do not belong to the same connected component of the moduli space.

In fact the equation of a generic Y is

$$\begin{cases} z^2 = f + wh & \text{with } f \in H^0(\mathcal{O}(40, 80)) \text{ and } h \in H^0(\mathcal{O}(18, 78)), \\ w^2 = g & \text{with } g \in H^0(\mathcal{O}(44, 4)), \end{cases}$$

where f, g, h are generic, and the arguments of [Ma3] show that the only automorphisms of Y are the identity and the "trivial" involution $z \mapsto -z$. Its quotient is the surface $Y_1 = \{w^2 = g\}$. Since the invariants of Y_1 are different from those of surfaces in $N(\mathbb{P}^2, L_2, L_3)$, the surface Y cannot belong to $N(\mathbb{P}^2, L_1, L_2, L_3)$.

Although explicitly finding simple iterated double covers of $\mathbb{P}^2$ having the same invariants is not easy, it is not difficult to use these surfaces to obtain again a lower bound of the form $\delta \geq (K^2)^{c \log K^2}$ for the number of connected components of moduli space, where c is a positive constant. In fact for sufficiently large n, if q_n is the number of sequences $l_1, \ldots, l_n$ such that $\sum l_i = T_n = 8 \cdot 3^n + 3$, $l_n \geq 5$ odd, l_i even for $i < n$ and $l_i > 2l_{i+1}$ then $\log q_n \geq a n^2$ for a positive constant a independent of n. For each of these q_n sequences, its quadratic sum $\sum l_i^2$ is smaller than T_n^2, and then there exists at least q_n/T_n^2 good sequences giving simple iterated double covers with the same invariants $K^2 = 2^n T_n^2$ and $I = 2^n (1 - \sum l_i^2)$. An easy computation along the lines of [Ma3, §5] gives the claimed lower bound for δ.

References

[Ba]   L. Bădescu: *Normal projective degenerations of rational and ruled surfaces.* Crelle Jour. **367** (1986) 76–89.

[B–S]   C. Bănică, O. Stănăşilă: *Méthodes algébrique dans la théorie globale des espaces complexes (2 vols).* Gauthier–Villars (1977).

[B–P–V] W. Barth, C. Peters, A. van de Ven: *Compact complex surfaces.* Springer Verlag Ergebnisse 4 (1984).

[Bri]   E. Brieskorn: *Rationale Singularitäten komplexer Flächen.* Inv. Math. **4** (1967/68) 336–358.

[B–W]   D. Burns, J. Wahl: *Local contribution to global deformations of surfaces.* Invent. Math. **26** (1974) 67–88.

[Ca1]   F. Catanese: *On the moduli spaces of surfaces of general type.* J. Diff. Geometry **19** (1984) 483–515.

[Ca2]   F. Catanese: *Moduli of surfaces of general type.* In: *Algebraic geometry: open problems (Ravello 1982).* Springer L.N.M. **997** (1983) 90–112.

[Ca3]   F. Catanese: *Automorphisms of rational double points and moduli spaces of surfaces of general type.* Comp. Math. **61** (1987) 81–102.

[Ca4]   F. Catanese: *Moduli of algebraic surfaces.* Springer L.N.M **1337** (1988) 1–83.

[E-V]   H. Esnault, E. Viehweg: *Two dimensional quotient singularities deform to quotient singularities.* Math. Ann. **271** (1985) 439–449.

[F-M]   B. Fantechi, M. Manetti: *Obstruction calculus for functors of Artin rings, I.* J. Algebra **202** (1998) 541–576.

[F-P]   B. Fantechi, R. Pardini: *Automorphisms and moduli spaces of varieties with ample canonical class via deformations of abelian covers.* Comm. in Algebra **25** (1997) 1413–1441.

[Fi]   G. Fischer: *Complex Analytic Geometry.* L.N.M. **538** Springer-Verlag (1976).

[Fl]   H. Flenner: *Über Deformationen holomorpher Abbildungen.* Habilitationsschrift, Osnabrück 1978.

[Gi]   D. Gieseker: *Global moduli for surfaces of general type.* Invent. Math. **43** (1977) 233–282.

[G-K]   G. M. Greuel, U. Karras: *Families of varieties with prescribed singularities.* Comp. Math. **69** (1989) 83–110.

[Ha1]   R. Hartshorne: *Algebraic geometry.* Springer Verlag GTM 52 (1977).

[Ha2]   R. Hartshorne: *Local Cohomology.* Springer L.N.M. **41** (1967).

[Hor1] E. Horikawa: *Algebraic surfaces of general type with small c_1^2. I,* Ann. Math. **104** (1976) 357–387; II, Inv. Math. **37** (1976) 121–155; III, Inv. Math. **47** (1978) 209–248; IV, Inv. Math. **50** (1979), 103–128.

[Hor2] E. Horikawa: *Surjectivity of the period map of K3 surfaces of degree 2.* Math. Ann. **228** (1977) 113–146.

[Kas]   A. Kas: *Ordinary double points and obstructed surfaces.* Topology **16** (1977) 51–64.

[Ko]   K. Kodaira: *Complex manifolds and deformation of complex structures.* Springer Verlag (1986).

[L-W]   E.J.N. Looijenga, J. Wahl: *Quadratic functions and smoothing surface singularities.* Topology **25** (1986) 261–291.

[Ma1]   M. Manetti: *Normal degenerations of the complex projective plane.* J. reine angew. Math. **419** (1991) 89–118.

[Ma2]   M. Manetti: *On some components of moduli space of surfaces of general type.* Comp. Math. **92** (1994) 285–297.

[Ma3]   M. Manetti: *Iterated double covers and connected components of moduli spaces.* Topology **36** (1996) 745–764.

[Ma4]   M. Manetti: *Degenerations of algebraic surfaces and applications to moduli problems.* Thesis, Scuola Normale Superiore, Pisa (1995).

[Ma5]   M. Manetti: *Automorphisms of generic cyclic covers.* Revista Matemática de la Universidad Complutense de Madrid **10** (1997) 149–156.

[Pa]   .V. P. Palamodov: *Deformations of complex spaces.* Uspekhi Mat. Nauk. **31:3** (1976) 129–194. Transl. Russian Math. Surveys **31:3** (1976) 129–197.

[Pi1]   H. Pinkham: *Singularités rationelles des surfaces (avec une appendice).* in Séminaire sur les singularités de surfaces, Lectures Notes in Math. **777** Springer Verlag (1980) 147–178.

[Pi2]   H. Pinkham: *Some local obstructions to deforming global surfaces.* Nova acta Leopoldina NF **52** Nr. 240 (1981) 173–178.

[Reid] M. Reid: *Young person's guide to canonical singularities.* Proc. Symp. in Pure Math. **46** part 1 (1987) A.M.S. 345–414.

[Rie]   O. Riemenschneider: *Deformationen von Quotientensingularitäten (nach zyklischen Gruppen).* Math. Ann. **209** (1974) 211–248.

[Sai]   B. Saint-Donat: *Projective models of K3 surfaces.* Amer. J. Math. **96** (1974) 602–639.

[Sal]   M. Salvetti: *On the number of non-equivalent differentiable structures on 4-manifolds.* Man. Math. **63** (1989) 157–171.

[Sch]   M. Schlessinger: *Functors of Artin rings.* Trans. Amer. Math. Soc. **130** (1968) 208–222.

[Ste]   J. H. M. Steenbrink: *Mixed hodge structure in the vanishing cohomology.* in *Real and complex singularities.* P.Holm editor, Oslo 1976, 525–564.

Marco Manetti,
Scuola Normale Superiore, P. Cavalieri 7,
I–56126 Pisa, Italy
e-mail: manetti@sabsns.sns.it

The geometry underlying mirror symmetry

David R. Morrison

Abstract

The recent result of Strominger, Yau and Zaslow relating mirror symmetry to the quantum field theory notion of T-duality is reinterpreted as providing a way of characterizing geometrically which Calabi–Yau manifolds have mirror partners. The geometric description is rather surprising: one Calabi–Yau manifold should serve as a compactified, complexified moduli space for special Lagrangian tori on the other. We formulate some precise mathematical conjectures concerning how these moduli spaces are to be compactified and complexified, as well as a definition of geometric mirror pairs (in arbitrary dimension) which is independent of those conjectures. We investigate how this new geometric description ought to be related to the mathematical statements which have previously been extracted from mirror symmetry. In particular, we discuss how the moduli spaces of the 'mirror' Calabi–Yau manifolds should be related to one another, and how appropriate subspaces of the homology groups of those manifolds could be related. We treat the case of K3 surfaces in some detail.

Precise mathematical formulations of the phenomenon in string theory known as "mirror symmetry" [21, 33, 19, 27] have proved elusive up until now, largely due to one of the more mysterious aspects of that symmetry: as traditionally formulated, mirror symmetry predicts an equivalence between physical theories associated to certain pairs of Calabi–Yau manifolds, but does not specify any geometric relationship between those manifolds. However, such a geometric relationship has recently been discovered in a beautiful paper of Strominger, Yau and Zaslow [53]. Briefly put, these authors find that the mirror partner X of a given Calabi–Yau threefold Y should be realized as the (compactified and complexified) moduli space for special Lagrangian tori on Y.

This relationship was derived in [53] from the assumption that the physical theories associated to the pair of Calabi–Yau threefolds satisfy a strong property called "quantum mirror symmetry" [52, 6, 12, 42]. In the present paper, we invert the logic, and use this geometric relationship as a *characterization* of mirror pairs, which we formulate in arbitrary dimension.[1] On the one hand, this characterization can be stated in purely mathematical terms, providing a criterion by which mathematicians can recognize mirror pairs. On the other hand, the characterization contains the essential ingredients needed to apply the quantum field theory argument known as "T-duality" which could in principle establish the equivalence of the associated string theories at the level of physical rigor (cf. [45, 53]).[2] This geometric characterization thus appears to capture the essence of mirror symmetry in mathematical terms.

This paper is organized as follows. In Section 1, we give a brief summary of quantum mirror symmetry and review the derivation of the geometric relationship given in [53]. In Section 2, we discuss the theory of special Lagrangian submanifolds [29] and their moduli spaces [37], and explain how these moduli spaces should be compactified and complexified (following [53]). In Section 3, we review in detail the topological and Hodge theoretic properties which have formed the basis for previous mathematical discussions of mirror symmetry. We then formulate in Section 4 our characterization of _geometric mirror pairs_, which we (conjecturally) relate to those topological and Hodge theoretic properties. In Section 5 we present some new results concerning the geometric mirror relationship, including a discussion of how it leads to a connection between certain subspaces of $H_n(Y)$ and $H_{\text{even}}(X)$, and in Section 6 we discuss geometric mirror symmetry for K3 surfaces.

1 Quantum mirror symmetry

Moduli spaces which occur in physics often differ somewhat between the classical and quantum versions of the same theory. For example, the essential mathematical data needed to specify the two dimensional conformal field theory associated to a Calabi–Yau manifold X consists of a Ricci flat metric g_{ij} on X and an $\mathbb{R}/\mathbb{Z}$-valued harmonic 2-form $B \in \mathcal{H}^2(X, \mathbb{R}/\mathbb{Z})$. The classical version of this theory is independent of B and invariant under rescaling the metric; one might thus call the set of all diffeomorphism classes of Ricci flat metrics of fixed volume on X the "classical moduli space" of the theory. The volume of the metric and the 2-form B must be included in the moduli space once quantum effects are taken into account; in a "semiclassical approximation" to the quantum moduli space, one treats the data (g_{ij}, B) (modulo diffeomorphism) as providing a complete description of that space. However,

---

[1]Our definition appears to produce valid mirror pairs of conformal field theories in any dimension, even though the string theoretic arguments of [53] cannot be directly extended to arbitrary dimension to conclude that all mirror pairs ought to arise in this fashion.

[2]There are some additional details which need to be understood before this can be regarded as fully established in physics.

a closer analysis of the physical theory reveals that this is indeed only an approximation to the quantum moduli space, with the necessary modifications becoming more and more significant as the volume is decreased. The ultimate source of these modifications—which are of a type referred to as "nonperturbative" in physics—is the set of holomorphic curves on X and their moduli spaces. A convenient mathematical way of describing how these modifications work is this: there are certain "correlation functions" of the physical theory, which are described near the large volume limit as power series whose coefficients are determined by the numbers of holomorphic 2-spheres on X.[3] The quantum moduli space should then be identified as the natural domain of definition for these correlation functions. To construct it starting from the semiclassical approximation, one first restricts to the open set in which the power series converge, and then extends by analytic continuation to find the complete moduli space.[4] We refer to this space as the quantum conformal field theory moduli space $\mathcal{M}_{\text{CFT}}(X)$. (When necessary, we use the notation $\mathcal{M}_{\text{CFT}}^{\text{sc}}(X)$ to refer to the semiclassical approximation to this space.)

A similar story has emerged within the last year concerning the moduli spaces for type IIA and IIB string theories compactified on a Calabi–Yau threefold X. The classical low energy physics derived from these string theories is determined by a quantum conformal field theory, so one might think of the quantum conformal field theory moduli space described above as being a "classical moduli space" for these theories.

In the semiclassical approximation to the *quantum* moduli spaces of these string theories, we encounter additional mathematical data which must be specified. In the case of the IIA theory, the new data consist of a choice of a nonzero complex number (called the "axion-dilaton expectation value"), together with an $\mathbb{R}/\mathbb{Z}$-valued harmonic 3-form $C \in \mathcal{H}^3(X, \mathbb{R}/\mathbb{Z})$. This last object has a familiar mathematical interpretation as a point in the intermediate Jacobian of X (taking a complex structure on X for which the metric is Kähler). In the case of the IIB theory on a Calabi–Yau threefold Y, the corresponding new data are a choice of nonzero "axion-dilaton expectation value" as before, together with what we might call a quantum $\mathbb{R}/\mathbb{Z}$-valued harmonic even class $C \in \mathcal{H}_Q^{\text{even}}(Y, \mathbb{R}/\mathbb{Z})$. The word "quantum" and the subscript Q here refer to the fact that we must use the quantum cohomology lattice rather than the ordinary cohomology lattice in determining when two

---

[3]There are several possible (equivalent) mathematical interpretations which can be given to these correlation functions: they can be interpreted as defining a new ring structure on the cohomology (defining the so-called quantum cohomology ring) or they can be regarded as defining a variation of Hodge structure over the moduli space. We review this in more detail in Section 3 below.

[4]There can also be modifications caused by higher genus curves [5], but these are less drastic and are not important for our purposes here.

harmonic classes C are equivalent. (The details of this difference are not important here; we refer the interested reader to [6, 42].) For both the IIA and IIB theories, a choice of such "data" as above can be used to describe a low energy supergravity theory in four dimensions.

Just as in the earlier example, there are additional corrections to the semiclassical description of the moduli space coming from "nonperturbative effects" [52, 26, 12]; some of these go by the name of "Dirichlet branes," or "D-branes" for short. The source of these D-brane corrections differs for the two string theories we are considering: in the type IIA theory, they come from moduli spaces of algebraic cycles on X equipped with flat $U(1)$-bundles, or more generally, from moduli spaces of coherent sheaves on X.[5] In the type IIB theory, the D-brane corrections come from (complexified) moduli spaces of so-called supersymmetric 3-cycles on Y, the mathematics of which we describe in the next section. Just as the correlation functions which we could use to determine the structure of the quantum conformal field theory moduli space involved a series expansion with contributions from the holomorphic spheres, the correlation functions in this theory will receive contributions from the coherent sheaves or supersymmetric 3-cycles, with the precise nature of the contribution arising from an integral over the corresponding moduli space.

Quantum mirror symmetry is the assertion that there should exist pairs of Calabi–Yau threefolds[6] (X, Y) such that the type IIA string theory compactified on X is isomorphic to the type IIB string theory compactified on Y; there should be compatible isomorphisms of both the classical and quantum theories. The isomorphism of the classical theories is the statement that the corresponding (quantum corrected) conformal field theories should be isomorphic. This is the version of mirror symmetry which was translated into mathematical terms some time ago, and leads to the surprising statements relating the quantum cohomology on X to the geometric variation of Hodge structure on Y (and *vice versa*).

On the other hand, the isomorphism of the quantum theories has only recently been explored.[7] At the semiclassical level, one infers isomorphisms

---

[5]A D-brane in type IIA theory is ordinarily described as a complex submanifold Z together with a flat $U(1)$-bundle on that submanifold; the associated holomorphic line bundle on Z can be extended by zero to give a coherent sheaf on X. The *arbitrary* coherent sheaves which we consider here correspond to what are called "bound states of D-branes" in physics. (This same observation has been independently made by Maxim Kontsevich, and by Jeff Harvey and Greg Moore.)

[6]There are versions of quantum mirror symmetry which can be formulated in other (low) dimensions, but since these are statements about compactifying ten dimensional string theories, they cannot be extended to arbitrarily high dimension.

[7]The speculation some time ago by Donagi and Markman [22] that some sort of Fourier transform should relate the continuous data provided by the intermediate Jacobian to the discrete data provided by the holomorphic curves is closely related to these isomorphisms

between the intermediate Jacobian of X (the 3-form discussed above), and an analogue of that intermediate Jacobian in quantum cohomology of Y. The full quantum isomorphism would involve properties of the coherent sheaves on X, as related to the supersymmetric 3-cycles on Y. In fact, there should be enough correlation functions in the quantum theory to fully measure the structure of the individual moduli spaces of these sheaves and cycles, so we should anticipate that the moduli spaces themselves are isomorphic.[8] It is this observation which was the key to the Strominger–Yau–Zaslow argument.

Strominger, Yau and Zaslow observe that the algebraic 0-cycles of length one on X (which can be thought of as torsion sheaves supported at a single point) have as their moduli space X itself. According to quantum mirror symmetry, then, there should be a supersymmetric 3-cycle M on Y with precisely the same moduli space, that is, the moduli space of M should be X. Since the complex dimension of the moduli space is three, it follows from a result of McLean [37] (see the next section) that $b_1(M) = 3$. Now as we explain in the next section, the complexified moduli space $\widehat{X}$ for the supersymmetric 3-cycles parameterizes both the choice of 3-cycle M and the choice of a flat $U(1)$-bundle on M. Fixing the cycle but varying the bundle gives a real 3-torus on $\widehat{X}$ (since $b_1(M) = 3$), which turns out to be a supersymmetric cycle on *that* space. This is the "inverse" mirror transform, based on a cycle $\widetilde{M}$ which is in fact a 3-torus. Thus, by applying mirror symmetry twice if necessary, we see that we can—without loss of generality—take the original supersymmetric 3-cycle M to be a 3-torus. In this case, we say that Y has a *supersymmetric T^3-fibration*; note that singular fibers must in general be allowed in such fibrations.

We have thus obtained the rough geometric characterization of the pair (X, Y) stated in the introduction: X should be the moduli space for supersymmetric 3-tori on Y. This characterization is "rough" due to technical difficulties involving both the compactifications of these moduli spaces, and the complex structures on them. We take a different path in Section 4 below, and give a precise geometric characterization which sidesteps these issues.

This line of argument can be pushed a bit farther, by considering the algebraic 3-cycle on X in the fundamental class equipped with a flat $U(1)$-bundle (which must be trivial, and corresponds to the coherent sheaf $\mathcal{O}_X$). There is precisely one of these, so we find a moduli space consisting of a single

---

of quantum theories.

[8] In the case of coherent sheaves, one should not use the usual moduli spaces from algebraic geometry, but rather some sort of "virtual fundamental cycle" on the algebraic geometric moduli space, whose dimension coincides with the "expected dimension" of the algebraic geometric moduli space as computed from the Riemann–Roch theorem. When the moduli problem is unobstructed, this virtual fundamental cycle should coincide with the usual fundamental cycle on the algebraic geometric moduli space.

point. Its mirror should then be a supersymmetric 3-cycle M' with $b_1(M') = 0$. Moreover, we should expect quantum mirror symmetry to preserve the intersection theory of the cycles represented by D-branes (up to sign), so that, since the 0-cycle and 6-cycle on X have intersection number one, we should expect M and M' to have intersection number one if M' is oriented properly. In other words, the supersymmetric T^3-fibration on Y should have a section,[9] and the base of the fibration should satisfy $b_1 = 0$.

The final step in the physics discussion given in [53] is to observe that given a Calabi-Yau threefold with a supersymmetric T^3-fibration and a mirror partner, the mirror partner can be recovered by dualizing the tori in the fibration, at least generically. This suggests that by applying an appropriate duality transformation to the path integral—in quantum field theory this is known as the "T-duality argument"—one should be able to conclude that mirror symmetry does indeed hold for the corresponding physical theories. Strominger, Yau and Zaslow take the first steps towards constructing such an argument, at appropriate limit points of the moduli space. To complete the argument and extend it to general points in the moduli space, one would need to understand the behavior of the T-duality transformations near the singular fibers; to this end, a detailed mathematical study of the possible singular fibers is needed. Some preliminary information about these singularities can be found in [29, 17] (see also [28]).

2   Moduli of special Lagrangian submanifolds

The structure of the supersymmetric 3-cycles which played a rôle in the previous section was determined in [12], where it was found that they are familiar mathematical objects known as special Lagrangian submanifolds. These are a particular class of submanifolds of Calabi-Yau manifolds first studied by Harvey and Lawson [29]. We proceed to the definitions.

A *Calabi-Yau manifold* is a compact connected orientable manifold Y of dimension $2n$ which admits Riemannian metrics whose (global) holonomy is contained in $SU(n)$. For any such metric, there is a complex structure on the manifold with respect to which the metric is Kähler, and a nowhere vanishing holomorphic n-form Ω (unique up to constant multiple). The complex structure, the n-form Ω and the Kähler form ω are all covariant constant with respect to the Levi-Civita connection of the Riemannian metric. This

---

[9]The existence of a section is also expected on other grounds: the set of flat $U(1)$-bundles on M has a distinguished element—the trivial bundle. This provides a section for the "dual" fibration, and suggests (by a double application of quantum mirror symmetry as above) that the original fibration could have been chosen to have a section, without loss of generality.

implies that the metric is Ricci flat, and that $\Omega \wedge \overline{\Omega}$ is a constant multiple of ω^n.

A *special Lagrangian submanifold* of Y is a compact real n-manifold M together with an immersion $f: M \to Y$ such that $f^*(\Omega_0)$ coincides with the induced volume form $d \operatorname{vol}_M$ for an appropriate choice of holomorphic n-form Ω_0. Equivalently [29], one can require that

(1) M is a Lagrangian submanifold with respect to the symplectic structure defined by ω, i.e., $f^*(\omega) = 0$, and

(2) $f^*(\operatorname{Im}\Omega_0) = 0$ for an appropriate Ω_0.

To state this second condition in a way which does not require that Ω_0 be specified, write an arbitrary holomorphic n-form Ω in the form $\Omega = c\,\Omega_0$, and note that

$$\int_M f^*(\Omega) = c \int_M f^*(\Omega_0) = c\,(\operatorname{vol} M).$$

Thus, the "appropriate" n-form is given by

$$\Omega_0 = \frac{(\operatorname{vol} M)\,\Omega}{\int_M f^*(\Omega)}$$

and we can replace condition (2) by

(2′) $f^* \left(\operatorname{Im} \left(\frac{\Omega}{\int_M f^*(\Omega)} \right) \right) = 0.$

(The factor of $\operatorname{vol} M$ is a real constant which can be omitted from this last condition.)

Very few explicit examples of special Lagrangian submanifolds are known. (This is largely due to our lack of detailed understanding of the Calabi–Yau metrics themselves.) One interesting class of examples due to Bryant [18] comes from Calabi–Yau manifolds which are complex algebraic varieties defined over the real numbers: the set of real points on the Calabi–Yau manifold is a special Lagrangian submanifold. Another interesting class of examples is the special Lagrangian submanifolds of a K3 surface, which we discuss in Section 6.

In general, special Lagrangian submanifolds can be deformed, and there will be a *moduli space* which describes the set of all special Lagrangian submanifolds in a given homology class. Given a special Lagrangian $f: M \to Y$ and a deformation of the map f, since $f^*(\omega) = 0$, the almost complex structure on Y induces a canonical identification between the normal bundle of M in Y and the tangent bundle of M. Thus, the normal vector field defined by the deformation can be identified with a 1-form on Y.

The key result concerning the moduli space is due to McLean.

Theorem   (McLean [37])

(1) *First order deformations of f are canonically identified with the space of harmonic 1-forms on Y.*

(2) *Every first order deformation of $f: M \to Y$ can be extended to an actual deformation. In particular, the moduli space $\mathcal{M}_{sL}(M,Y)$ of special Lagrangian maps from M to Y is a smooth manifold of dimension $b_1(M)$.*

We have in mind a global structure on $\mathcal{M}_{sL}(M,Y)$, in which two maps determine the same point in the moduli space if they differ by a diffeomorphism of Y. McLean also observes that $\mathcal{M} = \mathcal{M}_{sL}(M,Y)$ admits a natural n-form Θ defined by

$$\Theta(v_1,\ldots,v_n) = \int_M \theta_1 \wedge \cdots \wedge \theta_n$$

where θ_j is the harmonic 1-form associated to $v_j \in T_{\mathcal{M},f}$.

As discussed implicitly in the last section, the moduli spaces of interest in string theory contain additional pieces of data. To fully account for the "nonperturbative D-brane effects" in the physical theory (when $n = 3$), the moduli space we integrate over must include not only the choice of special Lagrangian submanifold, but also a choice of flat $U(1)$-bundle on it. If we pick a point b on a manifold M, then the space of flat $U(1)$-bundles on M is given by

$$\mathrm{Hom}(\pi_1(M,b), U(1)) \cong H^1(M,\mathbb{R})/H^1(M,\mathbb{Z}).$$

Thus, if we construct a universal family for our special Lagrangian submanifold problem, i.e., a diagram

$$\begin{array}{ccc} \mathcal{U} & \xrightarrow{f} & Y \\ {\scriptstyle p}\downarrow & & \\ \mathcal{M}_{sL}(M,Y) & & \end{array}$$

with the fibers of p diffeomorphic to M and $f_{|p^{-1}(m)}$ the map labeled by m, and if p has a section $s: \mathcal{M}_{sL}(M,Y) \to \mathcal{U}$, then we can define a moduli space including the data of a flat $U(1)$-bundle by setting

$$\mathcal{M}_D(M,Y) := R^1 p_* \mathbb{R}_{\mathcal{U}} / R^1 p_* \mathbb{Z}_{\mathcal{U}};$$

at each point $m \in \mathcal{M}_{sL}(M,Y)$, this specializes to

$$H^1(p^{-1}(m),\mathbb{R})/H^1(p^{-1}(m),\mathbb{Z}) \cong \mathrm{Hom}(\pi_1(p^{-1}(m), s(m)), U(1)).$$

(In the case $n = 3$, this is the "D-brane" moduli space, which motivates our notation.) Note that this space fibers naturally over $\mathcal{M}_{\mathrm{sL}}(M, Y)$, and that there is a section of the fibration, given by the trivial U(1)-bundles.

Both the base and the fiber of the fibration $\mathcal{M}_D(M, Y) \to \mathcal{M}_{\mathrm{sL}}(M, Y)$ have dimension $b_1(M)$, and the fibers are real tori. In fact, we expect from the physics that there will be a family of complex structures on $\mathcal{M}_D(M, Y)$ making it into a complex manifold of complex dimension $b_1(M)$. Roughly, the real tori should correspond to subspaces obtained by varying the arguments of the complex variables while keeping their norms fixed. It is expected from the physics that the complex structure should depend on the choice of both a Ricci flat metric on Y and also on an auxiliary harmonic 2-form B. (This would make $\mathcal{M}_D(M, Y)$ into a "complexification" of the moduli space $\mathcal{M}_{\mathrm{sL}}(M, Y)$ as mentioned in the introduction.) It it not clear at present precisely how those complex structures are to be constructed, although in the case $b_1(M) = n$, a method is sketched in [53] for producing an asymptotic formula for the Ricci flat metric which would exhibit the desired dependence on g_{ij} and B, and the first term in that formula is calculated.[10] The complex structure could in principle be inferred from the metric if it were known.

Motivated by the Strominger–Yau–Zaslow analysis, we now turn our attention to the case in which M is an n-torus. The earliest speculations that the special Lagrangian n-tori might play a distinguished rôle in studying Calabi–Yau manifolds were made by McLean [37], who pointed out that if $M = T^n$ then the deformations of M should locally foliate Y. (There should be no nearby selfintersections, because the harmonic 1-form corresponding to a first order deformation is expected to have no zeros if the metric on the torus is close to being flat.) By analogy with the K3 case, where such elliptic fibrations are well understood, McLean speculated that if certain degenerations were allowed, the deformations of M might fill out the whole of Y. We formulate this as a conjecture (essentially due to McLean).

Conjecture 1 *Suppose that $f \colon T^n \to Y$ is a special Lagrangian n-torus. Then there is a natural compactification $\overline{\mathcal{M}}_{\mathrm{sL}}(T^n, Y)$ of the moduli space $\mathcal{M}_{\mathrm{sL}}(T^n, Y)$ and a proper map $g \colon Y \to \overline{\mathcal{M}}_{\mathrm{sL}}(T^n, Y)$ such that*

$$g^{-1}(\mathcal{M}_{\mathrm{sL}}(M, Y)) \hookrightarrow Y$$

$$g \downarrow$$

$$\mathcal{M}_{\mathrm{sL}}(M, Y)$$

is a universal family of Lagrangian n-tori in the same homology class as f.

---

[10]In the language of [53], the "tree level" metric on the moduli space is computed, but the instanton corrections to that tree level metric are left unspecified.

Definition 1 *When the properties in Conjecture 1 hold, we say that Y has a special Lagrangian T^n-fibration.*

It is not clear at present what sort of structure should be required of $\overline{\mathcal{M}}_{sL}(T^n, Y)$: perhaps it should be a manifold with corners,[11] or perhaps some more exotic singularities should be allowed in the compactification. We will certainly want to require that the complex structures extend to the compactification, and that the section of the fibration extend to a map $\overline{\mathcal{M}}_{sL}(T^n, Y) \to Y$.

The mirror symmetry analysis of [53] as reviewed in the previous section suggests that the family of dual tori $\mathcal{M}_D(T^n, Y)$ can also be compactified, resulting in a space which is itself a Calabi–Yau manifold. We formalize this as a conjecture as well.

Conjecture 2 *The family $\mathcal{M}_D(T^n, Y)$ of dual tori over $\mathcal{M}_{sL}(T^n, Y)$ can be compactified to a manifold X with a proper map $\gamma \colon X \to \overline{\mathcal{M}}_{sL}(T^n, Y)$, such that X admits metrics with $\mathrm{SU}(n)$ holonomy for which the fibers of $\gamma|_{\gamma^{-1}(\mathcal{M}_{sL}(T^n, Y))}$ are special Lagrangian n-tori. Moreover, the fibration γ admits a section $\tau \colon \overline{\mathcal{M}}_{sL}(T^n, Y) \to X$ such that*

$$\tau(\mathcal{M}_{sL}(T^n, Y)) \subset \mathcal{M}_D(T^n, Y) \subset X$$

is the zero section.

It seems likely that for an appropriate holomorphic n-form Ω_0 on X, the pullback $\tau^*(\Omega_0)$ will coincide with McLean's n-form Θ when restricted to $\mathcal{M}_{sL}(T^n, Y)$.

The most accessible portion of these conjectures would be the following:

Sub-Conjecture *The family $\mathcal{M}_D(T^n, Y)$ of dual tori over $\mathcal{M}_{sL}(T^n, Y)$ admits complex structures and Ricci flat Kähler metrics. In particular, it has a nowhere vanishing holomorphic n-form.*

Strominger, Yau and Zaslow have obtained some partial results concerning this subconjecture, for which we refer the reader to [53]. It appears, for example, that the construction of the complex structure on the D-brane moduli space should be local around each torus in the torus fibration.

---

[11]This possibility is suggested by the structure of toric varieties, the moment maps for which express certain complex manifolds as T^n-fibrations over manifolds with corners (compact convex polyhedra).

3   Mathematical consequences of mirror symmetry

There is by now quite a long history of extracting mathematical statements from the physical notion of mirror symmetry. Many of these work in arbitrary dimension, where there is evidence in physics for mirror symmetry among conformal field theories [27, 25].[12] In this section, we review two of those mathematical statements, presented here as definitions. As the discussion is a bit technical, some readers may prefer to skip to the next section, where we formulate our new definition of *geometric mirror pairs* inspired by the Strominger–Yau–Zaslow analysis. Throughout this section, we let X and Y be Calabi–Yau manifolds of dimension n.

The first prediction one extracts from physics about a mirror pair is a simple equality of Hodge numbers.

Definition 2 *We say that the pair* (X, Y) *passes the topological mirror test if* $h^{n-1,1}(X) = h^{1,1}(Y)$ *and* $h^{1,1}(X) = h^{n-1,1}(Y)$.

Many examples of pairs passing this test are known; indeed, the observation of this "topological pairing" in a class of examples was one of the initial pieces of evidence in favor of mirror symmetry [19]. Subsequent constructions of Batyrev and Borisov [8, 16, 9] show that all Calabi–Yau complete intersections in toric varieties belong to pairs which pass this topological mirror test.

For simply connected Calabi–Yau threefolds, the Hodge numbers $h^{1,1}$ and $h^{n-1,1}$ determine all the others, but in higher dimension there are more. Naïvely one expects to find that $h^{p,q}(X) = h^{n-p,q}(Y)$. However, as was discovered by Batyrev and collaborators [11, 10], the proper interpretation of the numerical invariants of the physical theories requires a modified notion of "string theoretic Hodge numbers" $h_{st}^{p,q}$; once this modification has been made, these authors show that $h_{st}^{p,q}(X) = h_{st}^{n-p,q}(Y)$ for the Batyrev–Borisov pairs (X, Y) of complete intersections in toric varieties. The class of pairs for which this modification is needed includes some of those given by the Greene–Plesser construction [27] for which mirror symmetry of the conformal field theories has been firmly established in physics, so it would appear that this modification is truly necessary for a mathematical interpretation of mirror symmetry. Hopefully, it too will follow from the geometric characterization being formulated in this paper.

---

[12]In low dimension where a string theory interpretation is possible, this would become the "classical" mirror symmetry which one would also want to extend to a "quantum" mirror symmetry if possible.

Going beyond the simple topological properties, a more precise and detailed prediction arises from identifying the quantum cohomology of one Calabi–Yau manifold with the geometric variation of Hodge structure of the mirror partner (in the case that the Calabi–Yau manifolds have no holomorphic 2-forms). We discuss this prediction in considerable detail, in order to ensure that this paper has selfcontained statements of the conjectures being proposed within it (particularly those in Sections 4 and 5 below relating the "old" and "new" mathematical versions of mirror symmetry).

To formulate this precise prediction, let X be a Calabi–Yau manifold with $h^{2,0}(X) = 0$, and write $\widetilde{\mathcal{M}}^{\mathrm{sc}}_{\mathrm{CFT}}(X)$ for the moduli space of triples (g_{ij}, B, J) modulo diffeomorphism, where J is a complex structure for which the metric g_{ij} is Kähler. The map $\widetilde{\mathcal{M}}^{\mathrm{sc}}_{\mathrm{CFT}}(X) \to \mathcal{M}^{\mathrm{sc}}_{\mathrm{CFT}}(X)$ is finite-to-one, so this is another good approximation to the conformal field theory moduli space. Moreover, there is a natural map $\widetilde{\mathcal{M}}^{\mathrm{sc}}_{\mathrm{CFT}}(X) \to \mathcal{M}_{\mathrm{cx}}(X)$ to the moduli space of complex structures on X, whose fiber over J is $\mathcal{K}_{\mathbb{C}}(X_J)/\operatorname{Aut}(X_J)$, where

$$\mathcal{K}_{\mathbb{C}}(X_J) = \left\{ B + i\omega \in \mathcal{H}^2(X, \mathbb{C}/\mathbb{Z}) \mid \omega \in \mathcal{K}_J \right\}$$

is the *complexified Kähler cone*[13] of X_J ($\mathcal{K}_J$ is the usual Kähler cone), and $\operatorname{Aut}(X_J)$ the group of holomorphic automorphisms of X_J.

The moduli space of complex structures $\mathcal{M}_{\mathrm{cx}}(X)$ has a variation of Hodge structure defined on it which is of geometric origin: roughly speaking, one takes a universal family $\pi \colon \mathcal{X} \to \mathcal{M}_{\mathrm{cx}}(X)$ over the moduli space and constructs a variation of Hodge structure on the local system $R^n\pi_*\mathbb{Z}_{\mathcal{X}}$ by considering the varying Hodge decomposition of $H^n(X_J, \mathbb{C})$. The local system gives rise to a holomorphic vector bundle $\mathcal{F} := (R^n\pi_*\mathbb{Z}_{\mathcal{X}}) \otimes \mathcal{O}_{\mathcal{M}_{\mathrm{cx}}(X)}$ with a flat connection $\nabla \colon \mathcal{F} \to \Omega^1_{\mathcal{M}_{\mathrm{cx}}(X)} \otimes \mathcal{F}$ (whose flat sections are the sections of the local system), and the varying Hodge decompositions determine the *Hodge filtration*

$$\mathcal{F} = \mathcal{F}^0 \supset \mathcal{F}^1 \supset \cdots \supset \mathcal{F}^n \supset \{0\},$$

a filtration by holomorphic subbundles defined by

$$\mathcal{F}^p|_J = H^{n,0}(X_J) \oplus \cdots \oplus H^{p,n-p}(X_J),$$

which is known to satisfy *Griffiths transversality*

$$\nabla(\mathcal{F}^p) \subset \Omega^1_{\mathcal{M}_{\mathrm{cx}}(X)} \otimes \mathcal{F}^{p-1}.$$

Conversely, given the bundle with flat connection and filtration, the complexified local system $R^n\pi_*\mathbb{C}_{\mathcal{X}}$ can be recovered by taking (local) flat sections, and

---

[13]We are following the conventions of [40] rather than those of [38, 39].

the Hodge structures can be reconstructed from the filtration. However, the original local system of $\mathbb{Z}$-modules is additional data, and cannot be recovered from the bundle, connection and filtration alone.

The moduli space of complex structures $\mathcal{M}_{\mathrm{cx}}(X)$ can be compactified to a complex space $\overline{\mathcal{M}}$, to which the bundles $\mathcal{F}^p$ and the connection ∇ extend; however, the extended connection ∇ acquires *regular singular points* along the boundary $\mathcal{B}$, which means that it is a map

$$\nabla\colon \mathcal{F} \to \Omega^1(\log \mathcal{B}) \otimes \mathcal{F}.$$

The residues of ∇ along boundary components describe the *monodromy transformations* about those components, the same monodromy which defines the local system. At normal crossings boundary points there is always an associated *monodromy weight filtration*, which we take to be a filtration on the homology group $H_n(X)$.

The data of the flat connection and the Hodge filtration are encoded in the conformal field theory on X (at least for a sub-Hodge structure containing $\mathcal{F}^{n-1}$).[14] Since mirror symmetry reverses the rôles of base and fiber in the map

$$\widetilde{\mathcal{M}}^{\mathrm{sc}}_{\mathrm{CFT}}(X) \to \mathcal{M}_{\mathrm{cx}}(X),$$

one of the predictions of mirror symmetry will be an isomorphism between this structure and a similar structure on $\mathcal{K}_{\mathbb{C}}(X_J)/\operatorname{Aut}(X_J)$.

In fact, the conformal field theory naturally encodes a variation of Hodge structure on $\mathcal{K}_{\mathbb{C}}(X_J)/\operatorname{Aut}(X_J)$. To describe this mathematically, we must choose a *framing*, which is a choice of a cone

$$\sigma = \mathbb{R}_+ e^1 + \cdots + \mathbb{R}_+ e^r \subset H^2(X,\mathbb{R})$$

which is generated by a basis $e^1, \ldots, e^r$ of $H^2(X,\mathbb{Z})/$torsion and whose interior is contained in the Kähler cone of X. The complexified Kähler part of the semiclassical moduli space then contains as an open subset the space

$$\mathcal{M}_A(\sigma) := (H^2(X,\mathbb{R}) + i\sigma)/H^2(X,\mathbb{Z}),$$

elements of which can be expanded in the form $\sum \left(\frac{1}{2\pi i}\log q_j\right) e^j$, leading to the alternate description

$$\mathcal{M}_A(\sigma) = \{(q_1,\ldots,q_r) \mid 0 < |q_j| < 1\}.$$

---

[14]Note that $\mathcal{F}^n$ appears directly in the conformal field theory, and $\mathcal{F}^{n-1}/\mathcal{F}^n$ appears as a class of marginal operators in the conformal field theory. Thus, the conformal field theory contains *at least* as much of the Hodge theoretic data as is described by the smallest sub-Hodge structure containing $\mathcal{F}^{n-1}$, and quite possibly more.

The desired variation of Hodge structure will be defined on a partial compactification of this space, namely

$$\overline{\mathcal{M}}_A(\sigma) := \{(q_1, \ldots, q_r) \mid 0 \le |q_j| < 1\},$$

which has a distinguished boundary point $\vec{0} = (0, \ldots, 0)$.

The ingredients we need to define the variation of Hodge structure are the *fundamental Gromov–Witten invariants*[15] *of* X, which are trilinear maps

$$\Phi_\eta^0 \colon H^*(X, \mathbb{Q}) \oplus H^*(X, \mathbb{Q}) \oplus H^*(X, \mathbb{Q}) \to \mathbb{Q}.$$

Heuristically, when A, B and C are integral classes, $\Phi_\eta^0(A, B, C)$ should be the number of generically injective[16] holomorphic maps $\psi \colon \mathbb{CP}^1 \to X$ in class η, such that $\psi(0) \in Z_A$, $\psi(1) \in Z_B$, $\psi(\infty) \in Z_C$ for appropriate cycles Z_A, Z_B, Z_C Poincaré dual to the classes A, B, C. (The invariants vanish unless $\deg A + \deg B + \deg C = 2n$.) From these invariants we can define the *Gromov–Witten maps* $\Gamma_\eta \colon H^k(X) \to H^{k+2}(X)$ by requiring that

$$\Gamma_\eta(A) \cdot B_{|[X]} = \frac{\Phi_\eta^0(A, B, C)}{\eta \cdot C}$$

for $B \in H^{2n-k-2}(X)$, $C \in H^2(X)$. (This is independent of the choice of C.)

These invariants are usually assembled into the "quantum cohomology ring" of X, but here we present this structure in the equivalent form of a variation of Hodge structure over $\overline{\mathcal{M}}_A(\sigma)$ degenerating along the boundary. To do so, we define a holomorphic vector bundle $\mathcal{E} := \left(\bigoplus H^{\ell,\ell}(X)\right) \otimes \mathcal{O}_{\overline{\mathcal{M}}_A(\sigma)}$, and a flat[17] connection $\nabla \colon \mathcal{E} \to \Omega^1_{\overline{\mathcal{M}}}(\log \mathcal{B}) \otimes \mathcal{E}$ with regular singular points along the boundary $\mathcal{B} = \overline{\mathcal{M}}_A(\sigma) - \mathcal{M}_A(\sigma)$ by the formula[18]

$$\nabla := \frac{1}{2\pi i} \left(\sum_{j=1}^r d\log q_j \otimes \mathrm{ad}(e^j) + \sum_{0 \ne \eta \in H_2(X, \mathbb{Z})} d\log\left(\frac{1}{1 - q^\eta}\right) \otimes \Gamma_\eta \right),$$

where $q^\eta = \prod q_j^{e^j(\eta)}$, and where $\mathrm{ad}(e^j) \colon H^k(X) \to H^{k+2}(X)$ is the adjoint map of the cup product pairing, defined by $\mathrm{ad}(e^j)(A) = e^j \cup A$. We also define a "Hodge filtration"

$$\mathcal{E}^p := \left(\bigoplus_{0 \le \ell \le n-p} H^{\ell,\ell}(X) \right) \otimes \mathcal{O}_{\overline{\mathcal{M}}_A(\sigma)},$$

---

[15]These can be defined using techniques from symplectic geometry [49, 36, 50] or from algebraic geometry [32, 31, 15, 14, 13, 34].

[16]We have built the "multiple cover formula" [3, 35, 54] into our definitions.

[17]The flatness of this connection is equivalent to the associativity of the product in quantum cohomology.

[18]I am indebted to P. Deligne for advice [20] which led to this form of the formula.

which satisfies $\nabla(\mathcal{E}^p) \subset \Omega^1_{\overline{\mathcal{M}}}(\log \mathcal{B}) \otimes \mathcal{E}^{p-1}$. This describes a structure we call the *framed A-variation of Hodge structure with framing σ*. To be a bit more precise, we should refer to this as a "formally degenerating variation of Hodge structure," since the series used to define ∇ is only formal. (More details about such structures can be found in [43]; cf. also [20].) There are also some subtleties about passing from a local system of complex vector spaces to a local system of $\mathbb{Z}$-modules which we shall discuss in Section 5 below.

The residues of ∇ along the boundary components $q_j = 0$ are the adjoint maps $\mathrm{ad}(e^j)$; the corresponding monodromy weight filtration at $\vec{0}$ is simply

$$H_{0,0}(X) \subseteq H_{0,0}(X) \oplus H_{1,1}(X) \subseteq \cdots \subseteq (H_{0,0}(X) \oplus \cdots \oplus H_{n,n}(X)).$$

Under mirror symmetry, this maps to the geometric monodromy weight filtration at an appropriate "large complex structure limit" point in $\overline{\mathcal{M}}_{\mathrm{cx}}$ (see [41] and references therein). Note that the class of the 0-cycle is the monodromy-invariant class in $H_{\mathrm{even}}(X)$; thus, its mirror n-cycle will be the monodromy-invariant class in $H_n(Y)$.

Although the choice of a "framing" may look unnatural, the relationship between different choices of framing is completely understood [38] (modulo a conjecture about the action of the automorphism group on the Kähler cone). Varying the framing corresponds to varying which boundary point in the moduli space one is looking at, possibly after blowing up the original boundary of the moduli space in order to find an appropriate compactification containing the desired boundary point.

We finally come to the definition which contains our precise Hodge theoretic mirror prediction from physics.

Definition 3 *Let X and Y be Calabi–Yau manifolds with $h^{2,0}(X)$, $h^{2,0}(Y) = 0$. The pair (X, Y) passes the Hodge theoretic mirror test if there exists*

(1) a partial compactification $\overline{\mathcal{M}}_{\mathrm{cx}}(Y)$ of the complex structure moduli space of Y,

(2) a neighborhood $U \subset \overline{\mathcal{M}}_{\mathrm{cx}}(Y)$ of a boundary point P of $\overline{\mathcal{M}}_{\mathrm{cx}}(Y)$,

(3) a framing σ for $H^2(X)$, and

(4) a "mirror map" $\mu : U \to \mathcal{M}_A(\sigma)$ mapping P to $\vec{0}$,

such that μ^* induces an isomorphism between $\mathcal{E}^{n-1}$ and $\mathcal{F}^{n-1}$ extending to an isomorphism between subvariations of Hodge structure of the A-variation of Hodge structure with framing σ, and the geometric formally degenerating variation of Hodge structure at P.

The restriction to a subvariation of Hodge structure (which occurs only when the dimension of the Calabi–Yau manifold is greater than three) seems to be necessary in order to get an integer structure on the local system compatible with the complex variation of Hodge structure. (We will return to this issue in Section 5.) It seems likely that this is related to the need to pass to "string theoretic Hodge numbers," which may actually be measuring the Hodge numbers of the appropriate sub-Hodge structures.

The property described in the Hodge theoretic mirror test can be recast in terms of using the limiting variation of Hodge structure on Y to make predictions about enumerative geometry of holomorphic rational curves on X. In this sense, there is a great deal of evidence in particular cases (see [41, 25] and the references therein). There are also some specific connections which have been found between the variations of Hodge structure associated to mirror pairs of theories [44], as well as a recent theorem [23] which proves that the expected enumerative properties hold for an important class of Calabi–Yau manifolds.

Note that if (X, Y) passes the Hodge theoretic mirror test in both directions, then it passes the topological mirror test (essentially by definition, since the dimensions of the moduli spaces are given by the Hodge numbers $h^{1,1}$ and $h^{n-1,1}$).

4   Geometric mirror pairs

We now wish to translate the Strominger–Yau–Zaslow analysis into a definition of *geometric mirror pairs* (X, Y), which we formulate in arbitrary dimension. (As mentioned earlier, the arguments of [53] cannot be applied to conclude that all mirror pairs arise in this way, but it seems reasonable to suppose that a T-duality argument—applied to conformal field theories only—would continue to hold.) The most straightforward such definition would say that X is the compactification of the complexified moduli space of special Lagrangian n-tori on Y. However, as indicated by our conjectures of Section 2, at present we do not have adequate technical control over the compactification to see that it is a Calabi–Yau manifold. So we make instead an indirect definition, motivated by the following observation: if we had such a compactified moduli space X, then for generic $x \in X$ there would be a corresponding special Lagrangian n-torus $T_x \subset Y$, and we could define an *incidence correspondence*

$$Z = \text{closure of } \left\{ (x, y) \in X \times Y \mid y \in T_x \right\}.$$

By definition, the projection $Z \to X$ would have special Lagrangian n-tori as generic fibers. As we saw earlier, the analysis of [53] suggests that generic fibers of the other projection $Z \to Y$ will also be special Lagrangian n-tori. Furthermore, we should expect that as we vary the metrics on X and on Y, the fibrations by special Lagrangian n-tori can be deformed along with the metrics. (In fact, it is these dependencies on parameters which should lead to a "mirror map" between moduli spaces.) Thus, we formulate our definition using a *family* of correspondences depending on $t \in U$ for some (unspecified) parameter space U.

Definition 4 *A pair of Calabi–Yau manifolds (X, Y) is a geometric mirror pair if there is a parameter space U such that for each $t \in U$ there exist*

    (1) a correspondence $Z_t \subset (X \times Y)$ which is the closure of a submanifold of dimension $3n$,

    (2) maps $\tau_t \colon X \to Z_t$ and $\tilde{\tau}_t \colon Y \to Z_t$ which serve as sections for the projection maps $Z_t \to X$ and $Z_t \to Y$, respectively,

    (3) a Ricci flat metric $g_{ij}(t)$ on X with respect to which generic fibers of the projection map $Z_t \to Y$ are special Lagrangian n-tori, and

    (4) a Ricci flat metric $\tilde{g}_{ij}(t)$ on Y with respect to which generic fibers of the projection map $Z_t \to X$ are special Lagrangian n-tori.

Moreover, for generic $z \in Z_t$, the fibers through z of the two projection maps must be canonically dual as tori (with origins specified by τ_t and $\tilde{\tau}_t$).

In a somewhat stronger form of the definition, we might require that U be sufficiently large so that the images of the natural maps $U \to \mathcal{M}_{\mathrm{Ric}}(X)$ and $U \to \mathcal{M}_{\mathrm{Ric}}(Y)$ to the moduli spaces of Ricci flat metrics on X and on Y are open subsets of the respective moduli spaces. It is too much to hope that these maps would be surjective. The best picture we could hope for, in fact, would be a diagram of the form

$$\mathcal{M}_{\mathrm{Ric}}(X) \supseteq U_X \xleftarrow{\pi_X} U \xrightarrow{\pi_Y} U_Y \subseteq \mathcal{M}_{\mathrm{Ric}}(Y)$$

in which $U_X \subseteq \mathcal{M}_{\mathrm{Ric}}(X)$ and $U_Y \subseteq \mathcal{M}_{\mathrm{Ric}}(Y)$ are open subsets (near certain boundary points in a compactification, and contained within the set of metrics for which the semiclassical approximation is valid). The fibers of π_X will have dimension $h^{1,1}(X)$, and if the induced map is the mirror map each fiber of π_X must essentially be the set of B-fields on X, i.e., it must be a deformation of the real torus $H^2(X, \mathbb{R}/\mathbb{Z})$. This is compatible with the approximate formula[19] in [53] for a family of metrics on Y, produced by varying the B-field on X.

We expect that geometric mirror symmetry will be related to the earlier mathematical mirror symmetry properties in the following way.

---

[19]The "tree level" formula given in [53] is subject to unspecified instanton corrections.

Conjecture 3 *If (X, Y) is a geometric mirror pair, then the parameter space U and the data in the definition of the geometric mirror pair can be chosen so that*

(1) (X, Y) passes the topological mirror test;[20]

(2) $\pi_X : U \to \mathcal{M}_{\mathrm{Ric}}(X)$ lifts to a generically finite map $\tilde{\pi}_X : U \to \tilde{U}_X \subseteq \mathcal{M}_{\mathrm{CFT}}^{\mathrm{sc}}(X)$;

(3) $\pi_Y : U \to \mathcal{M}_{\mathrm{Ric}}(Y)$ lifts to a generically finite map $\tilde{\pi}_Y : U \to \tilde{U}_Y \subseteq \mathcal{M}_{\mathrm{CFT}}^{\mathrm{sc}}(Y)$; and

(4) if $h^{2,0}(X) = h^{2,0}(Y) = 0$, then there are boundary points $P \in \overline{\mathcal{M}}_{\mathrm{cx}}(Y)$, $P' \in \overline{\mathcal{M}}_{\mathrm{cx}}(X)$ and framings σ of $H^2(X)$ and σ' of $H^2(Y)$ with partial compactifications $\widetilde{\tilde{U}}_X \subset \mathcal{M}_A(\sigma) \times \overline{\mathcal{M}}_{\mathrm{cx}}(X)$ and $\widetilde{\tilde{U}}_Y \subset \overline{\mathcal{M}}_{\mathrm{cx}}(Y) \times \mathcal{M}_A(\sigma')$ such that the composite map $(\tilde{\pi}_X)_*(\tilde{\pi}_Y)^*$ extends to a map $\mu^{-1} \times \mu'$ which consists of mirror maps in both directions (in the sense of Definition 3). In particular, (X, Y) passes the Hodge theoretic mirror test.

Even in the case $h^{2,0}(X) \neq 0$, there is an induced map $(\tilde{\pi}_X)_*(\tilde{\pi}_Y)^*$ which should coincide with the mirror map between the moduli spaces.

If X has several birational models $X^{(j)}$, then all of the semiclassical moduli spaces $\mathcal{M}_{\mathrm{CFT}}^{\mathrm{sc}}(X^{(j)})$ give rise to a common conformal field theory moduli space (see [1], or for a more mathematical account, [39]). If we follow a path between the large radius limit points of two of these models, and reinterpret that path in the mirror moduli space, we find a path which leads from one large complex structure limit point of $\mathcal{M}_{\mathrm{cx}}(Y)$ to another. On the other hand, the calculation of [2] shows that the homology class of the torus[21] in a special Lagrangian T^n-fibration does not change when we move from one of these regions of $\mathcal{M}_{\mathrm{cx}}(Y)$ to another. Thus, the moduli space of special Lagrangian tori T^n must themselves change as we move from region to region. It will be interesting to investigate precisely how this change comes about.

5   Mirror cohomology and the weight filtration

The "duality" transformation which links the two members X and Y of a geometric mirror pair does not induce any obvious relationship between $H^{1,1}(X)$ and $H^{n-1,1}(Y)$, so it may be difficult to imagine how the topological mirror

---

[20](1) is a consequence of (4) if $h^{2,0}(X) = h^{2,0}(Y) = 0$.
[21]Recall that this is the monodromy-invariant cycle.

test can be passed by a geometric mirror pair. However, at least for a restricted class of topological cycles, such a relationship *can* be found, as part of a more general relationship between certain subspaces of $H_{\text{even}}(X)$ and $H_n(Y)$.

Fix a special Lagrangian T^n-fibration on Y with a special Lagrangian section, and consider n-cycles $W \subset Y$ with the property that W is the closure of a submanifold W_0 whose intersection with each nonsingular T^n in the fibration is either empty, or a subtorus of dimension $n - k$ (for some fixed integer $k \le n$). That is, we assume that W can be generically described as a T^{n-k}-bundle over a k-manifold, with the tori T^{n-k} linearly embedded in fibers of the given T^n-fibration. We call such n-cycles *pure*.

For any pure n-cycle $W \subset Y$, there is a *T-dual cycle*[22] $W^\vee \subset X$ $(= \overline{\mathcal{M}}_D(T^n, Y))$ defined as the closure of an n-manifold $W_0^\vee$ satisfying

$$W_0^\vee \cap (T^n)^* = \begin{cases} \text{the annihilator of } W \cap T^n \text{ in } (T^n)^* & \text{if } W \cap T^n \ne \emptyset, \\ \emptyset & \text{otherwise,} \end{cases}$$

for all smooth fibers $(T^n)^*$ in the dual fibration. Since the annihilator of an $(n - k)$-torus is a k-torus, we see that $W^\vee$ is generically described as a T^k-bundle over a k-manifold, and so it defines a class in $H_{2k}(X)$. This is our relationship between the space of pure n-cycles on Y and the even homology on X.

Taking the T-duality statements from physics very literally, we are led to the speculation that pure special Lagrangian n-cycles have as their T-duals certain algebraic cycles on X; moreover, the moduli spaces containing corresponding cycles should be isomorphic.[23] (Roughly speaking, the T^k-fibration on the corresponding algebraic k-cycle should be given by holding the norms of some system of complex coordinates on the k-cycle fixed, while varying their arguments.) We have already seen the simplest cases of this statement in the Strominger–Yau–Zaslow discussion: the special Lagrangian n-cycles which consist of a single fiber (i.e., $k = 0$) are T-dual to the 0-cycles of length one on X, while a special Lagrangian n-cycle which is the zero section of the fibration (i.e., $k = n$) is T-dual to the $2n$-cycle in the fundamental class. This new construction should extend that correspondence between cycles to a

---

[22]In physics, when a T-duality transformation is applied to a real torus, a D-brane supported on a subtorus is mapped to a D-brane supported on the "dual" subtorus (of complementary dimension); this can be mathematically identified as the annihilator. Here, we apply this principle to a family of subtori within a family of tori.

[23]As the referee has pointed out, our "purity" condition is probably too strong to be preserved under deformation, but one can hope that all nearby deformations of a (pure) special Lagrangian n-cycle are reflected in deformations of the corresponding algebraic cycle.

broader class (albeit still a somewhat narrow one, since pure cycles are quite special).

In fact, the correspondence should be even broader. If we begin with an arbitrary irreducible special Lagrangian n-cycle W on Y whose image in $\mathcal{M}_{\mathrm{sL}}(T^n, Y)$ has dimension k, then W can be generically described as a bundle of $(n - k)$-manifolds over the image k-manifold. The T-dual of such a cycle should be a coherent sheaf $\mathcal{E}$ on X having support a complex submanifold Z of dimension k whose image in $\mathcal{M}_{\mathrm{sL}}(T^n, Y)$ is that same k-manifold. Thus, to the homology class of W in $H_n(Y)$ we associate the total homology class in $H_{\mathrm{even}}(X)$ of the corresponding coherent sheaf.[24] Note that since the support has complex dimension k, this total homology class lies in $H_0(X) \oplus H_2(X) \oplus \cdots \oplus H_{2k}(X)$.

The homology class of the generic fiber of W within T^n should determine the subtori whose T-duals would sweep out Z; when that homology class is r times a primitive class, the corresponding coherent sheaf should have generic rank r along Z. For example, a multisection of the special Lagrangian T^n fibration which meets the fiber r times should correspond to a coherent sheaf whose support is all of X and whose rank is r.

We have thus found a mapping from the subspace $H_n^{\mathrm{sL}}(Y)$ of n-cycles with a special Lagrangian representative, to the subspace $H_{\mathrm{even}}^{\mathrm{alg}}(Y)$ of homology classes of algebraic cycles (and coherent sheaves). If we consider the Leray filtration on special Lagrangian n-cycles on Y

$$\mathbb{S}_k := \left\{ W \in H_n^{\mathrm{sL}}(Y) \mid \dim \mathrm{image}\, W \le k \right\},$$

then this will map to

$$H_0^{\mathrm{alg}}(X) \oplus H_2^{\mathrm{alg}}(X) \oplus \cdots \oplus H_{2k}^{\mathrm{alg}}(X)$$

(and the pure n-cycles on Y will map to homology classes of algebraic cycles on Y). But this latter filtration on $H_{\mathrm{even}}(X)$ is precisely the monodromy weight filtration of the A-variation of Hodge structures on X, which should be mirror to the geometric monodromy weight filtration on Y![25] We are thus led to the following refinement of Conjecture 3.

Conjecture 4 *If (X, Y) is a geometric mirror pair then there exists a large complex structure limit point $P \in \overline{\mathcal{M}}_{\mathrm{cx}}(Y)$ corresponding to the mirror partner X, and a subvariation of the geometric variation of Hodge structure defined on $H_n^{\mathrm{sL}}(Y)^*$ whose monodromy weight filtration at P coincides with the*

---

[24] It appears from both the K3 case discussed in the next section, and the analysis of [24] that the correct total homology class to use is the Poincaré dual of $\mathrm{ch}(\mathcal{E})\sqrt{\mathrm{td}\, Y}$.

[25] This property of the mapping of D-branes has also been observed by Ooguri, Oz and Yin [48].

Leray filtration for the special Lagrangian T^n-fibration on Y. Moreover, under the isomorphism of Conjecture 3, this maps to the subvariation of the A-variation of Hodge structure defined on $H^{\mathrm{alg}}_{\mathrm{even}}(X)^*$.

The difficulty in putting an integer structure on the A-variation of Hodge structure stems from the fact that $H^{p,p}(X)$ will in general not be generated by its intersection with $H^{2p}(X, \mathbb{Z})$. However, the *algebraic* cohomology $H^{\mathrm{alg}}_{\mathrm{even}}(X)^*$ does not suffer from this problem: its graded pieces are generated by integer (p,p)-classes. If Conjecture 4 holds, it explains why there is a corresponding subvariation of the geometric variation of Hodge structure on Y, also defined over the integers. We would thus get corresponding local systems over $\mathbb{Z}$ in addition to the isomorphisms of complex variations of Hodge structure.

6 Geometric mirror symmetry for K3 surfaces

The special Lagrangian submanifolds of a K3 surface can be studied directly, thanks to the following fact due to Harvey and Lawson [29]: given a Ricci flat metric on a K3 surface Y and a special Lagrangian submanifold M, there exists a complex structure on Y with respect to which the metric is Kähler, such that M is a complex submanifold of Y. This allows us to translate immediately the theory of special Lagrangian T^2-fibrations on Y to the standard theory of elliptic fibrations. In this section, we will discuss geometric mirror symmetry for K3 surfaces in some detail. (Some aspects of this case have also been worked out by Gross and Wilson [28], who went on to study geometric mirror symmetry for the Voisin–Borcea threefolds of the form $(K3 \times T^2)/\mathbb{Z}_2$.)

If we fix a cohomology class $\mu \in H^2(Y, \mathbb{Z})$ which is primitive (i.e., $\frac{1}{n}\mu \notin H^2(Y, \mathbb{Z})$ for $1 < n \in \mathbb{Z}$) and satisfies $\mu \cdot \mu = 0$, then for any Ricci flat metric we can find a compatible complex structure for which μ has type $(1,1)$ and $\kappa \cdot \mu > 0$ (κ being the Kähler form). The class μ is then represented by a complex curve, which moves in a one parameter family, defining the structure of an elliptic fibration. Thus, elliptic fibrations of this sort exist for every Ricci flat metric on a K3 surface.[26]

Our Conjecture 1 is easy to verify in this case: as is well known, the base of the elliptic fibration on a K3 surface can be completed to a 2-sphere, and

---

[26]They even exist—although possibly in degenerate form—for the "orbifold" metrics which occur at certain limit points of the moduli space: at those points, κ is only required to be semipositive, but by the index theorem $\kappa^\perp$ cannot contain an isotropic vector such as μ, so it is still possible to choose a complex structure such that $\kappa \cdot \mu > 0$.

the resulting map from K3 to S^2 is proper. In fact, the possible singular fibers are known very explicitly in this case [30].

To study Conjecture 2, we need to understand the structure of the "complexified" moduli space $\mathcal{M}_D(T^2, Y)$. Since a flat $U(1)$-bundle on an elliptic curve is equivalent to a holomorphic line bundle of degree zero, each point in $\mathcal{M}_D(T^2, Y)$ has a natural interpretation as such a bundle on some particular fiber of the elliptic fibration. Extending that bundle by zero, we can regard it as a sheaf $\mathcal{L}$ on Y, with $\mathrm{supp}(\mathcal{L}) = \mathrm{image}(f)$. We thus identify $\mathcal{M}_D(T^2, Y)$ as a moduli spaces of such sheaves.

Let us briefly recall the facts about the moduli spaces of simple sheaves on K3 surfaces, as worked out by Mukai [46, 47]. First, Mukai showed that for any simple sheaf $\mathcal{E}$ on Y, i.e., one without any nonconstant endomorphisms, the moduli space $\mathcal{M}_{\mathrm{simple}}$ is smooth at $[\mathcal{E}]$ of dimension $\dim \mathrm{Ext}^1(\mathcal{E}, \mathcal{E}) = 2 - \chi(\mathcal{E}, \mathcal{E})$. (The 2 in the formula arises from the spaces $\mathrm{Hom}(\mathcal{E}, \mathcal{E})$ and $\mathrm{Ext}^2(\mathcal{E}, \mathcal{E})$, each of which has dimension one, due to the constant endomorphisms in the first case, and their Kodaira–Serre duals in the second.)

Second, Mukai introduced an intersection pairing on $H^{\mathrm{ev}}(Y) = H^0(Y) \oplus H^2(Y) \oplus H^4(Y)$ defined by

$$(\alpha, \beta, \gamma) \cdot (\alpha', \beta', \gamma') = (\beta \cdot \beta' - \alpha \cdot \gamma' - \gamma \cdot \alpha')[Y],$$

and a slight modification of the usual Chern character $\mathrm{ch}(\mathcal{E})$, defined by

$$v(\mathcal{E}) = \mathrm{ch}(\mathcal{E})\sqrt{\mathrm{td}(Y)} = \left(\mathrm{rank}\,\mathcal{E}, c_1(\mathcal{E}), \mathrm{rank}\,\mathcal{E} + \frac{1}{2}(c_1(\mathcal{E})^2 - 2c_2(\mathcal{E}))\right),$$

so that the Riemann–Roch theorem reads

$$\chi(\mathcal{E}, \mathcal{F}) := \sum_{i=0}^{2}(-1)^i \dim \mathrm{Ext}^i(\mathcal{E}, \mathcal{F}) = v(\mathcal{E}) \cdot v(\mathcal{F}).$$

In particular, the moduli space $\mathcal{M}_{\mathrm{simple}}(v)$ of simple sheaves with $v(\mathcal{E}) = v$ has dimension

$$\dim \mathcal{M}_{\mathrm{simple}}(v) = 2 - \chi(\mathcal{E}, \mathcal{E}) = 2 - v \cdot v.$$

In the case of 2-dimensional moduli spaces $\mathcal{M}_{\mathrm{simple}}(v)$, Mukai goes on to show that whenever the space is compact, it must be a K3 surface.

The sheaves $\mathcal{L}$ with support on a curve from our elliptic fibration will have Mukai class $v(\mathcal{L}) = (0, \mu, 0)$ for which $v(\mathcal{L}) \cdot v(\mathcal{L}) = \mu \cdot \mu = 0$, so the moduli space has dimension two. That is, our moduli space $\mathcal{M}_D(T^2, Y)$ is contained in $\mathcal{M}_{\mathrm{simple}}(0, \mu, 0)$ as an open subset. Our second conjecture will follow if we can show that this latter space is compact, or at least admits a natural compactification. Whether this is true or not could in principle depend on the choice of Ricci flat metric on Y. If we restrict to metrics with the property that Y is algebraic when given the compatible complex structure for which μ defines an elliptic fibration (this is a dense set within the full moduli space), then techniques of algebraic geometry can be applied to this problem. General results of Simpson [51] imply that on an algebraic K3 surface, the set of semistable sheaves with a fixed Mukai vector v forms a projective variety. This applies to our situation with $v = (0, \mu, 0)$, and provides the desired compactification. It is to be hoped that compactifications such as this exist even for nonalgebraic K3 surfaces.

The Mukai class $v = (0, \mu, 0)$ should now be mapped under mirror symmetry to the class of a 0-cycle, or the corresponding sheaf $\mathcal{O}_P$; that Mukai class is $(0, 0, 1)$. In fact, the mirror map known in physics [4] does precisely that: given any primitive isotropic vector v in $H^{\text{ev}}(Y)$, there is a mirror map which takes it to the vector $(0, 0, 1)$. Moreover, it is easy to calculate how this mirror map affects complex structures, by specifying how it affects Hodge structures: if we put a Hodge structure on $H^{\text{ev}}(Y)$ in which H^0 and H^4 have been specified as type $(1, 1)$, then the corresponding Hodge structure at the mirror image point has $v^{\perp}/v$ as its H^2.

This is *precisely* the relationship between Hodge structures on Y and on $\mathcal{M}_{\text{simple}}(v)$ which was found by Mukai [47]! We can thus identify geometric mirror symmetry for K3 surfaces (which associates the moduli spaces of 0-cycles and special Lagrangian T^2's) with the mirror symmetry previously found in physics. It is amusing to note that in establishing this relationship, Mukai used elliptic fibrations and bundles on them in a crucial way.

As suggested in the previous section, such a mirror transformation should act on the totality of special Lagrangian 2-cycles. In fact, it is known that for at least some K3 surfaces, there is a *Fourier–Mukai transform* which associates sheaves on $\mathcal{M}_{\text{simple}}(v)$ to sheaves on Y [7]. The map between their homology classes is precisely the mirror map.[27] Thus, proving that there exists such a Fourier–Mukai transform for arbitrary K3 surfaces (including nonalgebraic ones) would establish a version of Conjecture 4 in this case.

Acknowledgments

It is a pleasure to thank Robbert Dijkgraaf, Ron Donagi, Brian Greene, Mark Gross, Paul Horja, Sheldon Katz, Greg Moore, Ronen Plesser, Yiannis Vlassopoulos, Pelham Wilson, Edward Witten, and especially Robert Bryant and Andy Strominger for useful discussions; I also thank Strominger for communicating the results of [53] prior to publication, and the referee for useful

---

[27]One example of this is given by a special Lagrangian section of the T^2-fibration, which will map to the class $(1, 0, 1)$ which is the Mukai vector of the fundamental cycle (i.e., of the structure sheaf) on the mirror.

remarks on the first version. I am grateful to the Rutgers physics department for hospitality and support during the early stages of this work, to the organizers of the European Algebraic Geometry Conference at the University of Warwick where this work was first presented, and to the Aspen Center for Physics where the writing was completed. This research was partially supported by the National Science Foundation under grant DMS–9401447.

References

[1] P. S. Aspinwall, B. R. Greene, and D. R. Morrison, *Calabi–Yau moduli space, mirror manifolds and spacetime topology change in string theory*, Nuclear Phys. B **416** (1994), 414–480, hep-th/9309097.

[2] _____, *Measuring small distances in N=2 sigma models*, Nuclear Phys. B **420** (1994), 184–242, hep-th/9311042.

[3] P. S. Aspinwall and D. R. Morrison, *Topological field theory and rational curves*, Comm. Math. Phys. **151** (1993), 245–262, hep-th/9110048.

[4] _____, *String theory on K3 surfaces*, Mirror Symmetry II (B. Greene and S.-T. Yau, eds.), International Press, Cambridge, 1997, pp. 703–716, hep-th/9404151.

[5] _____, *Chiral rings do not suffice: N=(2,2) theories with nonzero fundamental group*, Phys. Lett. B **334** (1994), 79–86, hep-th/9406032.

[6] _____, *U-duality and integral structures*, Phys. Lett. B **355** (1995), 141–149, hep-th/9505025.

[7] C. Bartocci, U. Bruzzo, and D. Hernández Ruipérez, *A Fourier–Mukai transform for stable bundles on K3 surfaces*, J. reine angew. Math. **486** (1997), 1–16, alg-geom/9405006.

[8] V. V. Batyrev, *Dual polyhedra and mirror symmetry for Calabi–Yau hypersurfaces in toric varieties*, J. Algebraic Geom. **3** (1994), 493–535, alg-geom/9310003.

[9] V. V. Batyrev and L. A. Borisov, *Dual cones and mirror symmetry for generalized Calabi–Yau manifolds*, Mirror Symmetry II (B. Greene and S.-T. Yau, eds.), International Press, Cambridge, 1997, pp. 71–86, alg-geom/9402002.

[10] _____, *Mirror duality and string-theoretic Hodge numbers*, Invent. Math. **126** (1996), 183–203, alg-geom/9509009.

[11] V. V. Batyrev and D. I. Dais, *Strong McKay correspondence, string-theoretic Hodge numbers and mirror symmetry*, Topology, **35** (1996), 901–929, alg-geom/9410001.

[12] K. Becker, M. Becker, and A. Strominger, *Fivebranes, membranes and non-perturbative string theory*, Nuclear Phys. B **456** (1995), 130–152, hep-th/9507158.

[13] K. Behrend, *Gromov–Witten invariants in algebraic geometry*, Invent. Math. **127** (1997), 601–617, alg-geom/9601011.

[14] K. Behrend and B. Fantechi, *The intrinsic normal cone*, Invent. Math. **128** (1997), 45–88, alg-geom/9601010.

[15] K. Behrend and Yu. Manin, *Stacks of stable maps and Gromov–Witten invariants*, Duke Math. J. **85** (1996), 1–60, alg-geom/9506023.

[16] L. Borisov, *Towards the mirror symmetry for Calabi–Yau complete intersections in toric Fano varieties*, alg-geom/9310001.

[17] R. L. Bryant, *Submanifolds and special structures on the octonians*, J. Differential Geom. **17** (1982), 185–232.

[18] _____, *Minimal Lagrangian submanifolds of Kähler–Einstein manifolds*, Differential Geometry and Differential Equations, Shanghai 1985, Lecture Notes in Math., vol. 1255, 1985, pp. 1–12.

[19] P. Candelas, M. Lynker, and R. Schimmrigk, *Calabi–Yau manifolds in weighted* $\mathbb{P}_4$, Nuclear Phys. B **341** (1990), 383–402.

[20] P. Deligne, *Local behavior of Hodge structures at infinity*, Mirror Symmetry II (B. Greene and S.-T. Yau, eds.), International Press, Cambridge, 1997, pp. 683–700.

[21] L. J. Dixon, *Some world-sheet properties of superstring compactifications, on orbifolds and otherwise*, Superstrings, Unified Theories, and Cosmology 1987 (G. Furlan et al., eds.), World Scientific, 1988, pp. 67–126.

[22] R. Donagi and E. Markman, *Cubics, integrable systems, and Calabi–Yau threefolds*, Proc. of the Hirzebruch 65 Conference on Algebraic Geometry (M. Teicher, ed.), Israel Math. Conf. Proc., vol. 9, Bar-Ilan University, 1996, pp. 199–221, alg-geom/9408004.

[23] A. B. Givental, *Equivariant Gromov–Witten invariants*, Internat. Math. Res. Notices (1996), 613–663, alg-geom/9603021.

[24] M. Green, J. A. Harvey, and G. Moore, *I-brane inflow and anomalous couplings on D-branes*, Class. Quant. Grav. **14** (1997), 47–52, hep-th/9605033.

[25] B. R. Greene, D. R. Morrison, and M. R. Plesser, *Mirror manifolds in higher dimension*, Comm. Math. Phys. **173** (1995), 559–598, hep-th/9402119.

[26] B. R. Greene, D. R. Morrison, and A. Strominger, *Black hole condensation and the unification of string vacua*, Nuclear Phys. B **451** (1995), 109–120, hep-th/9504145.

[27] B. R. Greene and M. R. Plesser, *Duality in Calabi–Yau moduli space*, Nuclear Phys. B **338** (1990), 15–37.

[28] M. Gross and P. M. H. Wilson, *Mirror symmetry via 3-tori for a class of Calabi–Yau threefolds*, Math. Ann., to appear, alg-geom/9608004.

[29] R. Harvey and H. B. Lawson, Jr., *Calibrated geometries*, Acta Math. **148** (1982), 47–157.

[30] K. Kodaira, *On compact analytic surfaces, II, III*, Ann. of Math. (2) **77** (1963), 563–626; **78** (1963), 1–40.

[31] M. Kontsevich, *Enumeration of rational curves via torus actions*, The moduli space of curves (R. Dijkgraaf, C. Faber, and G. van der Geer, eds.), Progress in Math., vol. 129, Birkhäuser, 1995, pp. 335–368, hep-th/9405035.

[32] M. Kontsevich and Yu. Manin, *Gromov–Witten classes, quantum cohomology, and enumerative geometry*, Comm. Math. Phys. **164** (1994), 525–562, hep-th/9402147.

[33] W. Lerche, C. Vafa, and N. P. Warner, *Chiral rings in $N=2$ superconformal theories*, Nuclear Phys. B **324** (1989), 427–474.

[34] J. Li and G. Tian, *Virtual moduli cycles and Gromov–Witten invariants of algebraic varieties*, alg-geom/9602007.

[35] Yu. I. Manin, *Generating functions in algebraic geometry and sums over trees*, The moduli space of curves (R. Dijkgraaf, C. Faber, and G. van der Geer, eds.), Progress in Math., vol. 129, Birkhäuser, 1995, pp. 401–417, alg-geom/9407005.

[36] D. McDuff and D. Salamon, *J-holomorphic curves and quantum cohomology*, University Lecture Series, vol. 6, American Mathematical Society, 1994.

[37] R. C. McLean, *Deformations of calibrated submanifolds*, (based on a 1991 Duke University Ph.D. thesis), Comm. Anal. Geom., to appear, preprint available at http://www.math.duke.edu/preprints/1996.html.

[38] D. R. Morrison, *Compactifications of moduli spaces inspired by mirror symmetry*, Journées de Géométrie Algébrique d'Orsay (Juillet 1992), Astérisque, vol. 218, Société Mathématique de France, 1993, pp. 243–271, alg-geom/9304007.

[39] _____, *Beyond the Kähler cone*, Proc. of the Hirzebruch 65 Conference on Algebraic Geometry (M. Teicher, ed.), Israel Math. Conf. Proc., vol. 9, Bar-Ilan University, 1996, pp. 361–376, alg-geom/9407007.

[40] _____, *Mirror symmetry and moduli spaces of superconformal field theories*, Proc. Internat. Congr. Math. Zürich 1994 (S. D. Chatterji, ed.), vol. 2, Birkhäuser Verlag, 1995, pp. 1304–1314, alg-geom/9411019.

[41] _____, *Making enumerative predictions by means of mirror symmetry*, Mirror Symmetry II (B. Greene and S.-T. Yau, eds.), International Press, Cambridge, 1997, pp. 457–482, alg-geom/9504013.

[42] _____, *Mirror symmetry and the type II string*, Trieste conference on S-duality and mirror symmetry, Nuclear Phys. B Proc. Suppl., vol. 46, 1996, pp. 146–155, hep-th/9512016.

[43] _____, *Mathematical aspects of mirror symmetry*, Complex algebraic geometry (J. Kollár, ed.), IAS/Park City Mathematics Series, vol. 3, 1997, pp. 265–340, alg-geom/9609021.

[44] D. R. Morrison and M. R. Plesser, *Summing the instantons: Quantum cohomology and mirror symmetry in toric varieties*, Nuclear Phys. B **440** (1995), 279–354, hep-th/9412236.

[45] _____, *Towards mirror symmetry as duality for two-dimensional abelian gauge theories*, Trieste conference on S-duality and mirror symmetry, Nuclear Phys. B Proc. Suppl., vol. 46, 1996, pp. 177–186, hep-th/9508107.

[46] S. Mukai, *Symplectic structure of the moduli space of sheaves on an abelian or K3 surface*, Invent. Math. **77** (1984), 101–116.

[47] _____, *On the moduli space of bundles on K3 surfaces, I*, Vector bundles on algebraic varieties, Bombay 1984, Oxford University Press, 1987, pp. 341–413.

[48] H. Ooguri, Y. Oz, and Z. Yin, *D-branes on Calabi–Yau spaces and their mirrors*, Nuclear Phys. B, **477** (1996), 407–430, hep-th/9606112.

[49] Y. Ruan, *Topological sigma model and Donaldson-type invariants in Gromov theory*, Duke Math. J. **83** (1996), 461–500.

[50] Y. Ruan and G. Tian, *A mathematical theory of quantum cohomology*, J. Differential Geom. **42** (1995), 259–367.

[51] C. T. Simpson, *Moduli of representations of the fundamental group of a smooth projective variety, I*, Inst. Hautes Études Sci. Publ. Math. **79** (1994), 47–129.

[52] A. Strominger, *Massless black holes and conifolds in string theory*, Nuclear Phys. B **451** (1995), 96–108, hep-th/9504090.

[53] A. Strominger, S.-T. Yau, and E. Zaslow, *Mirror symmetry is T-duality*, Nuclear Phys. B. **479** (1996), 243–259, hep-th/9606040.

[54] C. Voisin, *A mathematical proof of a formula of Aspinwall and Morrison*, Compositio Math. **104** (1996), 135–151.

David Morrison,
Department of Mathematics, Box 90320,
Duke University, Durham, NC 27708-0320, USA
e-mail drm@math.duke.edu

Duality of polarized K3 surfaces

Shigeru Mukai [*]

To the memory of Professor Hideyuki Matsumura

The dual $\widehat{A}$ of an abelian variety A is defined as the neutral component $\mathrm{Pic}^0 A$ of the moduli space of line bundles on A. Let $\mathcal{P}$ be a Poincaré bundle on the product $A \times \widehat{A}$ whose restriction to $0 \times \widehat{A}$ is trivial. Then the classification morphism in the opposite direction

$$A \to \mathrm{Pic}^0 \widehat{A} \quad \text{defined by} \quad x \mapsto \mathcal{P}|_{x \times \widehat{A}}$$

is also an isomorphism and gives the double duality $A \cong \widehat{\widehat{A}}$. Moreover, the construction of [5] gives an equivalence $\int_A \mathcal{P} \colon \mathbf{D}(\mathrm{Coh}\, A) \to \mathbf{D}(\mathrm{Coh}\, \widehat{A})$ of the derived categories of coherent sheaves on A and $\widehat{A}$; this equivalence of derived categories depends on $\mathcal{P}$, and we call it the *integral functor* with *kernel* $\mathcal{P}$ (it is sometimes called the *Fourier–Mukai transform*). See [13] for a systematic study of (auto-)equivalences of $\mathbf{D}(\mathrm{Coh}\, A)$ using universal families of semihomogeneous vector bundles. In this article we generalize these results to polarized K3 surfaces using the moduli space of semi-rigid sheaves studied in [7].

Let r and s be coprime positive integers and (S, h) a polarized K3 surface of degree $2rs$. We denote by $\widehat{S}$ the moduli space $M_S(r, h, s)$ of rank r stable sheaves E on S with $c_1(E) = h$ and $\chi(E) = r + s$ (cf. [2]). Then by the results of [6] and [7], $\widehat{S}$ is again a K3 surface and there exists a universal family $\mathcal{E}$ on the product $S \times \widehat{S}$. Let

$$c_1(\mathcal{E}) = h + \varphi \in H^2(S) \oplus H^2(\widehat{S}) \quad \text{and} \quad c_2^{\mathrm{mid}}(\mathcal{E}) \in H^2(S) \otimes H^2(\widehat{S}) \quad (1)$$

be the first Chern class of $\mathcal{E}$ and the middle Künneth component of the second Chern class, respectively. We define a class $\psi \in H^2(\widehat{S})$ by

$$h \cup c_2^{\mathrm{mid}}(\mathcal{E}) = p \otimes \psi \in H^4(S) \otimes H^2(\widehat{S}), \quad (2)$$

where p is the fundamental cohomology class of S. Both φ and ψ are algebraic by the Lefschetz type $(1,1)$ theorem.

---

[*]Partially supported by Grant-in-aid for Scientific Research (B) 0854004 of the Japanese Ministry of Education.

Proposition 1.1   *(i)* $\psi - 2(r-1)s\varphi$ *does not depend on the choice of the universal family* $\mathcal{E}$ *on* $S \times \widehat{S}$. *We denote it by* $\widehat{h}$.

(ii) $(\widehat{h})^2 = h^2$ *and* $\widehat{h} \equiv s\varphi \mod r$ *in* $\operatorname{Pic} \widehat{S}$.

Let v be an integer with $sv \equiv 1 \mod r$. By the proposition, there exists a divisor class λ on $\widehat{S}$ such that $v\widehat{h} = \varphi + r\lambda$. For this λ, $\mathcal{E} \otimes \pi_{\widehat{S}}^* \mathcal{O}_{\widehat{S}}(\lambda)$ is also a universal family and its first Chern class is $\pi_{\widehat{S}}^* h + v\pi_{\widehat{S}}^* \widehat{h}$. Such a universal family is said to be *normalized*. We denote by $\mathcal{F}_{rs+1}$ the moduli space of (quasi-)polarized K3 surfaces (S, h) of degree $2rs$. This is a quasiprojective algebraic variety of dimension 19.

Theorem 1.2 *Let* (S, h) *be a general member of* $\mathcal{F}_{rs+1}$ *and* $\mathcal{E}$ *a normalized universal sheaf on* $S \times \widehat{S}$, *where* $\widehat{S}$ *is the moduli space* $M_S(r, h, s)$. *If* $r \geq 2$, *then*

(i) $\widehat{h}$ *is primitive and ample;*

(ii) $\mathcal{E}$ *is locally free;*

(iii) *for every* $x \in S$, *the restriction* $\mathcal{E}|_{x \times \widehat{S}}$ *is stable with respect to* $\widehat{h}$ *and belongs to* $M_{\widehat{S}}(r, v\widehat{h}, v^2 s)$;

(iv) *the classification morphism in the opposite direction* $S \to M_{\widehat{S}}(r, v\widehat{h}, v^2 s)$ *given by* $x \mapsto \mathcal{E}|_{x \times \widehat{S}}$ *is also an isomorphism; and*

(v) *the* integral functor *with kernel* $\mathcal{E}$

$$\int_S \mathcal{E} \colon \mathbf{D}(\operatorname{Coh} S) \to \mathbf{D}(\operatorname{Coh} \widehat{S}) \quad \text{given by} \quad (?) \mapsto \mathbf{R}\pi_{T*}(\mathcal{E} \otimes \pi_S^*(?))$$

is an equivalence of categories between the derived categories $\mathbf{D}(\operatorname{Coh} S)$ *and* $\mathbf{D}(\operatorname{Coh} \widehat{S})$ *of coherent sheaves on* S *and* $\widehat{S}$; *here* (?) *stands for an object or morphism of* $\mathbf{D}(\operatorname{Coh} S)$ *and* $\mathbf{R}\pi_{T*}$ *for the derived functor.*

Example 1.3 $(r = 2, s = 3)$ Let (S, h) be a general member of $\mathcal{F}_7$. Then S is a complete linear section

$$\Sigma_{12}^{10} \cap H_1 \cap \cdots \cap H_8$$

of a 10-dimensional orthogonal Grassmannian $\Sigma_{12}^{10} \subset \mathbb{P}^{15}$, and h its hyperplane section. Here $\Sigma_{12}^{10} \subset \mathbb{P}^{15}$ is the Hermitian symmetric space $\mathrm{SO}(10, \mathbb{R})/\mathrm{U}(5)$ embedded into the projectivization of the 16-dimensional half spinor representation (cf. [8]). The discriminant variety $\check{\Sigma} \subset \check{\mathbb{P}}^{15}$ in the dual projective space is also a 10-dimensional orthogonal Grassmannian (cf. [9]). Let

$h_1, \ldots, h_8 \in \check{\mathbb{P}}^{15}$ be the points corresponding to the hyperplanes $H_1, \ldots, H_8$. Then the intersection

$$\check{\Sigma} \cap \langle h_1, \ldots, h_8 \rangle$$

is also a K3 surface, and is isomorphic to the moduli space $\widehat{S} = M_S(2, h, 3)$. In this case, the double duality (3) below also follows from the double duality $\Sigma \cong \check{\Sigma}$ of discriminant varieties.

The theorem also holds (as a tautology) in the case $r = 1$, provided that we modify (ii). In fact, every member of $M_S(1, h, s)$ is isomorphic to $\mathcal{O}_S(h) \otimes \mathfrak{m}_x$ for a point $x \in S$, and $M_S(1, h, s)$ is canonically isomorphic to S. The ideal sheaf of the diagonal $\Delta \subset S \times S$ tensored with $\mathcal{O}_S(h) \boxtimes \mathcal{O}_S(h)$ is a normalized universal family. The integral functor $\int_S \mathcal{E}$ is essentially the reflection studied in [7], §2.

When $s \equiv \pm 1$ modulo r, we can choose a universal bundle $\mathcal{E}$ such that $\det \mathcal{E} \cong \mathcal{O}_S(h) \boxtimes \mathcal{O}_{\widehat{S}}(\pm \widehat{h})$. This $\mathcal{E}$ gives the duality isomorphism

$$(S, h) \cong (\widehat{\widehat{S}}, \widehat{\widehat{h}}). \tag{3}$$

Even when s is general, we can show that $(S, h) \mapsto (\widehat{S}, \widehat{h})$ is an involution of the moduli space $\mathcal{F}_{rs+1}$ using the computation of periods in [7]. But this will be discussed elsewhere.

The most essential part of the theorem is the stability assertion in (iii), which we prove in §4 after examining $\widehat{S} = M_S(r, h, s)$ for K3 surfaces S with Picard number 2 in §3. We first show in §2 that (iv) and (v) follow from (iii).

In the talk at the Warwick conference, we explained a non-abelian analogue of Albanese maps for general polarized K3 curves of odd genus ≥ 11, which is an application of Theorem 1.2 (for $r = 2$) and non-abelian Brill–Noether theory. Consult [11] and [12] for this.

Notations and convention For simplicity, we consider algebraic varieties over the complex number field $\mathbb{C}$. For a coherent sheaf E, we denote its dual by $E^\vee$, its rank (at the generic point) by $r(E)$, and its Euler–Poincaré characteristic by $\chi(E) = \sum_i (-)^i h^i(E)$.

2 Moduli duality and integral functors

Let $\mathcal{L}$ be a line bundle on the product $A \times B$ of two abelian varieties A and B.

Proposition 2.1 ([14], §13) *The following three conditions are equivalent:*

(a) *the classification homomorphism* $A \to \mathrm{Pic}^0 B$ *given by* $a \mapsto \mathcal{L}_{|a \times B}$ *is an isomorphism;*

(b) *the classification homomorphism* $B \to \mathrm{Pic}^0 A$ *given by* $b \mapsto \mathcal{L}_{|A \times b}$ *is an isomorphism; and*

(c) $\chi(\pi_A^* L_A^{-1} \otimes \mathcal{L} \otimes \pi_B^* L_B^{-1}) = \pm 1$, *where* $L_A = \mathcal{L}_{|A \times 0}$ *and* $L_B = \mathcal{L}_{|0 \times B}$.

When these equivalent conditions hold, the integral functor with kernel $\mathcal{L}$

$$\mathbf{D}(\mathrm{Coh}\, A) \to \mathbf{D}(\mathrm{Coh}\, B) \quad \text{given by} \quad (?) \mapsto \mathbf{R}\pi_{B*}(\mathcal{L} \otimes \pi_A^*(?)),$$

essentially the Fourier functor, is an equivalence of categories ([5]), where (?) is an object or a morphism in $\mathbf{D}(\mathrm{Coh}\, A)$. We will generalize these results to K3 surfaces.

Let $\mathcal{E}$ be a vector bundle on the product $S \times T$ of a complete algebraic surface S and a scheme T. We consider the two integral functors

$$\int_S \mathcal{E} : \mathbf{D}(\mathrm{Coh}\, S) \to \mathbf{D}(\mathrm{Coh}\, T) \quad \text{given by} \quad (?) \mapsto \mathbf{R}\pi_{T*}(\mathcal{E} \otimes \pi_S^*(?))$$

with kernel $\mathcal{E}$ from S to T, and vice versa

$$\int_T \mathcal{E}^{\vee} : \mathbf{D}(\mathrm{Coh}\, T) \to \mathbf{D}(\mathrm{Coh}\, S) \quad \text{given by} \quad (?) \mapsto \mathbf{R}\pi_{S*}(\mathcal{E}^{\vee} \otimes \pi_T^*(?))$$

with kernel $\mathcal{E}^{\vee}$. By the Grothendieck Riemann–Roch theorem, the following diagram is commutative:

$$
\begin{array}{ccccc}
\mathbf{D}(\mathrm{Coh}\, T) & \xrightarrow{\int_T \mathcal{E}^{\vee}} & \mathbf{D}(\mathrm{Coh}\, S) & \xrightarrow{\int_S \mathcal{E}} & \mathbf{D}(\mathrm{Coh}\, T) \\
\mathrm{ch} \downarrow & & \mathrm{ch} \downarrow & & \mathrm{ch} \downarrow \\
H^*(T, \mathbb{Q}) & \xrightarrow[{[\mathcal{E}^{\vee}]_T}]{} & H^*(S, \mathbb{Q}) & \xrightarrow[{[\mathcal{E}]_S}]{} & H^*(T, \mathbb{Q}),
\end{array}
\qquad (4)
$$

where ch is the Chern character map, and $[\mathcal{E}]_S$, $[\mathcal{E}^{\vee}]_T$ the correspondences

$$
\begin{aligned}
\alpha &\mapsto \pi_{T*}(\mathrm{ch}\, \mathcal{E} \cup \pi_S^*(\alpha \cup \mathrm{td}_S)), \\
\beta &\mapsto \pi_{S*}(\mathrm{ch}\, \mathcal{E}^{\vee} \cup \pi_T^*(\beta \cup \mathrm{td}_T)),
\end{aligned}
$$

with kernels $\mathrm{ch}\, \mathcal{E} \cup \pi_S^* \mathrm{td}_S$ and $\mathrm{ch}\, \mathcal{E}^{\vee} \cup \pi_T^* \mathrm{td}_T$.

To generalize Proposition 2.1 to vector bundles, we need the following assumption:

$$\text{there exists an ample line bundle } \Lambda \text{ on } S \text{ such that}$$
$$\mathcal{E}_{|S \times t} \text{ is stable with respect to } \Lambda \text{ for every } t \in T. \qquad (\Sigma_{S,T})$$

Under this condition, we obtain a classification morphism $t \mapsto \mathcal{E}_{|S \times t}$ corresponding to $\mathcal{E}$ from T to the moduli M_S of stable vector bundles on S ([2]). We denote it by $\Phi : T \to M_S$.

Theorem 2.2 *Let S be a K3 surface, T a scheme, and $\mathcal{E}$ a vector bundle on $S \times T$ such that $(\Sigma_{S,T})$ holds. Then the following three conditions are equivalent:*

(i) the classification morphism $\Phi \colon T \to M_S$ is an isomorphism onto a connected component;

(ii) under the projection $\pi_{23} \colon S \times T \times T \to T \times T$, the direct images of the vector bundle $\pi_{12}^* \mathcal{E}^\vee \otimes \pi_{13}^* \mathcal{E}$ satisfy the following:

$$R^q \pi_{23*}(\pi_{12}^* \mathcal{E}^\vee \otimes \pi_{13}^* \mathcal{E}) = \begin{cases} \mathcal{O}_\Delta & \text{if } q = 2, \\ 0 & \text{otherwise,} \end{cases}$$

where Δ is the diagonal of $T \times T$; and

(iii) the composite $[\mathcal{E}]_S \circ [\mathcal{E}^\vee]_T$ in (4) is the identity map of $H^*(T, \mathbb{Q})$.

Proof (i) $\Rightarrow$ (ii) As proved in [7], Proposition 4.10, the second direct image is a line bundle, say L, on the diagonal $\Delta \subset T \times T$ and the other direct images are zero. Since the restriction $\mathcal{E}|_{S \times t}$ is a simple sheaf for every $t \in T$, the natural homomorphism

$$\mathcal{O}_T \to \pi_{T*}(\mathcal{E}^\vee \otimes \mathcal{E})$$

is an isomorphism. By Serre duality, $R^2 \pi_{T*}(\mathcal{E}^\vee \otimes \mathcal{E})$ is also isomorphic to $\mathcal{O}_T$. Since $\pi_{12}^* \mathcal{E}^\vee \otimes \pi_{13}^* \mathcal{E}$ restricted to $S \times \Delta$ is $\mathcal{E}^\vee \otimes \mathcal{E}$, the line bundle L is trivial by the base change theorem.

(ii) $\Rightarrow$ (iii) On the one hand, the composite $[\mathcal{E}]_S \circ [\mathcal{E}^\vee]_T \colon H^*(T, \mathbb{Q}) \to H^*(T, \mathbb{Q})$ is the correspondence given by the following cocycle in $H^*(T \times T, \mathbb{Q})$:

$$\begin{aligned} Z &= \pi_{23*}(\pi_{12}^* \operatorname{ch} \mathcal{E}^\vee \cup \pi_{23}^* \operatorname{ch} \mathcal{E} \cup \pi_1^* \operatorname{td}_S \cup \pi_2^* \operatorname{td}_T) \\ &= \pi_{23*}(\pi_{12}^* \operatorname{ch} \mathcal{E}^\vee \cup \pi_{23}^* \operatorname{ch} \mathcal{E} \cup \pi_1^* \operatorname{td}_S) \cup \pi_1^* \operatorname{td}_T. \end{aligned}$$

On the other hand, by (ii) and the Grothendieck Riemann–Roch theorem, we have

$$\pi_{23*}(\pi_{12}^* \operatorname{ch} \mathcal{E}^\vee \cup \pi_{23}^* \operatorname{ch} \mathcal{E} \cup \pi_1^* \operatorname{td}_S) = \operatorname{ch} \mathcal{O}_\Delta.$$

Hence $Z = \operatorname{ch} \mathcal{O}_\Delta \cup \pi_1^* \operatorname{td}_T = \Delta$ and the composite is the identity map.

(iii) $\Rightarrow$ (i) The condition (iii) implies

$$\pi_{23*}(\pi_{12}^* \operatorname{ch} \mathcal{E}^\vee \cup \pi_{23}^* \operatorname{ch} \mathcal{E} \cup \pi_1^* \operatorname{td}_S \cup \pi_2^* \operatorname{td}_T) = \Delta.$$

Comparing the H^0 component of both sides, we have

$$\varkappa_4(\operatorname{ch} E^\vee \cup \operatorname{ch} E \cup \operatorname{td}_T) = 0,$$

where E is the restriction of $\mathcal{E}$ to a fibre $S \times t$ and $\varkappa_4$ is the projection of $H^*(S, \mathbb{Q})$ to the direct summand $H^4(S, \mathbb{Q}) \cong \mathbb{Q}$. By the Riemann–Roch theorem, we have $\chi(E^\vee \otimes E) = 0$. Since S is a K3 surface and E is simple, we have $h^1(E^\vee \otimes E) = 2$. Therefore, the connected component $\widehat{S}$ of M_S containing the image $\Phi(T)$ is a surface.

Let K be the fibre of $\Phi \colon T \to M_S$ over $\Phi(y)$. As a set, K consists of the points $t \in T$ such that $\mathcal{E}_{|S \times t} \cong E$. By assumption $(\Sigma_{S,T})$, there exists a line bundle L on K such that $\mathcal{E}_{|S \times K} \cong E \boxtimes L$. Now apply the functor $\int_T \mathcal{E}^\vee$ to the structure sheaf $\mathcal{O}_K$. Then we have

$$R^q \pi_{S*}(\mathcal{E}^\vee \otimes \pi_T^* \mathcal{O}_K) = R^q \pi_{S*}(\mathcal{E}^\vee_{|S \times K})$$
$$\cong R^q \pi_{S*}(\pi_S^* E^\vee \otimes \pi_T^* L^{-1}) = E^\vee \otimes_{\mathbb{C}} H^q(S, L^{-1})$$

and

$$[\mathcal{E}^\vee]_T(\mathrm{ch}\, \mathcal{O}_K) = \chi(L^{-1}) \, \mathrm{ch}\, E^\vee.$$

Similarly, writing $k(t)$ for the sky scraper sheaf at $t \in T$, we have

$$\left(\int_T \mathcal{E}^\vee\right)(k(t)) = E^\vee \quad \text{and} \quad [\mathcal{E}^\vee]_T(p) = \mathrm{ch}\, E^\vee,$$

where $p \in H^4(S, \mathbb{Q})$ is the fundamental cohomology class of S. Since $[\mathcal{E}^\vee]_T$ is injective by (iii), we have $\mathrm{ch}\, \mathcal{O}_M = \chi(L^{-1})p$. Therefore, K is finite and $T \to \widehat{S} \subset M_S$ is finite. By the same argument as in [7], Proposition 4.10, the direct image $R^q \pi_{23*}(\pi_{12}^* \mathcal{E}^\vee \otimes \pi_{13}^* \mathcal{E})$ is zero for $q \neq 2$, and a line bundle on the pullback Δ' of the diagonal of $\widehat{S} \times \widehat{S}$ by the morphism $\Phi \times \Phi$ for $q = 2$. By (iii) and the Grothendieck Riemann–Roch formula, we have

$$\mathrm{ch}\big(R^2 \pi_{23*}(\pi_{12}^* \mathcal{E}^\vee \otimes \pi_{13}^* \mathcal{E})\big) = \mathrm{ch}\, \mathcal{O}_\Delta.$$

Therefore, Δ' coincides with Δ and $T \to \widehat{S}$ is an isomorphism. $\square$

When T is also a K3 surface, $[\mathcal{E}]_S \circ [\mathcal{E}^\vee]_T$ is the identity map if and only if $[\mathcal{E}]_T \circ [\mathcal{E}^\vee]_S$ is. Therefore, we have

Corollary 2.3 *Assume that both S and T are K3 surfaces and that a vector bundle $\mathcal{E}$ on $S \times T$ satisfies both $(\Sigma_{S,T})$ and $(\Sigma_{T,S})$. Then the classification morphism $T \to M_S$ is an isomorphism onto a connected component if and only if $S \to M_T$ is.*

Remark 2.4   (i) The Chern character $\mathrm{ch}\, \mathcal{E}$ is a cocycle with integral coefficients ([7], Lemma 4.1). Hence in the situation of the theorem, $[\mathcal{E}]_S$ is an isomorphism from $H^*(S, \mathbb{Z})$ to $H^*(T, \mathbb{Z})$.

(ii) We assume $\mathcal{E}$ locally free only to simplify the notation. The same holds for a coherent sheaf $\mathcal{E}$ on $S \times T$ that is both S-flat and T-flat.

The composite $\int_S \mathcal{E} \circ \int_T \mathcal{E}^\vee$ of two integral functors is also an integral functor. Moreover, its kernel is the composite $\mathbf{R}\pi_{23*}(\pi_{12}^*\mathcal{E}^\vee \otimes \pi_{13}^*\mathcal{E})$ of the two kernels, which is isomorphic to $\mathcal{O}_\Delta[2]$ under the equivalent conditions of Theorem 2.2 (where [2] denotes a shift in the derived category). Hence, we have

Theorem 2.5 *Under the same conditions as in Corollary 2.3, the integral functor with kernel $\mathcal{E}$*

$$\int_S \mathcal{E} \colon \mathbf{D}(\mathrm{Coh}\, S) \to \mathbf{D}(\mathrm{Coh}\, T)$$

is an equivalence of categories. Moreover, its inverse is $\int_T \mathcal{E}^\vee[-2]$, that is, the integral functor with kernel $\mathcal{E}^\vee$ in the opposite direction shifted by 2 to the right.

3 Construction of examples

Let r and s be integers with $r \geq 2$ and $s \geq 1$. In this section, we investigate the moduli space $\widehat{S} = M_S(r, h, s)$ for a polarized K3 surface (S, h) satisfying the following conditions:

$$h = e + rf, \text{ where } \mathrm{Pic}\, S = \mathbb{Z}e \oplus \mathbb{Z}f, \text{ and the divisor classes } e \text{ and } f \quad (*)$$
have intersection numbers $e^2 = -2r$, $e \cdot f = s+1$ and $f^2 = 0$.

By the surjectivity of the period map ([1], Chap. VIII), such polarized K3 surfaces (S, h) exist. Note that there is no -2-vector perpendicular to h. For integers m and n, we have

$$(me + nf)^2 = 2m\{-rm + (s+1)n\}$$
$$\text{and}\quad h \cdot (me + nf) = rsm + \{-rm + (s+1)n\}. \quad (5)$$

Lemma 3.1 (i) *If $s+1$ does not divide $r-1$, then $D^2 \geq 0$ for every effective divisor D on S.*

(ii) *Assume that $r - 1 = k(s+1)$, and put $e' = e + kf$. Then the divisor class e' contains an irreducible curve E' isomorphic to $\mathbb{P}^1$. Moreover, E' is the unique irreducible curve on S with negative self-intersection.*

Proof   Since the canonical class K_S of S is trivial, any irreducible curve C on S satisfies

$$C^2 = C \cdot (C + K_S) = 2p_a(C) - 2 \geq -2. \tag{6}$$

If $s+1$ does not divides $r-1$, then $C^2 = (me+nf)^2 \neq -2$ by (5). Therefore, we have (i).

Assume that $s+1$ divides $r-1$ and let C be an irreducible curve with negative self-intersection. By (5) and (6), C is linearly equivalent to $\pm e'$. Since $h \cdot C$ and $h \cdot e' = rs - 1$ are positive, C is linearly equivalent to e'. By the Riemann–Roch theorem, $\chi(\mathcal{O}_S(e')) = (e'^2)/2 + 2 = 1$ and e' contains an effective divisor E'. Since $(E'^2) < 0$, E' contains a curve C' with $(C'^2) < 0$ as an irreducible component. By what we have shown, C' is linearly equivalent to E'. Since $h^0(\mathcal{O}_S(C')) = 1$, we have $E' = C'$, which proves (ii). $\square$

By the lemma, we have

$$H^0(\mathcal{O}_S(e)) = H^0(\mathcal{O}_S(-e)) = 0 \tag{7}$$

since $(e^2) = -2r \leq -4$. By Serre duality and the Riemann–Roch theorem, we have

$$h^1(\mathcal{O}_S(\pm e)) = -\chi(\mathcal{O}_S(\pm e)) = r - 2.$$

Let $\mathfrak{m}_x$ be the maximal ideal of $\mathcal{O}_S$ at $x \in S$. By duality, we have

$$\dim \mathrm{Ext}^1\big(\mathfrak{m}_x\mathcal{O}_S(e+f), \mathcal{O}_S(f)\big) = h^1(\mathfrak{m}_x(-e)) = r - 1.$$

Hence we obtain the unique exact sequence

$$0 \to \mathcal{O}_S(f)^{\oplus(r-1)} \to E_x \to \mathfrak{m}_x(e+f) \to 0 \tag{8}$$

with the property $\mathrm{Hom}(E_x, \mathcal{O}_S(f)) = 0$. Since

$$\dim \mathrm{Ext}^1\big(\mathcal{O}_S(e+f), \mathcal{O}_S(f)\big) < \dim \mathrm{Ext}^1\big(\mathfrak{m}_x\mathcal{O}_S(e+f), \mathcal{O}_S(f)\big),$$

E_x is locally free of rank r. It is easy to verify that $\det E_x \cong \mathcal{O}_S(h)$ and

$$\chi(E_x) = (r-1)\chi(\mathcal{O}_S(f)) + \chi(\mathcal{O}_S(e+f)) - 1 = r + s. \tag{9}$$

We prove (i) and (iii) of Theorem 1.2 for this pair (S, h):

Theorem 3.2   *If (S, h) satisfies $(*)$, then*

   (i) for every $x \in S$, E_x is stable with respect to h and belongs to $M_S(r, h, s)$;

   (ii) if $x \neq y \in S$, then E_x and E_y are not isomorphic;

(iii) every member of $M_S(r, h, s)$ is locally free and isomorphic to E_x for some $x \in S$;

(iv) $\widehat{S} = M_S(r, h, s)$ is isomorphic to S; and

(v) there exists a vector bundle $\mathcal{E}$ on the product $S \times S$, a vector bundle G on S and an exact sequence

$$0 \to \mathcal{O}_S(f) \boxtimes G^{\vee} \to \mathcal{E} \to \mathcal{I}_{\Delta} \otimes \pi_1^* \mathcal{O}_S(e + f) \to 0 \qquad (10)$$

on $S \times S$ whose restriction to $S \times x$ is isomorphic to the exact sequence (8), where $\mathcal{I}_{\Delta}$ is the ideal sheaf of the diagonal $\Delta \subset S \times S$ and $\pi_1, \pi_2 \colon S \times S \to S$ are the projections to the two factors. Moreover, this vector bundle $\mathcal{E}$ is a universal family for $M_S(r, h, s)$.

In order to show the stability of E_x, we need an estimate of the degree $h \cdot D$ for an effective divisor D on S.

Lemma 3.3 *Let m and n be integers. If a divisor class $D = me + nf$ is effective, then $n \geq 0$ and $h \cdot D \geq (rs - 1)m$.*

Proof We first consider the case $D^2 \geq 0$. By (5), we have

$$m\{-rm + (s + 1)n\} \geq 0 \quad \text{and} \quad h \cdot D = rsm + \{-rm + (s + 1)n\} > 0.$$

Hence we have $m \geq 0$ and $-rm + (s+1)n \geq 0$. Therefore, we have $(s+1)n \geq rm \geq 0$ and $h \cdot D \geq rsm \geq (rs - 1)m$. This proves the lemma in the case where $s + 1$ does not divide $r - 1$. Assume that $s + 1$ divides $r - 1$. By (ii) of Lemma 3.1, D is linearly equivalent to the sum of an effective divisor D_+ with nonnegative self-intersection number and a multiple of E'. The lemma holds for both D_+ and E' since $E' = e + kf$ and $h \cdot E' = rs - 1$. Therefore, it also holds for D. $\square$

Proposition 3.4 *E_x is μ-stable with respect to h.*

Proof Let F be a torsion free quotient sheaf of $E_x(-f)$, which is neither zero nor $E_x(-f)$ itself. It suffices to show that

$$\mu(F) > \mu(E_x(-f)) = (h \cdot e)/r = s - 1.$$

Tensoring (8) with $\mathcal{O}_S(-f)$ gives the exact sequence

$$0 \to \mathcal{O}_S^{\oplus(r-1)} \to E_x(-f) \to \mathfrak{m}_x(e) \to 0.$$

with $\text{Hom}(E_x(-f), \mathcal{O}_S) = 0$. Let F_0 be the image of the composite

$$\mathcal{O}_S^{\oplus(r-1)} \hookrightarrow E_x(-f) \to F.$$

Then $\det F_0$ is effective. Since $\text{Hom}(F_0, \mathcal{O}_S) \subset \text{Hom}(E_x(-f), \mathcal{O}_S) = 0$, F_0 and hence $\det F_0$ is nontrivial.

If $\text{rank } F_0 < \text{rank } F$, then the quotient F/F_0 is isomorphic to $\mathfrak{m}_x(e)$. Hence $\det F \cong \det F_0 \otimes \mathcal{O}_S(e)$ and we have

$$\mu(F) \geq \frac{(h \cdot e)}{r(F)} > \frac{(h \cdot e)}{r} = \mu(E_x(-f)).$$

Assume that $\text{rank } F_0 = \text{rank } F$ and put $\det F_0 \cong \mathcal{O}_S(me + nf)$. Then, by the preceding lemma, we have

$$\mu(F) \geq \mu(F_0) = \frac{h \cdot (me + nf)}{r(F)} \geq \frac{m(rs - 1)}{r(F)} \geq ms.$$

Hence, we have $\mu(F) > s - 1$ if $m \geq 1$. So assume further that $m = 0$. Since a general member of $|f|$ is an (irreducible) elliptic curve, we have $h^0(\det F_0) = h^0(\mathcal{O}_S(nf)) = n + 1$. (Even when $s + 1$ divides $r - 1$, $|f|$ is base point free since $|f - \ell e'| = \emptyset$ for every integer $\ell \geq 1$.) There exists a trivial subbundle T of rank $r(F) + 1$ such that

$$\mathcal{O}_S^{\oplus r(F)+1} \cong T \to F_0$$

is surjective outside a finite set of points. Since its kernel is isomorphic to $(\det F_0)^{-1}$, we have

$$n + 1 = h^0(\det F) \geq r(F) + 1,$$

that is, $n \geq r(F)$. Hence, in this case, we also have

$$\mu(F_0) = \frac{(h \cdot nf)}{r(F)} \geq (f \cdot h) = s + 1 > \mu(E_x(-f)). \quad \square$$

(i) of Theorem 3.2 follows from this and the computation (9). It is easy to prove (ii):

Lemma 3.5 *If $E_x \cong E_y$, then $x = y$.*

Proof Let $f: E_x \to E_y$ be an isomorphism. Since $\text{Hom}(\mathcal{O}_S, \mathcal{O}_S(e)) = 0$ by (7), f maps the subsheaf $\mathcal{O}_S^{\oplus r-1} \subset E_x$ to the subsheaf $\mathcal{O}_S^{\oplus r-1} \subset E_y$. Therefore, f induces an isomorphism from $\mathfrak{m}_x(e)$ onto $\mathfrak{m}_y(e)$. Hence we have $x = y$. $\square$

The sheaf $I_\Delta \otimes \pi_1^* \mathcal{O}_S(e)$ is flat over S and its restriction to the fibre $S \times x$ is isomorphic to $\mathfrak{m}_x \otimes \mathcal{O}_S(e)$ for every $x \in S$. Let G be the sheaf on S associated to the presheaf

$$U \mapsto \mathcal{E}xt^1_{S \times U}(I_\Delta \otimes \pi_1^* \mathcal{O}_S(e)|_{S \times U}, \mathcal{O}_{S \times U})$$

$$\cong \mathcal{E}xt^1_{S \times U}(I_\Delta \otimes \pi_1^* \mathcal{O}_S(e+f)|_{S \times U}, \pi_1^* \mathcal{O}_S(f))$$

for U an open subset of S. Since $\text{Hom}(\mathfrak{m}_x(e), \mathcal{O}_S) = \text{Ext}^2(\mathfrak{m}_x(e), \mathcal{O}_S) = 0$ for every $x \in S$, G is locally free and we have

$$G \otimes k(x) \cong \text{Ext}^1(\mathfrak{m}_x(e), \mathcal{O}_S).$$

Precisely stated, the exact sequence (8) is

$$0 \to \mathcal{O}_S(f) \otimes_{\mathbb{C}} \text{Ext}^1(\mathfrak{m}_x(e), \mathcal{O}_S)^{\vee} \to E_x \to \mathfrak{m}_x(e+f) \to 0. \qquad (11)$$

Hence there exists an exact sequence

$$0 \to \pi_1^* \mathcal{O}_S(f) \otimes \pi_2^* G^{\vee} \to \mathcal{E} \to I_\Delta \otimes \pi_1^* \mathcal{O}_S(e+f) \to 0$$

over $S \times S$ whose restriction to $S \times x$ is (11) for every $x \in S$. This is the exact sequence (10) of Theorem 3.2. We have $G^{\vee} \cong R^1 \pi_{2*}(I_\Delta \otimes \pi_1^* \mathcal{O}(e))$ by (relative) Serre duality and obtain the exact sequence

$$0 \to \mathcal{O}_S(e) \to G^{\vee} \to \mathcal{O}_S \otimes_{\mathbb{C}} H^1(\mathcal{O}_S(e)) \to 0 \qquad (12)$$

from $0 \to I_\Delta \to \mathcal{O}_{S \times S} \to \mathcal{O}_\Delta \to 0$.

Let $S \to M_S(r, h, s)$ be the classification morphism of $\mathcal{E}$. Since $M_S(r, h, s)$ is smooth of dimension 2, it is an open embedding by Lemma 3.5; since $M_S(r, h, s)$ is connected, it is an isomorphism. So we have completed the proof of Theorem 3.2.

Now we compute the divisor class $\hat{h}$ on the moduli space $\hat{S} = M_S(r, h, s)$ defined in Proposition 1.1. We have

$$c_1(\mathcal{O}_S(f) \boxtimes G^{\vee}) = (r-1)\pi_1^* f + \pi_2^* e,$$
$$c_2(\mathcal{O}_S(f) \boxtimes G^{\vee}) = (r-2)\pi_1^* f \cup \pi_2^* e.$$

by the exact sequence (12), and

$$c_1(\mathcal{E}) = (r-1)\pi_1^* f + \pi_2^* e + \pi_1^*(e+f) = \pi_1^* h + \pi_2^* e,$$
$$c_2(\mathcal{E}) = c_1(\mathcal{O}_S(f) \boxtimes G^{\vee}) \cup \pi_1^*(e+f) + c_2(\mathcal{O}_S(f) \boxtimes G^{\vee}) + \Delta$$

by the exact sequence (10). Hence the middle Künneth component of $c_2(\mathcal{E})$ is

$$c_2^{\mathrm{mid}}(\mathcal{E}) = \pi_1^*(e + (r-1)f) \cup \pi_2^* e + \Delta \in H^2(S) \otimes H^2(S).$$

Therefore, by (1), (2) of the introduction and Proposition 1.1, we have

$$\varphi = e, \quad \psi = (h \cdot e + (r-1)f)e + h = h + (2rs - s - 1)e \quad \text{and}$$
$$\widehat{h} = \psi - 2(r-1)s\varphi = h + (s-1)e = se + rf = sh + r(1-s)f. \tag{13}$$

By our assumption $(*)$, $\{e, f\}$ is a free $\mathbb{Z}$-basis of Pic S. Hence we have

Lemma 3.6 $\widehat{h}$ *is primitive if and only if r and s are coprime.*

Since $(\widehat{h})^2 = 2rs > 0$ and $(\widehat{h} \cdot e') > 0$, $\widehat{h}$ is ample by Lemma 3.1. Since $2(h \cdot e)/(e^2) = s - 1$, we have

Lemma 3.7   *(i) $\widehat{h}$ is ample.*

(ii) $\widehat{h}$ is the reflection of h with respect to e, and hence $(\widehat{h})^2 = h^2$.

Let K_x be the restriction of the universal bundle $\mathcal{E}$ to a fibre $x \times S$ of the other direction. Restricting (10) to $x \times S$, we have an exact sequence

$$0 \to G^\vee \to K_x \to \mathfrak{m}_x \to 0. \tag{14}$$

Since K_x is locally free, this does not split.

Assume that r and s are coprime and fix a pair of integers (t, u) such that $ts + ur = 1$. The first Chern class of $\mathcal{E} \otimes \pi_2^* \mathcal{O}_S(-ue + tf)$ is $\pi_1^* h + t\pi_2^* \widehat{h}$. Hence this universal bundle is normalized and its restriction to $x \times S$ is $K_x(-ue + tf)$.

Lemma 3.8 *Assume that $(r, s) = 1$ and let F be a torsion free quotient sheaf of $K_x(-ue + tf)$, which is neither zero nor $K_x(-ue + tf)$ itself. If $c_1(F) = \ell\widehat{h}$ for an integer ℓ, then $\ell/r(F) > t/r$.*

Proof $F(ue - tf)$ is a quotient of K_x. By the exact sequences (12) and (14), either $c_1(F(ue - tf))$ or $c_1(F(ue - tf)) - e$ is effective. If we put $c_1(F(ue - tf)) = me + nf$, then we have $n \geq 0$ by Lemma 3.3. Since $me + nf - (ue - tf)q = \ell(se + rf)$ by our assumption, we have $n + tq = \ell r$, where we put $q = r(F)$. Since r and t are coprime and since $q < r$, $n = 0$ is impossible. Hence we have $\ell r - tq = n > 0$. $\square$

4   Proof of Proposition 1.1 and Theorem 1.2

First we prove (i) of Proposition 1.1, for which we do not need the results
of the preceding section. Let $\mathcal{E}$ be a universal sheaf on $S \times \widehat{S}$ with $\widehat{S} =
M_S(r, h, s)$. Any universal sheaf $\mathcal{E}'$ is isomorphic to $\mathcal{E} \otimes \pi_M^* \Lambda$ for a line bundle
Λ on $\widehat{S}$. Hence we have

$$c_1(\mathcal{E}') = c_1(\mathcal{E}) + r\pi_M^*\lambda,$$

$$c_2(\mathcal{E}') = c_2(\mathcal{E}) + (r-1)c_1(\mathcal{E}) \cup \pi_M^*\lambda + \frac{r(r-1)}{2}\pi_M^*\lambda^2, \quad \text{and}$$

$$c_2^{\mathrm{mid}}(\mathcal{E}') = c_2^{\mathrm{mid}}(\mathcal{E}) + (r-1)\pi_S^*h \cup \pi_M^*\lambda,$$

where we put $\lambda = c_1(\Lambda)$. Therefore, we have

$$\varphi' = \varphi + r\lambda, \quad \psi' = \psi + (r-1)h^2\lambda$$

and

$$\psi' - \frac{(r-1)h^2}{r}\varphi' = \psi - \frac{(r-1)h^2}{r}\varphi.$$

Since $h^2 = 2rs$, we have (i).

Now take a family $(\pi \colon \mathcal{S} \to T, \mathcal{L})$ of polarized K3 surfaces of degree $2rs$
with the following properties:

(a) The classification morphism

$$T \to \mathcal{F}_{rs+1} \quad \text{given by} \quad t \mapsto (S_t, L_t)$$

is dominant, where S_t is the fibre $\pi^{-1}(t)$ and L_t the restriction of $\mathcal{L}$ to
S_t.

(b) There exists a point $o \in T$ such that (S_o, L_o) satisfies the conditions $(*)$
at the beginning of §3.

For example, these are satisfied by an open subset of the Hilbert scheme
of a suitable projective space $\mathbb{P}^N$. By virtue of [3], there exists a scheme $\mathfrak{M}$
over T whose fibre over $t \in T$ is $\widehat{S}_t = M_{S_t}(r, L_t, s)$. By [6], $\mathfrak{M}$ is smooth
of dimension 2 over T. Since r and s are coprime, there exists a (coherent)
sheaf $\widetilde{\mathcal{E}}$ on the fibre product $\mathcal{S} \times_T \mathfrak{M}$ whose restriction to $S_t \times \widehat{S}_t$ is a universal
family $\mathcal{E}_t$ for every $t \in T$.

Define φ_t, ψ_t and $\widehat{h}_t \in H^{1,1}(\widehat{S}_t, \mathbb{Z})$ by

$$c_1(\mathcal{E}_t) = \pi_S^*c_1(L_t) + \pi_M^*\varphi_t, \quad c_1(L_t) \cup c_2^{\mathrm{mid}}(\mathcal{E}_t) = p \otimes \psi_t,$$

$$\text{and} \quad \widehat{h}_t = \psi_t - 2(r-1)s\varphi_t.$$

We have $(\widehat{h}_o^2) = 2rs$ by Lemma 3.7 and $\widehat{h}_o \equiv s\varphi_o$ mod r by (13). Hence we have $(\widehat{h}_t^2) = 2rs$ and $\widehat{h}_t \equiv s\varphi_t$ mod r for every t, which completes the proof of Proposition 1.1.

Now we prove Theorem 1.2. Let $(\mathcal{S}/T, \mathcal{L})$ and $\mathfrak{M}/T$ be as above. As we saw in the preceding section, $\widehat{h}_o$ is primitive and ample, and $\mathcal{E}_o$ is locally free. (See Lemmas 3.6, 3.7 and Theorem 3.2.) Hence, replacing T by an open subset containing o, we may assume that the following is satisfied for every $t \in T$:

1. the divisor class $\widehat{h}_t$ on $\widehat{S}_t$ is primitive and ample, and

2. $\mathcal{E}_t$ is locally free.

In particular, we have (i) and (ii) of Theorem 1.2.

Note that the isomorphism $[\mathcal{E}]_\mathcal{S} \colon H^*(S, \mathbb{Q}) \to H^*(T, \mathbb{Q})$ in the diagram (4) maps Hodge cocycles to Hodge cocycles since $\mathrm{ch}\,\mathcal{E}$ is algebraic. Therefore, S_t and the moduli space $\widehat{S}_t$ have the same Picard number (see [7], Theorem 1.4 for a more general result). Therefore, by our choice of $\mathcal{S}$, the set of points t for which $\widehat{S}_t$ is Picard general is dense in T. Hence we have

3. φ_t is a multiple of $\widehat{h}_t$ for every $t \in T$.

Claim 4.1 *There exists a point $t \in T$ such that $\mathcal{E}_t|_{x \times \widehat{S}}$ is stable with respect to $\widehat{h}_t$ for every $x \in S_t$.*

Proof   Assume the contrary and let Z be the set of points $x \in \mathcal{S}$ such that the restriction of $\mathcal{E}_{\pi(x)}$ to $x \times \widehat{S}_{\pi(x)}$ is unstable. This is a closed subvariety of $\mathcal{S}$ and we have $\pi(Z) = T$ by our assumption. Take a morphism $f \colon C \to Z$ from a curve C and two points p and $q \in C$ such that $\pi(f(p)) = o$ and $S_{\pi(f(q))}$ is Picard general. Let $\mathcal{E}_C$ be the pullback of $\widetilde{\mathcal{E}}$ to $C \times_T \mathfrak{M}$. Then $\mathcal{E}_C|_{x \times \widehat{S}}$ is unstable for every $x \in C$. Let $\mathcal{F}$ be a destabilizing subsheaf of $\mathcal{E}_C$, that is, an extension to $C \times_T \mathfrak{M}$ of the destabilizing subsheaf over the generic point of C (cf. [4]). Since $c_1(\mathcal{F}|_{q \times \widehat{S}})$ is an integral multiple of $\widehat{h}$, so is $c_1(\mathcal{F}|_{p \times \widehat{S}})$. But this contradicts Lemma 3.8.

By the claim and Corollary 2.3, if $S = S_t$ is general, then the classification morphism $S \to M_{\widehat{S}}$ is an isomorphism onto a connected component, which we denote by $M_{\widehat{S}}(r, v\widehat{h}, w)$ for integers v and w. Since $M_{\widehat{S}}(r, v\widehat{h}, w)$ is of dimension 2, we have $2rw = v^2(\widehat{h})^2$ and $w = v^2 s$ by the formula

$$h^1(\mathcal{E}nd\,E) - 2\dim\mathrm{End}\,E = -\chi(\mathcal{E}nd\,E) = c_1(E)^2 - 2r(E)(\chi(E) - r(E))$$

(cf. [6]). So we have proved (iii) and (iv) of Theorem 1.2. (v) is a consequence of (iv) and Theorem 2.5.

References

[1] Barth, W., Peters, C. and Van de Ven, A.: *Compact complex surfaces*, Springer Verlag, 1984.

[2] Gieseker, D.: On the moduli of vector bundles on an algebraic surface, Ann. of Math. **106** (1977), 45–60.

[3] Maruyama, M.: Moduli of stable sheaves, I, J. Math. Kyoto Univ. **17** (1977), 91–126.

[4] ——: Openness of a family of torsion free sheaves, J. Math. Kyoto Univ. **16** (1976), 627–637.

[5] Mukai, S.: Duality between $D(X)$ and $D(\widehat{X})$ with its application to Picard sheaves, Nagoya Math. J., **81** (1981), 153–175.

[6] ——: Symplectic structure of the moduli space of sheaves on an abelian or K3 surface, Invent. Math., **77** (1984), 101–116.

[7] ——: On the moduli space of vector bundles on a K3 surface, I, in *Vector Bundles on Algebraic Varieties*, Oxford University Press, 1987, pp. 341–413.

[8] ——: Curves, K3 surfaces and Fano manifolds of genus ≤ 10, in *Algebraic geometry and commutative algebra* in honor of Professor Masayoshi Nagata, Kinokuniya, Tokyo, 1988, pp. 357–377.

[9] ——: Curves and symmetric spaces, I, Amer. J. Math., **117** (1995), 1637–1644.

[10] ——: Vector bundles and Brill–Noether theory, Publ. Math. Sci. Res. Inst. **28** (1996), 145–158 (www.msri.org/Books/Book28).

[11] ——: Curves and K3 surfaces of genus eleven, '*Moduli of Vector Bundles*', (ed. M. Maruyama), Marcel Decker, 1996, pp. 189–197.

[12] ——: Non-Abelian Brill–Noether theory and Fano 3-folds, Sūgaku **49** (1997), no. 1, 1–24, English translation to appear in Sugaku Exposition: Duke file server, alg-geom/9704015, 36 pp.

[13] ——: Abelian variety and spin representation (in Japanese), Proceedings of symposium "Hodge theory and algebraic geometry (Sapporo, 1994)", pp. 110–135: English translation, Univ. of Warwick preprint, 13/1998.

[14] Mumford, D.: *Abelian Varieties*, Oxford University Press, 1970.

[15] Newstead, P.E.: Characteristic classes of stable bundles, Trans. Amer. Math. Soc., **169** (1972), 337–345.

Shigeru Mukai,
Graduate School of Mathematics,
Nagoya University,
Furō-chō, Chikusa-ku,
Nagoya, 464-01, Japan
e-mail: mukai@math.nagoya-u.ac.jp

On symplectic invariants of algebraic varieties coming from crepant contractions

Roberto Paoletti

1 Introduction

Let X be a smooth projective variety over the complex numbers; in standard notation, $N_1(X)$ denotes the finite dimensional vector space of 1-cycles on X modulo numerical equivalence. $N_1(X)$ contains the cone $\mathrm{NE}(X)$ of effective 1-cycles, generated by the classes of curves in X. In some geometric situations, the effective cone of X provides important deformation theoretic invariants of the manifold. For example, Wisniewski [Ws 2] observed that the halfcone $\mathrm{NE}(X) \cap (K_X z < 0)$ is invariant in smooth families of projective varieties. Recently, however, Ruan has used the theory of Gromov–Witten invariants to show that, at least in dimension two and three, Mori extremal rays provide invariants of algebraic manifolds in a stronger sense [Ru]. To describe his results, let Y denote a complete C^∞ manifold of real dimension $2n$, and let J_0 and J_1 be integrable complex structures on Y such that $Y_0 = (Y, J_0)$ and $Y_1 = (Y, J_1)$ are projective varieties. Then we say that Y_0 and Y_1 are *symplectically deformation equivalent* if there exists a family ω_t of symplectic forms on Y such that ω_0 is a Hodge form on Y_0 (that is, $i/2\pi$ times the curvature form of an ample line bundle) and ω_1 a Hodge form on Y_1. There is a family of almost complex structures J_t joining J_0 and J_1, such that J_t is compatible with ω_t for every t. If $n \leq 3$, Ruan proved that every Mori extremal ray has a nonzero Gromov–Witten invariant, and so is common to Y_0 and Y_1. Thus Mori extremal rays are symplectic deformation invariants of low dimensional projective manifolds.

In a related direction, Wilson [Wl 1] proved that the nef cone of a Calabi–Yau 3-fold is locally constant in its Kuranishi family if and only if X contains no quasi-ruled surface over an elliptic curve (see [Wl 2] for definitions). In [P 1], we proved a similar result for quasi-Fano 3-folds, that is, 3-dimensional projective manifolds for which $-K_X$ is nef and big: if X is a quasi-Fano 3-fold, then the nef cone of X is locally constant in the Kuranishi family if and only if X contains no quasi-ruled surface over a smooth rational curve, such that the canonical divisor K_X is trivial along the fibres, and descends to the canonical divisor of the base curve. The results of Ruan described above then lead to the question of the symplectic deformation invariance of the effective cone of these varieties. This problem was dealt with for Calabi–Yau 3-folds by Wilson in [Wl 2], where it was shown that two Calabi–Yau 3-folds that are symplectically deformation equivalent have the same nef cone unless one of them contains a quasi-ruled surface over an elliptic curve. Similarly, for quasi-Fano 3-folds, the effective cone is a symplectic invariant unless X contains a surface as above. More precisely, in this case, the effective cone of X is finite rational polyhedral, and each edge (extremal ray) has an associated nonzero Gromov–Witten invariant, except when the associated contraction φ is crepant, that is, the canonical divisor is φ-trivial, and the exceptional locus is a quasi-ruled surface over a smooth rational curve C, such that K_X descends to the canonical divisor on C. This statement follows from the arguments of [P 2].

In [P 2], we applied these methods to associate symplectic invariants to crepant extremal rays of 3-folds with K_X nef and big. For such 3-folds, the face $\mathrm{NE}(X) \cap (K_X z = 0)$ of the effective cone is finite rational polyhedral, and again, each edge of this part of the cone (crepant extremal ray) has an associated nonzero Gromov–Witten invariant, unless something special happens. Namely, the associated crepant contraction (see [Rd] for terminology) should have as its exceptional locus a quasi-ruled surface over a smooth curve C, with the property that the canonical divisor K_X descends to a divisor of degree $2g - 2$ on C, where g is the genus of C. In this case, the integer $d = \deg(\omega_C \otimes \omega_X^{-1})$ is the number of rational curves in the fibres of φ that deform with the almost complex structure. If the surface itself does not generically deform in the Kuranishi family, then (morally) this is the number of smooth rational curves in the exceptional locus of φ in a generic holomorphic deformation of X. This is illustrated by the examples of [P 1] and by the following.

Example 1.1 An element $t \in H^1(\Omega^1_{\mathbb{P}^1}) \cong \mathbb{C}$ corresponds to a short exact sequence $0 \to \mathcal{O}_{\mathbb{P}^1} \to \mathcal{F}_t \to \mathcal{O}_{\mathbb{P}^1}(2) \to 0$, and

$$\mathcal{F}_0 \cong \mathcal{O}_{\mathbb{P}^1} \oplus \mathcal{O}_{\mathbb{P}^1}(2), \quad \text{but} \quad \mathcal{F}_t \cong \mathcal{O}_{\mathbb{P}^1}(1)^{\oplus 2} \quad \text{for } t \neq 0.$$

Set $\mathcal{E}_t = \mathcal{O}_{\mathbb{P}^1}^{r-1} \oplus \mathcal{F}_t$ and write $X_t = \mathbb{P}(\mathcal{E}_t^*)$ for the corresponding $\mathbb{P}^1$-bundle. Then $\mathcal{O}_{X_t}(1)$ is nef and big for all t; indeed, it is globally generated because $\mathcal{E}_t$ is, and has top selfintersection number $\mathcal{O}_{X_t}(1)^{r+1} = 2$ by the Leray–Hirsch relation. Moreover, $K_{X_t} = \mathcal{O}_{X_t}(-(r+1))$, so that X_t is quasi-Fano for all t. Let $\Psi_t \colon X_t \to \overline{X}_t$ be the crepant birational contraction associated to $|-mK_{X_t}|$ for $m \gg 0$. Then Ψ_0 is a divisorial contraction, contracting $\mathbb{P}^1 \times \mathbb{P}^{r-1}$ to a subspace $\mathbb{P}^{r-1} \subset \overline{X}_0$ of canonical singularities, while Ψ_t is small, with exceptional locus $\mathbb{P}^1 \times \mathbb{P}^{r-2}$ contracted to a subspace $\mathbb{P}^{r-2} \subset \overline{X}_t$. For $m \gg 0$

and $i = 1, \ldots, r - 2$, let $V_{i,t} \in |-mK_{X_t}|$ be general, and set $V_t = \bigcap_{i=1}^{r-2} V_{i,t}$; the V_t form a smooth family of projective 3-folds with K_X nef and big, and $\varphi_t = \Psi_{t|V_t}$ is the pluricanonical contraction of V_t. For each t write $F_t \subset X_t$ for the exceptional locus of Ψ_t, and $E_t = F_t \cap V_t$ for that of φ_t. Then $E_0 \cong \mathbb{P}^1 \times C$, where C is a smooth complete intersection curve of type $(m, \ldots, m)$ in $\mathbb{P}^{r-1}$, and φ_0 contracts E_0 to C. However, $K_C = \mathcal{O}_C(m(r - 2) - r)$ and $K_{V_0|C} = K_C \otimes \mathcal{O}_C(-1)$. For $t \neq 0$, the contraction φ_t is small, with exceptional locus $d = \deg C$ copies of $\mathbb{P}^1$.

There may in principle be an exception to this, namely crepant contractions with exceptional locus a conic bundle surface whose generic fibre is a line pair, that degenerates somewhere to a double line. This is because in the case where K_X is nef and big one does not have such a good hold of the embedded deformation of subvarieties as when X is, say, a Calabi–Yau. I do not know of any example of this type associated with a discontinuity of the effective cone.

The aim of this paper is to make a few scattered remarks about this problem in the higher dimensional case. As above, we deal with varieties for which either K_X or $-K_X$ is nef and big, or K_X is trivial. Except in the case $K_X = 0$, there is a built-in crepant birational contraction $\Psi \colon X \to \overline{X}$, such that $\pm K_X$ descends to an ample $\mathbb{Q}$-divisor on $\overline{X}$, and thus the crepant effective cone is the cone $N_1(X/\overline{X}) \subset \mathrm{NE}(X)$ generated by classes of curves contracted by φ. In general, we define the crepant effective cone of X to be the subcone $\mathrm{NE}_{cr}(X) \subset \mathrm{NE}(X)$ generated by those curves contracting under some projective crepant contraction. This is a locally finite rational polyhedral cone, whose edges correspond to primitive contractions, and when $\pm K_X$ is nef and big we have $\mathrm{NE}_{cr}(X) = \mathrm{NE}(X/\overline{X})$. In this case, $\mathrm{NE}_{cr}(X)$ is the intersection of $\mathrm{NE}(X)$ with the hyperplane defined by K_X in $N_1(X)$. At any rate, $\mathrm{NE}_{cr}(X)$ is locally finite rational polyhedral, and we name the edges ℓ_i of this polyhedron the *crepant extremal rays* of X.

We remark that in Theorems 1.2, 1.3 and 1.4 below, the assumption on K_X is only needed to ensure that any given crepant extremal ray can be contracted. We could easily avoid this assumption by starting from an extremal contraction and its corresponding extremal ray.

We first look at the case that φ contracts a $\mathbb{P}^k$ to a point.

Theorem 1.2 *Let X be a smooth projective variety of dimension $n \geq 3$ for which $\pm K_X$ is nef and big or trivial. Suppose that $\ell \subset N_1(X)$ is an extremal ray, either crepant or of Mori type, with associated birational contraction $\varphi \colon X \to \overline{X}$. Suppose also that the exceptional locus $E \subset X$ of φ is isomorphic to $\mathbb{P}^k$, and that its normal bundle splits as a direct sum of line bundles: $N_{E/X} \cong \bigoplus_{i=1}^{n-k} \mathcal{O}_{\mathbb{P}^k}(-a_i)$. Then the ray ℓ is J-effective on all tamed almost* *complex deformations of X.*

For example, if X is 3-dimensional and a crepant birational contraction $\varphi\colon X \to \overline{X}$ has exceptional locus E isomorphic to $\mathbb{P}^1$, then any complex manifold which can be deformed to X through a family of tame almost complex structures has an effective curve homologous to E [Wl2]. Example 1.6 below shows that in general the assumption that the normal bundle splits as a direct sum of line bundles may not be omitted.

Theorem 1.3 *Let X be a smooth projective variety and $\varphi\colon X \to \overline{X}$ a divisorial contraction of an extremal ray ℓ of Mori type (that is, $-K_X$ is φ-ample [KMM]). Suppose that φ contracts a smooth $\mathbb{P}^1$-bundle $E \to B$ whose fibres generate the ray ℓ. If A is the homology class of a fibre of φ, then A is J-effective for any tamed almost complex deformation of the holomorphic structure of X.*

By [A], Theorem 2.3, the above contraction is simply a blowdown. The situation is more subtle for crepant contractions:

Theorem 1.4 *Let X be a smooth projective n-fold for which $\pm K_X$ is nef and big or trivial, $\ell \subset N_1(X)$ an extremal ray and $\varphi\colon X \to \overline{X}$ its crepant divisorial contraction. Assume that the exceptional locus $E \subset X$ is quasi-ruled via φ over $Z = \varphi(E) \subset X$, and that Z is smooth.*

  1. *If $c_1(\omega_{X|_E}) \neq (\varphi_{|_E})^*c_1(\omega_Z)$, then the ray ℓ remains effective on all tame almost complex deformations of the holomorphic structure of X.*

  2. *Suppose that E does not deform generically in the Kuranishi family of X. If $\omega_{X|_E} \cong (\varphi_{|_E})^*\omega_Z$, then the ray ℓ is not effective on a general small holomorphic deformation of X.*

We then look at the specific case $n = 4$:

Theorem 1.5 *Let X be a smooth projective 4-fold for which $\pm K_X$ is nef and big or trivial and $h^2(X, \mathcal{O}_X) = 0$, and suppose that the crepant effective cone of X is not locally constant in the Kuranishi family of X. Then X admits a primitive crepant birational contraction $\varphi\colon X \to \overline{X}$. If φ is divisorial, then φ contracts the exceptional locus $E \subset X$ (necessarily irreducible) to a surface $S \subset \overline{X}$ which is smooth in codimension one. In this case the following two possibilities hold:*

  (a) *away from finitely many fibres, E is quasi-ruled over S, meaning that the fibres are either all smooth conics or all reduced line pairs; if $S^* \subseteq S$ is the smooth locus, then $\omega_{S^*} \cong \omega_{X|_{S^*}}$; or*

(b) the general fibre of $E \rightarrow S$ is a reduced line pair, but there is a nonempty divisor $D \subset S$ over which the fibres of φ are double lines.

If on the other hand φ is a small contraction, then its exceptional locus is 2-dimensional and is contracted to a finite set in $\overline{X}$; furthermore the normalization of each irreducible component of the exceptional locus is isomorphic to $\mathbb{P}^2$.

This is proved by combining (a slight variation of) an argument of Wisniewski [Ws 2] with the results of [Wl 1], [P 1] and [P 2].

We illustrate the theorem by some examples:

Example 1.6 We give an example of a 4-fold admitting a crepant birational contraction whose exceptional locus is isomorphic to $\mathbb{P}^2$ and for which the crepant effective cone is not locally constant in the Kuranishi family. Let $\mathcal{G}_t$ denote the rank 3 vector bundle on $\mathbb{P}^2$ obtained as the extension

$$0 \rightarrow \mathcal{O}_{\mathbb{P}^2} \rightarrow \mathcal{G}_t \rightarrow T_{\mathbb{P}^2} \rightarrow 0 \quad \text{for } t \in H^1(\mathbb{P}^2, \Omega^1_{\mathbb{P}^2}) \cong \mathbb{C}.$$

For $t \neq 0$, this is the Euler sequence, so that $\mathcal{G}_t \cong \mathcal{O}_{\mathbb{P}^2}(1)^{\oplus 3}$ for $t \neq 0$, while $\mathcal{G}_0 \cong \mathcal{O}_{\mathbb{P}^2} \oplus T_{\mathbb{P}^2}$. Set $X_t = \mathbb{P}(\mathcal{G}_t^*)$, the relative projective space of lines in the dual vector bundle $\mathcal{G}_t^*$. This defines a smooth projective family $\mathcal{X} \rightarrow \mathbb{C}$, with $X_t \cong \mathbb{P}^2 \times \mathbb{P}^2$ if $t \neq 0$, and $X_0 \cong \mathbb{P}(\mathcal{O}_{\mathbb{P}^2} \oplus \Omega^1_{\mathbb{P}^2})$. Set $\xi_t = \mathcal{O}_{X_t}(1)$; then ξ_t is ample for $t \neq 0$ and globally generated for all $t \in \mathbb{C}$, since $\mathcal{G}_t$ is. Moreover, the Leray–Hirsch relation $\xi_t^3 - c_1(T_{\mathbb{P}^2})\xi_t^2 + c_2(T)\xi_t = 0$ gives $\xi_t^4 = 6$. The canonical line bundle of X_t is $\omega_{X_t} = \mathcal{O}_{X_t}(-3)$, and therefore X_t is a quasi-Fano 4-fold for all $t \in \mathbb{C}$, actually Fano for $t \neq 0$. It is, however, not Fano when $t = 0$: the direct summand $\mathcal{O}_{\mathbb{P}^2} \subset \mathcal{G}_0^*$ corresponds to a section $X_0 \supset F \cong \mathbb{P}^2$, such that $\mathcal{O}_F(\xi)$ is trivial. Large positive multiples of ξ_0 induce a crepant birational contraction of X_0, whose exceptional locus $F \cong \mathbb{P}^2$ is contracted to a point, and has normal bundle $N_{E/X_0} \cong \Omega^1_{\mathbb{P}^2}$. Note that by taking branched covers ramified along appropiate general multiples of ξ_t one can turn this into either a family of Calabi–Yau 4-folds or of 4-folds with nef and big canonical divisor.

Example 1.7 Let $P_i(t)$ for $i = 1, 2, 3$ be distinct points in the plane, in general position for $t \neq 0$ but collinear for $t = 0$, and denote by S_t the blowup of $\mathbb{P}^2$ in the $P_i(t)$; then set $X_t = S_t \times S$, where S is any fixed quasi-Fano surface. Then S_t is a del Pezzo surface for $t \neq 0$, while it is only quasi-Fano for $t = 0$, and if $\ell \subset \mathbb{P}^2$ is the line containing the $P_i(0)$ then its proper transform $\tilde{\ell} \subset S_0$ is a -2-curve. Let $S_0 \rightarrow \overline{S}_0$ denote the contraction of $\tilde{\ell}$ to a rational double point $Q \in \overline{S}_0$. Then X_t is Fano for $t \neq 0$, while X_0 is quasi-Fano but not Fano. Hence there is a discontinuity in the crepant effective cone, that is, some crepant extremal ray is effective for $t = 0$ but not for general t. In fact, the divisor $\tilde{\ell} \times S$ is the exceptional locus of a crepant contraction, where the factors $\ell \times \{P\}$, with $P \in S$, are the fibres, and the image is $\{Q\} \times S \subset \overline{S}_0 \times S$, a surface of cA_1 singularities. Note that similar examples may be constructed using K3 surfaces or minimal surfaces of general type, and then the families of 3-folds thus constructed are either Calabi–Yau or have nef and big canonical divisor.

Example 1.8 Let $\mathbb{P}^3 \supset \Lambda \cong \mathbb{P}^2$ be a plane and $D \subset \Lambda$ a smooth conic. Let $P(t) \in \mathbb{P}^3 \setminus D$ be a point, such that $P(t) \notin \Lambda$ for $t \neq 0$, while $P(0) \in \Lambda$. Denote by Y_t the blowup of $\mathbb{P}^3$ along C and $P(t)$. Then Y_t is a Fano 3-fold for $t \neq 0$, and $-K_{Y_0}$ is nef and big but not ample. Thus the crepant effective cone is not locally constant at the origin in the family. In fact, if $\tilde{\Lambda} \subset Y_0$ is the proper transform of Λ and $F \subset \tilde{\Lambda}$ is the exceptional divisor of the blowup of Λ at $P(0)$, then $\tilde{\Lambda}$ may be viewed as a rational ruled surface ($\cong \mathbb{F}_1$, with exceptional fibre F) of type $(0, 2)$ in the notation of [P 1]; it is the exceptional locus of a crepant divisorial contraction $X \to \overline{X}$, under which Λ is contracted to $F \subset \overline{X}$. We now set $X_t = Y_t \times \mathbb{P}^1$; then X_t is a Fano 4-fold when $t \neq 0$, but it is only quasi-Fano for $t = 0$. The divisor $\tilde{\Lambda} \times \mathbb{P}^1$ is the exceptional locus of a crepant morphism contraction, under which it contracts to $F \times \mathbb{P}^1 \subset \overline{X} \times \mathbb{P}^1$. Again, it is easy to modify this example so as to obtain families of 4-folds with $\pm K_X$ nef and big or trivial, for example, using a family of Calabi–Yau manifolds whose central fibre contains an elliptic ruled surface not deforming sideways in the family [Wl 1].

Notation   We recall some notation from the theory of J-holomorphic curves; a more adequate reference is [MS]. Let (M, ω) be any compact symplectic manifold. We say that an almost complex structure J on M is *compatible* with ω if it preserves ω and if $g_x(v, w) \overset{\text{def}}{=} \omega_x(v, Jw)$ (for $x \in M$ and $v, w \in T_x M$) is a Riemannian metric on M. The almost complex structure J is ω-*tame* if $\omega(v, J(v)) > 0$ for all tangent vectors $\neq 0$, and it is *tame* if it is ω-tame with respect to some symplectic structure ω. The space of all almost complex structures on M compatible with ω is contractible and nonempty; it is denoted by $\mathcal{J}(M, \omega)$. Having fixed J, and for $A \in H_2(M, \mathbb{Z})$ a spherical class, we denote by $\mathcal{M}(A, J)$ the space of all unparametrized J-holomorphic rational curves representing A. The almost complex structure is called A-*good* if for all J-holomorphic curves $f: \mathbb{P}^1 \to M$ representing the class A we have $H^{0,1}(f^*T_M) = 0$ and the space of cusp A-spheres has codimension at least two [MS], [Ru]; if J is A-good, it may be used to actually compute the Gromov–Witten invariants Φ_A. A ray ℓ in $H_2(X, \mathbb{R})$ is called J-*effective* if it is generated by an integral class for which the space of J-holomorphic curves is nonempty. Occasionally we say that a J-effective ray is a *jumping ray* if it does not remain effective under arbitrary small deformations of the almost complex structure on X. Accordingly, a *discontinuity* of the crepant effective cone will mean a jumping ray belonging to the crepant effective cone, and similarly for the Mori cone.

Finally, following Wilson [Wl 2], we say that a projective surface E is *quasi-ruled* if it admits a conic bundle structure $E \to C$ over a smooth curve C, all of whose fibres are either smooth conics or all line pairs.

Acknowledgments   I am grateful to the referee for various improvements in presentation, and for suggestions that led to the sharper version of Lemma 5.5 given here.

2   Proof of Theorem 1.2

We deal with the case $n > 3$, the case $n = 3$ having being treated in [Wl 2] (see [P 2] for some remarks in the non Calabi–Yau case). The strategy of the proof is first to show that under small deformations of the almost complex structure the excess dimensional moduli space of unparametrized J-holomorphic curves in the given homology class is replaced by a tractable geometric object: essentially the zero locus of a general section of the obstruction bundle on the moduli space itself. We then use this identification to prove that some associated Gromov–Witten invariant is nonzero, and therefore that the given homology class remains effective under arbitrary tamed almost complex deformations; this strategy is inspired by the work of Ruan [Ru] and Wilson [Wl 2]. In the present context all the relevant moduli spaces are *a priori* compact and smooth projective varieties, although our varieties are not necessarily weakly monotone.

Notice first that, either by an elementary direct computation using adjunction or by [Ws 1], Theorem 1.2, we have $2k \geq n - a$ with $a = 0$ in the Mori case and $a = 1$ in the crepant case. (Theorem 1.2 is stated for Mori extremal rays, but the proof works essentially just as well for crepant ones; see the more detailed discussion below.) Given that $n > 3$, then $k \geq 2$ and certainly $a_i > 0$ for all i; this is because under the hypothesis, $H^1(E, N) = 0$, so that the Hilbert scheme of X is smooth at E, and thus if $h^0(E, N) > 0$ then E must move in X, which is absurd. Let $L \subset E \cong \mathbb{P}^k$ be any line, and let $A \in H_2(X, \mathbb{Z})$ denote the homology class of L. We then have a short exact sequence of normal bundles

$$0 \to N_{L/E} \to N_{L/X} \to N_{E/X}|_L \to 0.$$

Given that $N_{L/E} \cong \mathcal{O}_L(1)^{k-1}$ and $N_{E/X} \cong \bigoplus_{i=1}^c \mathcal{O}_E(-a_i)$, this extension splits, that is, $N_{L/X} \cong \mathcal{O}_L(1)^{k-1} \oplus \bigoplus_{i=1}^c \mathcal{O}_E(-a_i)$ (here $c = n - k$). Therefore, if J denotes the complex structure of X, then J is not good for the class A in the sense of Gromov–Witten theory unless $a_i = 1$ for all i. The moduli space $\mathcal{M}(A, J)$ of all unparametrized J-holomorphic curves coincides with the Hilbert scheme of lines in $\mathbb{P}^k$, that is, with the Grassmannian $G = \mathrm{Grass}(2, k+1)$. Consider the incidence correspondence $F \subset G \times \mathbb{P}^k$, and let $p \colon F \to G$ and $q \colon F \to \mathbb{P}^k$ denote the projections. Then via p we may identify F with the relative projective space $\mathbb{P}_G(\mathcal{S})$ of lines in the tautological rank 2 bundle on G, and via q with $\mathbb{P}_{\mathbb{P}^k}(T_{\mathbb{P}^k})$.

For any given J-holomorphic curve L in the family (a line in $E \cong \mathbb{P}^k$), let $u \colon \mathbb{P}^1 \to X$ be a parametrization, and $D_u \colon \mathcal{A}(u^*T_M) \to \mathcal{A}^{0,1}(u^*T_M)$ the usual Fredholm operator of Gromov–Witten theory [MS], where $\mathcal{A}$ stands for C^∞ sections, and $\mathcal{A}^{0,1}(u^*T_X) = \mathcal{A}(\Omega^{0,1}_{\mathbb{P}^1} \otimes u^*(T_X))$. Since the almost complex structure of X is integrable, in a natural way $\ker(D_u) \cong H^0(\mathbb{P}^1, u^*T_X)$ and $\mathrm{coker}(D_u) \cong H^1(\mathbb{P}^1, u^*T_X) \cong H^1(\mathbb{P}^1, N_{L/X})$ [MS]. Thus,

$$\dim \mathrm{coker}(D_u) = \sum_{i=1}^{n-k} h^1(L, \mathcal{O}_L(-a_i)) = \sum_{i=1}^{n-k}(a_i - 1).$$

Besides, we have on the one hand $\deg(N_{L/X}) = -K_X \cdot L - 2$, and on the other $\deg(N_{L/X}) = -\sum_1^{n-k} a_i + k - 1$, and therefore $\sum_1^{n-k} a_i = K_X \cdot L + k + 1$. Hence, $\dim \mathrm{coker}(D_u) = K_X \cdot L + 2k + 1 - n$. Thus $n - K_X \cdot L - 3 + \dim \mathrm{coker}(D_u) = 2k - 2 = \dim \mathrm{Grass}(2, k+1)$. Therefore [Ru], Proposition 5.7 may be applied; in fact, although the variety in question is not necessarily weakly monotone, the moduli space at hand (the Grassmannian) is a priori smooth and compact. For the purpose of computing Gromov–Witten invariants, the moduli space of unparametrized J'-holomorphic curves, for J' a generic small deformation of J, may be identified with the zero locus of a transverse section of the obstruction bundle $\mathcal{E}_{\mathrm{ob}}$ on $G = \mathcal{M}(A, J)$. The latter, however, may in this case be described purely algebraically: on F there is a short exact sequence

$$0 \to T_{F/G} \to q^*T_X \to \mathcal{N} \to 0,$$

and $\mathcal{E}_{\mathrm{ob}} \cong R^1 p_*(\mathcal{N})$. We have the exact commutative diagram

$$
\begin{array}{ccccccccc}
& & & & 0 & & 0 & & \\
& & & & \downarrow & & \downarrow & & \\
0 & \to & T_{F/G} & \longrightarrow & q^*(T_{\mathbb{P}^k}) & \longrightarrow & \mathcal{R} & \to & 0 \\
& & \| & & \downarrow & & \downarrow & & \\
0 & \to & T_{F/G} & \longrightarrow & q^*(T_X) & \longrightarrow & \mathcal{N} & \to & 0 \\
& & & & \downarrow & & \downarrow & & \\
& & & & q^*(\oplus_1^{n-k}\mathcal{O}_{\mathbb{P}^k}(-a_i)) & = & q^*(\oplus_1^{n-k}\mathcal{O}_{\mathbb{P}^k}(-a_i)) & & \\
& & & & \downarrow & & \downarrow & & \\
& & & & 0 & & 0 & &
\end{array}
$$

and since $R^i p_*(\mathcal{R}) = 0$ for all $i > 0$, we have

$$\mathcal{E}_{ob} \cong R^1 p_* \left(q^* \bigoplus_{i=1}^{n-k} \mathcal{O}_{\mathbb{P}^k}(-a_i) \right) = \bigoplus_{i=1}^{n-k} R^1 p_* q^* \mathcal{O}_{\mathbb{P}^k}(-a_i).$$

On the other hand, with the identification $F = \mathbb{P}_G(\mathcal{S})$, we have $q^* \mathcal{O}_{\mathbb{P}^k}(-1) \cong \mathcal{O}_F(-1)$. Therefore,

$$\mathcal{E}_{ob} = \bigoplus_{i=1}^{n-k} R^1 p_* \mathcal{O}_F(-a_i) \cong \bigoplus_{i=1}^{n-k} \mathrm{Sym}^{a_i-2}(\mathcal{S}) \otimes \det(\mathcal{S}).$$

Now let $s \in \mathcal{A}(\mathcal{E}_{ob})$ be a transverse section. If s is nowhere zero, then $\mathcal{E}_{ob}$ has vanishing top Chern class: $c_r(\mathcal{E}_{ob}) = 0$, where $r = \sum_i (a_i - 1)$. This however implies that $c_r(\mathcal{E}_{ob}^*) = 0$, that is, $c_r \left(\bigoplus_{i=1}^{n-k} \mathrm{Sym}^{a_i-2}(\mathcal{S}^*) \otimes \det(\mathcal{S}^*) \right) = 0$. Set $H = \det(\mathcal{S}^*)$; this is an ample line bundle on G. Then

$$c_r \left(\bigoplus_{i=1}^{n-k} \mathrm{Sym}^{a_i-2}(\mathcal{S}^*) \otimes H \right) = c_1(H)^r + \sum_{k=1}^{r} c_1(H)^{r-k} c_k \left(\bigoplus_{i=1}^{n-k} \mathrm{Sym}^{a_i-2}(\mathcal{S}^*) \right);$$

however, since $\bigoplus_{i=1}^{n-k} \mathrm{Sym}^{a_i-2}(\mathcal{S}^*)$ is globally generated, each of its Chern classes may be represented by an effective cycle ([F], Chap. 14). Thus it is clear that the above Chern class is not zero. More succinctly, since $\mathcal{E}_{ob}^*$ is ample and globally generated, and has rank strictly less than the dimension of G, if $s \in H^0(\mathrm{Grass}(2, k+1), \mathcal{E}_{ob}^*)$ is a general section then its zero locus $Z = Z(s)$ is a nonempty, connected, smooth algebraic variety [F], [L]. This shows that the ray ℓ remains effective under small deformations of the almost complex structure of X. We claim, however, the stronger statement that the ray ℓ remains effective under arbitrary tamed almost complex deformations. To complete the proof, we now show that the Gromov–Witten invariant associated to A is nonzero.

Consider the case of smallest possible dimension of the exceptional locus: $2k = n - 1$. In this case $K_X \cdot A = 0$, that is, φ is a crepant contraction, and the normal bundle of E in X is $N \cong \mathcal{O}_{\mathbb{P}^k}(-1)^{\oplus(k+1)}$. Therefore, any line L in E has normal bundle

$$N_{L/X} = \mathcal{O}_L(1)^{\oplus(k-1)} \oplus \mathcal{O}_L(-1)^{\oplus(k+1)}.$$

In particular, the complex structure of X is A-good, and may be used to compute the invariant directly. To get a nonzero result we need to use a 2-point invariant, that is, we set $p = 2$ in [MS], Formula 7.1 on p. 93, and get $d = 4k + 4 = 2(2k + 2)$. Now let $H_i, H_i' \subset X$ for $i = 1, \ldots, k$ be general hyperplane sections of X, to fix ideas, in the same very ample linear series on X, and set $V = H_1 \cap \cdots \cap H_k$ and $V' = H'_1 \cap \cdots \cap H'_k$. Then $[V] = [V'] \in H_{2k+2}(X, \mathbb{Z})$, and $V \cap E$, $V' \cap E$ are two reduced finite subschemes, consisting of points $\{p_i\}$ and $\{p'_j\}$, say. The invariant $\Phi_A([V], [V])$ then counts the number of lines in E joining some p_i with some p'_j, and this is a positive number.

In all other cases we may write $n = 2k - h$, with $0 \leq h < k$. According to [MS], Formula 7.1, for $p = 2$ we have

$$d = 2(n-1) + 4 - 2c_1(A) = 2\left(k - h + \sum_{i=1}^{k-h} a_i\right).$$

Let $\Gamma \subset E$ be a linear subspace of dimension $c = n - k = k - h$. If $\Sigma \subset X$ is a general representative for the homology class $D \in H_{2(k-h)}(X, \mathbb{Z})$ of Γ, then Σ meets E in a finite number of points $\{p_i\}$, which counted with multiplicity is $(-1)^{k-h} \prod_{i=1}^{k-h} a_i$. Set $d_1 = 2(k-h)$ and $d_2 = d - d_1 = 2\sum_{i=1}^{k-h} a_i$. We need to produce a class of dimension d_2, that is, of codimension $2n - d_2 = 2\left(k - \sum_{i=1}^{k-h}(a_i - 1)\right)$. Let $s = k - \sum_{i=1}^{k-h}(a_i - 1)$, and $V = H_1 \cap \cdots \cap H_s$, where the H_i are general very ample divisors in X. Then $V \cap E$ is a smooth complete intersection subvariety of $E \cong \mathbb{P}^k$ of dimension $\sum_{i=1}^{k-h}(a_i - 1)$. If some $a_i > 1$, the structure is not good, and as above, we need to consider the zero locus of a transverse section s of the obstruction bundle. If $Z = Z(s)$ is such a zero locus, and $\widetilde{Z}$ the family of all parametrizations of lines in Z, there is a pseudo-cycle ([MS], Chap. 7) given by the 2-point evaluation map:

$$\mathrm{ev}_2 \colon \widetilde{Z} \times_{\mathbb{G}} (\mathbb{P}^1 \times \mathbb{P}^1) \to X \times X,$$

where $\mathbb{G} = \mathrm{PGL}(2, \mathbb{C})$. The Gromov–Witten invariant $\Phi_A([V], [D])$ may then be computed by taking suitable intersection numbers with this pseudo-cycle ([Ru], Proposition 5.7); in particular, Z may be replaced by any subvariety representing the same homology class. On the other hand, by changing the orientation of Z we change at most the sign of these intersection numbers, and therefore by identifying $\mathcal{E}_{\mathrm{ob}}$ (as a real bundle) with its dual by some choice of a Hermitian form, we may replace Z by the zero locus of a transversal section of $\mathcal{E}_{\mathrm{ob}}^*$. Then Z represents the top Chern class of $\mathcal{E}_{\mathrm{ob}}^*$, which decomposes as above as a sum of terms $c_1(H)^{r-k} c_k\left(\bigoplus_{i=1}^{n-k} \mathrm{Sym}^{a_i-2}(\mathcal{S}^*)\right)$. The relevant intersection numbers do not change if we replace Z by a sum of pseudo-cycles Z_k, with each Z_k representing one such term. Each Z_k may be represented as an intersection of Schubert cycles, and upon choosing the underlying flag sufficiently general we may assume that all relevant intersections are transversal, and then that the contribution of each Z_k to the total intersection number is a nonnegative multiple of $(-1)^{k-h} \prod_{i=1}^{k-h} a_i$. Hence to prove that the total invariant is nonzero it is sufficient to show that the contribution due to Z_0 does not vanish. Now

Z_0 represents $c_1(A)^r$, where $r = \sum_{i=1}^{k-h}(a_j - 1)$. Define $W = \mathbb{C}^{k+1}$ so that $E = \mathbb{P}(W)$, and let

$$W_*^{(i)} = \{W_1^{(i)} \subset W_2^{(i)} \subset \cdots \subset W_{k+1}^{(i)} = W\}$$

be general complete flags for $i = 1, \ldots, \sum_{i=1}^{k-h}(a_i - 1)$. We may then set

$$Z_0 = \sigma_1(W_*^{(1)}) \cap \cdots \cap \sigma_1(W_*^{(r)}),$$

where for each i we have

$$\sigma_1(W_*^{(i)}) = \{L \in \mathrm{Grass}(2, W) \mid L \cap \mathbb{P}(W_{k-1}^{(i)}) \neq \emptyset\}.$$

For every given point $p \in E$, the lines through p lying in the family Z_0 sweep out a linear subspace $\Pi \subset E$ of dimension $k - \sum_{i=1}^{k-h}(a_j - 1)$. Therefore Π meets $V \cap E$ in a positive number of points $\{q_j^{(i)}\}$. The contribution of Z_0 toward the invariant $\Phi_A([V], [D])$ is then $(-1)^{k-h} \prod_{i=1}^{k-h} a_i$ times the number of lines joining p_i with some $q_j^{(i)}$, and is therefore nonzero.   Q.E.D.

3   Proof of Theorem 1.3

Let $F \cong \mathbb{P}^1 \subset X$ be any fibre of E over its image $Z = \varphi(E) \subset \overline{X}$. Then we have the short exact sequence of normal bundles

$$0 \to \mathcal{O}_F^{n-2} \to N_{F/X} \to \mathcal{O}_F(E) \to 0;$$

on the other hand, letting $k = -K_X \cdot F > 0$, we have $\deg(N_{F/X}) = k - 2 \geq -1$. Hence, because $-E$ is φ-ample ([KMM], Cor. 0-3-5), we have $E \cdot F = -1$ and therefore the above sequence is split: $N_{F/X} \cong \mathcal{O}_F^{n-2} \cong \mathcal{O}_F(-1)$. In particular, $c_1(F) = 1$ and the holomorphic structure of X is A-good, where A is the homology class of F, and furthermore the moduli space of holomorphic curves in the class A is scheme theoretically isomorphic to the base manifold $\varphi(E)$. [MS], Formula 7.1 with $p = 2$ now gives $d = 2(n - 1) + 2$. Let $H \subset X$ be an ample divisor on X; then $\Phi_A([F], [H])$ computes the number of triples (u, z_1, z_2), where $u \colon \mathbb{P}^1 \to X$ is J-holomorphic and $z_1, z_2 \in \mathbb{P}^1$, $u(z_1) \in E$ and $u(z_2) \in H$, taken with multiplicity and modulo the action of $\mathrm{PGL}(2, \mathbb{C})$. These add up to the product of the intersection numbers, that is, $(F \cdot E)(F \cdot H) \neq 0$.   Q.E.D.

4   Proof of Theorem 1.4

Suppose that the crepant extremal ray $\ell \subset N_1(X)$ and its primitive contraction $\varphi \colon X \to \overline{X}$ are as in the statement of the theorem. Suppose first that the exceptional locus E is a $\mathbb{P}^1$-bundle over its image $Z = \varphi(E) \subset \overline{X}$, and let $A \in H_2(X, \mathbb{Z})$ denote the homology class of a fibre. If J_0 is the complex structure of X, then the moduli space $\mathcal{M}(A, J_0)$ of unparametrized holomorphic curves representing the class $[A]$ (that is, the component of the Hilbert scheme of X containing the point associated to a fibre of $E \to Z$) is isomorphic to Z. If, on the other hand, E is a conic bundle over Z all of whose fibres are reduced line pairs, then the normalization $\widetilde{E} \to E$ is a $\mathbb{P}^1$-bundle over an etale double cover $\widetilde{Z} \to Z$. In this case, let $A \in H_2(X, \mathbb{Z})$ be the homology class of the irreducible component of any fibre; then the moduli space $\mathcal{M}(A, J_0)$ is isomorphic to $\widetilde{Z}$. This is clear set theoretically, and from a scheme theoretic point of view it follows from the following:

Lemma 4.1 *Let $F \cong \mathbb{P}^1 \subset E$ be an irreducible component of any fibre of $\varphi_{|E}$. Then the normal bundle of F in X is $N_{F/X} \cong \mathcal{O}_F^{\oplus(n-2)} \oplus \mathcal{O}_F(-2)$.*

Proof Suppose first that E is a $\mathbb{P}^1$-bundle over $Z = \varphi(E)$, with Z smooth. Then we have the short exact sequence $0 \to N_{F/E} \to N_{F/X} \to N_{E/X}|_F \to 0$. Furthermore, $N_{F/E} \cong \mathcal{O}_F^{\oplus(n-2)}$ and $N_{E/X}|_F \cong \mathcal{O}_F(E) \cong \omega_F$; hence the above exact sequence is $0 \to \mathcal{O}_F^{\oplus(n-2)} \to N_{F/X} \to \mathcal{O}_F(-2) \to 0$, which is split. The argument in the case where all the fibres of E are reduced line pairs is similar.   Q.E.D.

This shows that $h^0(N_{F/X}) = n - 2$ for all irreducible components F of a fibre of $\varphi_{|E}$, and therefore that $\mathcal{M}(A, J_0)$ is indeed isomorphic to the smooth variety Z (or $\widetilde{Z}$) with its reduced structure. Thus $\mathcal{M}(A, J_0)$ does not have the expected dimension, and in fact the holomorphic structure of X is not A-good in the sense of [MS]. However, as in the proof of Theorem 1.2, we can rely on [Ru], Proposition 5.7 to relate the moduli space $\mathcal{M}(A, J)$ for generic J to the obstruction bundle on $\mathcal{M}(A, J_0)$. This is because on the one hand each fibre F maps immersively into X, and on the other, since $\mathrm{coker}(D_u) \cong H^1(F, N_{F/X})$, we have

$$2c_1(A) + 2n - 6 + \dim \mathrm{coker}(D_u) = 2n - 6 + 2 = 2(n-2) = \dim_{\mathbb{R}} \mathcal{M}(A, J_0),$$

so that the assumptions of [Ru], Proposition 5.7 are satisfied. Therefore, for a generic nearby almost complex structure J on X, it follows that $\mathcal{M}(A, J)$ is oriented cobordant to the zero set of a transverse section of the obstruction bundle $\mathcal{E}$. So as above our next step is to identify the obstruction bundle on Z (or $\widetilde{Z}$); this is a rank 2 real vector bundle over $\widetilde{Z}$.

Lemma 4.2 *Let $\widetilde{Z} = Z$ when E is a $\mathbb{P}^1$-bundle over Z, and let $\widetilde{Z}$ be the etale double cover of Z which is the base variety of $\widetilde{E}$ when the fibres of E*

are all line pairs; let $f\colon \widetilde{E} \to X$ be the induced immersive morphism. Note that $f^*\omega_X$ naturally descends to a holomorphic line bundle on $\widetilde{Z}$, that we still denote by ω_X. The obstruction bundle on $\widetilde{Z}$ is then given by $\mathcal{E}_{\mathrm{ob}} = \omega_{\widetilde{Z}} \otimes \omega_X^{-1}$.

Proof Let $T_{\widetilde{E}/\widetilde{Z}} \subset T_{\widetilde{E}}$ denote the subbundle of vertical tangent vectors, that is, $T_{\widetilde{E}/\widetilde{Z}} = \ker\{d\pi\colon T_{\widetilde{E}} \to \pi^*T_{\widetilde{Z}}\}$, where $\pi\colon \widetilde{E} \to \widetilde{Z}$ denotes the projection. For every fibre $F \subset \widetilde{E}$, clearly $T_{\widetilde{E}/\widetilde{Z}}|_F \cong T_F$. Let also $\widetilde{N}$ be the rank $n-1$ bundle on $\widetilde{Z}$ defined by the exact sequence $0 \to T_{\widetilde{E}/\widetilde{Z}} \to f^*T_X \to \widetilde{N} \to 0$, so that for each F there is an isomorphism $\widetilde{N}|_F \cong N_{F/X}$. The obstruction bundle on $\widetilde{Z}$ is then $R^1\pi_*\widetilde{N}$. On the other hand, it is easily seen that there is an exact sequence $0 \to \pi^*T_Z \to \widetilde{N} \to \omega_E \otimes \omega_X^{-1} \to 0$, from which it follows that $R^1\pi_*\widetilde{N} = \omega_{\widetilde{Z}} \otimes \omega_X^{-1}$. Q.E.D.

Thus if the first Chern class of the obstruction bundle is not zero, Ruan's result shows that the class A is represented by some J-holomorphic curve for a generic almost complex structure J on X which is sufficiently close to J_0. To complete the proof of (1), we show that a Gromov–Witten invariant associated to the ray ℓ is not zero. Since by assumption $c_1(\omega_Z) \neq c_1(\omega_X|_Z)$, there exists some irreducible curve $C \subset Z$ such that $c_1(\omega_Z) \cdot C \neq c_1(\omega_X|_Z) \cdot C$. Now let L be some sufficiently positive line bundle on Z such that both $\omega_Z \otimes L$ and $\omega_X|_Z \otimes L$ are very ample, and pick smooth divisors

$$V_1 \in |\omega_Z \otimes L| \quad \text{and} \quad V_2 \in |\omega_X|_Z \otimes L|;$$

moreover, let $\widetilde{V}_1$ and $\widetilde{V}_2 \subset \widetilde{\mathcal{M}}(A, J_0)$ be the family of all parametrizations of J_0-holomorphic curves contained in V_1 and V_2, respectively, and consider the surface $\widetilde{C} = \varphi^{-1}(C) \subset E$, whose homology class we denote by $D \in H_4(X, \mathbb{Z})$. Since $-E$ is φ-ample ([KMM], Cor. 0-3-5), a general representative $F \subset X$ of the homology class of the irreducible component of a fibre of E over Z meets E in a finite number of points, with multiplicities yielding a negative intersection number. If on the other hand we take a general representative of the homology class of $\widetilde{C}$, say $\Sigma \subset X$, then Σ meets E along a smooth subvariety C' of real dimension two, whose image in Z is homologous to a multiple of C, say $\varphi_*([C']) = a[C]$ for some $a \neq 0$. Consider the 2-point evaluation maps

$$\mathrm{ev}_2^{(i)}\colon \widetilde{V}_i \times_{\mathrm{PGL}(2)} (\mathbb{P}^1)^2 \to X \times X \quad \text{for } i = 1, 2,$$

each of which determines a pseudo-cycle in $X \times X$ ([MS], Chap. 7). Given that [MS], Formula 7.1 for $p = 2$ yields $d = 2(n-1) + 4$, we may consider the Gromov–Witten invariant $\Phi_A(D, [H])$, where A is the homology class of a fibre of E over Z, and $[H]$ denotes the class of some hyperplane section of X; our discussion shows that this invariant is determined by taking the difference of the intersection numbers related to the above pseudo-cycles. If $\overline{V}_i = \varphi^{-1}(V_i)$, then the image of $\mathrm{ev}_2^{(i)}$ is the fibre product $\overline{V}_i \times_Z \overline{V}_i$. The points of intersection between the pseudo-cycle $\mathrm{ev}_2^{(i)}$ and $\Sigma \times H$ are the pairs (z_1, z_2) such that $\varphi(z_1) = \varphi(z_2) \in V_i$ and $(z_1, z_2) \in C' \times H$. In particular, for a given $z_1 \in \overline{V}_i \cap C'$, the second component z_2 may range over the points of intersection of H with the fibre through z_2; furthermore, the intersection multiplicity at the pair (z_1, z_2) is given by the intersection multiplicity of Σ and $\overline{V}_i$, and this is the same as the intersection multiplicity (in E) of the Cartier divisor $\widetilde{V}_i$ and C'. Summing up, the value of the invariant is

$$\Phi_A(D, [H]) = a(F \cdot H)\Big(\big(c_1(\omega_Z) - c_1(\omega_{X|Z})\big) \cdot C\Big),$$

which is nonvanishing by construction.

To prove (2), one may argue as in [Wl 1], proof of Proposition 4.4 and erratum. For simplicity, let us consider the case where E is a $\mathbb{P}^1$-bundle and the first obstruction to the deformation of E in a given 1-parameter family of deformations of X, say $\mathcal{X} \to \Delta$, occurs at first order. Then the Kodaira–Spencer class $\theta \in H^1(X, T_X)$ of the given family has a nonzero image in $H^1(E, N_{E/X}) = H^1(X, \omega_E \otimes \omega_X^{-1})$. Let $\mathbb{P}^1 \cong F \subset E$ denote an arbitrary fibre of $\varphi_{|E}$; the claim will be proved if we can show that θ does not map to zero in $H^1(F, N_{F/X})$. We have however already seen that $N_{F/X} \cong N_{F/E} \oplus \mathcal{O}_F(E)$, so that $H^1(F, N_{F/X}) \cong H^1(F, N_{E/X|F})$. Hence, the map $H^1(X, T_X) \to H^1(N_{F/X})$ is in fact the composite $H^1(X, T_X) \to H^1(E, N_{E/X}) \to H^1(F, N_{E/X|F})$. Given the hypothesis, then, it is sufficient to show that $H^1(E, N_{E/X}) \to H^1(F, N_{E/X|F})$ is an isomorphism. But $H^1(F, N_{E/X|F}) \cong \mathcal{E}_{\mathrm{ob}}(x)$, where $x = \varphi(F) \in Z$ and $\mathcal{E}_{\mathrm{ob}} \cong \omega_Z \otimes \omega_X^{-1}$ is the obstruction bundle, while on the other hand, by relative duality $H^1(E, N_{E/X}) \cong H^1(E, \omega_E \otimes \omega_X^{-1}) \cong H^0(Z, \omega_Z \otimes \omega_X^{-1})$, and the map

$$H^1(E, N_{E/X}) \cong H^0(Z, \mathcal{E}_{\mathrm{ob}}) \to H^1(F, N_{E/X|F}) \cong \mathcal{E}_{\mathrm{ob}}(x)$$

is section evaluation. If φ is critical, that is, $\omega_Z \otimes \omega_X^{-1}$ is trivial, this is an isomorphism. Q.E.D.

5 Proof of Theorem 1.5

Any 1-parameter family of deformations $\mathcal{X} \to \Delta$ of $X = X_0$ may be assumed to be projective, perhaps after restricting to a smaller open disc. In fact, by [KS], Theorem 15, after restricting Δ to a smaller open disc we may assume that every fibre X_t is Kähler. The condition that $H^{0,2}(X) = 0$ implies $H^2(X_t, \mathbb{R}) = H^{1,1}_{\mathbb{R}}(X_t)$ for all t; therefore an integral class of type $(1,1)$ remains of the same type on X_t, for $t \in \Delta$. If H is a very ample line bundle on X with $H^i(X, H) = 0$ for $i > 0$, by taking a local section in a suitable Jacobian bundle we may extend H to a family of line bundles $\mathcal{H}$ over $\mathcal{X}$. By openness of the ample condition, H_t is very ample on X_t for sufficiently small t, with a linear system of the same dimension.

Let ℓ be a crepant extremal ray of X and $\varphi\colon X \to \overline{X}$ its contraction. If $E \subset X$ is the exceptional locus of φ, by [K], Theorem 1, E is covered by rational curves. We then write $\ell = \mathbb{R}_{\geq 0}[C]$, where $C \subset X$ is an unbreakable rational curve (see e.g., [BS] for terminology). If furthermore $\overline{A}$ is an ample line bundle on $\overline{X}$, then $A = \varphi^*(\overline{A})$ is a nef and big line bundle on X supporting φ; since $\overline{X}$ has only canonical, hence rational, singularities, we may assume that $H^i(X, A) = 0$ for all $i > 0$. Then by the above argument A deforms to a family of line bundles $\mathcal{A}$ on $\mathcal{X}$. Restricting the base of the deformation if necessary we then see that the linear series $|A_t|$ may be assumed to have the same dimension as $|A|$ for all $t \in \Delta$. Hence there is a family of morphisms $\Phi\colon \mathcal{X} \to \overline{\mathcal{X}}$ over Δ, which for $t = 0$ yields $\varphi\colon X \to \overline{X}$. By openness of the ample condition, the ray ℓ is not effective on X_t for $t \neq 0$ if and only if φ_t is an isomorphism for $t \neq 0$.

Since a 1-parameter family of deformations $\mathcal{X} \to \Delta$ of $X = X_0$ is necessarily differentiably locally trivial, perhaps after restriction, we may assume that $H^j(\mathcal{X}, \mathbb{Z}) \cong H^j(X, \mathbb{Z})$ and $H_j(\mathcal{X}, \mathbb{Z}) \cong H_j(X, \mathbb{Z})$. Given that $H^2(X, \mathcal{O}_X) = 0$, furthermore, we have $N^1(X) \cong H^2(X, \mathbb{R})$ and so, dually, $N_1(X) \cong H_2(X, \mathbb{R})$. Hence, there are isomorphisms $N_1(X) \cong N_1(\mathcal{X})$ and $N^1(\mathcal{X}) \cong N^1(X)$; the former induces an inclusion $\mathrm{NE}(X) \subseteq \mathrm{NE}(\mathcal{X}/\Delta)$. In this situation, the *relative* crepant effective cone $\mathrm{NE}_{\mathrm{cr}}(\mathcal{X}/\Delta) \subset \mathrm{NE}(\mathcal{X}/\Delta)$ is the subcone generated by those relative crepant rays ℓ for which there exists a crepant relative contraction $\varphi\colon \mathcal{X} \to \overline{\mathcal{X}}$ over Δ, such that $\mathrm{NE}(\mathcal{X}/\overline{\mathcal{X}}) = \mathbb{R}_{\geq 0}[\ell]$. Clearly, the above shows that $\mathrm{NE}_{\mathrm{cr}}(\mathcal{X}/\Delta) \supseteq \mathrm{NE}_{\mathrm{cr}}(X)$.

Lemma 5.1 *Let X be a smooth projective manifold and $\pi\colon \mathcal{X} \to \Delta$ a one parameter family of deformations of X. If $\pm K_X$ is nef and big then, perhaps after restricting the base to a smaller open disc containing the origin, $\mathrm{NE}_{\mathrm{cr}}(\mathcal{X}/\Delta) \subseteq \mathrm{NE}_{\mathrm{cr}}(X)$. If K_X is trivial, the same conclusion holds over the complement of at most countably many points in $\Delta \setminus \{0\}$.*

Thus the crepant effective cone of X is generically constant in the Kuranishi family if and only if each of the crepant extremal rays of X remains effective on the generic small deformation of X; in fact "generic" may be omitted when $\pm K_X$ is nef and big, while it stands for "on the complement of a countable union of subvarieties of $B \setminus \{0\}$" if K_X is trivial.

Proof Suppose that $\pm K_X$ is nef and big. After replacing B by a smaller open neighbourhood of the origin, we may assume that $\pm K_{X_b}$ is semiample and big, and $N_1(X_b) \cong H_2(X, \mathbb{R})$ for all $b \in B$. Fix an ample divisor $H \subset X$. Since $\pm K_X$ is nef and big, if $n \gg 0$ there exists an effective divisor $\Delta \subset X$ such that $\pm n K_X \equiv H + \Delta$. Thus $\ell \cdot \Delta < 0$ for any crepant extremal ray. After replacing Δ by a suitably small positive rational multiple, (X, Δ) has log terminal singularities ([KMM], §0). By [KMM], §4, the effective cone $\mathrm{NE}(X)$ is locally finite rational polyhedral in $\Sigma = \{\ell : \ell \cdot (K_X + \Delta) < 0\}$, and therefore finite rational polyhedral in a suitable neighbourhood of the hyperplane Λ of $H_2(X, \mathbb{R})$ defined by $[K_X]$. Thus, it is bounded in this neighbourhood by rational faces $F_1, \ldots, F_k$ of codimension one in $H_2(X, \mathbb{R})$. Each F_i is cut out on $\mathrm{NE}(X)$ by some nef divisor A_i; let $\Gamma_i = \{A_i \cdot z \geq 0\}$. Then $\mathrm{NE}_{\mathrm{cr}}(X) = \Lambda \cap \bigcap_i \Gamma_i$. Set $\widetilde{A}_i := A_i \pm r K_X$ for some $r \gg 0$; then $\widetilde{A}_i$ is big and semiample, and if $\widetilde{\Gamma}_i = \{\widetilde{A}_i \cdot z \geq 0\}$, then $\mathrm{NE}_{\mathrm{cr}}(X) = \Lambda \cap \bigcap_i \widetilde{\Gamma}_i$. For every i, a suitably large positive multiple of $\widetilde{A}_i$ extends to a family of relatively globally generated line bundles on the family. Hence for each i there is a relative crepant contraction $\varphi_i \colon \mathcal{X} \to \mathcal{X}_i$ over Δ extending φ. Therefore

$$\mathrm{NE}_{\mathrm{cr}}(\mathcal{X}/\Delta) \subseteq \bigcap_{i=1}^{k} H_i \cap \Lambda.$$

In particular, $\mathrm{NE}(X_b) \subset \widetilde{\Gamma}_i$ for all $b \in B$ sufficiently close to the origin, therefore $\mathrm{NE}_{\mathrm{cr}}(X_b) \subseteq \mathrm{NE}_{\mathrm{cr}}(X)$. The crepant effective cone is not locally constant in the family if and only if one of the edges (crepant extremal rays) $R_1, \ldots, R_l$ of $\mathrm{NE}_{\mathrm{cr}}(X)$ is not effective on X_b when $b \in B$ is general.

Suppose next that K_X is trivial. The statement then simply follows from the fact that if a class $\ell \in H_2(X)$ is not effective, then the locus in Δ over which it is effective is at most countable. Q.E.D.

We need the following slight variant of [Ws 1], Theorem 1.2:

Proposition 5.2 *Let X be a projective n-fold for which $\pm K_X$ is nef and big or trivial, $\ell \subset N_1(X)$ a crepant extremal ray of X and $\varphi \colon X \to \overline{X}$ its crepant primitive birational contraction. Let $\mathcal{X} \to \Delta$ be a 1-parameter family of holomorphic deformations of X, and $\Phi \colon \mathcal{X} \to \overline{\mathcal{X}}$ a crepant birational contraction over Δ extending φ. Let $E \subset \mathcal{X}$ be the exceptional locus of Φ, and $E_1 \subseteq E$ an irreducible component. Suppose that $F \subseteq E_1$ is an irreducible component of a fibre of the restriction $\varphi_{|E_1}$. Then*

$$\dim E_1 + \dim F \geq n. \tag{1}$$

The proof is identical to the proof given by Ionescu [Io] and Wisniewski [Ws 1] of the corresponding inequality

$$\dim E_1 + \dim F \geq n + \mathrm{length}\,\ell.$$

in Mori theory (here $n = \dim \mathcal{X} - 1$). Since however we need to tell when equality holds, we recall briefly the argument of [Ws 1], p. 144. Pick a general point $x \in F$ and let $F \supset C_0 \ni x$ be a rational curve minimal among the rational curves contained in F and passing through x, in the sense that it minimizes the intersection number with some fixed ample line bundle on X. Now let T be an irreducible variety parametrizing the embedded deformations of $C_0 \subset \mathcal{X}$, as constructed in [Io], and $V \subset \mathcal{X} \times T$ the incidence correspondence, with projections $p\colon V \to \mathcal{X}$ and $q\colon V \to T$. The standard Riemann–Roch estimate yields $\dim T \geq n - 2$ and therefore $\dim V \geq n - 1$. On the other hand, $\dim p^{-1}(x) \leq \dim F - 1$: as in [Ws 1], claim on p. 144, this follows from Mori's breaking lemma and the minimality assumption on C. Therefore, $\dim E_1 \geq \dim p(V) \geq \dim V - \dim p^{-1}(x) \geq n - \dim F$, and this is the claimed inequality. We see furthermore that equality holds in (1) if and only if $p(V) = E_1$ and the general fibre of $p\colon V \to X$ has dimension $\dim F - 1$.

We next recall [Ws 2], Key Lemma 1.1, which in the present context we may rephrase as follows:

Lemma 5.3 *Let X be a smooth projective variety for which $\pm K_X$ is nef and big or trivial and $h^2(X, \mathcal{O}_X) = 0$. Let ℓ be a jumping crepant extremal ray of X and $\varphi\colon X \to \overline{X}$ its crepant birational contraction, and let $E \subset X$ be any irreducible component of the exceptional locus of φ. Then*

$$2\dim E - \dim \varphi(E) \leq n.$$

By Proposition 5.2 and Lemma 5.3, we have:

Corollary 5.4 *Let X be a smooth projective n-fold for which $\pm K_X$ is nef and big or trivial and $h^2(X, \mathcal{O}_X) = 0$. Let ℓ be a jumping extremal ray of X and $\varphi\colon X \to \overline{X}$ its crepant birational contraction. If $E \subset X$ is any irreducible component of the exceptional locus of φ, then $2\dim E - \dim \varphi(E) = n$. In particular, if $n = 4$ then the exceptional locus of φ is either a union of surfaces mapping to a finite set, or a 3-fold mapping to a surface.*

Lemma 5.5 *Let X be a smooth projective 4-fold for which $\pm K_X$ is nef and big or trivial and $h^2(X, \mathcal{O}_X) = 0$. Let ℓ be a jumping ray on X, and assume that the associated crepant contraction $\varphi\colon X \to \overline{X}$ is small. Then the normalization of any irreducible component of the exceptional locus of φ is isomorphic to $\mathbb{P}^2$.*

Proof Let $F \subset E$ be any irreducible component and $x \in F$ a general point, lying in the smooth locus of F_{red}. Pick a rational curve $F \supset C_0 \ni x$ of minimal degree with respect to some fixed ample line bundle among curves in

F through x (see [K]). Since equality must hold in (1), in the notation used in the discussion following Proposition 5.2 we must have $\dim p^{-1}(x) = 1$, so that $\dim V = 3$. Setting $\delta_y = \dim p^{-1}(y)$ for $y \in F$, it thus follows that $\delta_y \geq 1$ for any $y \in F$. On the other hand, suppose that $\delta_y \geq 2$ for some $y \in F$; then $\dim V_y \geq 3$, and therefore every curve in the family T through x also passes through y, and by the nonbreaking lemma, this contradicts the assumption that C_0 is minimal. Thus $\delta_y = 1$ for every $y \in F$.

(I am indebted to the referee for the concise argument used in the rest of the proof.) Again referring to the notation used after Proposition 5.2, in this case $\dim T = 2$ and no point of F lies on all curves of the family. Let F', T', V' be the normalizations of F, T and V respectively, with the induced morphisms $q' : V' \to T'$ and $p' : V' \to F'$. If $F'' \to F'$ is a minimal desingularization of F, we have $H^i(F, \mathcal{O}_{F''}) = 0$ for $i = 1, 2$, so that algebraic and rational equivalence of divisors coincide on F''. Let $C' \subset F'$ be a curve from T' disjoint from the singular locus of F', and $C'' \subset F''$ its inverse image. By the above observation, the linear system $|C''|$ contains all deformations of $C' \subset F'$ disjoint from F'_{sing}; thus the map $|C''| \to T'$, which is birational, is an isomorphism, and therefore $|C''| \cong \mathbb{P}^2 \cong T'$. This linear system is base point free, and therefore its general representative is a smooth rational curve. By Riemann–Roch, since $h^i(F'', \mathcal{O}_{F''}(C'')) = 0$ for $i > 0$, $(C'', C'')_{F''} = 1$. Thus two general curves in the family T' are $\cong \mathbb{P}^1$ and meet at exactly one point. Now let $T'_x \subset T'$ be the subfamily parametrizing those curves through x; then $T'_x \cong |\mathcal{I}_x(C'')| \cong \mathbb{P}^1$. By minimality, the incidence correspondence $V' \supset V'_x \to T'_x$ is a $\mathbb{P}^1$-bundle. By the above, the induced map $V'_x \to F'$ is birational and contracts a section to x; since however x is a smooth point of F', we conclude $V'_x \cong \mathbb{F}_1$ and $F' \cong \mathbb{P}^2$.   Q.E.D.

The proof of the theorem is then completed by the following:

Proposition 5.6 *Let X be a 4-fold for which $\pm K_X$ is nef and big or trivial and $h^2(X, \mathcal{O}_X) = 0$. Suppose that ℓ is a crepant extremal ray whose contraction $\varphi \colon X \to \overline{X}$ is divisorial. If φ is not as in the critical cases (a) and (b) of Theorem 1.5, ℓ is not a jumping ray (that is, it remains effective on all small deformations of X).*

Proof Let $E \subset X$ denote the exceptional divisor of φ, which is an irreducible 3-fold, and set $S = \varphi(E) \subset \overline{X}$. Then S is a surface of canonical singularities of $\overline{X}$. Given a 1-parameter deformation $\alpha \colon \mathcal{X} \to \Delta$ of X, we prove that if φ is not critical then ℓ is effective on X_t, for all t close to the origin. Let $\overline{A} \in \mathrm{Pic}(\overline{X})$ be a sufficiently positive very ample line bundle such that $H^i(\overline{X}, \overline{A}) = 0$ for all $i > 0$, and set $A = \varphi^*(\overline{A})$. Fix a general section $\overline{\sigma} \in H^0(\overline{X}, \overline{A})$, and set $\sigma = \varphi^*\overline{\sigma} \in H^0(X, A)$. Set $\overline{Y} = \mathrm{div}(\overline{\sigma})$, $Y = \varphi^*\overline{Y}$. Then Y is a smooth 3-fold, with nef and big canonical divisor

$K_Y = (K_X + A)_{|Y}$, and the restriction $\varphi_{|Y}$ is a crepant birational contraction, whose exceptional locus is the surface $Z = E \cap Y$ and is contracted to the curve $C = S \cap \overline{Y}$. The ray ℓ is then the image of an effective ray ℓ' on Y, which is represented by the irreducible components of the fibres of the conic bundle $Z \to C$. Arguing as at the start of the proof of Theorem 1.5, perhaps after restricting the base Δ, we may extend A to a family of globally generated line bundles $\mathcal{A}$ on $\mathcal{X}$, such that $\dim |A_t|$ is constant. By taking a suitable local section of $\alpha_*(\mathcal{A})$, and the corresponding divisor, we obtain a smooth family of 3-folds $\mathcal{Y} \subset \mathcal{X}$ with $Y_0 = Y$. Therefore, in order to prove the statement it suffices to note that, by the results of [P 2], if φ is not critical then the ray ℓ' remains effective on all sufficiently small deformations of Y. Q.E.D.

References

[A]   T. Ando, *On extremal rays of the higher dimensional varieties*, Invent. Math. **81** (1985), 347-357

[BS]   M. Beltrametti, A. Sommese, *The adjunction theory of complex projective varieties*, De Gruyter Expositions in Mathematics **16**, Walter De Gruyter & co. Berlin 1995

[F]   W. Fulton, *Intersection theory*, Springer Verlag 1984

[Io]   P. Ionescu, *Generalized adjunction and applications*, Math. Proc. Camb. Phil. Soc. **99** (1986), 457–472

[K]   Y. Kawamata, *On the length of an extremal rational curve*, Invent. Math. **105** (1991), 609–611

[KMM]   Y. Kawamata, K. Matsuda, K. Matsuki, *Introduction to the minimal model program*, in Algebraic geometry (Sendai, 1985), Adv. Studies in Pure Math. **10**, Kinokuniya, 1987, pp. 283–360

[KS]   K. Kodaira, D. Spencer, *On deformations of complex analytic structures III. Stability theorems for complex analytic structures*, Ann. of Math. **71** (1960), 43–76

[L]   R. Lazarsfeld, *Some applications of the theory of positive vector bundles*, in S. Greco and R. Strano (Eds), *Complete intersections* (Acireale 1983), Lecture Notes in Mathematics **1092**, Springer Verlag 1984

[MS]   D. McDuff, D. Salamon, *J-holomorphic curves and quantum coho-mology*, Univ. Lect. Series **6**, Amer. Math. Soc. 1994

[P 1]   R. Paoletti, *The Kähler cone in families of quasi-Fano 3-folds*, Math. Zeit. **227** (1998), 45-68

[P 2]   R. Paoletti, *The crepant effective cone in families of 3-folds with big and nef canonical divisor*, preprint 1996

[Ran]   Z. Ran, *Deformations of manifolds with torsion or negative canonical bundle*, J. Alg. Geom. **1** (1992), 279–291

[Rd]   M. Reid, *Canonical 3-folds*, in A. Beauville, (Ed.), Géométrie algébrique (Angers 1979), Sijthoff and Noordhoff, Alphen 1980, pp. 273–310

[Ru]   Y. Ruan, *Symplectic topology and extremal rays*, Geom. and Funct. Analysis **3** (1994), 395–430

[Wl 1]   P.M.H. Wilson, *The Kähler cone on Calabi–Yau threefolds*, Invent. Math. **107** (1992), 561–583; *erratum*, Invent. Math. **114** (1993), 231–233

[Wl 2]   P.M.H. Wilson, *Symplectic deformations of Calabi–Yau threefolds*, J. Diff. Geom. **45** (1997), 611–637

[Wl 3]   P.M.H. Wilson, *Flops, Type III contractions and Gromov–Witten invariants on Calabi–Yau threefolds*, this volume, pp. 465–484

[Ws]   J. Wisniewski, *Length of extremal rays and generalized adjunction*, Math. Zeit. **200** (1989), 409–427

[Ws 1]   J. Wisniewski, *On contractions of extremal rays of Fano manifolds*, J. reine angew. Math. **417** (1991), 141–157

[Ws 2]   J. Wisniewski, *On deformations of nef values*, Duke Math. J. **64** (1991), 325–332

Roberto Paoletti,
Dipartimento di Matematica, Universitá di Pavia,
Via Abbiategrasso 215,
I-27100 Pavia, Italia
e-mail: rp@dragon.ian.pv.cnr.it

The Bogomolov–Pantev resolution, an expository account

Kapil H. Paranjape[*]

Statement of the result

Bogomolov and Pantev [3] have recently discovered a rather elegant geometric proof of the weak Hironaka theorem on resolution of singularities:

Theorem 0.1 *Let X be a projective variety and Z a proper Zariski closed subset of X. There is a projective birational map $\varepsilon\colon \widetilde{X} \to X$ such that $\widetilde{X}$ is smooth and the set theoretic inverse image $\varepsilon^{-1}(Z)$ is a divisor with simple normal crossings.*

Before their work, and that of Abramovich and de Jong [1] (appearing at roughly the same time) the only proof of this theorem was as a corollary of the famous result of Hironaka [5]. These new proofs were inspired by the recent work of de Jong [6], which Bogomolov and Pantev combine with a beautiful idea of Belyi [2] "simplifying" the ramification locus of a covering of $\mathbb{P}^1$ by successively folding up the $\mathbb{P}^1$ onto itself, over a fixed base. This latter step unfortunately only works in characteristic zero, limiting the scope of the argument (Abramovich and de Jong's paper gives some results even in characteristic p). Hence, we work over the field of complex numbers; the argument also works (with suitable modifications about rationality) over any field of characteristic zero.

The outline of the argument we follow is the same as that of the paper of Bogomolov and Pantev; however we offer different (and we hope simpler) proofs of the corresponding lemmas. To begin with, their argument using Grassmannians is replaced by an application of Noether normalisation in Section 1. Belyi's argument to reduce the degree of individual components of the ramification locus is presented in purely algebraic form in Section 1, Lemmas 1.3 and 1.4. The presentation of Bogomolov and Pantev refers to

---

[*]The author thanks the University of Warwick for hospitality during the period when this work was done.

"semi-stable families of pointed curves of genus 0"; the precise result required from that theory is proved here by means of blowups in Section 2. Finally we also give a summary (in Section 3) of the desingularisation of toroidal embeddings which is used by both the papers [3, 1]. To summarise, the aim of this account is to give all the details of the argument, so that it should be accessible to anyone with a basic knowledge of algebraic geometry (as contained for example in Mumford's book [8]).

Acknowledgments

Discussions with Lawrence Ein and D. S. Nagaraj have greatly helped me in understanding this proof better. V. S. Sunder patiently heard out my ramblings on "convex" barycentric subdivisions. M. V. Nori pointed out a flaw in an earlier argument in the section on genus 0 fibrations. I thank the referee for reading the paper very carefully and providing a number of suggestions. Of course, many thanks to Fedor Bogomolov and Tony Pantev for their beautiful proof!

Trivial reductions

We begin the proof with some trivial reductions.

0.1 We may assume that X is normal.

0.2 Any normal curve is smooth and any proper closed subset of a smooth curve is a simple normal crossing divisor (!). Hence the result is true in dimension 1.

0.3 We prove the theorem by induction on the dimension of X. Thus we may assume that $n = \dim X > 1$ and that the result holds for all varieties of smaller dimension than X.

0.4 We can blow up X along Z if necessary, and so we may assume that Z is of codimension 1 in X.

1 $\mathbb{P}^1$-bundles

We will prove the following:

Claim 1.1 *After replacing X by X', where the latter is the blow up of X at a finite set of smooth points we have:*

1. *There is a finite surjective map $f\colon X \to \mathbb{P}_Y(F)$, where Y is smooth and F is a rank 2 vector bundle on Y.*

2. *The branch locus of f is contained in the union $B = \bigcup_{i=1}^{n} s_i(Y)$ of a finite number of sections $s_i \colon Y \to \mathbb{P}_Y(F)$ of the $\mathbb{P}^1$-bundle $\mathbb{P}_Y(F)$.*

3. *The image $f(Z)$ of Z under f is a divisor contained in B.*

Lemma 1.2 *After replacing X by its blowup at a suitable finite set of smooth points, we have the following situation. There is a smooth variety Y, a line bundle L and a finite surjective map $g \colon X \to Q = \mathbb{P}_Y(\mathcal{O} \oplus L)$ such that if E denotes the section corresponding to the quotient homomorphism $\mathcal{O} \oplus L \to \mathcal{O}$, then there is a divisor $B \subset Q$ disjoint from E and containing $g(Z)$ and the branch locus of g.*

Proof This is essentially just Noether normalisation. Let $X \hookrightarrow \mathbb{P}^N$ be a projective embedding of X and $L = \mathbb{P}^{N-n-1}$ a linear subspace in $\mathbb{P}^N$ that does not meet X, where $n = \dim X$. The projection from L gives a finite map $X \to \mathbb{P}^n$. Let B be the hypersurface in $\mathbb{P}^n$ that contains the image of Z and the branch locus of the map. Let p be a point of $\mathbb{P}^n$ not on B. Replace X by its blowup at the points lying over p (which are all smooth) and let $Y = \mathbb{P}^{n-1}$ with $L = \mathcal{O}(1)$. The blowup of $\mathbb{P}^n$ at p is naturally isomorphic to $\mathbb{P}_Y(\mathcal{O} \oplus L)$ and the resulting morphism from X to this $\mathbb{P}^1$-bundle has the required form. $\square$

Now B is a divisor in the $\mathbb{P}^1$-bundle $Q \to Y$ that does not meet a section E. Thus the projection $B \to Y$ is finite and flat. In fact we have:

Lemma 1.3 *In the above situation, if $\mathcal{O}_Q(1)$ denotes the universal quotient bundle, then $\mathcal{O}_Q(B) = \mathcal{O}_Q(b)$ where b is the degree of the map $B \to Y$.*

Proof Consider the line bundle $\mathcal{O}_Q(B) \otimes \mathcal{O}_Q(-b)$. Since this line bundle is trivial on the fibres of $Q \to Y$, it is the pullback of a line bundle from Y. But it restricts to the trivial bundle on E, which is a section of $Q \to Y$. Hence it is the trivial line bundle. $\square$

Set $Q = \mathbb{P}_Y(\mathcal{O} \oplus L)$, and let B be a divisor in Q which is finite over Y; assume moreover that $B = B_1 \cup \bigcup_{i=1}^{r} s_i(Y)$, where s_i are sections, and B_1 does not meet E. Now let $d(B)$ denote the maximum degree over Y of any irreducible component of B_1 and $m(B)$ the number of components of this degree. We wish to construct a new map $X \to Q' = \mathbb{P}_Y(\mathcal{O} \oplus L^N)$ for which B_1 is empty. Thus we may assume that $d(B) > 1$ and $m(B) > 0$. The following lemma shows that we can arrange for at least one of these numbers to drop, and that completes the required inductive step.

Lemma 1.4 *There is a map $h \colon Q \to Q' = \mathbb{P}_Y(\mathcal{O} \oplus L^d)$ such that if B' is the union of $h(B)$ and the branch locus of h then $(d(B), m(B)) > (d(B'), m(B'))$ in the lexicographic ordering.*

Proof Let A be an irreducible component of B_1 of degree $d = d(B)$ and consider the two maps $\mathcal{O}_Q \to \mathcal{O}_Q(d)$ and $L_Q^d \to \mathcal{O}_Q(d)$ given by A and $d \cdot E$. Since these two divisors do not meet, the direct sum $\mathcal{O}_Q \oplus L_Q^d \to \mathcal{O}_Q(d)$ is surjective. Thus, by the universal property of Q' we obtain a morphism $Q \to Q'$ which on every fibre is a map $\mathbb{P}^1 \to \mathbb{P}^1$ of degree d. Its ramification divisor R has the form $R = (d-1)E + R'$ for some divisor R' that does not meet E. Moreover, the Hurwitz formula on the general fibre $\mathbb{P}^1$ (equivalently, computing the canonical divisors) gives $\mathcal{O}_Q(R') = \mathcal{O}_Q(d-1)$. Thus we have the required result. $\square$

We compose the morphism obtained from Lemma 1.2 with a succession of morphisms obtained from Lemma 1.4. By the latter lemma the pair $(d(B), m(B))$ can be reduced until eventually B_1 becomes empty and thus the composite morphism f is as stated in Claim 1.1.

2   Genus 0 fibrations

We will prove the following:

Claim 2.1 *We can replace X by a blow up X' so that:*

1. *There is a finite map $f\colon X \to W$ with W smooth.*

2. *The union of the image $f(Z)$ of Z and the branch locus of f is contained in a divisor D with simple normal crossings (or strict normal crossings).*

For completeness, we recall the definition of a simple normal crossing divisor D in a smooth projective variety. If $D = \bigcup_{i=1}^{n} D_i$ is the decomposition of the divisor into irreducible components, then for each $I \subset \{1, \dots, n\}$ the scheme theoretic intersection $D_I = \bigcap_{i \in I} D_i$ is reduced and smooth of codimension equal to the cardinality $\#I$ of the set I.

Write $g\colon X \to P = \mathbb{P}_Y(F)$ for the map constructed in Section 1 and let $\{s_i\}_{i=1}^{m}$ be sections of $P \to Y$ such that the union $\bigcup_{i=1}^{m} s_i(Y)$ contains $g(Z)$ and the branch locus of g. Let

$$Z_1 = \bigcup_{1 \le i < j \le m} p_Y(s_i(Y) \cap s_j(Y))$$

be the divisor in Y obtained as the image of the pairwise intersections of the sections. We apply the induction hypothesis 0.3 to obtain a map $\varepsilon_Y\colon \widetilde{Y} \to Y$ such that the inverse image $Z_1 = \varepsilon_Y^{-1}(Z)$ is a divisor with simple normal crossings in $\widetilde{Y}$. Replacing Y by $\widetilde{Y}$ and P by its pullback, we obtain a configuration with the following properties (A):

(1) $p: P \to Y$ is a flat morphism of smooth varieties whose reduced fibres are trees of projective lines $\mathbb{P}^1$ (in other words, p is a *genus 0 fibration*);

(2) we have a finite collection $\{s_i\}_{i=1}^m$ of sections of p such that

$$\bigcup_{1 \leq i < j \leq m} p_Y(s_i(Y) \cap s_j(Y)) \subset Z_1,$$

where Z_1 is a divisor with simple normal crossings in Y;

(3) the morphism p is smooth outside $p^{-1}(Z_1)$ and the latter is a divisor with simple normal crossings.

We wish to perform a succession of blowups resulting in a configuration still having the same properties, but with the sections disjoint.

For each component C of Z_1 and each section s_i, the image $C' = s_i(C) \subset P$ is a codimension 2 subvariety. Write $n(C')$ for the number of j such that $s_j(C) = C'$, and n_P for the maximum of such $n(C')$.

Lemma 2.2 *Let C' be so chosen that $n(C') = n_P$. Then for any j, either $s_j(C) = C'$ or $s_j(Y) \cap C' = \emptyset$.*

Proof Suppose j is such that $s_j(Y) \cap C'$ is a proper nonempty subset of C', and let C''' be an irreducible component of this intersection. Choose i so that $s_i(C) = C'$, let D' be a component of $s_i(Y) \cap s_j(Y)$ containing C''', and denote by D the image $p(D')$. On the one hand we have

$$s_j(C) \cap C' \subset s_j(Y) \cap C' \subsetneq C' = s_i(C),$$

while $D' = s_i(D) = s_j(D)$. Thus we see that D and C are different components of Z_1 containing $p(C''')$. But $C''' \subset C'$ has codimension 3 in P, and p restricts to an isomorphism from C' to C. Thus $p(C''')$ is of codimension 2 in Y; hence it is an irreducible component of $D \cap C$. Since Z_1 is a simple normal crossing divisor we see that D and C are the *only* components of Z_1 that contain $p(C''')$.

Take any k such that $s_k(C) = C'$; then $s_k(Y) \cap s_j(Y)$ contains a component E' which contains C'''. Then by the above reasoning we must have $D = p(E')$, and thus

$$D' = s_j(D) = s_j(p(E')) = E' = s_k(p(E')) = s_k(D)$$

But then we get $s_k(D) = D'$ for all k such that $s_k(C) = C'$, and in addition, $s_j(D) = D'$ so that $n(D') = n(C') + 1$. This contradicts the maximality of $n(C')$. $\square$

Together with the preceding lemma, the following result shows that blowing up C' with $n(C') = n_P$ again leads to a configuration of type (A).

Lemma 2.3 *Let $p: P \to Y$ be a genus 0 fibration which is smooth outside Z_1 so that the inverse image $p^{-1}(Z_1)$ of Z_1 is a simple normal crossing divisor in P. For $s: Y \to P$ a section and C a component of Z_1, consider the blowup $P' \to P$ along $s(C)$.*

Then $p': P' \to Y$ is again a genus 0 fibration which is smooth outside Z_1 so that the inverse image $p'^{-1}(Z_1)$ of Z_1 is a simple normal crossing divisor, and the birational transform (or strict transform) of $s(Y)$ gives a section of P' over Y. Moreover, if t is a section of p which is disjoint from $s(C)$ then the birational transform of $t(Y)$ continues to be a section of p'.

Proof   The first two statements and the final statement about P' are obvious. The union of $s(Y)$ with $p^{-1}(Z_1)$ is a simple normal crossing divisor since s is a section. Moreover, $s(C)$ is the locus of intersection of $s(Y)$ and one of the components of $p^{-1}(C)$. Thus the blowup preserves the property of being a simple normal crossing divisor.

Suppose that $n_P > 1$ and let N_P be the number of C' attaining this maximum. For each such C' and each pair i, j such that $s_i(C) = s_j(C) = C'$ let $m(i, j, C')$ denote the multiplicity of intersection of $s_i(Y)$ and $s_j(Y)$ along C'. We are in the situation of the following lemma

Lemma 2.4 *Let P be a smooth variety, D_1 and D_2 smooth divisors meeting along a smooth codimension 2 locus B with multiplicity $m > 0$. Let P' be the blowup of P along C', and let D_1' and D_2' be the birational transforms of D_1 and D_2 respectively. Then the multiplicity of intersection of D_1' and D_2' is $m - 1$ along a codimension two locus B' lying over B.*

Proof   In a neighbourhood of the generic point of B, the given condition can be written as follows

$$\mathcal{O}_P(D_1) \otimes \mathcal{O}_{D_2} = \mathcal{O}_{D_2}(m \cdot B).$$

Let E denote the exceptional divisor of the blowup $\varepsilon: P' \to P$ and let B' be the intersection $D_2' \cap E$; this is a section of the $\mathbb{P}^1$-bundle $E \to B$. Now the birational transform of D_1 represents the divisor class $\varepsilon^*(D_1) - E$ and thus we see that

$$\mathcal{O}_{P'}(D_1') \otimes \mathcal{O}_{D_2'} = (\mathcal{O}_P(D_1) \otimes \mathcal{O}_{D_2'}) \otimes \mathcal{O}_{D_2'}(-B') = \mathcal{O}_{D_2'}((m-1) \cdot B'),$$

which proves the result.

From this lemma, we see that the multiplicity of intersection of the birational transforms of $s_i(Y)$ and $s_j(Y)$ is $m(i, j, C') - 1$. If this becomes zero then $n(C')$ drops, hence either N_P decreases or n_P does so. This completes the argument by induction since all these numbers are positive and we wish to obtain the situation where $n_P = 1$.

Now we replace X by the normalisation of the fibre product $X \times_{\mathbb{P}_Y(F)} P$, where $P \to Y$ is the configuration of type (A) with $n_P = 1$ obtained above. We then have a finite morphism $h\colon X \to P$. The image $h(Z)$ of Z and the branch locus of h are contained in the simple normal crossing divisor consisting of the finitely many disjoint sections of the configuration (A) and $p^{-1}(Z_1)$. This proves Claim 2.1.

3   Toric singularities

As a last step we must prove the result in the following situation. There is a finite map $f\colon X \to W$ with W smooth and X normal and $D = \bigcup_{i=1}^{n} D_i$ a divisor with simple normal crossings so that f is étale outside D; moreover Z is a union of (some of) the components of $f^{-1}(D)$. So we need to construct a birational morphism $\varepsilon\colon \widetilde{X} \to X$ so that $\widetilde{X}$ is smooth and $\varepsilon^{-1}f^{-1}(D)$ is a divisor with simple normal crossings in $\widetilde{X}$.

We will show (see Lemma 3.1 below) that the inclusion of $X \setminus f^{-1}(D)$ in X is a strict toroidal embedding in the sense of [7], where the desingular-isation problem for such embeddings has been studied and solved. For the sake of completeness we also give a brief summary of their method. Many statements below are given without detailed proofs—readers are invited to complete the arguments on their own or look for proofs in one of the books on toric geometry such as Oda [9], [10] or Fulton [4].

3.1   Affine toric singularities

Let us first consider the simple situation where $W = \mathbb{A}^n$ and D is a union of coordinate hyperplanes. In this case $W \setminus D$ is isomorphic to $\mathbb{G}_m^r \times \mathbb{A}^{n-r}$ where r is the number of components of D. Hence (because we are working over the complex numbers $\mathbb{C}$) its fundamental group is a product of infinite cyclic groups and the finite cover $X \setminus f^{-1}(D)$ is also isomorphic to $\mathbb{G}_m^r \times \mathbb{A}^{n-r}$. In fact, the homomorphism f^* on rings has the form

$$\mathbb{C}[z_1, z_1^{-1}, \ldots, z_r, z_r^{-1}, z_{r+1}, \ldots, z_n] \;\to\; \mathbb{C}[t_1, t_1^{-1}, \ldots, t_r, t_r^{-1}, t_{r+1}, \ldots, t_n]$$
$$z_i \;\mapsto\; \begin{cases} m_i & \text{for } i \le r \\ t_i & \text{for } i > r \end{cases} \tag{1}$$

where the m_i are certain monomials in $t_1, \ldots, t_r$ (with negative powers allowed). The natural action of the torus $\mathbb{G}_m^n$ on $X \setminus f^{-1}(D)$ then descends to an action via f making this map equivariant. Moreover, since X is the normalisation of W in $X \setminus f^{-1}(D)$, the action extends to X. Since the map f is equivariant there are only finitely many orbits for the torus action on X; in other words, X is an (affine) *toric variety*.

An explicit description of X can be given as follows. Let M be the free Abelian group of all monomials in the variables $t_1, \ldots, t_r$. Let M^+ be the *saturated* submonoid of M generated by the m_i. Then $X = \operatorname{Spec} \mathbb{C}[M^+] \times \mathbf{A}^{n-r}$. For future reference, we note that there is a unique closed orbit in X which maps isomorphically to the closed orbit $0 \times \mathbf{A}^{n-r}$ in W. Moreover, M is the group of Cartier divisors supported on $f^{-1}(D)$ and M^+ is the submonoid of effective Cartier divisors.

Let N be the dual Abelian group to M and

$$N^+ = \left\{ n \mid n(m) \geq 0 \text{ for all } m \in M^+ \right\}$$

the "dual" monoid to M^+. Then N^+ is a finitely generated saturated monoid in N (like M^+ in M). Let σ be any finitely generated submonoid of N^+ and

$$M^\sigma = \left\{ m \mid n(m) \geq 0 \text{ for all } n \in N^+ \right\}$$

the dual monoid in M (which contains M^+). Then $X(\sigma) = \operatorname{Spec} \mathbb{C}[M^\sigma] \times \mathbf{A}^{n-r}$ is an affine toric variety and the natural morphism $X(\sigma) \to X$ is birational and equivariant for the torus action. Moreover, we see that $X(\sigma)$ is nonsingular and the pullback to $X(\sigma)$ of D is a simple normal crossing divisor if and only if the monoid σ is generated over nonnegative integers by a (sub-)basis of N; such a monoid is called *simplicial*.

Let $\Sigma = \{\sigma_i\}_{i=1}^n$ be a collection of finitely generated saturated submonoids of N^+ which give a subdivision of N^+; i.e., N^+ is the union of all the σ_i and for any pair σ_i, σ_j their intersection is a σ_k for some k. We then obtain a collection of equivariant birational morphisms $X_i = X(\sigma_i) \to X$ so that if $\sigma_j \subset \sigma_i$ then $X_j \subset X_i$ in a natural way. Thus we can patch together the X_i to obtain $X_\Sigma \to X$ which is birational and equivariant (but X_Σ need not be affine any more). Moreover, the condition that the σ_i cover N^+ implies that $X_\Sigma \to X$ is proper.

Thus to obtain a desingularisation of X it is enough to find a subdivision of N^+ consisting entirely of simplicial monoids; an easy enough combinatorial problem solved by barycentric subdivision. The intrepid reader is warned that proving that the resulting morphism is projective is a little intricate since an arbitrary simplicial subdivision *need not* result in a projective morphism; however, the barycentric subdivision does yield a projective morphism.

3.2   Local toric singularities

Now we examine the general case locally. Let $x \in X$ be any point such that $w = f(x)$ lies in D. There is an analytic neighbourhood U of w in W

and coordinates on U so that $D \cap U$ is given by the vanishing of a product of coordinate functions; by further shrinking U we can assume that U is a polydisk in these coordinates. Let V be the component of $f^{-1}(U)$ which contains x. The normality of X implies the normality of the analytic space V. Hence, the open subset $V \setminus f^{-1}(D)$ is connected. So it is a topological cover of $U \setminus D$. The coordinate functions give an inclusion $U \hookrightarrow W' = \mathbf{A}^n$ so that $D \cap U$ is the restriction to U of a union of coordinate hyperplanes in W'. The resulting inclusion $U \setminus D \hookrightarrow \mathbb{G}_m^r \times \mathbf{A}^{n-r}$ induces an isomorphism of fundamental groups. Thus there is a Cartesian square in which the horizontal maps are inclusions:

$$
\begin{array}{ccc}
V \setminus f^{-1}(D) & \longrightarrow & \mathbb{G}_m^r \times \mathbf{A}^{n-r} \\
f \downarrow & & \downarrow f' \\
U \setminus D & \longrightarrow & \mathbb{G}_m^r \times \mathbf{A}^{n-r}
\end{array}
$$

and f' is a covering of the form (1) above. Let X' denote the normalisation of W' in the cover f'. By the normality of X (and hence the normality of V as an analytic space) it follows that we obtain a commutative diagram

$$
\begin{array}{ccc}
V & \longrightarrow & X' \\
f \downarrow & & \downarrow f' \\
U & \longrightarrow & W'
\end{array}
$$

so that V is isomorphic to an analytic open neighbourhood of a point in the toric variety X'. By means of the two diagrams above we can carry over the desingularisation of 3.1 to the local analytic space V.

3.3   General toroidal embeddings

The local description 3.2 can be repeated in a suitable neighbourhood of any point $x \in X$. This shows that the inclusion of $X \setminus f^{-1}(D)$ in X is a *toroidal embedding*. The desingularisations obtained locally need to be constructed in a coherent manner so that they "patch up".

Let us stratify W by connected components of intersections of the form

$$
(D_{i_1} \cap \cdots \cap D_{i_r}) \setminus \bigcup_{j \neq i_s} D_j. \tag{2}
$$

If U and V are as in 3.2, then there is a unique stratum S of W so that $S \cap U$ is closed. Under the inclusions of U in W' and V in X' of 3.2 the strata correspond to the orbits of the torus action. As we have noted in

3.1 the unique closed strata in U and V then become isomorphic. Thus, if $T = f^{-1}(S)$ then $T \cap V$ is closed in V and the morphism $T \cap V \to S \cap U$ is an isomorphism; indeed $T \cap V$ and $S \cap U$ are the restrictions of the closed orbits in X' and W' respectively. Thus we see that if S is *any* stratum in W and T a connected component of $f^{-1}(S)$ then $T \to S$ is étale and proper. We thus stratify X by connected components of the inverse images of strata in W under f. Let $\{T_a\}_{a \in A}$ denote this stratification; by abuse of notation we define $S_a = f(T_a)$.

Let $E = \bigcup_{j=1}^m E_j$ be the decomposition into irreducible components of the inverse image $f^{-1}(D)$ of D. Then there is a function $i \colon \{1, \dots, m\} \to \{1, \dots, n\}$ so that $f(E_j) = D_{i(j)}$ for all j. For any stratum T_a let X^a denote the complement in X of all those E_j that do not meet T_a. Similarly, we denote by W^a the complement in W of all D_i that do not meet S_a. The morphism f clearly maps X^a to W^a.

Lemma 3.1 *Let M_a be the Abelian group of all Cartier divisors in X^a with support in $E \cap X^a$; let M_a^+ be the submonoid consisting of effective Cartier divisors. Then M_a has rank $r_a = \mathrm{codim}_X T_a$. The distinct analytic components of E in a neighbourhood of any point x of X are precisely the algebraic components; i.e., $X \setminus E \hookrightarrow X$ is a strict toroidal embedding.*

Proof Let M_a' be the group of Cartier divisors on W^a with support on $D \cap W^a$. Clearly, this is the free Abelian group on those D_i which contain S_a. Since S_a is a connected component of an intersection of the type (2), there are exactly $r_a = \mathrm{codim}_W S_a$ of such D_i. The pullback homomorphism makes M_a' a subgroup of M_a. On the other hand, consider an analytic neighbourhood $V = f^{-1}(U)$ of some point $x \in T_a$ as in 3.2. If M_x'' denotes the group of all Cartier divisors on V supported on $E \cap V$; then we have noted in 3.1 that M_x'' is a free Abelian group of rank r_a. Moreover, M_a is included as a subgroup of M_x'' under the restriction from X_a to V.

It follows that the homomorphism $M_a \to M_x''$ has finite cokernel; but then the normality of X means that it is surjective. In particular, we see that the distinct analytic components of $E \cap V$ correspond to distinct algebraic components of E. This concludes the proof. $\square$

Let M_a^+ denote the monoid of effective divisors in M_a; under the isomorphism $M_a \to M_x'' = M$ this maps isomorphically onto the submonoid M^+ considered in 3.1. Thus X_a is smooth (and the divisor E is a simple normal crossing divisor) if and only if M_a^+ is simplicial. If T_b lies in the closure of T_a then X_a is an open subset of X_b. This induces by restriction a (surjective) homomorphism $M_b \to M_a$ which further restricts to a surjection $M_b^+ \to M_a^+$. Thus we have a (finite) projective system of monoids.

Let N_a and N_a^+ be the dual objects as defined in 3.1. These form a finite injective system of monoids. By a compatible family S of subdivisions we mean a subdivision Σ_a of N_a^+ for each a so that the subdivision Σ_b restricts to the subdivision Σ_a on the submonoid N_a^+ of N_b^+. We then obtain a proper birational morphism $X_{\Sigma_b} \to X_b$ for each b which restricts to $X_{\Sigma_a} \to X_a$ on the open subset X_a of X_b. Thus, we see that any such compatible family leads to a proper birational morphism $X_S \to X$.

Thus, in order to desingularise, we have to find a compatible family of subdivisions so that each of the new monoids is simplicial. This is achieved by the barycentric subdivision. As seen earlier, this ensures that the morphism $X_S \to X$ is locally projective. Since X is projective, this morphism is indeed projective.

References

[1] D. Abramovich and A. J. de Jong, *Smoothness, semistability and toroidal geometry*, J. Algebraic Geom. **6** (1997), 789–801, preprint alg-geom/9603018, 1996.

[2] G. V. Belyi, *Galois extensions of a maximal cyclotomic field*, Izv. Akad. Nauk SSSR Ser. Math. **43** (1979), no. 2, 267–276.

[3] F. Bogomolov and T. Pantev, *Weak Hironaka theorem*, alg-geom/9603019, 11 pp.

[4] W. Fulton, *Introduction to toric varieties*, Ann. of Math Studies, Vol. 151, Princeton University Press, Princeton, USA, 1993.

[5] H. Hironaka, *Resolution of singularities of an algebraic variety over a field of characteristic zero: I, II*, Ann. of Math.(2) **79** (1964), 109–326.

[6] A. J. de Jong, *Smoothness, semistability and alterations*, Publ. Math. IHES **83** (1996), 51–93.

[7] G. Kempf, F. Knudsen, D. Mumford, and B. Saint-Donat, *Toroidal embeddings. I*, Lecture Notes in Mathematics, no. 339, Springer-Verlag, Berlin-Heidelberg-New York, 1973.

[8] D. Mumford, *The red book of varieties and schemes*, LNM **1358** Springer-Verlag, Berlin–New York, 1988 (Indian edition: Narosa Publishers, New Delhi, India, 1996).

[9] T. Oda, *Lectures on torus embeddings and applications*, TIFR Lecture Notes, Tata Institute, Bombay, India, 1982.

[10] T. Oda, *Convex bodies and algebraic geometry. An introduction to the theory of toric varieties*, Ergebnisse (3) **15**, Springer-Verlag, Berlin–New York, 1988.

Kapil H. Paranjape,
Institute of Mathematical Sciences,
CIT Campus, Tharamani, Chennai,
600 113, INDIA
e-mail: kapil@imsc.ernet.in

Mordell–Weil lattices for
higher genus fibration over a curve

Tetsuji Shioda

To Sōichi Kawai for his 60th birthday

1 Introduction, notation

Let $K = k(C)$ be the function field of an algebraic curve C over an algebraically closed ground field k. Let Γ/K be a smooth projective curve of genus $g > 0$ with a K-rational point $O \in \Gamma(K)$, and let J/K denote the Jacobian variety of Γ/K. Further let (τ, B) be the K/k-trace of J (see §2 below and [4]).

Then the Mordell–Weil theorem (in the function field case) states that the group of K-rational points $J(K)$ modulo the subgroup $\tau B(k)$ is a finitely generated Abelian group.

Now, given Γ/K, there is a smooth projective algebraic surface S with genus g fibration $f\colon S \to C$ which has Γ as its generic fibre and which is relatively minimal in the sense that no fibres contain an exceptional curve of the first kind (-1-curve). It is known that the correspondence $\Gamma/K \leftrightarrow (S, f)$ is bijective up to isomorphisms (cf. [7], [5]).

The main purpose of this paper is to give the Mordell–Weil group $M = J(K)/\tau B(k)$ (modulo torsion) the structure of Euclidean lattice via intersection theory on the algebraic surface S. The resulting lattice is the *Mordell–Weil lattice* (MWL) of the Jacobian variety J/K, which we sometimes call MWL of the curve Γ/K or of the fibration $f\colon S \to C$.

For this, we first establish the relationship between the Mordell–Weil group and the Néron–Severi group $\mathrm{NS}(S)$ of S (Theorem 1, stated in §2 and proved in §3). Then (in §4) we introduce the structure of lattice on the Mordell–Weil group by defining a natural pairing in terms of the intersection pairing on $\mathrm{NS}(S)$. Thus the basic idea is the same as the elliptic case $g = 1$ (cf. [8]).

We think that Theorem 1 must have been known to mathematicians such as Lang, Néron, Tate, Weil, ... who did the foundational work on the Mordell–Weil theorem and the theory of canonical height. But we could not

359

locate the statement of Theorem 1 (as given below) in the literature; compare e.g., [16], [4, Chap. 6, §6], [14, §4].

For the sake of completeness, we give a proof of Theorem 1 in §3. In the proof, we use (a consequence of) Theorem 2 due to Raynaud which describes a precise relationship between the K/k-trace of J/K, the Picard variety of S and the Jacobian variety of C. As far as we know, Theorem 2 is new, in that it takes care of the infinitesimal part, which is a delicate point in dealing with the K/k-trace. Actually, to prove Theorem 1, we do not need the full strength of Theorem 2; its consequence on the k-valued points (i.e., ignoring the infinitesimal part) is sufficient. This was how we proved Theorem 1 in the original version of this paper. But it is much nicer to have a statement as definitive as Theorem 2.

The main results of this paper were announced in our short note [11] under the assumption that the K/k-trace in question is trivial. We apologize for the delay in providing a detailed, generalized version. During this delay, two important results have come to light; one is Raynaud's theorem mentioned above, and the other is a new viewpoint to use the canonical height for Abelian varieties over higher dimensional function fields (cf. [13]).

I am very grateful to M. Raynaud for informing me of his proof of Theorem 2, and for allowing me to include it here. Also I would like to thank many people for giving me the opportunity to reconsider the subject in a stimulating atmosphere, especially J. Coates (Cambridge Univ.), M. Reid (Warwick Univ.), S. Ramanan (Tata Inst.), F. Hirzebruch (Max-Planck Inst.) and others. The final version was completed during my stay at Johns Hopkins Univ. (JAMI).

List of Notation

- k: an algebraically closed field
  C or C/k: a smooth, projective curve defined over k
  $K = k(C)$: the function field of C/k
  $J_C = J_{C/k}$: the Jacobian variety of C

- Γ or Γ/K: a smooth, projective curve of genus $g > 0$ defined over K
  $J = J_{\Gamma/K}$: the Jacobian variety of Γ/K
  (τ, B): the K/k-trace of J

- $f: S \to C$: relatively minimal genus g fibration with generic fibre Γ
  S or S/k: a smooth projective surface defined over k
  $\Gamma(K)$: the set of K-rational points of Γ, identified with the set of sections of $f: S \to C$
  (P) (for $P \in \Gamma(K)$): the image of $P: C \to S$ (a curve on S)
  $\mathrm{Pic}_{S/k}$: the Picard scheme of S/k

PicVar$_S$: the Picard variety of S/k
Pic$_{S/C}$: the relative Picard scheme of S/C (see [6])
Pic$(S) := $ Pic$_{S/k}(k)$; PicVar$(S) := $ PicVar$_S(k)$
NS(S): the Néron–Severi group of S
$\rho(S)$: the Picard number of S $(=$ rank NS$(S))$

- $R := \{v \in C | f^{-1}(v) = \sum_{i=0}^{m_v-1} \mu_{v,i} \Theta_{v,i}$ is a reducible fibre$\}$
  $m_v (v \in R)$: the number of irreducible components of $f^{-1}(v)$
  $\Theta_{v,i}$ (for $0 \leq i \leq m_v - 1$): irreducible components
  $\Theta_{v,0}$: the identity components of $f^{-1}(v)$ (with $\mu_{v,0} = 1$)
  T_v: the subgroup of NS(S) generated by $\Theta_{v,i}$ (for $1 \leq i \leq m_v - 1$)
  F: the class of any fibre of f in NS(S)
  T: the subgroup of NS(S) generated by $(O), F$ and T_v (for $v \in R$)

2 K/k-trace, Mordell–Weil group and Néron–Severi group

We use the notation introduced above. Thus J is the Jacobian variety of a curve Γ defined over $K = k(C)$, C being a curve over an algebraically closed field k. Unless otherwise mentioned, we always assume:

$$\Gamma \text{ has a } K\text{-rational point } O. \qquad (A_0)$$

According to the Mordell–Weil theorem for the function field case, the group of K-rational points $J(K)$ modulo the subgroup $\tau B(k)$, i.e., the quotient group $M = J(K)/\tau B(k)$, is a finitely generated Abelian group, where (τ, B) denotes the K/k-trace of J/K (cf. [4] for this and the following). We call M the *Mordell–Weil group* of J/K or of Γ/K.

For the convenience of the reader, let us recall here the definition of the K/k-trace of an Abelian variety A defined over K. It is a pair (τ, B) in which B is an Abelian variety defined over k and $\tau \colon B \to A$ is a homomorphism of Abelian varieties over K with the following universal mapping property: given any pair (τ', B') as above, there is a unique homomorphism $\psi \colon B' \to B$ such that $\tau' = \tau \circ \psi$. It is known that the K/k-trace exists (Chow) and that $\tau \colon B \to A$ is a radical map (that is, it is injective on points, but it may have an infinitesimal kernel in case of characteristic $p > 0$).

Now, letting $A = J$ be a Jacobian variety as before, we state the following basic theorems.

Theorem 1 *Under (A_0), the Mordell–Weil group $M = J(K)/\tau B(k)$ is naturally isomorphic to the quotient group of* NS(S) *by T:*

$$J(K)/\tau B(k) \simeq \text{NS}(S)/T. \qquad (1)$$

Theorem 2 (Raynaud) *There is a short exact sequence of Abelian varieties over k:*

$$0 \longrightarrow J_C \longrightarrow \text{PicVar}_S \longrightarrow B \longrightarrow 0 \qquad (2)$$

The maps in these theorems are induced from the natural maps: restricting line bundles on S to the generic fibre Γ and taking the pullback (via $f \colon S \to C$) of line bundles on the base curve C to S.

Postponing the proof of Theorems 1 and 2 to the next section, let us deduce some obvious consequences, which will be the most interesting case for application.

Theorem 3 *The following conditions are equivalent:*

(i) The K/k-trace of J/K is trivial, i.e., $B = 0$.

(ii) $f^* \colon J_C \to \text{PicVar}_S$ is an isomorphism.

(iii) The irregularity of S is equal to the genus of C.

If any (hence all) of these conditions holds, then $J(K)$ is a finitely generated Abelian group such that

$$J(K) \simeq \text{NS}(S)/T. \qquad (3)$$

In particular, we have

$$\text{rank } J(K) = \rho(S) - 2 - \sum_{v \in R}(m_v - 1) \qquad (4)$$

(cf. the notation in §1 and §4, Proposition 5).

The irregularity of a surface is, by definition, the dimension of its Picard variety. A surface with irregularity zero is classically called a *regular* surface. Examples of regular surfaces include rational surfaces, K3 surfaces and smooth surfaces in $\mathbb{P}^3$.

Corollary 4 *Suppose that S is a regular surface with a given genus g fibration $f \colon S \to C = \mathbb{P}^1$. Then the K/k-trace of its generic fibre Γ is trivial, and the conclusion of Theorem 3 holds.*

3 Proof of Theorems 1 and 2

First we prove Theorem 1, assuming Theorem 2. Restricting a line bundle on S to the generic fibre Γ defines a group homomorphism

$$\text{Pic}(S) := \text{Pic}_{S/k}(k) \longrightarrow \text{Pic}_{\Gamma/K}(K). \tag{5}$$

In terms of divisors, it associates with each divisor class $\text{cl}(D)$ on S the class of the divisor $D|_\Gamma = D \cdot \Gamma$ on the generic fibre Γ. By adjusting the image to have degree zero using the given rational point $O \in \Gamma(K)$, we get a homomorphism

$$\varphi\colon \text{Pic}(S) \longrightarrow J(K) := \text{Pic}^0_{\Gamma/K}(K) \tag{6}$$

which sends $\text{cl}(D)$ to $\text{cl}(\delta)$ with $\delta = D \cdot \Gamma - (D \cdot \Gamma)O$.

Obviously (5) and (6) are surjective, because given any K-rational divisor, say δ', on Γ, we can take the k-loci (schematic closure) of the components of δ' to obtain a divisor D' on S such that $D' \cdot \Gamma = \delta'$.

Next we determine the kernel of φ. Let $\widetilde{T}$ denote the subgroup of $\text{Pic}(S)$ generated by the divisor classes of irreducible components of fibres and the zero section (O). It is obvious that $\widetilde{T}$ is contained in $\text{Ker}(\varphi)$. Let us show that $\widetilde{T} = \text{Ker}(\varphi)$. Take any element $\xi = \text{cl}(D) \in \text{Ker}(\varphi)$. By definition, $D|_\Gamma$ is linearly equivalent to zero on Γ, and hence it is equal to a principal divisor (h) for some rational function $h \in K(\Gamma)$. Since the function field $K(\Gamma)$ can be identified with the function field $k(S)$, we can find some function $H \in k(S)$ such that $(H)|_\Gamma = (h)$ (note that H is unique up to multiplication by elements in $K^\times = k(C)^\times$). Then every component of the divisor $D' := D - (H)$ on S must be contained in some fibre; in other words, D' is a linear combination of irreducible components of fibres. Hence $\xi = \text{cl}(D)$ belongs to $\widetilde{T}$.

Therefore we have an exact sequence (of Abelian groups):

$$0 \longrightarrow \widetilde{T} \longrightarrow \text{Pic}(S) \longrightarrow J(K) \longrightarrow 0 \tag{7}$$

Furthermore, the restriction of the map φ to the Picard variety $\text{PicVar}(S) \subset \text{Pic}(S)$ factors through the K/k-trace B of J/K (see the proof of Theorem 2 below), which fits in the exact sequence of k-points of (2):

$$0 \longrightarrow J_C(k) \longrightarrow \text{PicVar}(S) \longrightarrow B(k) \longrightarrow 0 \tag{8}$$

From the above two sequences, we obtain a surjective map:

$$\widetilde{\varphi}\colon \text{NS}(S) := \text{Pic}(S)/\text{PicVar}(S) \longrightarrow J(K)/\tau B(k) \tag{9}$$

which induces an isomorphism (1) in Theorem 1, since T is nothing but the image of $\widetilde{T}$ in $\text{NS}(S)$. This proves Theorem 1 (assuming Theorem 2).

Proof of Theorem 2　Following Raynaud, we prove Theorem 2 under the assumption:

$$\text{For every fibre } f^{-1}(v) \text{ of } f\colon S \to C, \text{ the g.c.d. of} \atop \text{multiplicities of irreducible components is equal to 1.} \qquad (A_1)$$

This condition is weaker than our assumption (A_0) that Γ has a K-rational point O, because the latter implies that every fibre has at least one component with multiplicity 1, namely, the (identity) component intersecting the zero section (O).

As proved in [6] (see also [1]), under (A_1), the sheaf $R^1 f_* \mathcal{O}_S$ is locally free of rank g. The Leray spectral sequence for $\mathcal{O}_S$ gives a short exact sequence

$$0 \longrightarrow H^1(C, \mathcal{O}_C) \xrightarrow{f^*} H^1(S, \mathcal{O}_S) \longrightarrow H^0(C, R^1 f_* \mathcal{O}_S) \qquad (10)$$

(N.B. We can add the arrow $\to 0$ at the end under the stronger assumption (A_0), but this is not needed for the argument below.)

At the level of Picard schemes, we have an exact sequence

$$0 \longrightarrow \operatorname{Pic}_{C/k} \longrightarrow \operatorname{Pic}_{S/k} \longrightarrow \operatorname{Pic}_{S/C} \qquad (11)$$

in which $\operatorname{Pic}_{S/C}$ is the relative Picard scheme representing $R^1 f_* \mathcal{O}_S$ (see [6]), and (10) corresponds to taking the Lie algebras of (11), or more precisely to the Lie algebras of the identity components of group schemes in (11):

$$0 \longrightarrow \operatorname{Pic}^0_{C/k} \longrightarrow \operatorname{Pic}^0_{S/k} \longrightarrow \operatorname{Pic}^0_{S/C} \qquad (12)$$

Note that the image of the map $J_C = \operatorname{Pic}^0_{C/k} \to \operatorname{Pic}^0_{S/k}$ lies in $(\operatorname{Pic}^0_{S/k})_{\mathrm{red}}$, since smooth part is mapped into smooth part, and that we have $\operatorname{PicVar}_S = (\operatorname{Pic}^0_{S/k})_{\mathrm{red}}$ by definition.

On the other hand, the restriction to the generic fibre gives a map

$$\operatorname{Pic}^0_{S/C} \longrightarrow J_{\Gamma/K} \qquad (13)$$

such that the corresponding map of Lie algebras

$$H^0(C, R^1 f_* \mathcal{O}_S) \longrightarrow \operatorname{Lie}(J_{\Gamma/K}) \qquad (14)$$

is injective, since $R^1 f_* \mathcal{O}_S$ is locally free under (A_1) as remarked before.

By (12) and (13), we have

$$0 \longrightarrow J_C \xrightarrow{\alpha} \operatorname{PicVar}_S \xrightarrow{\gamma} J = J_{\Gamma/K}. \qquad (15)$$

The map γ factors through the K/k-trace (τ, B) by the universal mapping property: $\gamma = \tau \circ \beta$ for a unique homomorphism $\beta\colon \operatorname{PicVar}_S \to B$.

Now we claim that

$$0 \longrightarrow J_C \xrightarrow{\alpha} \mathrm{PicVar}_S \xrightarrow{\beta} B \longrightarrow 0 \tag{16}$$

is an exact sequence of Abelian varieties over k, which will prove Theorem 2.

The most crucial point is the exactness at the middle. To take care of the infinitesimal part, look at the corresponding Lie algebras:

$$
\begin{array}{ccccc}
\mathrm{Lie}(J_C) & \xrightarrow{\alpha'} & \mathrm{Lie}(\mathrm{PicVar}_S) & \xrightarrow{\beta'} & \mathrm{Lie}(B) \\
\| & & \cap & & \downarrow \tau' \\
H^1(C, \mathcal{O}_C) & \xrightarrow{f^*} & H^1(S, \mathcal{O}_S) & \longrightarrow & \mathrm{Lie}(J).
\end{array}
\tag{17}
$$

The bottom row is (10), combined with (14); hence it is exact.

Now we observe that τ' is an injective map (of a k-vector space into a K-vector space). Indeed, if $\mathrm{Ker}(\tau') \neq 0$, then there would be some radicial subgroup scheme $G \neq 0$ of B giving rise to a homomorphism $B' = B/G \to J$, which contradicts the universality of the trace (τ, B).

It follows that the top row in (17) is exact (at the middle).

Going back to (16), we show the exactness at the middle. First we note that $\beta \circ \alpha = 0$. For, otherwise, $\mathrm{Im}(\beta \circ \alpha) \neq 0$ will be an Abelian subvariety of B which is mapped to zero under $\tau \colon B \to J$; again this contradicts the universality of τ.

To prove $\mathrm{Ker}(\beta) = \mathrm{Im}(\alpha)$, it is now enough to test it at the k-valued points. Thus, take any $\xi \in \mathrm{Ker}(\beta)$. Then we have $\xi = \mathrm{cl}(D)$ for some divisor D with $D|_\Gamma = 0$ (cf. (7)); hence it is a linear combination of components of fibres. Thus we can write $D = \sum_v D_v$ (a finite sum) where D_v has support in $F_v = f^{-1}(v)$ (for $v \in C$) and $(D_v \cdot \Theta) = 0$ for every irreducible component Θ of F_v. We want to see that D_v is an integral multiple of F_v: $D_v = n_v F_v$ (with $n_v \in \mathbb{Z}$). There is no problem at good fibres. At a bad fibre, we know in general from [6] that dD_v is an integral multiple of F_v if d denotes the g.c.d. of multiplicities of irreducible components of F_v. Hence, under the assumption (A_1), we conclude that $D = f^{-1}(\sum_v n_v v)$, and hence ξ belongs to $\mathrm{Im}(\alpha)$.

To complete the proof of Theorem 2, we should also see that, in (16), α is injective and β is surjective. The injectivity of α follows from the fact that $f \colon S \to C$ is a flat morphism with $f_* \mathcal{O}_S = \mathcal{O}_C$.

For the surjectivity of β, it is enough to show that $\tau(B(k)) \subset \tau(\beta(A(k)))$. Let b be a generic point of B over K (in Weil's language [15] which will be used in this section). Because the homomorphism $\tau \colon B \to J$ is K-rational, $b' := \tau(b)$ is a $K(b)$-rational point of J. Taking a generic point t of C over k, we identify the function field $K = k(C)$ with $k(t)$. Then we have

$$K' := K(b) = k(t)(b) = k(b)(t). \tag{18}$$

Let δ be a K'-rational divisor on Γ of degree 0 representing b', and let D be a $k(b)$-rational divisor on S such that $D|_\Gamma = \delta$. We have (cf. (6))

$$\varphi(\mathrm{cl}(D)) = b' = \tau(b). \tag{19}$$

Now D defines an algebraic family of divisors $\{D_u | u \in W\}$ on S parametrized by some open subset W of B; D_u is a specialization of D over the specialization $b \to u$ over k. Thus we have $\varphi(\mathrm{cl}(D_u)) = \tau(u)$. For any $u_1, u_2 \in W$, $D_{u_1} - D_{u_2}$ is algebraically equivalent to zero, which defines a point, say a, of the Picard variety $A = \mathrm{PicVar}_S$. Then

$$\tau(\beta(a)) = \varphi(\mathrm{cl}(D_{u_1} - D_{u_2})) = \tau(u_1 - u_2). \tag{20}$$

Fixing $u_2 \in W(k)$, let W' be the translate of W by $-u_2$ in B. Then the above shows that $\tau(W'(k)) \subset \tau(\beta(A(k)))$. Since $B(k)$ is obviously generated by $W'(k)$ as a group, we conclude that $\tau(B(k)) \subset \tau(\beta(A(k)))$, i.e., that $\beta \colon A \to B$ is surjective.   *q.e.d*

4   Mordell–Weil lattices

In Theorem 1, we established an isomorphism of the Mordell–Weil group $M = J(K)/\tau B(k)$ with $\mathrm{NS}(S)/T$. Once this is done, we can introduce a natural pairing $\langle \, , \, \rangle$ on M by "splitting" this isomorphism and using the intersection pairing $(\,\cdot\,)$ on $\mathrm{NS}(S)$. This simple idea has proved to be fruitful in the elliptic case ([8]) and we follow it here.

   Let $N := \mathrm{NS}(S)$ and $\overline{N} := N/N_{\mathrm{tor}}$. Then $\overline{N}$ is an integral lattice with respect to the intersection pairing, and it has signature $(1, \rho - 1)$ by the Hodge index theorem, where $\rho = \rho(S)$ is the Picard number of S (cf. [3]). We can regard T as a sublattice of $\overline{N}$, since it is torsion-free by the following well-known result (see e.g., [2, Chap. 9]).

Proposition 5 *The lattice T has an orthogonal decomposition:*

$$T = U \oplus \sum_{v \in R} T_v, \tag{21}$$

where U is the rank 2 unimodular lattice spanned by $(O), F$ and T_v is the negative-definite sublattice of rank $m_v - 1$ spanned by the irreducible components $\Theta_{v,i}$ (with $i > 0$), other than the identity component, of a reducible fibre $f^{-1}(v)$ (for $v \in R$).

   Let us denote by

$$\overline{\varphi} \colon M \to N/T \tag{22}$$

the map giving the isomorphism (1) of Theorem 1.

Lemma 6 *There is a unique map*

$$\varphi \colon M \longrightarrow N \otimes \mathbb{Q} \qquad (23)$$

such that (a) it splits $\overline{\varphi}$ in the sense that

$$\varphi(x) \bmod T \otimes \mathbb{Q} = \overline{\varphi}(x) \text{ in } (N/T) \otimes \mathbb{Q} \quad \text{for any } x \in M \qquad (24)$$

and (b) the image of φ is orthogonal to T w.r.t. $(\,\cdot\,)$:

$$\varphi(x) \perp T \quad \text{for any } x \in M. \qquad (25)$$

Moreover, φ is a group homomorphism such that

$$\mathrm{Ker}(\varphi) = M_{\mathrm{tor}}. \qquad (26)$$

Proof This is essentially the same as in the elliptic case (cf. [8, §8]). First we assume the existence of the map φ, and prove (i) its uniqueness, (ii) that φ is a group homomorphism, and (iii) $\mathrm{Ker}(\varphi) = M_{\mathrm{tor}}$.

(i) Suppose that both φ_1 and φ_2 are maps satisfying the conditions (a) and (b). Take any $x \in M$. Then, by (a), $t = \varphi_1(x) - \varphi_2(x)$ belongs to $T \otimes \mathbb{Q}$, while (b) implies $t \perp T$. Using Proposition 5, it is easy to show that $t = 0$, that is, $\varphi_1(x) = \varphi_2(x)$. Hence $\varphi_1 = \varphi_2$.

(ii) Take any $x, y \in M$ and let $z = x + y$. We show that $\varphi(x) + \varphi(y) = \varphi(z)$. Indeed, noting that $\overline{\varphi}$ is a homomorphism, we have by (a)

$$(\varphi(x) + \varphi(y)) \bmod T \otimes \mathbb{Q} = \overline{\varphi}(x) + \overline{\varphi}(y) = \overline{\varphi}(z).$$

On the other hand, $\varphi(x) + \varphi(y)$ is obviously orthogonal to T by (b). Hence the uniqueness (i) implies $\varphi(x) + \varphi(y) = \varphi(z)$, proving (ii).

(iii) Suppose that $x \in M$ belongs to $\mathrm{Ker}(\varphi)$. Then $\overline{\varphi}(x)$ is an element of N/T which is killed in $(N/T) \otimes \mathbb{Q}$. This happens precisely when $\overline{\varphi}(x) \in (N/T)_{\mathrm{tor}}$. Since $M \simeq N/T$, we conclude that $\mathrm{Ker}(\varphi) = M_{\mathrm{tor}}$.

Now we prove the existence of the map φ satisfying (a), (b). This will be done elementwise. Thus fix any $x \in M$. Take an element $\xi \in N$ such that $\xi \bmod T = \overline{\varphi}(x)$. We can choose ξ so that $(\xi \cdot F) = 0$ by using $(O), F \in T$. To satisfy the condition (a), we write $\varphi(x) = \xi + t$ for some $t \in T \otimes \mathbb{Q}$, and express t as a $\mathbb{Q}$-linear combination of the free generators $(O), F, \Theta_{v,i}$ $(v \in R, i > 0)$ of T. Then by Proposition 5, this expression has a unique solution satisfying the condition (b). Explicitly, we have

$$\varphi(x) = \xi - (\xi \cdot (O))F$$
$$+ \sum_{v \in R} (\Theta_{v,1}, \cdots, \Theta_{v,m_v-1})(-I_v^{-1})^t((\xi \cdot \Theta_{v,1}), \cdots, (\xi \cdot \Theta_{v,m_v-1})), \qquad (27)$$

where I_v denotes the intersection matrix of $\Theta_{v,i}$ (for $1 \le i \le m_v - 1$) and ^{t} denotes transpose. *q.e.d*

Theorem 7 *For any $P \in J(K)$, denote by $\overline{P}$ its image in $M = J(K)/\tau B(k)$. Define a symmetric bilinear form on $J(K)$ by*

$$\langle P, Q \rangle = -(\varphi(\overline{P}) \cdot \varphi(\overline{Q})) \quad (P, Q \in J(K)). \tag{28}$$

Then it defines the structure of a positive definite lattice on

$$M/M_{\mathrm{tor}} = J(K)/(J(K)_{\mathrm{tor}} + \tau B(k)). \tag{29}$$

This will be called the Mordell–Weil lattice of J/K.

The pairing $\langle \, , \, \rangle$ is called the height pairing. The formula (27) gives an explicit formula for the height pairing. If we take a representative divisor class $\mathrm{cl}(D_P)$ of $\overline{P} \in M \simeq \mathrm{NS}(S)/T$ for each $P \in J(K)$, then it reads as

$$\langle P, Q \rangle = -(D_P \cdot D_Q) - \sum_{v \in R} \mathrm{contr}_v(P, Q) \tag{30}$$

where the local contribution term is defined by

$$\mathrm{contr}_v(P, Q) = ((D_P \cdot \Theta_{v,1}), \cdots, (D_P \cdot \Theta_{v,m_v-1}))(-I_v^{-1})$$
$$\times {}^t((D_Q \cdot \Theta_{v,1}), \cdots, (D_Q \cdot \Theta_{v,m_v-1})) \tag{31}$$

(cf. [8]).

In the following, we assume that the K/k-trace of J/K is trivial: $B = 0$, and hence $M = J(K)$ itself is finitely generated (cf. Theorem 3 in §2). Further, for the sake of simplicity, we also assume that $\mathrm{NS}(S)$ is torsion-free.

Then the above process defines the structure of lattice on $J(K)/J(K)_{\mathrm{tor}}$, the Mordell–Weil lattice of J/K. To define the narrow Mordell–Weil lattice of J/K, let $J(K)^0$ be the subgroup of $J(K)$ which is the image of $L = T^{\perp} \subset \mathrm{NS}(S)$ under (1). For $P \in J(K)^0$, we obviously have $\mathrm{contr}_v(P, Q) = 0$ for any $Q \in J(K)$. In particular, we have $\langle P, P \rangle = -(D_P^2)$ by (30). Since we can take $D_P \in L$ by definition, and L is negative definite under our assumption, we obtain:

Theorem 8 *Assume that (i) the K/k-trace of J/K is trivial and (ii) $\mathrm{NS}(S)$ is torsion-free. Then $J(K)^0$ is a positive definite integral lattice, isomorphic to the opposite lattice L^- of L. It will be called the narrow Mordell–Weil lattice of J/K.*

As in the elliptic case, the Mordell–Weil lattice embeds into the dual lattice of the narrow one, and we have the following:

Theorem 9 *In addition to the assumptions (i) and (ii) above, suppose that the Néron–Severi lattice of S is unimodular, that is, $\det \mathrm{NS}(S)/(\mathrm{tor}) = \pm 1$. Then the Mordell–Weil lattice of J/K is isomorphic to the dual lattice of the narrow Mordell–Weil lattice $J(K)^0$.*

The proof is the same as in [8, §9], and is omitted. This theorem applies, for example, when S is a rational surface (see §2, Corollary 4).

5 Examples

5.1 Milnor lattice of the A_n singularities

For any positive integer $n > 1$, consider the hyperelliptic curve $\Gamma = \Gamma_\lambda$ over $K = k(t)$, defined by the equation:

$$y^2 = x^{n+1} + p_2 x^{n-1} + \cdots + p_n x + p_{n+1} + t^2, \quad \lambda = (p_2, \ldots, p_{n+1}) \in k^n. \quad (32)$$

Its genus g is given by $n = 2g$ or $n = 2g + 1$ according to the parity of n. For simplicity, assume that k has characteristic 0.

This equation is a familiar one in the theory of rational singularities; it defines a "semi-universal deformation" of the A_n singularity $y^2 = x^{n+1} + t^2$, with parameter λ. We are interested in comparing the Mordell–Weil lattice arising from this situation with the Milnor lattice which, for the A_n singularity, is known to be the root lattice A_n. Such a viewpoint led to rather interesting results in the framework of MWL of elliptic surfaces, covering the exceptional case E_6, E_7, E_8 (cf. [9], [10]).

Let $u_1, \ldots, u_n, u_{n+1}$ be the roots of the algebraic equation

$$\Phi(x) = \Phi_\lambda(x) = x^{n+1} + p_2 x^{n-1} + \cdots + p_n x + p_{n+1} = 0. \quad (33)$$

Corresponding to them, there are K-rational points of Γ:

$$P_i : (x = u_i, y = t) \quad \text{and} \quad P_i' : (x = u_i, y = -t) \quad \text{for } i = 1, \ldots, n + 1. \quad (34)$$

(I) Assume first that $n = 2g$ is even and that the u_i are all distinct. Let $O \in \Gamma(K)$ be the unique point at infinity, and we take an embedding Γ into J sending O to the origin of J. Then the $2(2g+1)$ points above satisfy the following relation in $J(K)$:

$$\sum_{i=1}^{2g+1} P_i = 0, \quad P_i' = -P_i. \quad (35)$$

(For the former, look at the divisor of the function $h = y - t$ on Γ.)

Theorem 10 *With the above notation, assume that $n = 2g$ and the u_i are mutually distinct. Then the Mordell–Weil group $J(K)$ is a torsion-free Abelian group of rank $r = 2g$. More precisely, the narrow MWL $J(K)^0$ is* *isomorphic to the root lattice A_{2g}, and the full MWL $J(K)$ is isomorphic to its dual lattice A_{2g}^*:*

$$J(K) \quad \simeq \quad A_{2g}^*$$
$$\cup \qquad \cup \qquad \textit{index } 2g + 1. \tag{36}$$
$$J(K)^0 \quad \simeq \quad A_{2g}$$

Moreover $\{P_i, P_i'\}$ correspond to the minimal vectors of A_{2g}^* (with minimal norm $2g/(2g+1)$); in particular, $\{P_1, \ldots, P_{2g}\}$ forms a set of free generators of $J(K)$.

The algebraic surface S associated with (32) is a rational surface, and so we can use Theorem 9. It has a unique reducible fibre at $t = \infty$, and we have

$$\operatorname{rank} T_\infty = 2g + 4, \quad \det T_\infty = 2g + 1. \tag{37}$$

The case $g = 1$ reduces to the known result (Case (A_2) in [9]); in that case, we have $T_\infty^- \simeq E_6$. For $g > 1$, we need to resolve the singularity. Omitting the details, we find that T_∞^- is not a root lattice, but its dual graph can be described as follows: first take the Dynkin graph of type D_{2g+2} and then adjoin a new vertex with norm $g + 1$ to each extreme vertex of the two short branches in D_{2g+2}. The sections (P_i) (or (P_i')) pass through the irreducible component corresponding to one (or the other) of the two new vertices.

(II) Still assuming that $n = 2g$ is even, we consider the general case where $\Phi_\lambda(x)$ has multiple roots. Let

$$\Phi_\lambda(x) = (x - u_1)^{n_1+1} \cdots (x - u_l)^{n_l+1} \tag{38}$$

with $u_1, \ldots, u_l$ distinct, $n_1, \ldots, n_l \geq 0$ and $\sum_j (n_j + 1) = n + 1$. In this case we have

Theorem 11 *Assume $n = 2g$, and denote by L the orthogonal complement of $T_{\text{fin}} = A_{n_1} \oplus \cdots \oplus A_{n_l}$ in A_n; let L^* be its dual lattice. Then, the narrow MWL $J(K)^0$ is isomorphic to L, and $J(K) \simeq L^* \oplus (torsion)$. In particular, $J(K)$ has rank $r = n - \sum n_j = l - 1$.*

It is easy to see that the trivial lattice T is equal to $U \oplus T_{\text{fin}} \oplus T_\infty$ where T_{fin} is as above and T_∞ is the same as in (I). By the theory of simultaneous resolution of rational singularities (see the references in [10]), the surfaces S_λ form a smooth family when a suitable change of parameter is chosen. In particular, $N = \operatorname{NS}(S_\lambda)$ is constant over such a parameter space. Then the orthogonal complement of T in N is equal to that of T_{fin} in A_{2g} by (I), hence the assertion follows from Theorem 9.

As a consequence, we have $r \leq 2g$ where equality holds only for the case treated in (I). On the other hand, we have $r = 0$ iff $l = 1$, which corresponds to the totally degenerate case $\lambda = 0$ with $\Phi_\lambda(x) = x^{2g+1}$. In this case, $J(K)$ is a finite cyclic group of order $2g + 1$, i.e., $J(K) \simeq A_{2g}^*/A_{2g} \simeq \mathbb{Z}/(2g+1)\mathbb{Z}$.

(III) In case $n = 2g + 1$ is odd, the structure of MWL depends more on each individual parameter λ. In any case, we choose one of the two rational points over $x = \infty$ as the origin O of J.

Theorem 12 *Assume $n = 2g + 1$. In the general case where there are no reducible fibres other than $f^{-1}(\infty)$, $\{P_i\}$ generate a lattice isomorphic to the standard unimodular lattice $\mathbb{Z}^{2g+2}$. The narrow MWL $J(K)^0$ is a sublattice of index $g + 1$, consisting of $(x_i) \in \mathbb{Z}^{2g+2}$ satisfying $\sum x_i \equiv 0(g+1)$, and $J(K)$ is isomorphic to its dual lattice.*

Even if $\Phi_\lambda(x)$ has no multiple roots, the structure of MWL can be different from the above general case. For instance, for $\Phi_\lambda(x) = x^{n+1}-1(n+1 = 2g+2)$, there are two more reducible fibres over $t = \pm 1$. In this case, we can prove the following:

$$J(K) \simeq A_g^* \oplus A_g^* \oplus \mathbb{Z}/(g+1)\mathbb{Z}$$
$$\cup \qquad\qquad \cup \qquad\qquad \text{index } (g+1)^3. \qquad (39)$$
$$J(K)^0 \simeq \qquad A_g \oplus A_g$$

Here the torsion part is generated by Q which is the other point at $x = \infty$ other than the chosen origin O. We have $P_i + P_i' = Q$ for all i. The $g + 1$ points P_i corresponding to the $g+1$ roots of $x^{g+1} = 1$ (or $x^{g+1} = -1$) generate a sublattice of $J(K)$ isomorphic to A_g^*.

5.2 The Jacobian of a pencil of hyperplane sections of a regular surface

The following result was announced in [12]:

Theorem 13 *Let X be a regular surface such that $NS(X)$ is torsion-free and suppose that it is embedded as a surface of degree d in a projective space $\mathbf{P}^N$. Let $\{\Gamma_t|t \in \mathbf{P}^1\}$ be a linear pencil of hyperplane sections of X such that every member is irreducible. Let Γ be the generic member of this pencil and let J be* *its Jacobian variety defined over the rational function field $K = k(t)$. Then $J(K)$ is a positive definite integral lattice of rank*

$$r = \rho(X) + d - 2$$

whose determinant is equal to $|\det \mathrm{NS}(X)|$.

Proof By Bezout's theorem, the linear pencil has d base points. By blowing up X at these points, we obtain a regular surface S with a fibration $f \colon S \to \mathbb{P}^1$ whose generic fibre is Γ and which has no reducible fibres. We have d sections arising from the base points; choose one of them as the zero section O. The Néron–Severi group $\mathrm{NS}(S)$ is torsion-free and has rank $\rho(S) = \rho(X) + d$, and the Néron–Severi lattice $\mathrm{NS}(S)$ has $|\det \mathrm{NS}(S)| = |\det \mathrm{NS}(X)|$. The trivial lattice T is equal to the rank 2 unimodular lattice U generated by (O) and F (a fibre), and we have $\mathrm{NS}(S) = U \oplus L$ with $L = T^\perp$. Therefore, by Theorem 3 (cf. Cor. 4) in §2 and Theorem 8 above, we have $J(K) = J(K)^0 \simeq L^-$. It has rank $\rho(S) - 2$ and determinant $\det L = \det \mathrm{NS}(S)$ up to sign.   q.e.d

As noted in [12], an example is given by a linear pencil of degree m curves in a projective plane all of whose members are irreducible. More precisely, take X to be the isomorphic image of the projective plane $\mathbb{P}^2$ under the embedding given by the complete linear system $|mH|$ (H: a line). In this case, we have $d = m^2$ and $J(K)$ is a unimodular lattice of rank $r = m^2 - 1$. This gives a generalization of the Manin–Shafarevich theorem for $m = 3$ saying that, if we choose one of the base points as the origin, the remaining $r = m^2 - 1$ points are linearly independent and generate a subgroup of finite index m in $J(K)$.

References

[1] Artin, M., Winters, G.: Degenerate fibres and stable reduction of curves, Topology 10, 373–383 (1971).

[2] Bosch, S., Lütkebohmert, W., Raynaud, M.: Néron models, Springer-Verlag (1990).

[3] Hartshorne, R.: Algebraic geometry, Springer-Verlag (1977).

[4] Lang, S.: Fundamentals of Diophantine geometry, Springer-Verlag (1983).

[5] Lichtenbaum, S.: Curves over discrete valuation rings, Amer. J. Math. 90, 380–404 (1968).

[6] Raynaud, M.: Spécialisation du foncteur de Picard, Publ. Math. IHES 38, 27–76 (1970).

[7] Shafarevich, I.R.: Lectures on minimal models and birational transformations of two dimensional schemes, Tata Inst. Fund. Res. (1966).

[8] Shioda, T.: On the Mordell–Weil lattices, Comment. Math. Univ. St. Pauli 39, 211–240 (1990).

[9] — : Construction of elliptic curves with high rank via the invariants of the Weyl groups, J. Math. Soc. Japan 43, 673–719 (1991).

[10] — : Mordell–Weil lattices of type E_8 and deformation of singularities, in: Lecture Notes in Math. 1468, 177–202 (1991).

[11] — : Mordell–Weil lattices for higher genus fibration, Proc. Japan Acad. 68A, 247–250 (1992).

[12] — : Generalization of a theorem of Manin–Shafarevich, Proc. Japan Acad. 69A, 10–12 (1993).

[13] — : Constructing curves with high rank via symmetry, Amer. J. Math. **120**, 551–566 (1998).

[14] Tate, J.: On the conjectures of Birch and Swinnerton-Dyer and a geometric analog, Sém. Bourbaki 1965/66, n^0306; in Dix exposés sur la cohomologie des schémas, Masson & CIE; North-Holland Pub. Co. (1968).

[15] Weil, A.: Foundation of algebraic geometry, 2nd ed., AMS (1962).

[16] — : Sur les critères d'équivalence en géométrie algébrique, Math. Ann. 128, 95–127 (1954); Collected papers, II, Springer, 127–159 (1980).

T. Shioda,
Department of Mathematics,
Rikkyo University, Nishi-Ikebukuro,
Tokyo 171, Japan
e-mail shioda@rikkyo.ac.jp

Symplectic Gromov–Witten invariants

Bernd Siebert

Contents

Introduction

Gromov–Witten (GW) invariants "count" algebraic curves on a smooth projective variety X with certain incidence conditions, but in a rather refined way. They originated in the realm of symplectic rather than algebraic geometry. Salient features are

(1) in unobstructed situations, i.e., if the relevant moduli spaces of algebraic curves are smooth of the expected dimension ("expected" by looking at the Riemann–Roch theorem), one obtains the number that one would naively expect from algebraic geometry. A typical such example is the number of rational curves of degree d in the plane passing through $3d-1$ generic closed points, which is in fact a finite number.

(2) GW invariants are constant under (smooth projective) deformations of the variety.

For the original definition, one deforms X as an *almost complex* manifold and replaces algebraic by pseudoholomorphic curves (i.e., holomorphic with respect to the almost complex structure). For a generic choice of almost complex structure on X, the relevant moduli spaces of pseudoholomorphic curves are oriented manifolds of the expected dimension, and GW invariants can be defined by naive counting. Not every almost complex structure J is admitted though, but (for compactness results) only those *tamed* by a symplectic form ω, which by definition means that $\omega(v, Jv) > 0$ for any nonzero tangent vector $v \in T_X$. In the algebraic case, if J is sufficiently close to the integrable structure, ω may be chosen as the pullback of the Fubini–Study form. It turns out that GW invariants really depend only on the symplectic structure (or its deformation class), hence are symplectic in nature. Since the original definition basically neglects singular curves, GW invariants were restricted to projective manifolds with numerically effective anticanonical bundle.

More recently the situation has changed with the advent of a beautiful, purely algebraic and completely general theory of GW invariants based on an idea of Li and Tian [Be1], [BeFa], [LiTi1]. This development is surveyed in [Be2]. Due to the independent effort of many people there is now also a completely general definition of symplectic GW invariants available [FkOn], [LiTi2], [Ru2], [Si1]. The purpose of the present paper is to supplement Behrend's contribution to this volume [Be2] by the symplectic point of view. We also sketch our own more recent proof of equivalence of symplectic and algebraic GW invariants for projective manifolds.

While it is perfect to have a purely algebraic theory, I believe that the symplectic point of view is still rewarding, even if one is not interested in symplectic questions: apart from the aesthetic appeal of the interplay between geometric and algebraic methods, symplectic techniques are sometimes easier and more instructive to use (if only as a preparation for an algebraic treatment). In [Si3], Proposition 1.1, I gave an example of GW invariants of certain projective bundles that are much more readily accessible by symplectic techniques. I also find that the properties of GW invariants, especially deformation invariance, are more intuitively obvious from the symplectic point of view, cf. also Section 4.2; this may just be a matter of taste. More philosophically, the symplectic nature of enumerative invariants in algebraic geometry should mean something, especially in view of their appearance in mirror symmetry. Finally, it is important to establish algebraic techniques to compute symplectic invariants. In fact, a closed formula for GW invariants, holding in even the most degenerate situations, can easily be derived from the definition, cf. [Si2]. The formula involves only Fulton's canonical class of the moduli space and the Chern class of a virtual bundle.

Gromov–Witten invariants have a rather interesting and involved history, with connections to gauge theory, quantum field theory, symplectic geometry and algebraic geometry. I include some remarks on this, following the referee's suggestion; however, I should point out that I concentrate only on the history of how these invariants were *defined*, rather than their computation and their many interesting applications.

The story begins with Gromov's seminal paper of 1985 [Gv]. In this paper Gromov laid the foundations for a theory of (pseudo-) holomorphic curves in almost complex manifolds. Of course, a notion of holomorphic maps between almost complex manifolds existed already for a long time. However, Gromov's points were that

(1) while higher dimensional almost complex submanifolds or holomorphic functions might not exist even locally, there are always many local holomorphic curves;

(2) the local theory of curves in almost complex manifolds largely parallels the theory in the integrable case, i.e., on $\mathbb{C}^n$ with the standard complex structure (Riemann removable singularities theorem, isolatedness of singular points and intersections, identity theorem);

(3) to get good global properties one should require the existence of a "*taming*" symplectic form ω (a closed, non-degenerate two-form) with $\omega(v, Jv) > 0$ for any nonzero tangent vector v (where J is the almost complex structure).

In fact, in the tamed setting, Gromov proves a compactness result for spaces of pseudoholomorphic curves in a fixed homology class. At first sight, the requirement of a taming symplectic form seems to be merely technical. However, Gromov turned this around and observed that given a symplectic manifold (M, ω), the space of almost complex structures tamed by ω is always nonempty and connected. With the then recent ideas of gauge theory, Gromov studied moduli spaces of pseudoholomorphic curves in some simple cases for generic tamed almost complex structures. One such case was pseudoholomorphic curves homologous to $\mathbb{P}^1 \times \{\text{pt}\}$ on $\mathbb{P}^1 \times T$ with T an (compact) n-dimensional complex torus. He shows that for any almost complex structure on $\mathbb{P}^1 \times T$ tamed by the product symplectic structure, such pseudoholomorphic curves exist. In modern terms, he shows that the associated GW invariant is nonzero. This can then be used to prove his famous squeezing theorem: the symplectic ball of radius r cannot be symplectically embedded into the cylinder $B_R^2 \times \mathbb{C}^n$ for $R < r$.

Several more applications of pseudoholomorphic curves to the global structure of symplectic manifolds were given already in Gromov's paper, and many more have appeared since. Probably the most striking of these is due to Floer [Fl]. He interpreted the Cauchy–Riemann equation of pseudoholomorphic curves as flow lines of a functional on a space of maps from the circle S^1 to the manifold. He can then do Morse theory on this space of maps. The homology of the associated Morse complex is the celebrated (symplectic) *Floer homology*, which has been used to solve the Arnol'd conjecture on fixed points of nondegenerate Hamiltonian symplectomorphisms. I mention Floer's work also because it is in the (rather extended) introduction to [Fl] that a (quantum) product structure on the cohomology of a symplectic manifold makes its first appearance (and is worked out for $\mathbb{P}^n$). As we now (almost) know [RuTi2], [PSaSc] this agrees with the product structure defined via GW invariants, i.e., quantum cohomology.

An entirely different, albeit related, development took place in physics. Starting from Floer's *instanton homology*, a homology theory developed by Floer in analogy to the symplectic case for gauge theory on three-manifolds, Witten [Wi1] observed that one can formulate supersymmetric gauge theory on closed four-manifolds, provided one changes the definition of the fields in an appropriate way ("twisting procedure"). The result is a physical theory that reproduces Donaldson's polynomial invariants as correlation functions. Because the latter are (differential) topological invariants, the twisted theory is referred to as a *topological quantum field theory*. In [Wi2] Witten applied the twisting procedure to nonlinear sigma models instead of gauge theory. Such a theory is modelled on maps from a Riemann surface to a closed, almost complex manifold. The classical extrema of the action functional are then pseudoholomorphic maps. The correlation functions of the theory are physical analogues of GW invariants. Witten was the first to observe much of the rich algebraic structure that one expects for these correlation functions from degenerations of Riemann surfaces [Wi3].

It is curious that, while simple versions of GW invariants were used as a tool in symplectic topology, and the technical prerequisites for a systematic treatment along the lines of Donaldson theory were all available (notably through the work of McDuff, the compactness theorem by Gromov, Pansu, Parker–Wolfson and Ye), it was only in 1993 that Ruan tied up the loose ends [Ru1] and defined symplectic invariants based on moduli spaces of pseudoholomorphic (rational) curves, mostly for positive symplectic manifolds. It was quickly pointed out to him that one of his invariants was the mathematical analogue of correlation functions in Witten's topological sigma model.

At the end of 1993 the breakthrough in the mathematical development of GW invariants and their relations was achieved by Ruan and Tian in the important paper [RuTi1]. Apart from special cases (complex homogeneous manifolds), until recently, the results of Ruan and Tian were the only methods available to make precise sense of GW invariants for a large class of (semipositive) manifolds, including Fano and Calabi–Yau manifolds, and to establish relations between them, notably the associativity of the quantum product and the WDVV equation. Many of the deeper developments in GW theory used these methods, including Taubes' relationship between GW invariants and Seiberg–Witten invariants of symplectic four-manifolds [Ta], as well as Givental's proof of the mirror conjecture for the quintic via equivariant GW invariants [Gi]. For the case of positive symplectic manifolds, proofs for the glueing theorem for two rational pseudoholomorphic curves, which is the reason for associativity of the quantum product, were also given by different methods in the Ph.D. thesis of G. Liu [Liu] and in the lecture notes [McSa].

Early in 1994, Kontsevich and Manin [KoMa] advanced the theory in a different direction: rather than proving the relations among GW invariants, they formulated them as axioms and investigated their formal behaviour. They introduced a rather big compactification of the moduli space of maps from a Riemann surface by "stable maps" (cf. Definition 1.1 below). With this choice, all relations coming from degenerations of domains can be formulated in a rather regular and neat way. In the algebraic setting, spaces of stable maps have projective algebraic coarse moduli spaces [FuPa]; fine moduli spaces exist in the category of Deligne–Mumford stacks [BeMa]. Another plus is the regular combinatorial structure that allows the use of methods of graph theory to compute GW invariants in certain cases. No suggestion was made, however, of how to address the problem of degeneracy of moduli spaces, that in Ruan and Tian's approach, applied to projective algebraic manifolds, forces the use of general almost complex structures rather than the integrable one.

This problem was only solved in the more recent references given above, first in the algebraic and finally in the symplectic category, by constructing virtual fundamental classes on spaces of stable maps.

Here is an outline of the paper: Chapter 1 starts with a simple model case to discuss both the traditional approach and the basic ideas of [Si1]. Chapter 2 is devoted to the most technical part of my approach, the construction of a Banach orbifold containing the moduli space of pseudoholomorphic curves. The ambient Banach orbifold is used in Chapter 3 to construct the virtual fundamental class on the moduli space. Chapter 4 discusses the properties of GW invariants that one obtains easily from the virtual fundamental class. We follow here the same framework as in [Be2], so a comparison is easily possible. A fairly detailed sketch of the equivalence with the algebraic definition is given in the last chapter. The proof shows that the obstruction theory chosen in the algebraic context is natural also from the symplectic point of view. For this chapter we assume some understanding of the algebraic definition.

After finishing this survey, I received a similar survey by Li and Tian [LiTi3], in which they also announce a proof of equivalence of symplectic and algebraic Gromov–Witten invariants.[1]

A little warning is in order: the symplectic definition of GW invariants is more involved than the algebraic one. Modulo checking the axioms and the formal apparatus needed to do things properly, the latter can be given a rather concise treatment, cf. [Si2]. But as long as symplectic GW invariants are based on pseudoholomorphic curves, even to find local embeddings of the moduli space into finite dimensional manifolds ("Kuranishi model") means a considerable amount of technical work. In this survey I tried to emphasize ideas and the reasons for doing things in a particular way, but at the same time keep the presentation as nontechnical as possible. While we do not assume any knowledge of symplectic geometry or GW theory, the ideal reader would have some basic acquaintance with the traditional approach, e.g., from [McSa]. The reader who feels uneasy with symplectic manifolds is invited to replace the word "symplectic" by "Kähler".

I thank the referee and Miles Reid for pointing out some unclear points and M.R. for carefully editing the whole text.

1   Setting up the problem

1.1   The traditional approach

The purpose of this section is to present Ruan's approach to GW invariants on a simple example. We refer to the lecture notes [McSa] for background information and a more careful exposition. Let Σ be a closed Riemann surface with complex structure j, and (M, J) an almost complex manifold. We also fix some $k \geq 1$ and $\alpha \in (0, 1)$. The space $\mathcal{B} := C^{k,\alpha}(\Sigma; M)$ of k-times

---

[1]Note added in proof: Both papers are meanwhile available as [LiTi4], [Si4].

differentiable maps from Σ to M with kth derivative of Hölder class α is a Banach manifold. Charts at $\varphi\colon \Sigma \to M$ are given, for example, by

$$C^{k,\alpha}(\Sigma; \varphi^*T_M) \supset V \longrightarrow C^{k,\alpha}(\Sigma; M)$$
$$v \longmapsto \left(z \mapsto \exp_{\varphi(z)} v(z)\right).$$

Here the exponential map is with respect to some fixed connection on M, and V is a sufficiently small open neighbourhood of 0 where the map is injective. The equation for φ to be J-holomorphic is

$$\bar{\partial}_J\varphi = 0, \quad \bar{\partial}_J\varphi = \frac{1}{2}(D\varphi + J \circ D\varphi \circ j) \in C^{k-1,\alpha}(\Sigma; \varphi^*T_M \otimes_{\mathbb{C}} \overline{\Omega}).$$

Here we write $\overline{\Omega} = \Lambda^{0,1}T_\Sigma^*$ for the bundle of $(0,1)$-forms on Σ. The equation $\bar{\partial}_J\varphi \circ j = -\varphi^*J \circ \bar{\partial}_J\varphi$ in the space of homomorphisms between the complex vector bundles (T_Σ, j) and $(\varphi^*T_M, \varphi^*J)$ shows that, viewed as section of φ^*T_M, $\bar{\partial}_J\varphi$ is indeed $(0,1)$-form valued. Intrinsically, these equations fit together to a section $s_{\bar{\partial},J}$ of the Banach bundle $\mathcal{E} \downarrow \mathcal{B}$ with fibers

$$\mathcal{E}_\varphi = C^{k-1,\alpha}(\Sigma; \varphi^*T_M \otimes \overline{\Omega}).$$

Local trivializations of $\mathcal{E}$ over the above charts can easily be constructed by (the complex linear part of) parallel transport of vector fields along the family of closed geodesics $\gamma_z(t)$ with $\gamma_z(0) = \varphi(z)$, $\dot{\gamma}_z(0) = v(z)$.

Obviously, $s_{\bar{\partial},J}$ is differentiable. As for any differentiable section of a vector bundle, its linearization *at a point of the zero locus*, which is a linear map $T_{\mathcal{B},\varphi} \to \mathcal{E}_\varphi$, is independent of the choice of local trivialization. Thus over the zero locus Z of $s_{\bar{\partial},J}$ the linearization induces a section σ of $\mathrm{Hom}(T_\mathcal{B}, \mathcal{E})$. One computes for the linearization at J-holomorphic φ:

$$\sigma_\varphi := D_\varphi s_{\bar{\partial},J} \colon T_{\mathcal{B},\varphi} = C^{k,\alpha}(\Sigma; \varphi^*T_M) \longrightarrow \mathcal{E}_\varphi = C^{k-1,\alpha}(\Sigma; \varphi^*T_M \otimes \overline{\Omega})$$
$$v \longmapsto \bar{\partial}_J^\varphi v + N_J(v, D\varphi).$$

Here N_J is the Nijenhuis tensor or almost complex torsion of (M, J), a certain tensor on M depending only on J that vanishes identically if and only if J is integrable, see, e.g., [KoNo], Chap. IX, §2; and $\bar{\partial}_J^\varphi$ is the $\bar{\partial}$ operator belonging to a natural holomorphic structure on φ^*T_M. Concerning the latter, one actually *defines* $\bar{\partial}_J^\varphi$ as the $(0,1)$-part of the linearization of $s_{\bar{\partial},J}$. The integrability condition being void in dimension one, this complex linear partial connection defines a holomorphic structure on φ^*T_M, cf. Section 2.4.

We see that up to a zeroth order differential operator, σ_φ is just the Cauchy–Riemann operator of a holomorphic vector bundle over Σ of rank $n = \dim_{\mathbb{C}} M$. But $\bar{\partial}_J^\varphi$ is a Fredholm operator on appropriate spaces of sections, i.e., it has finite dimensional kernel and cokernel. It is crucial at this point to work with Hölder spaces: the Fredholm property is false for $\alpha = 0$. Alternatively, as below, one may work with Sobolev spaces. Now lower order perturbations are compact operators by the Arzela–Ascoli compactness theorem, and the Fredholm property and the index (the dimension of the kernel minus that of the cokernel) do not change under adding a compact operator. This shows that σ_φ is a Fredholm operator whose index is given by

$$\frac{1}{2}\operatorname{ind}\sigma_\varphi \;=\; \operatorname{ind}_\mathbb{C}\overline{\partial}_J^\varphi \;=\; \chi(\varphi^*T_M,\overline{\partial}_J^\varphi),$$

the holomorphic Euler characteristic of φ^*T_M. The latter can be computed by the ordinary Riemann–Roch theorem to be

$$\deg(\varphi^*T_M,\varphi^*J) + (1-g)\dim_\mathbb{C}M \;=\; c_1(M,J)\cdot\varphi_*[\Sigma] + (1-g)n.$$

If $s_{\overline{\partial},J}$ is transverse at φ, which by definition means that σ_φ is surjective, then an application of the implicit function theorem shows that in a neighbourhood of φ, the space $\mathcal{C}^{\mathrm{hol}}(\Sigma,M,J)$ of J-holomorphic maps $\Sigma \to M$ is a differentiable manifold of dimension $\operatorname{ind}\sigma_\varphi$. Moreover, near such points, $\mathcal{C}^{\mathrm{hol}}(\Sigma,M,J)$ is naturally oriented by complex linearity of $\overline{\partial}_J^\varphi$. Ignoring questions of compactness for the moment, if transversality is true everywhere, $\mathcal{C}^{\mathrm{hol}}(\Sigma,M,J)$ is a good moduli space for enumerative purposes involving J-holomorphic curves, i.e., GW invariants. In the integrable situation (i.e., M a complex manifold), transversality at φ means that the deformation theory of φ is unobstructed: φ deforms both under deformations of J and j. By the same token we see that the same statement holds even under deformations of J as almost complex structure.

Using the Sard–Smale theorem one can make $s_{\overline{\partial},J}$ transverse everywhere except possibly at so-called *multiple cover curves*, simply by a generic choice of J. A multiple cover curve is a $\varphi\colon\Sigma\to M$ that factors over a holomorphic map of Riemann surfaces of higher degree. For J-holomorphic curves this is equivalent to saying that φ is not generically injective. The reason is that the Sard–Smale theorem requires that perturbations of J generate the tangent space of the ambient space, and this may fail if φ is not generically injective. For certain (positive) manifolds the bad locus of compactifying and multiple cover curves can be proved to be of lower dimension for generic J and thus to be ignorable for enumerative questions. This is the original, very successful, approach of Ruan to GW theory [Ru1], following a similar scheme in gauge theory.

In the general case there are two ways of proceeding. If one wants to stick to manifolds one can introduce an *abstract perturbation term*, which is just a section $\nu \in C^1(\mathcal{B};\mathcal{E})$, and consider solutions of the perturbed equation

$$\overline{\partial}_J\varphi \;=\; \nu(\varphi),$$

i.e., look at the zero locus of the perturbed section $s_{\bar{\partial},J} - \nu$. Again by the Sard–Smale theorem, for generic choice of ν, $\mathcal{C}^{\mathrm{hol}}(\Sigma, M, J, \nu) := Z(s_{\bar{\partial},J} - \nu)$ is a canonically oriented manifold of dimension $\mathrm{ind}\,\sigma_\varphi$. So one replaces the possibly singular, wrong dimensional $\mathcal{C}^{\mathrm{hol}}(\Sigma, M, J)$ by an approximating manifold inside $\mathcal{B}$. In GW theory this has again been pioneered by Ruan to extend the range of the previous approach to semipositive manifolds by removing nongenericity of multiple cover curves (the perturbation term unfortunately vanishes at "bubbling components" of the domain, cf. Section 1.3, that have to be included on the compactifying part; so this method still does not work generally). The idea of associating (cobordism classes of) finite dimensional submanifolds to Fredholm maps by perturbations goes back (at least) to Smale [Sm].

The other approach, that we follow for the most part, is to replace the manifold by a homology class located on $\mathcal{C}^{\mathrm{hol}}(\Sigma, M, J)$. The homology class should be thought of as the limit of the fundamental classes of the perturbed manifolds $Z(s_{\bar{\partial},J} - \nu)$ as ν tends to zero. Because its image in $H_*(\mathcal{B})$ represents any of these fundamental classes, the homology class is called the *virtual fundamental class* of $\mathcal{C}^{\mathrm{hol}}(\Sigma, M, J)$. For conceptual clarity let us discuss this topic in an abstract setting.

1.2 Localized Euler classes in finite and infinite dimensions

First a note on homology theories. While cohomology has good properties on a large class of spaces making it essentially unique, (singular) homology behaves well only on compact spaces. Several extensions to noncompact spaces are possible. Since we need fundamental classes of noncompact oriented manifolds, the natural choice is singular homology of the second kind, i.e., with only *locally finite* singular chains, or Borel–Moore homology with coefficients in the ring $\mathbb{Z}$ (or later also $\mathbb{Q}$). These two homology theories are isomorphic under fairly mild conditions on the spaces, that are fulfilled in cases of our interest, cf. [Sk]. Note that this homology theory has restriction homomorphisms to open sets, obeys invariance only under *proper* homotopy, and pushes forward only under *proper* morphisms. General references are [Br],[Iv] and [Sk].

Given a Hausdorff space X with a closed subspace Z the localized cap products are homomorphisms

$$\cap\colon H_n(X) \otimes H_Z^k(X) \longrightarrow H_{n-k}(Z),$$

where, as mostly in the sequel, we suppressed coefficient rings. If s is a section of an oriented topological vector bundle E of finite rank r over X, the Euler class of E can be localized on the zero locus Z of s. Namely, let $\Theta_E \in H_X^r(E)$ be the Thom class of E. Locally, Θ_E is of the form $\pi^* \delta_0$, where $\pi \colon E|_U \to \mathbb{R}^r$ is a local trivialization and $\delta_0 \in H_{\{0\}}^r(\mathbb{R}^r, \mathbb{Z})$ is the unique generator compatible with the orientation. Then $s^* \Theta_E \in H_Z^r(X)$ represents the Euler class of E. And if X is an oriented topological manifold one may pair $s^* \Theta_E$ with the fundamental class $[X]$ using the localized cap product to arrive at a homology class on Z which is Poincaré dual on X to the Euler class of E.

In a differentiable situation, i.e., E a differentiable vector bundle over an oriented differentiable manifold X, let s_i be (differentiable) transversal sections converging to s. Then $[X] \cap s_i^* \Theta_E = [Z(s_i)]$, the fundamental class of the naturally oriented manifold $Z(s_i)$. These converge to $[X] \cap s^* \Theta_E \in H_{n-r}(Z)$, $n = \dim X$, which may thus be viewed as natural homological replacement for the zero loci of generic perturbations of s.

In the infinite dimensional setting of the previous section, neither $\Theta_{\mathcal{E}}$ nor $[\mathcal{B}]$ make sense. But if s is differentiable with Fredholm linearizations, if $Z = Z(s)$ is compact and if $\mathcal{B}$ admits differentiable bump functions, we may do the following: by hypothesis it is possible to construct a homomorphism from a trivial vector bundle $\tau \colon F = \mathbb{R}^r \to \mathcal{E}$ so that for any $x \in Z(s)$, $\tau_x + D_x s \colon \mathbb{R}^r \oplus T_{\mathcal{B},x} \to \mathcal{E}_x$ is surjective. Replace $\mathcal{B}$ by the total space of F (which is just $\mathcal{B} \times \mathbb{R}^r$, but we will need nontrivial bundles later), and consider the section $\tilde{s} := q^* s + \tau$ of $q^* \mathcal{E}$, where $q \colon F \to \mathcal{B}$ is the bundle projection. Note that if we identify $\mathcal{B}$ with the zero section of F then $\tilde{s}|_{\mathcal{B}} = s$. $\tau_x + D_x s$ to be surjective means that $\tilde{s}$ is a transverse section, at least in a neighbourhood of $Z = \tilde{Z} \cap \mathcal{B}$, $\tilde{Z} = Z(\tilde{s})$. So $\tilde{Z} \subset F$ is a manifold near the zero section of F. And F being of finite rank it has a Thom class, no matter the base is infinite dimensional. Ignoring questions of orientation we may then define

$$[\mathcal{E}, s] := [\tilde{Z}] \cap \Theta_F \in H_*(Z).$$

It is not hard to check independence of choices and coincidence with $[Z]$ in transverse situations. The dimension of $[\mathcal{E}, s]$ is locally constant and equals the index of the linearization of s.

Similar ideas have been applied in certain cases to compute both Donaldson and GW invariants, notably if the dimension of the cokernel is constant. In the presented generality this is due to Brussee who used it to study Seiberg–Witten theory in degenerate situations [Bs].

I should also point out that, locally, $[\mathcal{E}, s]$ is uniquely determined by a *Kuranishi model* for Z: let s be given locally by a differentiable Fredholm map $f \colon U \to E = \mathcal{E}_x$, $U \subset \mathcal{B}$ open. Let $q \colon E \to Q$ be a projection with finite dimensional kernel C ("C" for cokernel) such that $D(q \circ f) \colon T_{\mathcal{B},x} \to Q$ is surjective. Possibly after shrinking U, $\tilde{Z} = (q \circ f)^{-1}(0)$ is a manifold of dimension $\mathrm{ind}\, f + r$, $r = \dim C$. Then $f|_{\tilde{Z}} \colon \tilde{Z} \to C$ is a differentiable map between finite dimensional manifolds. Again ignoring questions of orientation, we observe

$$[\mathcal{E}, s]\big|_U = [\widetilde{Z}] \cap (f|_{\widetilde{z}})^* \delta_0 \in H_*(Z \cap U),$$

$\delta_0 \in H^r_{\{0\}}(C, \mathbb{Z})$ the positive generator. However, unless Z is of *expected dimension* ind s, the glueing of the local classes may not be unique. So the knowledge of Kuranishi models covering $Z(s)$ may not be enough to determine the class $[\mathcal{E}, s]$. It would be interesting to understand precisely what additional datum is needed to globalize these classes. In a sense this is the question how topological a theory of localized Euler classes of *differentiable* Fredholm sections can be made.

1.3 The compactification problem: stable J-curves

Since we used compactness of Z for the construction of $[\mathcal{E}, s]$, the method of the last section applies directly to our model in 1.1 only on compact components of the space of J-holomorphic maps from Σ to M. This never holds for the important case of $\Sigma = \mathbb{P}^1$ because of noncompactness of $\mathrm{Aut}(\mathbb{P}^1) = \mathrm{PGL}(2)$, which acts on these spaces by reparametrization. This trivial cause of noncompactness could be avoided by factoring out the connected component of the identity $\mathrm{Aut}^0(\Sigma)$. More fundamentally though, the space of J-holomorphic maps is not compact if so-called *bubbles* appear in limits of sequences of such maps. If $\varphi_i \colon \Sigma \to M$ is a sequence of J-holomorphic maps, a bubble is a J-holomorphic rational curve $\psi \colon \mathbb{P}^1 \to M$ obtained as the limit of rescalings of φ_i near a sequence of points $P_i \to P \in \Sigma$ with $|D\varphi_i(P_i)|$ unbounded. A simple example of bubbling off in algebraic geometry is the degeneration of a family of conics in the plane to a line pair. The content of the Gromov compactness theorem is that this phenomenon is the precise reason for noncompactness of moduli spaces of J-holomorphic curves of bounded volume, cf. Theorem 1.2. As one knows from examples in algebraic geometry this happens quite often unless $(\varphi_i)_*[\Sigma]$ (constant on connected components of $\mathcal{C}^{\mathrm{hol}}(\Sigma, M, J)$) is indecomposable in the cone in $H_2(M, \mathbb{Z})$ spanned by classes representable by J-holomorphic curves.

We are thus lead to the problem of introducing an appropriate compactification of $\mathcal{C}^{\mathrm{hol}}(\Sigma, M, J)$.[2] This is due to Gromov [Gv], and Parker and Wolfson from a different point of view [PrWo], but has been put into its final form by Kontsevich through the notion of stable map [KoMa]. It is convenient to also incorporate marked points on Σ now.

Definition 1.1 $(C, \mathbf{x}, \varphi)$ is called a *stable J-holomorphic curve* if

---

[2]Another reason compactification is essential is of course that we need a degree map to extract well-defined numbers out of the homology class, cf. 4.1

- C is a connected, reduced, complete complex algebraic curve with at most ordinary double points (a "Riemann surface with nodes");

- $\mathbf{x} = (x_1, \ldots, x_k)$ with pairwise distinct $x_i \in C_{\text{reg}}$;

- for any irreducible component $D \subset C$, $\varphi|_C$ is J-holomorphic;

- $\text{Aut}(C, \mathbf{x}, \varphi) := \{\sigma \colon C \to C \text{ biregular} \mid \varphi \circ \sigma = \varphi\}$ is finite. $\square$

The first two conditions are sometimes summarized by saying that $(C, \mathbf{x})$ is a *pre-stable (marked) curve*. The condition on finiteness of the automorphism group is the *stability condition*. It can be rephrased by saying that the normalization (=desingularization) D of any rational component of C contracted by φ contains at least three special points (marked points or preimages of nodes under $D \to C$). Note that by putting $M = \{\text{pt}\}$ one retrieves the definition of stable algebraic curves with marked points due to Deligne and Mumford [DeMu] and Knudson [Kn]. So the notion of stable J-holomorphic curve should be viewed as natural extension of the concept of Deligne–Mumford stable curve to the situation relative to M rather than the spectrum of a field. The *genus* $g(C)$ of $(C, \mathbf{x}, \varphi)$ is by definition the arithmetic genus $h^1(C, \mathcal{O}_C)$ of C. Since $h^1(C, \mathcal{O}_C) = 1 - \chi(C, \mathcal{O}_C)$ is constant in flat families, $g(C)$ could alternatively be defined as the genus of a smooth fiber of a deformation of C, i.e., the genus of the closed surface obtained from C by replacing each double point $x \cdot y = 0$ by a cylinder $x \cdot y = \varepsilon$.

How does this concept incorporate bubbling phenomena, say in our model of maps $\varphi \colon \Sigma \to M$? After rescaling at P_i in such a way that the differentials become uniformly bounded at P_i, there might be another sequence of points with unbounded differentials. So we may end up in the limit with a *tree* $\psi \colon B \to M$ of J-holomorphic rational curves at P. To be a tree means that B is simply connected and has at worst ordinary double points. To achieve the latter, one might have to introduce more rational components than necessary to make sense of a limiting map, i.e., ψ might be trivial on some irreducible components $D \subset B$, but only if D contains at least three double points. Because the only marked Riemann surface with infinitesimal automorphisms fixing one more point (a double point making the whole curve connected) is $\mathbb{P}^1$ with fewer than two marked points, this is the stability condition! So in this case, the domain of the limiting map will be $(C, \mathbf{x}) = (\Sigma \cup_P B, \emptyset)$, or more generally, several trees $B_1, \ldots, B_b$ of rational curves attached to Σ at several points.

Conversely, if $(C, \mathbf{x}, \varphi)$ is a stable J-holomorphic curve, there is always an associated stable curve $(C, \mathbf{x})^{\text{st}} = (C_{\text{st}}, \mathbf{x}_{\text{st}})$ so that $(C, \mathbf{x}, \varphi)$ looks like the curve obtained starting from a sequence of J-holomorphic curves by bubbling off (in reality this deformation problem might be obstructed). $(C, \mathbf{x})^{\text{st}}$

is just the *stabilization* of the abstract curve $(C, \mathbf{x})$ obtained by successive contraction of (absolutely) unstable components. The latter are by definition rational components $D \subset C$ with $\#\{x_i \in D\} + \#C_{\text{sing}} \cap D < 3$. Equivalently, $\text{Aut}^0(C, \mathbf{x})$ acts nontrivially on D. After contracting all unstable components of $(C, \mathbf{x})$, some previously stable components may become unstable. The process is then repeated until the result is a Deligne–Mumford stable curve $(C, \mathbf{x})^{\text{st}}$. Note that the genus does not change under this process.

By this picture it is natural to distinguish *bubbling* and *principal* components of the domain of stable J-holomorphic curves $(C, \mathbf{x}, \varphi)$, depending on whether or not the component gets contracted under the *stabilization map* $(C, \mathbf{x}) \to (C, \mathbf{x})^{\text{st}}$. Note that if $(C, \mathbf{x})^{\text{st}}$ is singular, there is another possible type of bubbling off, different from that discussed above, introducing a *chain* of $\mathbb{P}^1$s joining the two branches over an ordinary double point.

One subtlety in the discussion of stable J-curves is that their domains are only pre-stable curves, which do not in general possess decent moduli spaces. To explain this, recall the local description of $\mathcal{M}_{g,k}$, the coarse moduli space of (Deligne–Mumford) stable curves of genus g with k marked points. For later use, it is better to work complex analytically now. If $(C, \mathbf{x}) \in \mathcal{M}_{g,k}$ there is an open subset $S \subset \text{Ext}^1(\Omega_C(x_1 + \cdots + x_k), \mathcal{O}_C) \simeq \mathbb{C}^{3g-3+k}$, a flat family $q \colon \mathcal{C} \to S$ (with $\mathcal{C}$ smooth) of pre-stable curves with k sections $\underline{x} \colon S \to \mathcal{C} \times_S \cdots \times_S \mathcal{C}$, such that the germ of $(\mathcal{C} \to S, \underline{x})$ at $0 \in S$ is an analytically universal deformation of $(C, \mathbf{x})$. This means that the germ of any flat family of marked stable curves with central fiber $(C, \mathbf{x})$ is (canonically isomorphic to) the pullback of $(\mathcal{C} \to S, \underline{x})$ under a map from the parameter space to S. If $(C, \mathbf{x})$ has nontrivial automorphisms, the action on the central fiber extends to an action on (the germ at $0 \in S$ of) $\mathcal{C}$ and S making q and $\underline{x}$ equivariant. After possibly shrinking S, one may also assume that s and $s' \in S$ parametrize isomorphic marked stable curves if and only if there exists an automorphism of $(C, \mathbf{x})$ carrying s to s'. Since $\text{Aut}(C, \mathbf{x})$ is finite we may assume that the action is in fact well defined on all of $\mathcal{C}$ and S. The quotient $S/\text{Aut}(C, \mathbf{x})$ exists as complex space and is a neighbourhood of $(C, \mathbf{x})$ in $\mathcal{M}_{g,k}$.

If $(C, \mathbf{x})$ is just pre-stable we still have a pair $(\mathcal{C} \to S, \underline{x})$. But now $\text{Aut}(C, \mathbf{x})$ is higher dimensional and $\dim S = 3g - 3 + k + \dim \text{Aut}(C, \mathbf{x})$. There is the germ of an action of $\text{Aut}^0(C, \mathbf{x})$ on $\mathcal{C} \to S$, which is a map from a neighbourhood of $\{\text{Id}\} \times \mathcal{C}$ in $\text{Aut}^0(C, \mathbf{x}) \times \mathcal{C}$ to $\mathcal{C}$ (respectively, from a neighbourhood of $(\text{Id}, 0) \in \text{Aut}^0(C, \mathbf{x}) \times S$ to S). Now $(\mathcal{C} \to S, \underline{x})$ is no longer a universal deformation, but only semiuniversal, which means that uniqueness holds only on the level of tangent maps at $0 \in S$. The moduli "space" $\mathfrak{F}_{g,k}$ of pre-stable curves of genus $g = g(C)$ with $k = \#\underline{x}$ marked points should locally around $(C, \mathbf{x})$ be thought of as the quotient of S by the analytic equivalence relation generated by this action. Now $\text{Aut}^0(C, \mathbf{x})$ decomposes into a product with factors $\mathbb{C}^*$ for each bubbling component with only two special (i.e., marked or singular) points and $\mathbb{C} \rtimes \mathbb{C}^*$ for each bubbling component with one special point. In appropriate coordinates, the restriction of the action to one such $\mathbb{C}^*$ looks like the standard $\mathbb{C}^*$ action on $\mathbb{C}$ cross a trivial factor. So a quotient does not even exist as a Hausdorff topological space, let alone analytic space or scheme.

Despite this, $\mathfrak{F}_{g,k}$ behaves in many respects like a scheme. It has a structure of what is called an *Artin stack*. We do not go into details of this, but instead keep in mind the local description as a quotient of the base S of a semiuniversal deformation of $(C, \mathbf{x})$ by the analytic equivalence relation generated by $\mathrm{Aut}(C, \mathbf{x})$.

It is also useful to observe that the semiuniversal deformation $\mathcal{C} \to S$ of $(C, \mathbf{x})$ fibers over the universal deformation $\overline{\mathcal{C}} \to \overline{S}$ of its stabilization $(C, \mathbf{x})^{\mathrm{st}}$. The map $S \to \overline{S}$ is smooth (a linear projection in appropriate coordinates) unless $(C, \mathbf{x})$ has bubble chains (bubbles inserted at a double point of $(C, \mathbf{x})^{\mathrm{st}}$), in which case it is only of complete intersection type (with factors of the form $(x_1, \ldots, x_r) \mapsto x_1 \cdots \cdots x_r$ in appropriate coordinates). This is important in the proof of the isogeny axiom of GW invariants in Section 4.2.

There is a natural topology on the set of stable J-holomorphic curves, the *Gromov topology* [Gv], §1.5 cf. also [Pn], Def. 2.12. We do not give the definition here because it will become obvious once we introduce local charts for the ambient Banach manifold in Chapter 3. To state the compactness theorem, let $R \in H_2(M, \mathbb{Z})$ and $g, k \geq 0$.

Theorem 1.2 ([Gv], [Pn], [PrWo], [RuTi1], [Ye]) *Assume that the almost complex structure J is tamed by some symplectic form ω on M, i.e., $\omega(X, JX) > 0$ for all $X \in T_M \setminus \{0\}$. Then the space*

$$C_{R,g,k}^{\mathrm{hol}}(M, J) := \left\{ \begin{array}{c|c} (C, \mathbf{x}, \varphi) \text{ is a stable} & \varphi_*[C] = R, \#\mathbf{x} = k, \\ J\text{-holomorphic curve} & \text{and } g(C) = g \end{array} \right\} \Big/ \mathrm{iso}$$

with the Gromov topology is compact and Hausdorff. $\square$

The taming condition allows us to bound the volume of J-holomorphic curves in terms of its homology class R by the analogue of the Wirtinger theorem. In fact, Ye's method of proof uses only a bound on the volume. That such a bound is crucial in compactness results has been understood in complex analysis since [Bi]. The Hausdorff property is not proved in the given references but requires some additional arguments as given in any of [FkOn], [LiTi2], [Ru2], [Si1].

We also need to enlarge the ambient Banach manifold. We discuss this in a separate chapter since it involves a number of subtleties.

2   The ambient space

To carry out the program of Section 1.2 for the spaces $\mathcal{C}^{\text{hol}}_{R,g,k}(M, J)$ of stable J-holomorphic curves in an almost complex manifold tamed by some symplectic form ω, we want to construct a Banach manifold into which $\mathcal{C}^{\text{hol}}_{R,g,k}(M, J)$ embeds. Obvious choices are spaces of triples $(C, \mathbf{x}, \varphi)$ with $(C, \mathbf{x})$ a k-marked pre-stable curve of genus g and $\varphi\colon C \to M$ just a continuous map such that $\varphi_*[C] = R$, together with some kind of regularity. We will see that requiring φ to be of Sobolev class L^p_1 with $2 < p < \infty$, i.e., with one distributional derivative in L^p, is a very natural condition. The measure is with respect to a metric on C with certain weights at the singular points. For general facts on Sobolev spaces we refer to the books [Ad] (for Sobolev spaces on $\mathbb{R}^n$) and [Au] for the case of manifolds. Most standard textbooks on partial differential equations also contain ample information on Sobolev spaces. Note that since the domain is two dimensional, L^2_1 is a critical case of the Sobolev embedding theorem: there exist discontinuous L^2_1-functions on $\mathbb{R}^2$, but functions in L^p_1 with $p > 2$ always have continuous representatives. Thus L^p_1 with $p > 2$ is the minimal possible regularity for a sensible formulation of the $\bar{\partial}_J$-equation. Conversely, it is unreasonable to expect maps of higher regularity to give rise to a smooth total space, as will be clear from Section 2.4.

2.1   Charts

To produce charts, observe that, intuitively, a small deformation of $(C, \mathbf{x}, \varphi)$ can be split into

(1) a deformation of the domain $(C, \mathbf{x})$ as pre-stable curve, arriving at a possibly less singular curve C', and

(2) a deformation of a pullback $\varphi \circ \kappa$, where $\kappa\colon C' \to C$ is some comparison map that is a diffeomorphism away from the singularities of C.

As in our discussion of the Artin stack $\mathfrak{F}_{g,k}$ in Section 1.3, let $(q\colon \mathcal{C} \to S, \underline{\mathbf{x}})$ be an analytically semiuniversal deformation of $(C, \mathbf{x}) = (q^{-1}(0), \underline{\mathbf{x}}(0))$. Let us write C_s for $q^{-1}(s)$. If $\varphi\colon C \to M$ is L^p_1, we want $\varphi \circ \kappa$ also to be L^p_1. Since $\bigcap_{p>2} L^p_1 = L^\infty_1$ (the Sobolev space of functions with essentially bounded first derivative) and $L^\infty_1 = C^{0,1}$ is the space of Lipschitz maps, a good choice that works for all p should be Lipschitz. Using an analytic description of $q\colon \mathcal{C} \to S$, it is not hard to construct a retraction

$$\kappa\colon \mathcal{C} \longrightarrow C_0 = C,$$

which, when restricted to C_s, is a diffeomorphism away from C_{sing}, and which near the smoothing $zw = t$ of a node $zw = 0$ is given by a linear rescaling

$$(z = re^{i\varphi}) \longmapsto z = \frac{r - |t|^{1/2}}{1 - |t|^{1/2}} e^{i\varphi} \quad \text{if } |z| \geq |w|,$$

and similarly for w if $|w| \geq |z|$. In particular the circle $|z| = |w| = |t|^{1/2}$ is contracted to the node and κ is Lipschitz (note that $\mathcal{C}$ is also smooth).

Next, we define our weighted Sobolev spaces. The choice is distinguished by the fact that we want κ to induce isomorphisms of L^p-spaces. Since the ordinary Lebesgue measure on a nonsingular C_s corresponds to the finite cylindrical measure $drd\varphi = r^{-1}dxdy$ on each branch of C near a singular point (where $z = re^{i\varphi} = x + iy$), our measure μ on C is required to be of this type near C_{sing} and locally equivalent to Lebesgue measure away from this set. We write

$$\check{L}^p(C, \mathbb{R}) \ := \ L^p(C, \mathbb{R}; \mu)$$

and $\check{L}^p_1(C, \mathbb{R}) \subset L^p_1(C, \mathbb{R})$ for the functions possessing one weak derivative in $\check{L}^p(C, \mathbb{R})$ (on each irreducible component of C). Since $\check{L}^p(C, \mathbb{R}) \subset L^p(C, \mathbb{R})$, by the Sobolev embedding theorem there is an embedding of $\check{L}^p_1(C, \mathbb{R})$ into the space of continuous functions $C^0(C, \mathbb{R})$ for $p > 2$. We adopt the usual abuse of notation and identify $\check{L}^p_1(C, \mathbb{R})$ with its image in the space of continuous functions, i.e., we take the unique continuous representative of any class in $\check{L}^p_1(C, \mathbb{R})$. Note that in general there is no distinguished choice of metric on C, so these spaces are well defined only as topological vector spaces, not as normed spaces.

For vector bundles E over C one defines similarly $\check{L}^p(C, E)$ and $\check{L}^p_1(C, E)$. And as usual, spaces of maps $\check{L}^p_1(C, M)$ can be defined either by embedding M into some $\mathbb{R}^N$ and requiring the Sobolev property componentwise, or by taking local coordinates on M and require composition with the coordinate functions to be $\check{L}^p_1$. This is well defined for $p > 2$ by continuity.

Here is the definition of our ambient space. We fix once and for all some p with $2 < p < \infty$.

Definition 2.1 $(C, \mathbf{x}, \varphi)$ is a *stable complex curve* in M of Sobolev class L^p_1 if and only if

- $(C, \mathbf{x})$ is a pre-stable marked curve;

- $\varphi \in \check{L}^p_1(C, M)$;

- for any unstable component D of $(C, \mathbf{x})$, $\varphi|_D$ is homotopically non-trivial. $\square$

We use $\mathcal{C}(M; p)$ to denote the set of isomorphism classes of such curves and

$$\mathcal{C}_{R,g,k}(M;p) := \left\{ (C, \mathbf{x}, \varphi) \in \mathcal{C}(M;p) \mid \varphi_*[C] = R, \#\mathbf{x} = k, g(C) = g \right\} \Big/ \text{iso.}$$

By abuse of notation, we write $(C, \mathbf{x}, \varphi) \in \mathcal{C}(M;p)$ to mean a representative of the isomorphism class.

By construction $\kappa_s^*\colon \check{L}^p(C, \mathbb{R}) \to \check{L}^p(C_s, \mathbb{R})$ is an isomorphism for any $s \in S$. On $\check{L}_1^p$-spaces pullback is also well defined, because κ_s is Lipschitz, but since $\varphi \circ \kappa_s$ is constant on the contracted circles, κ_s^* is certainly not surjective. What we are interested in for the construction of charts are identifications $\Pi_s\colon \check{L}_1^p(C, \varphi^*T_M) \to \check{L}_1^p(C_s; (\varphi \circ \kappa_s)^*T_M)$, i.e., a structure of Banach bundle on $\coprod_s \check{L}_1^p(C_s; (\varphi \circ \kappa_s)^*T_M)$. The latter space will be denoted $q_*^{1,p}(\kappa^*\varphi^*T_M)$, which captures the idea of being the direct image of a sheaf of sections of $(\varphi \circ \kappa)^*T_M$ that are fiberwise locally of class $\check{L}_1^p$. We will sketch a proof of the following result in Section 2.4:

Theorem 2.2 *Let* $(C, \mathbf{x}, \varphi : C \to M)$ *be a stable map of class* L_1^p *and* $(q : C \to S, \underline{\mathbf{x}})$ *an analytically semiuniversal deformation of* $(C, \mathbf{x})$*. Thus* S *is an open set in* $\mathbb{C}^N$*, where* $N = 3g - 3 + k + \dim \operatorname{Aut}(C, \mathbf{x})$*. Let* $\kappa : C \to C_0 = C$ *be the Lipschitz retraction to the central fiber as above. Then there exists a family of isomorphisms*

$$\Pi_s\colon \check{L}_1^p(C, \varphi^*T_M) \to \check{L}_1^p(C_s; (\varphi \circ \kappa_s)^*T_M).$$

Π_s *enjoys the following continuity property: The composition with pullback to the central fibre*

$$(\kappa_s^*)^{-1} \circ \Pi_s\colon \check{L}_1^p(C, \varphi^*T_M) \to L^\infty(C, \varphi^*T_M)$$

is a norm continuous family of operators (between fixed Banach spaces!) parametrized by S*.*

Here the stated norm continuity implies that for $v \in \check{L}_1^p(C, \varphi^*T_M)$, small changes of s lead to small pointwise changes of $\Pi_s v$ in an intuitive sense. Thus the graphs of $\Pi_s v$ in the total space of $\kappa^*\varphi^*T_M$ fit together continuously. More regularity properties of Π_s will be discussed later.

Given Π_s we just need to write down the analogue of the charts for fixed domains (cf. 1.1) to get charts for $\mathcal{C}(M;p)$:

$$\Phi\colon S \times \check{L}_1^p(C, \varphi^*T_M) \supset S \times V \longrightarrow \mathcal{C}(M;p), \quad (s, v) \longmapsto \varphi(s, v)$$

$$\varphi(s, v)(z) := \exp_{\varphi \circ \kappa_s(z)} \left(\Pi_s v \right)(z).$$

2.2   Automorphisms and differentiable structure

There are still two problems with the proposed charts Φ of 3.1. First, since S does not in general parametrize pre-stable curves near $(C, \mathbf{x})$ effectively, Φ need not be injective. Not every $\Psi \in \mathrm{Aut}(C, \mathbf{x})$ has $\mathrm{im}\,\Phi \cap \mathrm{im}(\Psi^* \circ \Phi) \neq \emptyset$, but this is certainly the case for Ψ close to $\mathrm{Aut}(C, \mathbf{x}, \varphi) \subset \mathrm{Aut}(C, \mathbf{x})$. Here we write Ψ both for the automorphism of the central fibre of the deformation and for the germ of the action on the total space $\mathcal{C}$. Thus the best we can hope for is the structure of a Banach *orbifold* on $\mathcal{C}(M; p)$.

Orbifolds are a generalization of the notion of manifolds where we replace open subsets of vector spaces as local models by quotients of such open subsets by linear actions of finite groups. More precisely, one defines

Definition 2.3 A *local uniformizing system* (of Banach orbifolds) is a triple $(q \colon \widehat{U} \to U, G, \alpha)$ where:

(1) α is a continuous linear action of the finite group G on some Banach space T;

(2) $\widehat{U}$ is a G-invariant open subset of T;

(3) q induces a homeomorphism $\widehat{U}/G \to U$.

For a more intuitive notation we often write $U = \widehat{U}/G$ instead of (q, G, α). $\square$

Compatibility of local uniformizing systems is defined through the notion of *open embeddings*: let $V = \widehat{V}/G'$, $U = \widehat{U}/G$ be two local uniformizing systems. An open embedding $V = \widehat{V}/G' \hookrightarrow U = \widehat{U}/G$ is a γ-equivariant open embedding $\widehat{f} \colon \widehat{V} \to \widehat{U}$ for a monomorphism $\gamma \colon G' \to G$. Thus this induces an open embedding of the quotient spaces $f \colon V \hookrightarrow U$. If the actions of the groups are not effective one should also require a maximality condition for γ, namely

$$\mathrm{im}\,\gamma \ = \ \{g \in G \mid \widehat{f}(\widehat{V}) \cap g \cdot \widehat{f}(\widehat{V}) \neq \emptyset\}.$$

This ensures that for any $\widehat{X} \in \widehat{V}$, γ induces an isomorphism of stabilizers $G'_{\widehat{X}} \simeq G_{\widehat{f}(\widehat{X})}$.[3]

Recall that a covering $\{U_i\}_{i \in I}$ of a set X is called *fine* if for any $i, j \in I$ with $U_i \cap U_j \neq \emptyset$ there exists $k \in I$ with $U_k \subset U_i \cap U_j$. An *atlas* for the structure of Banach orbifold on a Hausdorff space X is now a fine covering of X by local

---

[3]The group actions are traditionally required to be effective. This is too restrictive for our purposes: because any curve of genus 2 has a hyperelliptic involution, $\mathcal{M}_2$ is an instance of an orbifold with $\mathbb{Z}_2$-kernel of the action everywhere, cf. below.

uniformizing systems $\{U_i = \widehat{U}_i/G_i\}_{i\in I}$ (i.e., $\{U_i\}$ is an open covering of X) such that for any $i, j \in I$ there is a $k \in I$ and open embeddings

$$U_k = \widehat{U}_k/G_k \hookrightarrow U_i = \widehat{U}_i/G_i \quad \text{and} \quad U_k = \widehat{U}_k/G_k \hookrightarrow U_j = \widehat{U}_j/G_j.$$

It is in general not possible to find an open embedding of the restriction of $U_i = \widehat{U}_i/G_i$ to $U_i \cap U_j$ into $U_j = \widehat{U}_j/G_j$. Consider, for example, the orbifold structure on S^2 with cyclic quotient singularities of orders 2 and 3 at the poles. This orbifold can be covered by two local uniformizing systems $\mathbb{R}^2 = \mathbb{C}/\mathbb{Z}^2$, $\mathbb{R}^2 = \mathbb{C}/\mathbb{Z}^3$ via stereographic projection from the poles. So unlike the case of manifolds, one has to restrict to sufficiently small open sets to compare two local uniformizing systems.

As usual a *(topological) Banach orbifold* is defined as an equivalence class of atlases (or as a maximal atlas). If one requires all open embeddings $\widehat{f}$ to be differentiable or holomorphic immersions, one arrives at *differentiable* and *complex* Banach orbifolds. In the latter case, the representations $G \to \mathrm{GL}(T)$ should of course also be complex.

Quotients of manifolds by finite groups are examples of orbifolds, but the whole point of the concept of orbifold is that not every orbifold is of this form. An easy example is S^2 with a $\mathbb{Z}_m$-quotient singularity at one point P: if $q\colon X \to S^2$ were a global cover (connected, w.l.o.g.), $X \setminus q^{-1}(P) \to S^2 \setminus \{P\}$ is an unbranched cover, hence trivial by simply connectivity of the base, hence bijective. But any local uniformizer at P is m to one. Contradiction, such a global cover X does not exist! According to Thurston, orbifolds are called *good* or *bad* depending on whether or not they are globally covered by manifolds.

Note that to build up an orbifold from a set of local uniformizing systems $\{U_i = \widehat{U}_i/G_i\}$ through glueing by open embeddings, we need only require the cocycle condition to hold on the level of the underlying sets U_i; on the level of uniformizers $\widehat{U}_i$, it needs to hold only up to the group actions.

Note also that any $x \in X$ has an associated group G_x, the isomorphism class of the stabilizer $G_{\widehat{x}}$ of a lift $\widehat{x}$ of x to any local uniformizing system containing x. It is the smallest group of a local uniformizing system containing x. Thus the concept of orbifold incorporates groups intrinsically associated to the points of X.

Natural examples of orbifolds are thus moduli spaces of objects in complex analysis with finite automorphism groups and unobstructed deformation theory (the latter to ensure smoothness of the local covers). As we saw in the last section, moduli spaces $\mathcal{M}_g$ (or $\mathcal{M}_{g,k}$) of Deligne–Mumford stable curves of fixed genus are examples of this: local uniformizers at C are of the form $S \to S/\mathrm{Aut}(C)$, where $S \subset \mathrm{Ext}^1(\Omega_C, \mathcal{O}_C)$ is an open set. What is nice about viewing $\mathcal{M}_g$ as an orbifold is that, unlike the underlying scheme or complex space, the orbifold is a fine moduli space, i.e., it wears a universal family of stable curves. The latter is the orbifold $\mathcal{M}_{g,1}$ of stable 1-pointed curves $(C, \mathbf{x})$ fibered over $\mathcal{M}_g$ via the forgetful and stabilization map $(C, x) \to C^{\text{st}}$: any family $X \to T$ of stable curves of genus g over a complex manifold, say, is isomorphic to the pullback family $T \times_{\mathcal{M}_g} \mathcal{M}_{g,1} \to T$ for some morphism of orbifolds $T \to \mathcal{M}_g$ (where T is viewed as orbifold with trivial groups). Similarly for $\mathcal{M}_{g,k}$.

Such a *morphism of orbifolds* is just a continuous map of the underlying topological spaces with compatible lifts to local uniformizers. We also need the notion of *(vector) orbibundle*. This is a morphism of orbifolds $\pi \colon E \to X$ that is locally uniformized by projections $E_0 \times \widehat{U} \to \widehat{U}$, with E_0 a Banach space. If the local groups are G^E and G for $E|_U$ and U then the action of G^E on $E_0 \times \widehat{U}$ is required to be diagonal via a linear representation of G^E on E_0 and an epimorphism $G^E \to G$. Open embeddings of E have to be linear on the fibers E_0. Note that the topological fiber $\pi^{-1}(x)$ is isomorphic to E_0/G_x^E and thus does not have an additive structure in general. Tangent bundles of differentiable orbifolds are examples of orbibundles (with $G^E = G$ everywhere).

We now proceed with our discussion of charts for $\mathcal{C}(M; p)$. We will see that if the domain $(C, \mathbf{x})$ of $(C, \mathbf{x}, \varphi)$ is stable as an abstract curve then the map $\Phi \colon S \times V \to \mathcal{C}(M; p)$ from the end of Section 2.1 will indeed provide a local uniformizing system at $(C, \mathbf{x}, \varphi)$. But if $(C, \mathbf{x})$ is not stable as an abstract curve, i.e., if $(C, \mathbf{x}, \varphi)$ has bubbles, then $\dim \operatorname{Aut}(C, \mathbf{x}) > 0$ and any $\eta \in \operatorname{Lie} \operatorname{Aut}(C, \mathbf{x})$ induces a vector field v_η on S in such a way that the pre-stable curves are mutually isomorphic along any integral curve of v_η. We will see in the next section how to deal with this problem by taking a slice of the induced equivalence relation on $S \times V$. The slice will again be a family of Banach manifolds over S but with tangent space at $(0, 0) \in S \times V$ of the form $S \times \overline{V}$, where $\overline{V} \subset \check{L}_1^p(C, \varphi^* T_M)$ is a linear subspace of codimension equal to the dimension of $\operatorname{Aut}(C, \mathbf{x})$. Moreover, the slice can be taken to be $\operatorname{Aut}(C, \mathbf{x}, \varphi)$-invariant.

The existence of the slice is not merely a simple application of the implicit function theorem; this is related to the second problem that we face with our charts: the action of the group of self-diffeomorphisms of C (and even of $\operatorname{Aut}^0(C, \mathbf{x})$) on $\check{L}_1^p(C, M)$ is only continuous, not differentiable. In fact, the differential with respect to a one parameter group of diffeomorphisms would mean applying the corresponding vector field to the maps $\varphi \colon C \to M$, and this costs one derivative. So looking at the simple case of nonsingular C, two choices of retraction $\kappa, \kappa' \colon C \to C$ will lead to two different structures of differentiable Banach orbifold near $(C, \mathbf{x}, \varphi)$: the coordinate change need not be differentiable. The solution to the problem is that, locally, the differentiable structure *relative* to S is well defined. Since S is finite dimensional, this is enough to make the implicit function theorem work, albeit locally in a version relative to S.

From a categorical point of view, we are thus led to a category of topological Banach orbifolds that locally have a well-defined differentiable structure relative to some finite dimensional spaces. From a point of view closer to algebraic geometry, one might alternatively view our Banach orbifold as "fibered in differentiable Banach orbifolds over the Artin stack $\mathfrak{F}_{g,k}$".

2.3 Slices

According to the discussion in the last section, when bubbles are present in $(C, \mathbf{x}, \varphi)$, we need to find slices of the equivalence relation generated by the germ of the action of $\mathrm{Aut}^0(C, \mathbf{x})$ on $S \times V \subset S \times \check{L}_1^p(C, \varphi^* T_M)$. The usual method, applied both in algebraic geometric and analytic approaches, is *rigidification*. This means adding enough points $\mathbf{y} = (y_1, \dots, y_l)$ to $(C, \mathbf{x})$ to make $(C, \mathbf{x} \vee \mathbf{y})$ stable as an abstract curve; here $\mathbf{x} \vee \mathbf{y} = (x_1, \dots, x_k, y_1, \dots, y_l)$ is the concatenation of $\mathbf{x}$ and $\mathbf{y}$. Explicitly, this means that we add at least $3 - i$ points to each rational component having only i special points, $i \in \{1, 2\}$. By stability (!) the y_i can be chosen in such a way that φ is locally injective there. Choose locally closed submanifolds $H_1, \dots, H_l \subset M$ of real codimension two and transversal to φ through $\varphi(y_1), \dots, \varphi(y_l)$. The slice is

$$\gamma = \left\{ (s, v) \in S \times V \;\middle|\; \varphi(s, v)(y_i) \in H_i \right\}.$$

This will be a submanifold at $(0, 0) \in S \times V$ if transversality to H_i is an open condition in the function spaces we employ. This is indeed the case in function spaces with at least one continuous derivative. So the idea of rigidification is to let the map rule the deformation of the points we add.

Unfortunately, this method does not work in our case, because local injectivity is not an open condition in L_1^p. The way out is an integral version of rigidification: let $z \colon U \to \mathbb{C}$ restrict to holomorphic coordinates on $U_s = U \cap C_s$, where $U \subset C$ is an open set with U_0 contained in a bubble we want to rigidify. By stability, if U_0 is sufficiently large, there are differentiable bump functions ρ on M with $\varphi^* \rho|_{U_0}$ nontrivial and having compact support. Consider

$$\lambda(s, v) = \frac{\int_{U_s} z \cdot \varphi(s, v)^* \rho \, d\mu(z)}{\int_{U_s} \varphi(s, v)^* \rho \, d\mu(z)},$$

which computes the center of gravity of $\varphi^* \rho$ in the coordinate $z(s)$ on U_s. Assembling one (respectively, two different) such λ for each unstable component of $(C, \mathbf{x})$ with two (respectively, one) special points into a vector valued function $\Lambda \colon S \times V \to \mathbb{C}^b$, where $b = \dim \mathrm{Aut}^0(C, \mathbf{x})$, our candidate for a slice is

$$\gamma = \Lambda^{-1}(\lambda_0), \quad \lambda_0 = \Lambda(0,0).$$

Due to the lack of differentiability it seems hard to prove that this is in fact a slice, i.e., induces a local homeomorphism $\mathrm{Aut}^0(C, \mathbf{x}) \times \gamma \to S \times V$ on appropriate domains of definition. However, this is easy if we choose z to be a *linear* coordinate, i.e., such that the action of $\mathrm{Aut}^0(C, \mathbf{x})$ is affine linear. Such coordinates can be constructed explicitly, cf. [Sil]. Then the implicit function theorem allows us to change coordinates on $S \times V$ relative to S in such a way that $\gamma = S \times \overline{V}$, with $\overline{V} \subset V$ a linear subspace of codimension equal to $\dim \mathrm{Aut}(C, \mathbf{x})$. Moreover, the slice can be chosen $\mathrm{Aut}(C, \mathbf{x}, \varphi)$-invariant.

An alternative in the differentiable setting is to take directly linear slices of the form $S \times \overline{V}$, with $\overline{V} \subset V$ complementary to the finite dimensional subspace spanned by the action of $\mathrm{Lie}\,\mathrm{Aut}(C, \mathbf{x})$. This is the approach of [FkOn]. Since in the final analysis we are only interested in the germ of $\mathcal{C}(M; p)$ along $\mathcal{C}^{\mathrm{hol}}(M, J)$, we may assume the centers $(C, \mathbf{x}, \varphi)$ of our charts are J-holomorphic. The map φ is then smooth by elliptic regularity. So $D\varphi$ maps $\mathrm{Lie}\,\mathrm{Aut}(C, \mathbf{x})$ to a finite dimensional subspace in $\check{L}_1^p(C, \varphi^* T_M)$, to which we may choose a complementary subspace $\overline{V}$. However, in our setting, it still seems hard to prove that $S \times \overline{V}$ is in fact a slice.

2.4   Trivializing the relative tangent bundle

What is still missing is a structure of Banach bundle on

$$q_*^{1,p}(\kappa^* \varphi^* T_M) = \coprod_{s \in S} \check{L}_1^p(C_s; (\varphi \circ \kappa_s)^* T_M),$$

which should be viewed as the *tangent bundle* of $\coprod_s \check{L}_1^p(C_s; M)$ *relative to* S, restricted to $\{(C_s, \mathbf{x}_s, \varphi \circ \kappa_s) \mid s \in S\}$. This problem is at the heart of our approach to symplectic GW invariants. Our solution has three ingredients:

(1) For any $\varphi \in \check{L}_1^p(C, M)$ there is a natural structure of *holomorphic* vector bundle on the complex vector bundle $(\varphi^* T_M, \varphi^* J)$, even though φ is only L_1^p. In particular, we get a $\varphi^* J$-linear first order linear differential operator

$$\overline{\partial}_J^\varphi \colon \check{L}_1^p(C, \varphi^* T_M) \longrightarrow \check{L}^p(C, \varphi^* T_M \otimes_{\mathbb{C}} \overline{\Omega}).$$

Here $\overline{\Omega} = \Lambda^{0,1}$ is a bundle only away from C_{sing}, and the right-hand side is defined using frames of the form $d\overline{z}$ on a branch of C near $P \in C_{\mathrm{sing}}$, where z is a holomorphic coordinate of this branch at P.[4]

---

[4]Equivalently, one may use the relative dualizing bundle $\omega_{C/S}$, which is more natural from the point of view of algebraic geometry. Local frames near a singularity of C are now of the form dz/z, which requires insertion of the p-dependent weight $|z|^p$ in the definition of the measure μ near C_{sing}.

(2) Prove the Poincaré Lemma for the above $\overline{\partial}$ operator. Then the sequence of coherent sheaves on C

$$0 \to \mathcal{O}(\varphi^*T_M) \longrightarrow \check{\mathcal{L}}_1^p(\varphi^*T_M) \xrightarrow{\overline{\partial}_J^\varphi} \check{\mathcal{L}}^p(\varphi^*T_M \otimes \overline{\Omega}) \to 0 \qquad (*)$$

is exact (this is well known for smooth C).

(3) Use (2), plus the trivialization of $\check{L}^p$-spaces via pullback by κ, plus a Čech construction for the holomorphic part to exhibit the Banach bundle structure on $q_*^{1,p}(\kappa^*\varphi^*T_M)$.

Informally speaking, the $\overline{\partial}$ operator is used to reduce the nonholomorphic part to the simple case of $\check{L}^p$-spaces, while the holomorphic part is taken care of by a Čech construction. We should remark that this kind of argument does not work for $\check{L}_k^p$ with $k > 1$ because only a subspace of $\check{L}_k^p$ is mapped to $\check{L}_{k-1}^p$ by $\overline{\partial}$. This is due to the continuity at the node imposed on sections of the sheaf $\check{\mathcal{L}}_{k-1}^p(\varphi^*T_M \otimes \overline{\Omega})$. Restricting to this subspace would mean a higher tangency condition of the two branches at the nodes which is not what we want in the application to J-holomorphic curves. This dictates the choice of L_1^p as modelling space in our approach.

The rest of this section is devoted to filling in the details of the above steps.

Holomorphic structure on φ^*T_M

The logic here is actually the opposite of that presented above; namely, the operator $\overline{\partial}_J^\varphi$ is constructed first. There are various ways to do this, but at J-holomorphic φ the choice should reduce to the φ^*J-linear part of the linearization of the $\overline{\partial}_J$ operator (which is independent of choices of local trivialization). Letting ∇ be the Levi-Civita connection on M with respect to some fixed Riemannian metric, $\nabla^\varphi = \varphi^*\nabla$ the induced connection on φ^*T_M, our choice for $(\overline{\partial}_J^\varphi)_\xi v$ is the φ^*J-linear part of

$$\frac{1}{2}\Big(\nabla_\xi^\varphi v + J\nabla_{j(\xi)}^\varphi v + (\nabla_v J)\partial_J\varphi(j(\xi))\Big),$$

for $\xi \in \Gamma(T_C)$ and $v \in \check{L}_1^p(C, \varphi^*T_M)$; here $\partial_J\varphi := \frac{1}{2}(D\varphi - J \circ D\varphi \circ j)$ and j is the complex structure on C. Note that since we assumed φ to be only of class $\check{L}_1^p$, this expression does not make sense pointwise, but only as $\check{L}^p$-section of φ^*T_M, itself only a complex vector bundle of class $\check{L}_1^p$. Nevertheless, an application of the implicit function theorem shows that local solutions of $\overline{\partial}_J^\varphi$ define a locally free coherent sheaf on C, i.e., induce the structure of a holomorphic vector bundle on φ^*T_M, cf. [HLiSk] and [IvSh], Lemma 6.1.1.

Poincaré Lemma for weighted Sobolev spaces

Away from the singularities of C, exactness of the stated sequence of sheaves is well known. What is left at a node is to prove surjectivity of the restriction of the ordinary $\bar{\partial}$ operator to each branch in nonstandard Sobolev spaces:

$$\bar{\partial} \colon \check{L}_1^p(\Delta) \to \check{L}^p(\Delta).$$

These spaces can be identified with Sobolev spaces on a half-infinite cylinder with exponential weights $e^{-\mu s}$ by the identification

$$\Delta^* \longrightarrow \mathbb{R}_{>0} \times S^1, \quad re^{i\varphi} \longmapsto (s, \psi) = (-\log r, \varphi),$$

under which the $\bar{\partial}$ operator transforms to an operator of the form $e^{-s} \cdot (\partial_s + i\partial_\psi)$. For such linear elliptic differential operators on manifolds with cylindrical ends ($\mathbb{R}_{>0} \times N$ with compact N) there does exist a general theory, which implies the result we need [LcMc].

Alternatively, and maybe even more enlightening than invoking general theory, one may employ the explicit right inverse to the $\bar{\partial}$ operator on the disk given by the Cauchy integral operator

$$T(g\mathrm{d}\bar{z})(z) \;=\; \frac{1}{2\pi i} \int_\Delta \frac{g(w)}{w - z} \mathrm{d}w \wedge \mathrm{d}\bar{w}.$$

To show that T does indeed map $\check{L}^p$ to $\check{L}_1^p$, one just needs to estimate $\partial \circ T$. The latter equals a singular integral operator

$$S(g\mathrm{d}\bar{z})(z) \;=\; \frac{1}{2\pi i} \left(\lim_{\varepsilon \to 0} \int_{\Delta \setminus B_\varepsilon(z)} \frac{g(w)}{(w - z)^2} \mathrm{d}w \wedge \mathrm{d}\bar{w} \right) \mathrm{d}z.$$

The classical Calderon–Zygmund inequality says that S is a continuous endomorphism of $L^p(\Delta)$. The corresponding statement for the relevant case with weights can be found in [CfFe].

The Čech construction

Because $(*)$ is a soft resolution of $\mathcal{O}(\varphi^* T_M)$, the long exact cohomology sequence reads

$$0 \to H^0(\varphi^* T_M) \longrightarrow \check{L}_1^p(\varphi^* T_M) \xrightarrow{\bar{\partial}_J^\varphi} \check{L}^p(\varphi^* T_M \otimes \overline{\Omega}) \longrightarrow H^1(\varphi^* T_M) \to 0,$$

where all sections are understood over the domain C of φ. Now let $\mathcal{U} = \{U_i\}_{i=0,\dots,d}$ be a finite open covering of C with the properties:

(1) U_0 has components conformally equivalent to the unit disk minus a number of pairwise disjoint closed disks in its interior (so is an open Riemann surface of genus 0);

(2) each U_i for $i > 0$ is conformally equivalent to (possibly degenerate) cylinders $Z_t = \{(z, w) \in \Delta^2 \mid zw = t\}$, and such that $U_i \cap U_j \cap U_k = \emptyset$ for any three pairwise different indices i, j, k.

Write $C^i(\mathcal{U}; \mathcal{O}(\varphi^* T_M))$ for the ith (alternating) Čech cochains of holomorphic sections extending continuously to the boundary. These are Banach spaces under the sup norm. Since $\mathcal{U}$ is a Stein (hence acyclic) cover, there is a similar sequence

$$0 \to H^0(\varphi^* T_M) \to C^0(\mathcal{U}; \mathcal{O}(\varphi^* T_M)) \xrightarrow{\check{d}} C^1(\mathcal{U}; \mathcal{O}(\varphi^* T_M)) \to H^1(\varphi^* T_M) \to 0,$$

where $\check{d}$ is the Čech coboundary operator. Note that $C^1(\mathcal{U}; \mathcal{O}(\varphi^* T_M))$ consists of cochains rather than cocycles, because by our choice of $\mathcal{U}$, triple intersections are empty. To find an explicit quasi-isomorphism between the two middle arrows $\overline{\partial}_J^\varphi$ and $\check{d}$, we just need to go through standard constructions of cohomology theory: define maps

$$\Theta \colon \check{L}_1^p(\varphi^* T_M) \to C^0(\mathcal{U}; \mathcal{O}(\varphi^* T_M)) \quad \text{by} \quad f \mapsto \left(f|_{U_i} - T^i(\overline{\partial} f|_{U_i}) \right)_i,$$

and

$$\Lambda \colon \check{L}^p(\varphi^* T_M \otimes \overline{\Omega}) \to C^1(\mathcal{U}; \mathcal{O}(\varphi^* T_M)) \quad \text{by} \quad \alpha \mapsto \left(T^j(\alpha|_{U_j}) - T^i(\alpha|_{U_i}) \right)_{ij}.$$

where $T^i \colon \check{L}^p(U_i; \varphi^* T_M \otimes \overline{\Omega}) \to \check{L}_1^p(U_i; \varphi^* T_M)$ is a right inverse of the above $\overline{\partial}_J^\varphi$. One can show that Θ and Λ induce isomorphisms on kernels and cokernels of $\overline{\partial}_J^\varphi$ and $\check{d}$. This is equivalent to exactness of the associated mapping cone

$$0 \to \check{L}_1^p(\varphi^* T_M) \xrightarrow{\binom{\overline{\partial}_J^\varphi}{\Theta}} \begin{matrix} \check{L}^p(\varphi^* T_M \otimes \overline{\Omega}) \\ \oplus \\ C^0(\mathcal{U}; \mathcal{O}(\varphi^* T_M)) \end{matrix} \xrightarrow{(\Lambda, -\check{d})} C^1(\mathcal{U}; \mathcal{O}(\varphi^* T_M)) \to 0.$$

We have thus exhibited the $\check{L}_1^p$-spaces as kernels of epimorphisms of Banach spaces, and we can reasonably hope to trivialize these in families. Alternatively, since all maps have right inverses, one may use a similar sequence with the arrows reversed, cf. [Si1].

Introducing parameters

So far we have discussed the situation at a fixed curve $(C, \mathbf{x}, \varphi)$. For the purpose of producing charts we wanted to identify $\check{L}_1^p(C_s; (\varphi \circ \kappa_s)^* T_M)$ with $\check{L}_1^p(C, \varphi^* T_M)$, where $q \colon \mathcal{C} \to S$ (together with $\underline{\mathbf{x}} \colon S \to \mathcal{C} \times_S \cdots \times_S \mathcal{C}$, which is not of interest here) is a semiuniversal deformation of $(C, \mathbf{x})$, and $\kappa \colon \mathcal{C} \to C$ is a Lipschitz retraction to the central fiber as in 2.1, $C_s = q^{-1}(s)$, $\kappa_s = \kappa|_{C_s}$.

Using κ_s we may identify $\check{L}^p(C, \varphi^* T_M \otimes \overline{\Omega})$ with $\check{L}^p(C_s; (\varphi \circ \kappa_s)^* T_M \otimes \overline{\Omega})$. In view of (a parametrized version of) the exact sequence of Banach spaces in the last paragraph, it remains to trivialize spaces of Čech cochains. We choose an open covering $\mathcal{U} = \{U_i\}_{i=0,\dots,d}$ of the total space $\mathcal{C}$ in such a way that on $\overline{U}_i$ there are holomorphic functions z, w identifying $U_i(s) := U_i \cap C_s$, $i > 0$, with possibly degenerate cylinders $Z_{t_i(s)}$, $t_i \in \mathcal{O}(S)$, while on $\overline{U}_0$ there is just one holomorphic function z identifying $U_0(s)$ with a union of plane open sets as above. (Holomorphic *relative* to S would suffice for what follows.) Together with continuously varying holomorphic trivializations of $(\varphi \circ \kappa)^* T_M|_{U_i(s)}$ we are left to find isomorphisms between $\mathcal{O}(Z_t) \cap C^0(\overline{Z}_t)$ for different t. This can be done by observing that these spaces are given (up to constants) by positive Fourier series $\sum_{n>0} a_n e^{in\varphi}$ on the two boundary circles $|z| = 1$ or $|w| = 1$ via

$$(a_n, b_n) \longmapsto \sum_{n>0} a_n z^n + \sum_{n>0} b_n w^n.$$

A similar method works for U_0.

2.5   Summary

We summarize our discussion in the following

Theorem 2.4 *Let (M, J) be an almost complex manifold. Then the space $\mathcal{C}(M; p)$ of stable complex curves in M of Sobolev class $\check{L}_1^p$ has the structure of a weakly differentiable Banach orbifold with local group $\mathrm{Aut}(C, \mathbf{x}, \varphi)$ at $(C, \mathbf{x}, \varphi)$.*

Moreover, $\mathcal{C}(M; p)$ has a weakly differentiable Banach orbibundle $\mathcal{E}$ with fibers $\mathcal{E}_{(C, \mathbf{x}, \varphi)}$ uniformized by $\widehat{\mathcal{E}}_{(C, \mathbf{x}, \varphi)} = \check{L}^p(C, \varphi^* T_M \otimes \overline{\Omega})$. $\mathcal{E}$ has a weakly differentiable orbibundle section $s_{\overline{\partial}, J}$ sending $(C, \mathbf{x}, \varphi)$ to $\overline{\partial}_J \varphi$. Its zero locus $Z(s_{\overline{\partial}, J})$ is the space $\mathcal{C}^{\mathrm{hol}}(M, J)$ of stable J-holomorphic curves with the Gromov topology. $\square$

Here "weak differentiability" means that the differentiable structure is well defined locally only relative to some finite dimensional space. The differentiability properties of the section $s_{\overline{\partial}, J}$ will be discussed further in 3.1. The construction of $\mathcal{E}$ is straightforward.

3   Construction of the virtual fundamental class

In this chapter we outline our construction of the virtual fundamental class along the lines of Chapter 1 inside the ambient space of Chapter 2.

3.1 Local transversality

We first show how to solve the problem locally, i.e., how to construct a manifold $\widetilde{Z}$ containing a neighbourhood of $(C, \mathbf{x}, \varphi)$ in $\mathcal{C}^{\mathrm{hol}}(M, J)$ as the zero set of a map to a finite dimensional space. In view of the analogy with the construction of germs of moduli spaces of complex manifolds by Kuranishi [Ku], this data is often called a *Kuranishi model* (here for $\mathcal{C}^{\mathrm{hol}}(M, J)$ at $(C, \mathbf{x}, \varphi)$). Since the construction relies on the implicit function theorem we must now discuss the regularity properties of $s_{\bar{\partial}, J}$.

Recall that charts at $(C, \mathbf{x}, \varphi)$ are of the form $S \times \overline{V} \hookrightarrow S \times \check{L}_1^p(C, \varphi^* T_M)$, where $\overline{V}$ is of finite codimension in $\check{L}_1^p(C, \varphi^* T_M)$, of positive codimension whenever $(C, \mathbf{x}, \varphi)$ has bubbles. Fixing s means fixing the domain $(C, \mathbf{x})$, which implies differentiability of all objects involved, including the local uniformizers $\widehat{\mathcal{E}}$ of $\mathcal{E}$ and $\widehat{s}_{\bar{\partial}, J}$ of $s_{\bar{\partial}, J}$. However, due to the phenomenon discussed in Section 2.2, one cannot even expect the differential σ of $\widehat{s}_{\bar{\partial}, J}$ *relative to* S to be uniformly continuous. But since we used the $\varphi^* J$-linear part of σ as $\bar{\partial}$ operator to trivialize $q_*^{1,p}(\kappa^* \varphi^* T_M)$, σ turns out to be uniformly continuous at the center of our charts. This is just enough to apply the implicit function theorem in a version relative to S.

Now let $(C, \mathbf{x}, \varphi)$ be J-holomorphic and let σ_0 be the differential of $\widehat{s}_{\bar{\partial}, J}$ relative to S at $(0, 0) \in S \times \overline{V}$. Then

$$\sigma_0(w) = \bar{\partial}_J^\varphi w + \varphi^* N_J(w, D\varphi) \quad \text{for } w \in T_0 \overline{V} \subset \check{L}_1^p(C, \varphi^* T_M).$$

So possibly up to a zeroth order term and restriction to a subspace of finite codimension in $\check{L}_1^p(C, \varphi^* T_M)$, σ_0 is just the $\bar{\partial}$ operator on $\varphi^* T_M$. By the results of Section 2.4, the latter is a Fredholm operator to $\check{L}^p(C, \varphi^* T_M \otimes \overline{\Omega})$, and so is σ_0. Moreover, it is not hard to see that when restricted to sufficiently small neighbourhoods of C_{sing}, the corresponding operators are surjective. Therefore we may choose $\alpha_1, \ldots, \alpha_c \in \check{L}^p(C, \varphi^* T_M \otimes \overline{\Omega})$ supported away from C_{sing}, $c = \dim \operatorname{coker} \sigma_0$, such that $\operatorname{im} \sigma_0 + \sum_i \mathbb{C} \alpha_i = \check{L}^p(C, \varphi^* T_M)$. We say that the α_i *span* $\operatorname{coker} \sigma_0$. Define a morphism τ from a trivial bundle $F = \mathbb{R}^c$ over $S \times \overline{V}$ to $\widehat{\mathcal{E}}$ by sending the ith standard section to the parallel transport of α_i. Then an application of the implicit function theorem relative to S to the section $\widetilde{s} := q^* s + \tau$ of $q^* \mathcal{E}$ ($q \colon F \to S \times \overline{V}$ the bundle projection), viewed as map from $S \times \overline{V} \times \mathbb{R}^c$ to $\widehat{\mathcal{E}}_{(C, \mathbf{x}, \varphi)} = \check{L}^p(C, \varphi^* T_M \otimes \overline{\Omega})$, shows that $\widetilde{Z} = Z(\widetilde{s})$ is a topological submanifold of F of expected dimension plus rank F. The restriction of $q^* F$ to $\widetilde{Z}$ has a tautological section s_{can} (mapping $f \in F$ to f). A germ of $\mathcal{C}^{\mathrm{hol}}(M, J)$ at $(C, \mathbf{x}, \varphi)$ is given by the zero locus of s_{can}.

If $(C, \mathbf{x}, \varphi)$ has nontrivial automorphisms we would like to make the Kuranishi model $\operatorname{Aut}(C, \mathbf{x}, \varphi)$-equivariant. Since it is not always possible to span $\operatorname{coker} \sigma_0$ by $\operatorname{Aut}(C, \mathbf{x}, \varphi)$-invariant sections (this is the notorious obstruction to transversality under the presence of multiply covered components) this inevitably forces a nontrivial action of $G = \mathrm{Aut}(C, \mathbf{x}, \varphi)$ on the fibers of F. The easiest way to make τ equivariant is then to replace F by F^G (#G copies of F) and define $\tau^G \colon F^G \to \widehat{\mathcal{E}}$ on the Ψth copy of F, $\Psi \in \mathrm{Aut}(C, \mathbf{x}, \varphi)$, by $(\Psi^{-1})^* \circ \tau$.

Choosing α to have support away from C_{sing} will be convenient in going over to other charts that we will need below.

3.2   Globalization

To globalize we would like:

- to extend F^G to an orbibundle over $\mathcal{C}(M; p)$;

- to extend τ by multiplication with a bump function that is differentiable relative to S in *any* chart $S \times \overline{V}$.

Neither of these problems is immediate. On Banach orbifolds the existence of finite rank orbibundles with effective actions of the local groups on the fibers, say on a neighbourhood of a compact set, seems to be a nontrivial condition. The general solution to this question given in a previous version of [Si1] is insufficient, because the cocycle condition cannot be verified. Fortunately, such orbibundles do exist on $\mathcal{C}(M; p)$ by a method similar to that given in [Be1], Prop. 5.

To this end, we now assume J tamed by some symplectic form ω. By slightly deforming ω and taking a large multiple, we may assume ω to represent an integral de Rham class. Then there exists a U(1)-bundle L over M with $[\omega] = c_1(L)$. L is a substitute for an ample line bundle in the algebraic setting. Let ∇ be a U(1)-connection on L. Let $\pi \colon \Gamma \to \mathcal{C}(M; p)$ be the universal curve and ev$\colon \Gamma \to M$ the evaluation map sending $p \in C$ over $(C, \mathbf{x}, \varphi) \in \mathcal{C}(M; p)$ to $\varphi(p)$. Thus π is a morphism of topological orbifolds with fiber over $(C, \mathbf{x}, \varphi)$ equal to the complex analytic orbispace $C / \mathrm{Aut}(C, \mathbf{x}, \varphi)$. As in Section 2.4 one shows that via ∇, ev$^*L$ has naturally the structure of a continuously varying family of holomorphic line bundles over the fibers of π. And since $[\omega]$ evaluates positively on any nonconstant J-holomorphic curve, φ^*L is ample on any bubbling component. To achieve ampleness on the other components we just need to tensor with $\omega_C(x_1 + \cdots + x_k)$, which is the sheaf of meromorphic 1-forms on C with at most simple poles at C_{sing} and the marked points x_i. These sheaves again fit into a continuously varying family of holomorphic line bundles $\omega_\pi(\mathbf{x})$ over the fibers of π. Then ev$^*L \otimes \omega_\pi(\mathbf{x})$ is π-ample (i.e., ample on each fiber), hence a sufficiently large multiple (power $\geq N_0$, say) has vanishing H^1 on any fiber of π. Let $N > N_0$ be an even bigger natural number such that for any #$\mathrm{Aut}(C, \mathbf{x}, \varphi)$ points on

C there exist a section of $(\varphi^L \otimes \omega_C(\mathbf{x}))^{\otimes N}$ vanishing at all but one point. We consider

$$\pi_*(\mathrm{ev}^*L \otimes \omega_\pi(\underline{\mathbf{x}}))^{\otimes N} := \coprod_{(C,\mathbf{x},\varphi) \in \mathcal{C}(M;p)} \Gamma(C, \mathrm{ev}^*L^{\otimes N} \otimes \omega_\pi^{\otimes N}(\underline{\mathbf{x}})).$$

Using a Čech construction as in Section 2.4, one shows that this is locally uniformized by the kernel of a Fredholm epimorphism of Banach bundles (of Čech cocycles) over $\mathcal{C}(M;p)$ and hence glues to an orbibundle F of finite rank. Moreover, by the choice of N, for any section α of $\check{L}^p(C, \varphi^*T_M \otimes \overline{\Omega})$ with sufficiently small support there exists a vector v of the fiber of F at $(C, \mathbf{x}, \varphi)$ such that the dimensions of the linear subspace spanned by the $\mathrm{Aut}(C, \mathbf{x}, \varphi)$-orbits of v in F and of α in $\check{L}^p(C, \varphi^*T_M \otimes \overline{\Omega})$ coincide. Direct sums of bundles of this type allow us to extend F^G on a neighbourhood of $\mathcal{C}^{\mathrm{hol}}(M, J) \subset \mathcal{C}(M;p)$.

As for extending the morphism τ, one might try to use parallel transports of differentiable bump functions on $\check{L}_1^p(C, \varphi^*T_M)$, which do in fact exist provided p is even. This will be insufficient for our purposes though. The problem is that if we look at such bump functions constructed at a curve with bubbles (with deformation space S say) from a chart centered at a curve without bubbles (with deformation space $\overline{S}$ say) then, locally, S fibers over $\overline{S}$. Differentiability holds relative to S but will fail relative to $\overline{S}$. The way out is to take a "bump function" χ which only takes into account the behaviour on an open set $U \subset C \setminus C_{\mathrm{sing}}$. U has to be chosen in such a way that the coordinates of S ruling the deformations of nodes belonging to the bubbles do not influence the trivialization of $q_*^{1,p}(\kappa^*\varphi^*T_M)$ over U. χ will not have bounded support on $\mathcal{C}(M;p)$, but its restriction to $\mathcal{C}^{\mathrm{hol}}(M, J)$ does. This is enough for extending τ along a neighbourhood of $\mathcal{C}^{\mathrm{hol}}(M, J)$ in $\mathcal{C}(M;p)$. Note that by choosing $\mathrm{supp}\,\alpha_i$ inside U, τ will be differentiable even relative to $\mathcal{M}_{g,k}$ in any appropriately chosen coordinate chart.

3.3 The Main Theorem

Since we need compactness (and for the construction of F) we further assume J tamed by some symplectic form ω. Fix $R \in H_2(M, \mathbb{Z})$, g, k. Then $\mathcal{C}_{R,g,k}^{\mathrm{hol}}(M, J)$ is compact. The direct sum of finitely many morphisms to $\mathcal{E}$ as in 3.2 yields a morphism $\tau : F \to \mathcal{E}$ spanning the cokernels of the differentials of $s_{\overline{\partial},J}$ relative to S in any chart $S \times \overline{V}$ centered at J-holomorphic $(C, \mathbf{x}, \varphi)$. Thus $\widetilde{Z} = Z(\widetilde{s})$, where $\widetilde{s} = q^*s + \tau$, $q : F \to \mathcal{C}(M;p)$, is a finite dimensional (topological) suborbifold of the total space of F. It is also not hard to see that $\widetilde{Z}$ can be naturally oriented by complex linearity of $\overline{\partial}_J^\varphi$, provided that F is also oriented. The latter can be achieved by taking $F \oplus F$ if necessary (this is just a matter of convenience; what one needs is a relative orientation of q^*F over $\widetilde{Z}$). Let Θ_F be the Thom class of F. For the virtual fundamental class of $C^{\text{hol}}_{R,g,k}(M,J)$, we set

$$\mathcal{GW}^{M,J}_{R,g,k} := [\widetilde{Z}] \cap \Theta_F \in H_{2d(M,R,g,k)}(C^{\text{hol}}_{R,g,k}(M,J)),$$

where $d(M,R,g,k) = \dim_{\mathbb{C}} \mathcal{M}_{g,k} + c_1(M,J) \cdot R + (1-g)\dim_{\mathbb{C}} M$ is computed by the Riemann–Roch theorem to be the index of $\bar{\partial}^{\varphi}_J$ plus $\dim \mathcal{M}_{g,k}$ (this needs to be corrected if $2g + k < 3$).

Theorem 3.1 *The class* $\mathcal{GW}^{M,J}_{R,g,k}$ *is independent of the choices made. Its image in* $H_*(C(M;p))$ *depends only on the symplectic deformation class of* ω. $\square$

Independence of choices (of τ, and of the Sobolev index p) is easy to establish. The second claim asserts independence under deformations of J inside the space of almost complex structures tamed by some symplectic form. To this end, one sets up a family version of the approach with fixed J, from which independence of the image in $C(M;p)$ follows immediately. For details we refer to [Si1].

3.4 Alternative approaches

The purpose of this section is to discuss, in a rather sketchy way, various other approaches to the construction of virtual fundamental classes, as given by Fukaya and Ono [FkOn], Li and Tian [LiTi2] and Ruan [Ru2]. Still another definition can be extracted from a paper of Liu and Tian [LiuTi] on a solution to the closely related Arnol'd conjecture on nondegenerate exact symplectomorphisms (the latter is also covered in [FkOn] and [Ru2]).

Recall that in formulating our problem as that of constructing a localized Euler class of a section of a Banach orbibundle over a Banach orbifold we had to pay the following price:

(1) working in spaces of maps with very weak differentiability (this caused problems in the slice theorem, cf. 2.3),

(2) the loss of differentiability in a finite dimensional direction (which made the construction of τ more subtle, cf. 3.1), and

(3) having to construct a finite dimensional orbibundle F with effective actions of the local groups on the fibers, cf. 3.2.

But what we are finally interested in is the zero locus $\widetilde{Z} \subset F$ of a perturbed section $\widetilde{s} = q^*s + \tau$. As a set, $\widetilde{Z}$ consists of (isomorphism classes of) tuples

$(C, \mathbf{x}, \varphi, f)$ with $(C, \mathbf{x}, \varphi) \in \mathcal{C}(M; p)$ and $f \in F_{(C, \mathbf{x}, \varphi)}$ such that $\overline{\partial}_J \varphi = \tau(f)$. Thus if the sections $\alpha \in \check{L}^p(C'; (\varphi')^* T_M \otimes \overline{\Omega})$ spanning the cokernel at various $(C', \mathbf{x}', \varphi') \in \mathcal{C}^{\mathrm{hol}}(M, J)$ used to construct τ, are chosen to be smooth, solutions φ of $\overline{\partial}_J \varphi = \tau(f)$ will also be smooth by elliptic regularity. That is, in constructing $\widetilde{Z}$ we may safely restrict to spaces of smooth maps. A common feature of the other approaches to GW theory is that they work in ambient spaces of C^∞ maps, and that $\widetilde{Z}$ is first constructed locally for any local, finite dimensional perturbation. The problem is then to find a global object that matches up the local perturbations.

The local construction of $\widetilde{Z}$ can be done by more or less straightforward modifications of the known glueing constructions for J-holomorphic curves in generic situations (i.e., when the linearization of the relevant Fredholm operator is already surjective), as given in [RuTi1], [Liu], [McSa]. "Glueing" means the following: given a nodal J-holomorphic curve $\varphi \colon C \to M$ and a family $\{C_s\}_{s \in S}$ of deformations of C as a pre-stable curve, one wants to deform φ to a family of J-holomorphic curves $\varphi_s \colon C_s \to M$. This is achieved by first constructing φ_s approximately by some kind of differentiable glueing construction involving bump functions. The $\overline{\partial}_J$ operators on the C_s set up a family of elliptic problems, albeit with varying Banach spaces (here: versions of L_1^p and L^p). The basic analytic problem is to establish a uniform estimate on the norm of the inverse of the linearized problem. Here one has to assume that the linearization is invertible at $s = 0$, which is true for generic situations as in op. cit. The inverse of the linearized problem enters into effective versions of the implicit function theorem, which can then be applied to identify the solution set as a manifold. In the nongeneric case one can consider a perturbed problem by introducing abstract perturbation terms spanning the cokernel. A solution to the latter problem will yield an ambient smooth space into which the original solution set is embedded as the zero set of finitely many functions, i.e., a Kuranishi model for $\mathcal{C}^{\mathrm{hol}}(M, J)$ at $(C, \mathbf{x}, \varphi)$. Several choices of spaces, differentiable glueing and deformation of abstract perturbation terms are possible, cf. op. cit. Notice that if the perturbations are chosen to be smooth, then the solutions of the perturbed equation will also be smooth by elliptic regularity.

The problem of globalization of local transversality in this setting (in particular in the presence of local automorphisms) is new. This is where the approaches differ most.

1. Fukaya and Ono let the dimension of the perturbation space (rank F in our setting) and hence also the dimensions of the manifolds containing $Z = \mathcal{C}_{R,g,k}^{\mathrm{hol}}(M, J)$ locally (dim $\widetilde{Z}$ in our setting) vary along a finite open cover of Z. The result is that Z is the zero locus of a section s' of a strange fiber space $F \to \widetilde{Z}$. Locally, the fiber space is a finite union of orbibundles of finite ranks over finite dimensional orbifolds fitting together nicely, but of jumping ranks and dimensions. The basic observation of [FkOn] is that while it is usually impossible to make orbifold sections transverse by perturbation, one may do so by going over to sufficiently high multivalued sections ("multisections"). These are sections of a symmetric product $S^l F$ that locally lift to $F^{\oplus l}$. And transversality means transversality of *each* component of a lift, i.e., of each branch of the multisection. The zero locus of a multisection is defined as the union of the zero loci of its branches. A generic perturbation of $\widetilde{s}$ will thus have a zero locus which locally is the finite union of (oriented) orbifolds of the expected dimension. The sums of the fundamental classes of these orbifolds, appropriately normalized, glue to a homology class on the base. The same works for sections of the strange fiber space $F \to \widetilde{Z}$. The homology class on $\widetilde{Z}$ thus obtained is the virtual fundamental class of Z. Note that if one insists on a class localized on Z one might take a limit of these classes as the perturbations tend to zero. But since the maps $\mathrm{ev}\colon Z \to M^k$ and $q\colon Z \to \mathcal{M}_{g,k}$ extend to $\widetilde{Z}$, this is not important for GW theory.

2. Li and Tian also describe Z as the zero locus of a section of a fiber space $F \to \widetilde{Z}$ with jumping dimensions as in (1). But instead of trying to perturb the section, they show how to glue cycles representing the Euler class and supported on $\widetilde{Z}$ directly.

3. Ruan works inside the stratified Fréchet orbifold of C^∞-stable complex curves in M. This is a topological space, but locally stratified into finitely many Fréchet orbifolds, depending on the combinatorial type of the curve. Nevertheless, by the glueing construction, it suffices to work within this space. The argument proceeds in an analogous way to Section 1.2, i.e., one constructs the perturbation as a morphism from a stratified orbibundle F of finite rank over the ambient space to a Banach bundle. Ruan claimed that one may take a trivial orbibundle of the form (base space) $\times\ (\mathbb{R}^N/G)$, where G is the product of the local groups of finitely many Kuranishi models covering $Z = \mathcal{C}^{\mathrm{hol}}_{R,g,k}(M, J)$. This is not in general possible. The argument is however right if one takes a nontrivial orbibundle, e.g., as in Section 3.2.

4. Another method, due to Liu and Tian, uses a compatible system of perturbation terms in the following sense: Z can be covered by finitely many local uniformizers $\{V_I = \widehat{V}_I/\Gamma_I\}_I$, $I = \{i_1, \dots, i_k\}$, $i_\nu \in \{1, \dots, n\}$, $k \leq m$, with

- $V_I \cap V_J = \emptyset$ if $\#I = \#J$ and $I \neq J$

- whenever $I \subset J$ there are morphisms $\pi_{IJ} : V_J = \widehat{V}_J/\Gamma_J \to V_I = \widehat{V}_I/\Gamma_I$

uniformizing open embeddings $V_J \subset V_I$, and these are compatible in the obvious way.

The $V_I = \widehat{V}_I/\Gamma_I$ are open sets in fibered products $\widehat{U}_{i_1} \times_\mathcal{B} \cdots \times_\mathcal{B} \widehat{U}_{i_k}$, and so are not smooth for $\#I > 1$, but finite unions of manifolds. Using the π_{IJ} one may compare perturbation terms over different V_I and thus define compatible systems $\{v_I\}_I$ of perturbation terms. For a generic choice of $\{v_I\}$ the zero loci of the perturbed section $\{(\widehat{s}_{\overline{\partial},J})_I - v_I\}$ form a compatible system of finite dimensional oriented orbifolds $\widehat{Z}_I^\nu \subset \widehat{V}_I$. This is enough to produce a homology class on the underlying space of the expected dimension of $Z^\nu \subset \mathcal{B}$.

While it might be somewhat tedious to do this in detail, it is rather obvious that all these definitions lead to the same homology class in an appropriate common ambient space, say $\mathcal{C}(M; p)$. In fact, in all these approaches one might take the Kuranishi model to be the restriction of our embedding $Z = Z(s_{\text{can}}) \subset \widetilde{Z}$ to open sets, at least if we choose our perturbations α sufficiently smooth. The problem is then essentially reduced to comparing various constructions of Euler classes for orbibundles in finite dimensions.

4 Axioms for GW invariants

4.1 GW invariants

There are several ways of extracting symplectic invariants from the virtual fundamental classes $\mathcal{GW}_{R,g,k}^{M,J} \in H_*(\mathcal{C}_{R,g,k}^{\text{hol}}(M, J))$. Assume that $2g + k \geq 3$. Then $\mathcal{M}_{g,k}$ exists as orbifold and, in particular, it satisfies rational Poincaré duality. There are diagrams

$$
\begin{array}{ccc}
\mathcal{C}_{R,g,k}^{\text{hol}}(M, J) & \xrightarrow{\text{ev}} & M^k \\
{\scriptstyle p}\downarrow & & \\
\mathcal{M}_{g,k} & &
\end{array}
$$

with ev and p the evaluation map and the forgetful map, sending $(C, \mathbf{x}, \varphi)$ respectively to $(\varphi(x_1), \ldots, \varphi(x_k))$ and to the stabilization of $(C, \mathbf{x})$. Note that both maps extend to $\mathcal{C}_{R,g,k}(M; p)$. By Theorem 3.1, we conclude the following:

Proposition 4.1 *The associated* GW *correspondence*

$$
\mathrm{GW}_{R,g,k}^M \colon H^*(M)^{\otimes k} \longrightarrow H_*(\mathcal{M}_{g,k}),
$$

$$
\alpha_1 \otimes \cdots \otimes \alpha_k \longmapsto p_*\Big(\mathcal{GW}_{R,g,k}^{M,J} \cap ev^*(\alpha_1 \times \cdots \times \alpha_k)\Big)
$$

is invariant under deformations of J *inside the space of almost complex structures tamed by some symplectic form. In particular,* $\mathrm{GW}_{R,g,k}^M$ *is an invariant of the symplectic deformation type of* (M, ω). $\square$

The following equivalent objects also occur commonly:

- composition of $\mathrm{GW}^M_{R,g,k}$ with Poincaré duality $H_*(\mathcal{M}_{g,k}) \to H^*(\mathcal{M}_{g,k})$;

- the associated homomorphism $H_*(\mathcal{M}_{g,k}) \otimes H_*(M)^k \to \mathbb{Q}$;

- the cycle $(p \times \mathrm{ev})_* \mathcal{GW}^M_{R,g,k} \in H_*(\mathcal{M}_{g,k} \times M^k)$.

Of these the second one is perhaps the most intuitive. For cycles $K \subset \mathcal{M}_{g,k}$, $A_1, \ldots, A_k \subset M$, it counts the "ideal" number of k-marked stable J-holomorphic curves $(C, \mathbf{x}, \varphi)$ in M of genus g with $(C, \mathbf{x})^{\mathrm{st}} \in K$ and the ith point mapping to A_i. "Ideal" means that this agrees with the actual (signed) number only in nice situations, say when $\mathcal{C}^{\mathrm{hol}}_{R,g,k}(M, J)$ is indeed an orbifold of the expected dimension which is transversal to $K \times A_1 \times \cdots \times A_k$ under $p \times \mathrm{ev}$. I prefer to reserve the name *GW invariant* for such numbers, i.e., by applying the second map to a product of cycles.

As already pointed out in [RuTi1], Remark 7.1 one can also define invariants by restricting the domain to certain singular curves and requiring homological conditions for the restriction of the maps to subcurves. The full perspective of this point of view has been given in [BeMa], where *marked modular graphs* τ are introduced as a bookkeeping device for the combinatorial data, cf. [Be2], Definition 1.2 (we adopt the abuse of notation and use τ both for the marked modular graph and the associated stable modular graph, i.e., with the marking omitted). If τ has n vertices and l edges, the corresponding moduli spaces $\mathcal{C}_\tau(M) = \mathcal{C}^{\mathrm{hol}}_\tau(M, J)$ (which corresponds to $\overline{M}(M, \tau)$ in [Be2]) are constructed as the fiber over a product of diagonals $\Delta^l \subset M^{2l}$ of a partial evaluation map

$$\mathrm{pev}: \prod_{i=1}^{n} \mathcal{C}^{\mathrm{hol}}_{R_i, g_i, k_i}(M, J) \longrightarrow M^{\Sigma k_i} \longrightarrow M^{2l}.$$

The meaning of this is that any edge of τ implements the requirement that the two marked points of the subcurves (= vertices of τ) bounding the edge, map to the same point in M. We refer to [Be2] for details of this concept.

To define virtual fundamental classes on these more general moduli spaces, let $\delta_{\Delta^l} \in H^*_{\Delta^l}(M^{2l})$ be the Poincaré dual class of Δ^l. We may then set

$$\mathcal{GW}^M_\tau := \left(\prod_i \mathcal{GW}^M_{R_i, g_i, k_i} \right) \cap \mathrm{pev}^* \delta_{\Delta^l} \in H_*(\mathcal{C}_\tau(M))$$

as the virtual fundamental class of $\mathcal{C}_\tau(M)$ (this corresponds to $J(M, \tau)$ in [Be2]). As above, we get an associated GW correspondence

$$\mathrm{GW}^M_\tau : H^*(M)^{\otimes \# S_\tau} \longrightarrow H_*\left(\mathcal{M}_\tau := \prod_i \mathcal{M}_{g_i, k_i} \right),$$

where S_τ is the set of tails of τ (which encode the positions of marked points).

4.2   Properties

From the intuitive geometric meaning one expects GW invariants to have a number of properties. These turned up as proved identities for a restricted class of varieties [RuTi1], Theorem A and Prop. 2.5, and in [RuTi3], or as axioms for $GW_{R,g,k}^M$ in [KoMa]. The corresponding axioms for the system of GW correspondences parametrized by marked modular graphs are given in [BeMa], cf. also [Be2]. As the presentation in the last reference is quite appropriate we just indicate what is to be added to establish Axioms I–V in op. cit. in the symplectic context. For statement and geometric explanation of the axioms we mostly refer to op. cit.

One should probably add to the axioms the important property of invariance under deformations of the (tamed) almost complex structure (respectively, under smooth projective deformations in the algebraic setting) that we have already commented on.

I. Mapping to point   This is the case $R = 0$. Since by the Wirtinger inequality, $\omega(\varphi_*[C]) = 0$ for a connected J-holomorphic curve $\varphi\colon C \to M$ implies $\varphi \equiv \mathrm{const}$, we get

$$\mathcal{C}_{0,g,k}^{\mathrm{hol}} = \mathcal{M}_{g,k} \times M$$

with universal curve $\pi = p \times \mathrm{Id}\colon \mathcal{M}_{g,k+1} \times M \to \mathcal{M}_{g,k} \times M$. This is an orbifold, but possibly of the wrong dimension. In fact, the cokernels of the linearization of $s_{\bar{\partial},J}$ glue to $R^1\pi_*\mathrm{ev}^*T_M = R^1p_*\mathcal{O} \boxtimes T_M$ (that we view as orbibundle, rather than its orbisheaf of sections). We set $F = R^1\widetilde{\pi}_*\widetilde{\mathrm{ev}}^*T_M$ and define $\tau\colon F \to \mathcal{E} = \widetilde{\pi}_*^p(\widetilde{\mathrm{ev}}^*T_M \otimes \overline{\Omega})$ in such a way that it restricts to a lift of this identification. For clarity, this time we write $\widetilde{\pi}, \widetilde{\mathrm{ev}}$ for the extensions of π, ev to $\mathcal{C}_{0,g,k}(M;p)$. Then $\widetilde{Z} = Z(\widetilde{s} = q^*s + \tau) \subset F$ is nothing other than $\mathcal{C}_{0,g,k}^{\mathrm{hol}}(M,J)$, and $[\widetilde{Z}] \cap \Theta_F$ computes the Euler class of $R^1\pi_*\mathrm{ev}^*T_M = R^1p_*\mathcal{O} \boxtimes T_M$ as claimed in the mapping to point axiom.

II. Products   This axiom forced our definition of the virtual fundamental class for nonconnected τ.

III. Glueing tails/cutting edges   Again, this axiom follows directly from our definition of virtual fundamental classes, now for connected components of τ with more than one vertex.

IV. Forgetting tails   Forgetting tails in a marked modular graph means omitting marked points from a stable complex curve and stabilizing (as a complex curve in M). Let us restrict to $\tau = (R, g, k)$, from which the general case follows easily. In view of the analogous fact for $\mathcal{M}_{g,k}$ and our construction of charts for $\mathcal{C}_{R,g,k}(M;p)$, it is easy to check that the corresponding map

$$\Phi \colon \mathcal{C}_{R,g,k+1}(M;p) \longrightarrow \mathcal{C}_{R,g,k}(M;p)$$

is the universal curve. If $\mathcal{E}_k$ and $s_k = s_{\bar{\partial},J}$ are the Banach bundle and section over $\mathcal{C}_{R,g,k}(M;p)$, then $\Phi^*\mathcal{E}_k$, Φ^*s_k can be identified with the bundle and section $\mathcal{E}_{k+1}$, s_{k+1} over $\mathcal{C}_{R,g,k+1}(M;p)$. Let $\tau \colon F \to \mathcal{E}_k$ span the cokernel of the (relative) linearization of s_k. Then $\Phi^*\tau$ will span the cokernel of the (relative) linearization of s_{k+1}. We obtain

$$\widetilde{Z}_{k+1} \;=\; Z(\widetilde{s}_{k+1} = \Phi^*\widetilde{s}_k) \;=\; \Phi^{-1}\Big(Z(\widetilde{s}_k = q^*s_k + \tau)\Big) \;=\; \Phi^{-1}(\widetilde{Z}_k),$$

and hence

$$\mathcal{GW}^M_{R,g,k+1} \;=\; [\widetilde{Z}_{k+1}] \cap \Theta_{\Phi^*F} \;=\; \Phi^{!}\Big([\widetilde{Z}_k] \cap \Theta_F\Big) \;=\; \Phi^{!}\mathcal{GW}^M_{R,g,k}.$$

V. Isogenies   Among the axioms this is the most interesting, having as consequence for instance the associativity of quantum products. The axiom comprises those modifications of marked modular graphs that do not change its genus. There are four basic cases:

(1) Contraction of a loop: omitting a loop, i.e., an edge connecting a vertex with itself, from a modular graph corresponds to dropping the requirement that a certain subcurve has a nondisconnecting double point. In a sense this case says something about potential smoothings of such double points of the domain.

(2) Contraction of a nonlooping edge: nonlooping edges correspond to disconnecting double points of the curve. Contraction of such an edge means that we consider two adjacent subcurves of genera g_1, g_2 as one subcurve of genus $g_1 + g_2$. So here we deal with potential smoothings of disconnecting double points.

(3) Forgetting a tail: as in axiom IV, but the conclusion is different.

(4) Relabelling: this treats isomorphisms of marked modular graphs, which in particular covers renumberings of the set of marked points.

Let τ be the marked modular graph obtained from σ by any of the operations (1–4). There is an embedding $\mathcal{C}_\sigma(M) \hookrightarrow \mathcal{C}_\tau(M)$ over the closed embedding of moduli spaces of curves $\mathcal{M}_\sigma \hookrightarrow \mathcal{M}_\tau$. The latter is divisorial in the first three cases and an isomorphism in the last case. Except possibly in (2), the choice of τ and the underlying modular graph of σ determine the marking of σ and the diagram

$$
\begin{array}{ccc}
\mathcal{C}_\sigma(M) & \longrightarrow & \mathcal{C}_\tau(M) \\
{\scriptstyle q_\sigma}\downarrow & & \downarrow{\scriptstyle q_\tau} \\
\mathcal{M}_\sigma & \longrightarrow & \mathcal{M}_\tau
\end{array}
$$

is Cartesian. Let $\delta_{\mathcal{M}_\sigma} \in H^*_{\mathcal{M}_\sigma}(\mathcal{M}_\tau)$ be Poincaré dual to $\mathcal{M}_\sigma$. The axiom can then be stated by requiring

$$\mathcal{GW}^M_\sigma = \mathcal{GW}^M_\tau \cap q_\tau^* \delta_{\mathcal{M}_\sigma}.$$

In case (2) the homology class R of the joined subcurve of τ can be arbitrarily distributed to the two adjacent subcurves of σ. We get a proper surjection

$$h: \coprod_{R=R_1+R_2} \mathcal{C}_{\sigma=\sigma(R_1,R_2)}(M) \longrightarrow q_\tau^{-1}(\mathcal{M}_\sigma) = \mathcal{M}_\sigma \times_{\mathcal{M}_\tau} \mathcal{C}_\tau(M).$$

Note that h is not injective if there are J-holomorphic curves with bubbles inserted at the double points. The claim is

$$\sum_{R=R_1+R_2} h_*\left(\mathcal{GW}^M_\sigma\right) = \mathcal{GW}^M_\tau \cap q_\tau^* \delta_{\mathcal{M}_\sigma}.$$

Except in the obvious case (4), the proof runs as follows. We again restrict to the basic case $\tau = (R, g, k)$. The embedding $\mathcal{M}_\sigma \hookrightarrow \mathcal{M}_\tau$ identifies $\mathcal{M}_\sigma$ with a divisor parametrizing singular curves or curves with two infinitely near marked points. By the form of our charts it is not hard to see that the Kuranishi space $\widetilde{Z}_\tau \subset F$ for $\mathcal{C}_\tau(M)$ intersected with $q_\tau^{-1}(\mathcal{M}_\sigma)$ is a union of suborbifolds $\widetilde{Z}_\sigma$ that can be identified with Kuranishi spaces for the components of $q_\tau^{-1}(\mathcal{M}_\sigma)$. Capping with the Thom class of F yields the result.

5 Comparison with algebraic GW invariants

For a smooth complex projective variety $M \subset \mathbb{P}^N$ we have now two definitions of virtual fundamental classes fulfilling the list of axioms plus deformation invariance: the symplectic ones $\mathcal{GW}^M_\sigma$ discussed so far (where $J = I$ is the integrable complex structure tamed by the Fubini–Study form), and the algebraic ones $J(M, \sigma)$ discussed in [Be2]. The latter are taken here in $H_*(\mathcal{C}_\sigma(M))$ by sending the analogous Chow class in the Deligne–Mumford stack $\overline{M}(M, \tau)$ to its homology class on the underlying complex space. It is natural to expect

Theorem 5.1 [Si4] *For any marked modular graph* σ

$$\mathcal{GW}_\sigma^M = J(M, \sigma). \quad \square$$

It is of course enough to treat the case $\sigma = (R, g, k)$. To compare the two definitions it is most convenient to work in the category of complex orbispaces, which are defined analogously to complex orbifolds, but with local models taken as finite group quotients of complex spaces (the underlying space will also be a complex space, but we want to keep in mind the group actions).

We first present an argument that does not work as stated, but where the basic reason for this equivalence is clear, and then outline the actual proof.

5.1 A model argument

Let us pretend that we can find $\tau\colon F \to \mathcal{E}$ such that

- $\widetilde{Z} = Z(\widetilde{s} = q^*s + \tau) \subset F$ is a *complex* suborbifold and $\widetilde{F} = q^*F|_{\widetilde{Z}}$ is a *holomorphic* vector bundle with *holomorphic* tautological section $s_{\mathrm{can}}\colon \widetilde{Z} \to \widetilde{F}$.

- The induced structure of complex orbispace on $C_\sigma(M) = Z = Z(s_{\mathrm{can}})$ is the right one (coming from the notion of holomorphic families of stable holomorphic curves in M).

According to [Fu], §14.1, the Euler class of $\widetilde{F}$ can be expressed in terms of the normal cone $C_{Z|\widetilde{Z}}$ of Z in $\widetilde{Z}$ and the total Chern class of $\widetilde{F}|_Z$ by

$$\left\{ c(\widetilde{F}) \cap s(C_{Z|\widetilde{Z}}) \right\}_d.$$

Here $d = \dim \widetilde{Z} - \operatorname{rank} \widetilde{F} = d(M, R, g, k)$ is the expected dimension of Z and $s(C_{Z|\widetilde{Z}})$ is the Segre class of $C_{Z|\widetilde{Z}}$. By construction, $\widetilde{Z}$ is smooth over the Artin stack $\mathfrak{F}_{g,k}$ of pre-stable curves (we should work with the analytic analogue here). Let $T_{\widetilde{Z}|\mathfrak{F}_{g,k}}$ be the relative tangent bundle, which is in fact an ordinary vector bundle over $\widetilde{Z}$, cf. below for an explicit construction. Next observe that $c_F(Z/\mathfrak{F}_{g,k}) := c(T_{\widetilde{Z}|\mathfrak{F}_{g,k}}) \cap s(C_{Z|\widetilde{Z}})$ is a class intrinsically associated to $Z \to \mathfrak{F}_{g,k}$, i.e., does not depend on the choice of embedding into a space smooth over $\mathfrak{F}_{g,k}$. This is a relative version of Fulton's *canonical class* [Fu], Expl. 4.2.6. We may thus write

$$\mathcal{GW}_\sigma^M = \left\{ c(\widetilde{F} - T_{\widetilde{Z}/\mathfrak{F}_{g,k}}) \cap c_F(Z/\mathfrak{F}_{g,k}) \right\}_d.$$

There is a quasi-isomorphism

$$
\begin{array}{ccc}
T_{\widetilde{Z}/\mathfrak{F}_{g,k}} & \xrightarrow{q_*} & T_{C_\sigma(M)/\mathfrak{F}_{g,k}} = \pi_*^{1,p}\mathrm{ev}^*T_M \\
{\scriptstyle Ds_{\mathrm{can}}}\downarrow & & \downarrow{\scriptstyle \bar{\partial}=D_{\mathfrak{F}_{g,k}}s_{\bar{\partial},I}} \\
\widetilde{F} & \xrightarrow{\tau} & \mathcal{E} = \pi_*^p(\mathrm{ev}^*T_M \otimes \overline{\Omega})
\end{array}
$$

where q_* and the tangent bundle and differential relative to $\mathfrak{F}_{g,k}$ are only meant formally, but are defined directly. The right-hand vertical arrow in turn is quasi-isomorphic to $R\pi_*\text{ev}^*T_M$ (to make this precise one should represent $R\pi_*\text{ev}^*T_M$ by a morphism of vector bundles, cf. below). We obtain

$$\mathcal{GW}_\sigma^M = \left\{ c(R\pi_*\text{ev}^*T_M) \cap c_F(Z/\mathfrak{F}_{g,k}) \right\}_d.$$

This is exactly the formula that one can derive for $J(M,\sigma)$ [Si3], for Behrend's relative obstruction theory is of the form $(R\pi_*\text{ev}^*T_M)^\vee \to L^\bullet_{C_\sigma(M)/\mathfrak{F}_{g,k}}$.

This argument is of course somewhat ad hoc and does not work as stated, because a τ with the required properties does not in general exist. But it shows already the basic reason behind the equivalence of the two theories: $R\pi_*\text{ev}^*T_M$ can be interpreted as the *virtual tangent bundle* of $C_\sigma(M)$ relative to $\mathfrak{F}_{g,k}$ when viewed either as zero locus of $s_{\overline{\partial},I}$ or as equipped with an obstruction theory relative to $\mathfrak{F}_{g,k}$.

5.2   Sketch proof by comparing cones

Recall that the construction of $J(M,\sigma)$ worked by writing $R^1\pi_*\text{ev}^*T_M$ as homomorphism of vector bundles $[G \to H]$, constructing a cone $C^H \subset H$ invariant under the additive action of G and intersecting $[C^H]$ with the zero section of H. Our proof that this class coincides with the symplectic virtual fundamental class is divided into three steps:

(1) The local construction of complex analytic Kuranishi models;

(2) a limit construction to obtain a (bundle of) cone(s) $C(\tau) \subset F|_Z$ of dimension $d + \text{rank}\, F$ and supporting a homology class $[C(\tau)]$ of the same dimension;

(3) finding an inclusion of vector bundles $\mu \colon H \hookrightarrow F|_Z$ with $\mu^![C(\tau)] = [C^H]$. Here $\mu^!$ is defined as the cap product with the pullback of the Thom class of $(F|_Z)/H$.

The theorem then follows from

$$\mathcal{GW}_\sigma(M) = [\widetilde{Z}] \cap \Theta_F = [C(\tau)] \cap \Theta_F = [C^H] \cap \Theta_H = J(M,\sigma).$$

Analytic Kuranishi models

Finding Kuranishi models in an integrable situation is actually easier than generally, because we may restrict to holomorphic maps near the double points. Let $(C, \mathbf{x}, \varphi)$ be a stable holomorphic curve, i.e., the map $\varphi \colon C \to M$ is holomorphic. As in Section 2.4, let $(q \colon \mathcal{C} \to S, \mathbf{x})$ be a semiuniversal deformation of $(C, \mathbf{x})$. If C_i are the irreducible components of C, we choose this time an open covering $\mathcal{U} = \{\mathcal{U}_i\}_{i=0,\ldots,d}$ of $\mathcal{C}$ with the following properties:

- for $i > 0$ there are holomorphic maps

$$z_i : U_i \longrightarrow \Delta$$

  extending holomorphically to $\overline{U_i}$ and inducing isomorphisms $U_i(s) := U_i \cap q^{-1}(s) \to \Delta$ for any $s \in S$;

- $U_i(0) \subset C_i$ and $U_i \cap U_j = \emptyset$ for $i, j > 0$;

- $U_0 = C \setminus \bigcup_{i>0} z_i^{-1}(\overline{\Delta_{1/2}})$, $\Delta_{1/2} = \{z \in \mathbb{C} \mid |z| < 1/2\}$;

- for $i > 0$ there are holomorphic charts

$$M \supset W_i \xrightarrow{\gamma_i} \mathbb{C}^n$$

with $\varphi(\overline{U_i(0)}) \subset W_i$.

The part over U_0 is dealt with by the space $\overline{\mathrm{Hom}}_S(U_0; M)$, that as a set consists of holomorphic maps $U_0(s) \to M$ extending continuously to $\overline{U_0(s)}$. Using a Čech construction together with the fact that open Riemann surfaces have vanishing higher coherent cohomology (they are Stein), one can show

Proposition 5.2 $\overline{\mathrm{Hom}}_S(U_0; M)$ *is a complex Banach manifold mapping submersively onto* S. $\square$

By this we mean of course that this complex Banach manifold represents a certain functor. The functor sends a morphism $\varphi : T \to S$ to the set of holomorphic maps from $T \times_S U_0$ to M that extend continuously to $T \times_S \overline{U_0}$.

For $i > 0$ we may identify (an open set in) $L_1^p(U_i(s); M)$ with $L_1^p(\Delta; W_i)$ via z_i and γ_i, and $L_1^p(U_0(s) \cap U_i(s); M)$ with $L_1^p(A_{1/2}; W_i)$, $A_{1/2} = \Delta \setminus \overline{\Delta_{1/2}}$. Consider the differentiable map of complex Banach manifolds

$$H : \overline{\mathrm{Hom}}_S(U_0; M) \times \prod_{i>0} L_1^p(\Delta; \gamma_i(W_i)) \longrightarrow \prod_{i>0} L_1^p(A_{1/2}; \mathbb{C}^n)$$

$$\left(\psi_0 : U_0(s) \to M; \psi_i \right) \longmapsto \left(\psi_i - \gamma_i \circ \psi_0 \circ z_i^{-1} \right).$$

$H^{-1}(0)$ can be identified with an open neighbourhood of φ in the space of L_1^p maps $\psi : C_s \to M$, some $s \in S$, that are holomorphic on $U_0(s)$. H is a split submersion along $H^{-1}(0)$. Hence

Proposition 5.3 $\mathcal{B} := H^{-1}(0)$ *is a complex Banach manifold.* $\square$

The $\overline{\partial}$ operator can now be viewed as a holomorphic map

$$G : \mathcal{B} \longrightarrow \prod_{i>0} L^p(A_{1/2}; \mathbb{C}^n),$$

and this induces the complex analytic structure on $\operatorname{Hom}_S(\mathcal{C}; M) = G^{-1}(0)$. An embedding of $\operatorname{Hom}_S(\mathcal{C}; M)$ into a finite dimensional complex manifold submerging onto S can be found as follows: let $Q \subset \prod_{i>0} L^p(A_{1/2}; \mathbb{C}^n)$ be a finite dimensional linear subspace spanning the cokernel of the linearization of G at some holomorphic φ. Q exists by the Stein property of $U_0(0)$. Then $G^{-1}(Q)$ is the desired finite dimensional complex manifold containing (an open part of) $\operatorname{Hom}_S(\mathcal{C}; M)$ as a closed complex subspace.

Note that by taking a basis of Q as perturbation terms α and a trivialization of $\mathcal{E}$ compatible with the complex analytic structure over W_i in the construction of $\tau\colon F \to \mathcal{E}$ (Section 3.2), we can achieve:

> Let $(C, \mathbf{x}, \varphi) \in \mathcal{C}_\sigma(M)$. Then, locally, there is a complex sub-bundle $F^h \subset F$ such that $\tau^h := \tau|_{F^h}$ spans the cokernel of the linearization of $s_{\overline{\partial}}$ relative to $\mathfrak{F}_{g,k}$ and $\widetilde{Z}^h := \widetilde{Z} \cap F^h$ is a *complex* orbifold.

For this purpose let $\overline{\mathcal{B}}$ be the image of $\mathcal{B}$ in $\mathcal{C}(M; p)$. Then by the choice of τ, $\widetilde{Z}^h$ must be a subset of $F^h|_{\overline{\mathcal{B}}}$, while over $\overline{\mathcal{B}}$ a uniformizer of $\widetilde{s}^h$ factorizes over G. Note that $\widetilde{Z}^h$ is $Z(\widetilde{s}^h)$ with $\widetilde{s}^h = (q^h)^* s + \tau^h$, $q^h = q|_{F^h}$.

The limit cone

The next step concerns the construction of the cone $C(\tau) \subset F|_Z$ that we get as the limit of $t \cdot \widetilde{Z} \subset F$ as t tends to infinity. This has nothing to do with holomorphicity.

We start with any $\tau\colon F \to \mathcal{E}$ over our Banach orbifold $\mathcal{C}_\sigma(M; p)$ spanning the cokernel of σ and write as usual $q\colon F \to \mathcal{C}_\sigma(M; p)$ for the bundle projection. For any $l > 0$,

$$F \times \mathbb{R}^l \ni (f, v) \longmapsto s_{\overline{\partial}}(q(f)) + |v|^2 \cdot \tau(f) \in \mathcal{E}$$

defines a section $\widetilde{s}_l$ of $q_l^* \mathcal{E}$, where $q_l = q \circ \operatorname{pr}_1$ is the projection from $F \times \mathbb{R}^l$ to $\mathcal{C}_\sigma(M; p)$. $\widetilde{s}_l$ is constant on spheres $\{f\} \times S_t^{l-1}(0)$. For $t \neq 0$ the zero locus $\widetilde{Z}_l$ of $\widetilde{s}_l$ restricted to $F \times S_t^{l-1}(0)$ is just $(t \cdot \widetilde{Z}) \times S_t^{l-1}(0)$, while $\widetilde{Z}_l \cap (F \times \{0\}) = F|_{\mathcal{C}_\sigma(M)}$.

Definition 5.4 Set $A = \widetilde{Z}_l \cap (F \times (\mathbb{R}^l \setminus \{0\}))$ and write $\overline{A}$ for its closure in $F \times \mathbb{R}^l$. The *limit cone* $C(\tau) \subset F$ of $s_{\overline{\partial}}$ with respect to τ is defined to be $\overline{A} \cap (F \times \{0\})$. $\square$

$C(\tau)$ is the set theoretic limit of $t \cdot \widetilde{Z}$ as t tends to infinity. As such:

(1) it does not depend on l; and

(2) it lies over the zero locus $C_\sigma(M)$ of $s_{\overline{\sigma}}$.

The reason for introducing l is the exact sequence of homology groups (of the second kind, see the discussion in 1.2)

$$H_{l+d+r}(C(\tau)) \longrightarrow H_{l+d+r}(\overline{A}) \longrightarrow H_{l+d+r}(A) \longrightarrow H_{l+d+r-1}(C(\tau)).$$

Here $r = \operatorname{rank} F$ and $d = d(M, R, g, k)$. The fundamental class $[A]$ of the oriented manifold A extends uniquely to a $(l+d+r)$-homology class (conveniently denoted $[\overline{A}]$ by abuse of notation) on $\overline{A}$, provided that $l+d+r-1 > \dim C(\tau)$. This uses the general vanishing theorem for homology, cf. [Iv], IX.1, Prop. 1.6. But from $C(\tau) \subset F|_{C_\sigma(M)}$, it follows that $\dim C(\tau) < r + \dim C_\sigma(M)$ is always finite, so the inequality can be satisfied by choosing l large enough. We can now define a homology class on $C(\tau)$ that is the limit of $[t \cdot \widetilde{Z}]$.

Proposition 5.5 *Let* $\delta_0 \in H^l_{\{0\}}(\mathbb{R}^l)$ *be the Poincaré dual of* $\{0\} \subset \mathbb{R}^l$. *Then*

$$[C(\tau)] := [\overline{A}] \cap \delta_0 \ \in H_{d+r}(C(\tau))$$

is independent of l *and homologous to* $[\widetilde{Z}]$ *as class on* F. $\square$

Note that the construction of $C(\tau)$ and $[C(\tau)]$ actually happens in finite dimensions. This is more apparent if we work in q^*F over the fixed finite dimensional orbifold $\widetilde{Z}$. Let s_{can} be the tautological section of q^*F. The natural map $q^*F \to F$ identifies the graph $\Gamma_{t \cdot s_{\mathrm{can}}}$ of $t \cdot s_{\mathrm{can}}$ with $t \cdot \widetilde{Z}$, and we may as well work with these graphs.

In a holomorphic situation we retrieve the following familiar picture [Fu], §14.1 (the following construction also works for singular spaces): let E be a holomorphic vector bundle over a complex manifold N, and let Z be the zero locus of a holomorphic section s of E. The differential of s induces a closed embedding of the *normal bundle* $N_{Z|N}$ of Z in N into E. $N_{Z|N}$ is the linear fiber space over Z associated to the conormal sheaf $\mathcal{I}/\mathcal{I}^2$ (the analytic analogue of $\operatorname{Spec}_Z S^\bullet \mathcal{I}/\mathcal{I}^2$ in the algebraic situation), $\mathcal{I}$ the ideal sheaf of Z in N. The *normal cone* $C_{Z|N}$ (the analytic analogue of $\operatorname{Spec}_Z \oplus_{d \geq 0} \mathcal{I}^d/\mathcal{I}^{d+1}$) is a closed subspace of $N_{Z|N}$. Let $\iota: C_{Z|N} \hookrightarrow E$ be the induced closed embedding. Take the identity morphism $E \to E$ for τ. Then $t \cdot \widetilde{Z}$ is the graph of $t \cdot s$. One can show ([Fu], Remark 5.1.1) that

$$C(\tau) = \iota(C_{Z|N}) \quad (\text{as spaces}) \quad \text{and} \quad [C(\tau)] = \iota_*[C_{Z|N}] = [\iota(C_{Z|N})].$$

This is used below to identify $C(\tau^h)$ with the image in F^h of the normal cone of $Z = C_\sigma(M)$ in $\widetilde{Z}^h$. The use of s^h_{can} is equivalent to the present use of the identity morphism.

The other ingredient is the following method to get rid of a nonholomorphic part of τ locally.

Proposition 5.6 *Let $F = F^h \oplus \overline{F}$ be a decomposition such that*

- $\tau^h := \tau|_{F^h}$ *spans the cokernel of the linearization σ along $Z = Z(s)$ and has the regularity properties of τ;*

- $\overline{\tau} := \tau|_{\overline{F}}$ *maps to $\mathrm{im}\,\sigma$ along Z.*

Then $C(\tau) = C(\tau^h) \oplus \overline{F}$ and $[C(\tau)] = [C(\tau^h)] \oplus [\overline{F}]$. □

The *proof* runs by considering the two parameter family $\widetilde{s}_{t,u} := q^*s + t \cdot \tau^h + u \cdot \overline{\tau}$ of perturbed sections with $|u| \leq |t|$. This interpolates between the original family $\widetilde{s}_t = \widetilde{s}_{t,t}$ and the family $\widetilde{s}_{t,0}$ having $\overline{F}$ added as trivial factor. As long as $t \neq 0$, $\widetilde{Z}_{t,u} = Z(\widetilde{s}_{t,u})$ is a suborbifold of F. The essential point is this:

Lemma 5.7 *The set theoretic limit of $\widetilde{Z}_{t,u}$ as $t, u \to 0$, $|u| \leq |t|$, equals $C(\tau^h) \oplus \overline{F}$. More precisely,*

$$\mathrm{cl}\left(\bigcup_{\substack{t \neq 0 \\ |u| \leq |t|}} \widetilde{Z}_{t,u} \times (t,u) \right) \cap (F \times (0,0)) = \left\{ (f,g) \in F^h \oplus \overline{F} \,\middle|\, f \in C(\tau^h) \right\}.$$

Intuitively if slightly imprecisely, we write $[C(\tau)] = \lim_{t \to 0}[\widetilde{Z}_t]$, to indicate both set theoretical and homological convergence. The proposition follows from

$$
\begin{aligned}
[C(\tau)] &= \lim_{t \to 0}[\widetilde{Z}_{t,t}] = \lim_{\substack{t,u \to 0 \\ |u| \leq |t|}} [\widetilde{Z}_{t,u}] \\
&= \lim_{t \to 0}[\widetilde{Z}_{t,0}] = \lim_{t \to 0}[\widetilde{Z}_t^h] \oplus [\overline{F}] = [C(\tau^h)] \oplus [\overline{F}].
\end{aligned}
$$

As for the lemma we may restrict attention to a fixed fiber F_z over $z \in C_\sigma(M)$. One may then use uniform continuity of the relative differential σ at centers of local uniformizing systems in connection with the implicit function theorem to modify sequences $(f_\nu, g_\nu) \in \widetilde{Z}_{t_\nu, u_\nu}$ with limit $(f,g) \in C(\tau)$ to $(f'_\nu, g'_\nu) \in \widetilde{Z}_{t_\nu} \oplus \overline{F}$ with the same limit. This shows $C(\tau) \subset C(\tau^h) \oplus \overline{F}$. The converse inclusion is obvious. □

The following will also be used.

Lemma 5.8 *Let χ be a continuous function on $C_\sigma(M; p)$ without zeros on an open set U. Then*

$$C(\tau)|_U = C(\chi \cdot \tau|_U), \quad [C(\tau)]|_U = [C(\chi \cdot \tau|_U)].$$

This is because multiplication by χ on the fibers of F induces an isomorphism from $Z(q^*s + t \cdot \chi \cdot \tau)$ to $Z(q^*s + t \cdot \tau)$. □

Global comparison

We begin by recalling the global free resolution of $R\pi_*\text{ev}^*T_M$ used by Behrend [Be1], Prop. 5. Let $\pi\colon \Gamma \to \mathcal{C}_\sigma(M)$ be the universal family, $\text{ev}\colon \Gamma \to M$ the universal morphism. By a twisting procedure with a relatively ample line bundle one obtains a sequence of holomorphic vector bundles

$$0 \to K \longrightarrow N \longrightarrow \text{ev}^*T_M \to 0$$

with $\pi_*K = \pi_*N = 0$. Then $R\pi_*\text{ev}^*T_M$ is (up to unique isomorphism) given by the homomorphism of vector bundles $G := R^1\pi_*K \to H := R^1\pi_*N$, viewed as a complex in degrees 0 and 1, as element of the derived category. The latter vector bundles can be described as cokernels of $\bar\partial$ operators obtained by resolving the above sequence by sheaves of fiberwise Sobolev sections and pushing forward. We get a diagram of complex (rather than holomorphic) Banach bundles

$$
\begin{array}{ccccccccc}
 & & 0 & & 0 & & & & \\
 & & \downarrow & & \downarrow & & & & \\
0 & \to & \pi_*^{1,p}K & \longrightarrow & \pi_*^{1,p}N & \longrightarrow & \pi_*^{1,p}\text{ev}^*T_M & =: \mathcal{T} & \to 0 \\
 & & \bar\partial_K \downarrow & & \downarrow \bar\partial_N & & \downarrow \bar\partial & & \\
0 & \to & \pi_*^{p}(K \otimes \overline{\Omega}) & \longrightarrow & \pi_*^{p}(N \otimes \overline{\Omega}) & \longrightarrow & \pi_*^{p}(\text{ev}^*T_M \otimes \overline{\Omega}) & = \mathcal{E} & \to 0 \\
 & & \downarrow & & \downarrow & & & & \\
 & & G & \overset{d}{\longrightarrow} & H & & & & \\
 & & \downarrow & & \downarrow & & & & \\
 & & 0 & & 0 & & & &
\end{array}
$$

As we occasionally do in the sequel, we omit to indicate some restrictions to $\mathcal{C}_\sigma(M)$. Similarly to the above construction of τ^h, we may now construct local homomorphisms $\tau_i\colon H \to \mathcal{E}$ that come from lifts to $\pi_*^{p}(N \otimes \overline{\Omega})$ of local holomorphic sections of H with support away from the singular locus of π. τ_i is easily seen to span the cokernel of $\sigma = \bar\partial$. In fact, locally, we even obtain a Cartesian diagram of vector bundles

$$
\begin{array}{ccc}
G & \longrightarrow & \mathcal{T} \\
d \downarrow & & \downarrow \bar\partial \\
H & \overset{\tau_i}{\longrightarrow} & \mathcal{E}
\end{array}
$$

K and N extend naturally to $\mathcal{C}_\sigma(M;p)$, as do H, G and τ_i. Keeping the notations H, G and τ_i for the extended objects we may set

$$F := H^{\oplus l} \quad \text{and} \quad \tau = \sum_i \chi_i\tau^i\colon F \longrightarrow \mathcal{E},$$

where we now insist that the bump functions χ_i form a partition of unity along $C_\sigma(M)$ (this can be done by going over to $\chi_i / \sum_j \chi_j$). Then τ composed with the diagonal embedding $H \hookrightarrow F$ spans the cokernel of σ along all of $C_\sigma(M)$. To compare Behrend's cone $C^H \subset H$ and $C(\tau)$ we embed H in F diagonally:

$$\mu \colon H \longrightarrow F, \quad h \longmapsto (h, \ldots, h).$$

Over an open set where τ_i spans the cokernel, put $F^h := F_i$ and $\iota_i \colon F_i \hookrightarrow F$ the embedding. Then, up to a harmless scaling factor, $\tau|_{F_i} = \chi_i \tau_i$ is of the form as given in the construction of analytic Kuranishi models. Put $\tau^h := \tau_i$. To find the complementary subbundle $\overline{F}$, let $\widetilde{T} := F \oplus_\tau T = \{(f, v) \mid \tau(f) = \sigma(v)\}$. $\widetilde{T}$ should be viewed as the tangent bundle of $\widetilde{Z}$ relative to $\mathfrak{F}_{g,k}$. Both F_i and $\operatorname{im} \mu$ span the cokernel of the projection $\widetilde{T} \to F$. A linear algebra argument gives:

Lemma 5.9 *Over the open set under consideration there exists a (continuous) suborbibundle $P \subset \widetilde{T}$ with $\overline{F} := \sigma(P)$ complementary to both $\mu(H)$ and $\iota_i(F_i)$.* $\square$

Proposition 5.6 applied to $F = F^h \oplus \overline{F}$ now shows that

$$C(\tau) = C(\chi_i \tau_i) \oplus \overline{F} \quad \text{and} \quad [C(\tau)] = [C(\chi_i \tau_i)] \oplus [\overline{F}].$$

By Lemma 5.8 we may also replace $\chi_i \tau_i$ by $\tau^h = \tau_i$. Let $\rho \colon F \to Q$ be the cokernel of μ. By transversality of $\overline{F}$ to $\operatorname{im} \mu$ we may identify $\overline{F}$ with Q via ρ. Let Θ_Q be the Thom class of Q. Then

$$\mu^!{[C(\tau)]} = [C(\tau)] \cap \rho^* \Theta_Q = [C(\tau^h)].$$

It thus remains to show that $C(\tau^h)$ coincides with $C^H \subset H$. To this end note that the morphism

$$\varphi^\bullet \colon [\mathcal{F}_i \to \Omega_{\widetilde{Z}^h/\mathfrak{F}}|_Z] \longrightarrow [\mathcal{I}/\mathcal{I}^2 \to \Omega_{\widetilde{Z}^h/\mathfrak{F}}|_Z],$$

that we obtain from the description of Z as zero locus of s_{can}^h in $\widetilde{Z}^h$ (with ideal sheaf $\mathcal{I}$), is a *(global resolution of a perfect) obstruction theory* as defined in [BeFa], or a *free global normal space* in the language of [Si2]. Note that the right-hand side of $\varphi^\bullet$ is isomorphic to the truncated cotangent complex $\tau_{\geq -1} L_Z^\bullet$ of Z. And $C(\tau^h)$ is exactly the closed subcone of $F^h|_Z$ obtained from this obstruction theory.

Let $\mathcal{G} = \mathcal{O}(G^\vee)$, $\mathcal{H} = \mathcal{O}(H^\vee)$ be the sheaves corresponding to G and H. Then $\mathcal{H} = \mathcal{F}_i$ and $\mathcal{G}$ can be identified with $\Omega_{\widetilde{Z}^h/\mathfrak{F}}|_Z$. Let

$$\psi^\bullet \colon [\mathcal{H} \to \mathcal{G}] \longrightarrow [\mathcal{I}/\mathcal{I}^2 \to \Omega_{\widetilde{Z}^h/\mathfrak{F}}|_Z]$$

be the obstruction theory used for algebraic GW invariants in [Be1], cf. [BeFa] before Prop. 6.2. The central result is

Proposition 5.10 *With the identifications $\mathcal{H} = \mathcal{F}_i$ and $\mathcal{G} = \Omega_{\widetilde{Z}^h/\widetilde{S}}|_Z$, the morphisms $\varphi^\bullet$ and $\psi^\bullet$ are locally homotopic, i.e., equal as morphisms in the derived category.* □

Since the cone belonging to an obstruction theory depends only on the morphism in the derived category, this shows that $C^H = C(\tau^h)$ as complex subspaces of H. So the proposition will finish the proof of Theorem 5.1.

To prove the proposition it suffices to check equality of the maps in cohomology, because we are dealing here with locally split two-term complexes [Si2], Lemma 2.4. $\psi^\bullet$ is constructed from the morphisms of the universal curve over $\mathcal{C}^{\mathrm{hol}}_\sigma(M)$ to M (evaluation map) and to $\mathcal{C}^{\mathrm{hol}}_\sigma(M)$ (projection) by constructions in the derived category. The difficulty in proving the proposition is to make the abstract constructions in derived categories explicit in a way suitable for comparison with the $\overline{\partial}$ operator. Let us just briefly indicate here how the $\overline{\partial}$ operator shows up, which is the key part.

First note that it suffices to work with truncated cotangent complexes $\tau_{\geq -1} L^\bullet$. By embedding into smooth spaces these can always be expressed in the form "conormal sheaf maps to restriction of cotangent sheaf of ambient smooth space". The smooth spaces we take are of course $\widetilde{Z}^h$ and the universal curve $\widetilde{\Gamma}$ over $\widetilde{Z}^h$. The holomorphic evaluation map from the universal curve Γ over Z does not in general extend holomorphically to $\widetilde{\Gamma}$. The point is that $\widetilde{\mathrm{ev}}$ provides a *differentiable* extension. The defect in holomorphicity leads to the $\overline{\partial}$ operator in the following explicit description of the map

$$\ker(\mathcal{H} \to \mathcal{G}) \simeq \pi_*(\mathrm{ev}^* \Omega_M \otimes \omega) \longrightarrow \mathcal{I}/\mathcal{I}^2.$$

Namely, for $U \subset \widetilde{Z}$ an open set, we send $\alpha \in (\mathrm{ev}^* \Omega_M \otimes \omega)(\pi^{-1} U)$ to $f_\alpha \in \mathcal{I}(U)$ by

$$U \ni z \longmapsto \int_{\widetilde{\Gamma}_z} \alpha(\overline{\partial} \widetilde{\varphi}_z),$$

where $\widetilde{\varphi}_z \colon \widetilde{\Gamma}_z \to M$ is the curve parametrized by z, and where we apply the dual pairing $\Omega_M \otimes T_M \to \mathbb{C}$ to make $\alpha(\overline{\partial} \widetilde{\varphi}_z)$ a $(1,1)$-form on $\widetilde{\Gamma}_z$. Note that $\widetilde{\varphi}_z$ is holomorphic near the singularities, so this form is smooth (in contrast to α, which may have poles at the singularities of $\widetilde{\Gamma}_z$). It should be more or less clear, and can be checked easily that this is exactly $H^{-1}(\varphi i^\bullet)$, the map induced by s^h_{can}. Similarly for the cokernels of $\varphi^\bullet$ and $\psi^\bullet$.

One final remark concerning rigidification: the limit cones that one obtains over an unrigidified chart $S \times V$ are invariant under the automorphism group of $(C, \mathbf{x})$ and hence restricts to the limit cone uniformizing $C(\tau)$ on the actual, rigidified chart $S \times \overline{V}$. A similar statement holds for the algebraic cones.

References

[Ad]   R. Adams: *Sobolev spaces*, Academic Press 1975

[Au]   T. Aubin: *Nonlinear analysis on manifolds*, Springer 1982

[Be1]   K. Behrend: *GW invariants in algebraic geometry*, Inv. Math. **127** (1997) 601–617

[Be2]   K. Behrend: *Algebraic Gromov–Witten invariants*, this volume, pp. 19–70

[BeFa] K. Behrend, B. Fantechi: *The intrinsic normal cone*, Inv. Math. **128** (1997) 45–88

[BeMa] K. Behrend, Y. Manin: *Stacks of stable maps and Gromov–Witten invariants*, Duke Math. J. **85** (1996) 1–60

[Bi]   E. Bishop: *Conditions for the analyticity of certain sets*, Michigan Math. J. **11** (1964) 289–304

[Br]   G. E. Bredon: *Sheaf theory*, McGraw-Hill 1965

[Bs]   R. Brussee: *The canonical class and the C∞-properties of Kähler surfaces*, New York J. Math. **2** (1996) 103–146 (available from http://nyjm.albany.edu/)

[CfFe] R. R. Coifman, C. Fefferman: *Weighted norm inequalities for maximal functions and singular integrals*, Studia Math. **51** (1974) 241–250

[Co]   J. Conway: *Functions of one complex variable II*, Springer 1995

[DeMu] P. Deligne, D. Mumford: *The irreducibility of the space of curves of given genus*, Publ. Math. IHES **36** (1996) 75–110

[Fi]   G. Fischer: *Complex analytic geometry*, LNM 538, Springer 1976

[FkOn] K. Fukaya, K. Ono: *Arnol'd conjecture and Gromov–Witten invariant*, Warwick preprint 29/1996

[Fl]   A. Floer: *Symplectic fixed points and holomorphic spheres*, Comm. Math. Phys. **120** (1989), 575–611

[Fu]   W. Fulton: *Intersection theory*, Springer 1984

[Gi]   A. Givental: *Equivariant Gromov–Witten invariants*, Intern. Math. Res. Notices 1996, 613–663

[Gv]   M. Gromov: *Pseudo-holomorphic curves in symplectic manifolds*, Inv. Math. **82** (1985) 307–347

[HLiSk] H. Hofer, V. Lizan, J.-C. Sikorav: *On genericity for holomorphic curves in four-dimensional almost-complex manifolds*, J. Geom. Anal. **7** (1997) 149–159

[IvSh] S. Ivashkovich, V. Shevchishin: *Pseudo-holomorphic curves and envelopes of meromorphy of two-spheres in* $\mathbf{CP}^2$, preprint Bochum 1995, revision math.CV/980401

[Iv]   B. Iversen: *Cohomology of sheaves*, Springer 1986

[Kn]   F. Knudsen: *The projectivity of the moduli space of stable curves, II: The stacks* $M_{g,n}$, Math. Scand. **52** (1983) 161–199

[KoMa] M. Kontsevich, Y. Manin: *Gromov–Witten classes, quantum cohomology, and enumerative geometry*, Comm. Math. Phys. **164** (1994) 525–562

[KoNo] S. Kobayashi, K. Nomizu: *Foundations of differential geometry*, Wiley 1969

[Ku]   M. Kuranishi: *New proof for the existence of locally complete families of complex structures*, in Proceedings of the conference on complex analysis, Minneapolis 1964, A. Aeppli, E. Calabi, H. Röhrl (eds.), Springer 1965

[LcMc] R. Lockhardt, R. Mc Owen: *Elliptic differential operators on noncompact manifolds*, Ann. Sc. Norm. Sup. Pisa **12** (1985) 409–447

[LiTi1] J. Li, G. Tian: *Virtual moduli cycles and GW invariants of algebraic varieties*, J. Amer. Math. Soc. **11** (1998) 119–174

[LiTi2] J. Li, G. Tian: *Virtual moduli cycles and Gromov–Witten invariants of general symplectic manifolds*, in Topics in symplectic 4-manifolds (Irvine, CA, 1996), R. Stern (ed.), First Int. Press Lect. Ser., I, Internat. Press, Cambridge, MA, 1998, 47–83

[LiTi3] J. Li, G. Tian: *Algebraic and symplectic geometry of Gromov–Witten invariants*, in Algebraic geometry—Santa Cruz 1995, 143–170, Proc. Sympos. Pure Math. **62**, Part 2, Amer. Math. Soc. 1997

[LiTi4]   J. Li, G. Tian: *Comparison of the algebraic and the symplectic Gromov-Witten invariants*, preprint `alg-geom/9712035`

[Liu]   G. Liu, PhD thesis, Stony Brook 1994

[LiuTi]   G. Liu, G. Tian: *Floer homology and Arnol'd conjecture*, to appear in J. Diff. Geom.

[McSa]   D. McDuff, D. Salamon: *J-holomorphic curves and quantum cohomology*, Amer. Math. Soc. 1994

[PSaSc]   S. Piunikhin, D. Salamon, M. Schwarz: *Symplectic Floer-Donaldson theory and quantum cohomology*, in Contact and symplectic geometry (Cambridge, 1994), 171–200, Publ. Newton Inst. 8, Cambridge Univ. Press 1996

[Pn]   P. Pansu: *Compactness*, in Holomorphic curves in symplectic geometry, M. Audin, J. Lafontaine (eds.), Birkhäuser 1994

[PrWo]   T. Parker, J. Wolfson: *Pseudoholomorphic maps and bubble trees*, J. Geom. Anal. **3** (1993) 63–98

[Ru1]   Y. Ruan: *Topological sigma model and Donaldson type invariants in Gromov theory*, Duke Math. J. **83** (1996) 461–500

[Ru2]   Y. Ruan: *Virtual neighbourhoods and pseudo-holomorphic curves*, preprint `alg-geom 9611021`

[RuTi1]   Y. Ruan, G. Tian: *A mathematical theory of quantum cohomology*, J. Diff. Geom. **42** (1995) 259–367 (announced in Math. Res. Lett. **1** (1994) 269–278)

[RuTi2]   Y. Ruan, G. Tian: *Bott-type Floer cohomology and its multiplication structures*, Math. Res. Lett. **2** (1995) 203–219

[RuTi3]   Y. Ruan, G. Tian: *Higher genus symplectic invariants and sigma model coupled with gravity*, Invent. Math. **130** (1997) 455–516

[Si1]   B. Siebert: *Gromov-Witten invariants for general symplectic manifolds*, preprint `dg-ga 9608005`, revised 12/97

[Si2]   B. Siebert: *Virtual fundamental classes, global normal cones and Fulton's canonical classes*, preprint 1997

[Si3]   B. Siebert: *An update on (small) quantum cohomology*, Proceedings of the conference on Geometry and Physics, Montreal 1995

[Si4]   B. Siebert: *Symplectic and algebraic Gromov–Witten invariants co-incide*, preprint math.AG/9804108

[Sk]   E. G. Sklyarenko: *Homology and cohomology theories of general spaces*, in *General topology II*, A. V. Arhangel'skii (ed.), Encyclope-dia of mathematical sciences, Springer 1996

[SeSi]   R. Seeley, I. Singer: *Extending $\bar{\partial}$ to singular Riemann surfaces*, J. Geom. Phys. **5** (1988) 121–136

[Sm]   S. Smale: *An infinite dimensional version of Sard's theorem*, Amer. J. Math. **87** (1965) 861–866

[Ta]   C. Taubes: SW → Gr: *from the Seiberg–Witten equations to pseudo-holomorphic curves*, J. Amer. Math. Soc. **9** (1996) 845–918

[Vi]   A. Vistoli: *Intersection theory on algebraic stacks and on their mod-uli spaces*, Inv. Math. **97** (1989) 613–670

[Wi1]   E. Witten: *Topological quantum field theory*, Comm. Math. Phys. **117** (1988) 353–386

[Wi2]   E. Witten: *Topological sigma models*, Comm. Math. Phys. **118** (1988) 411–449

[Wi3]   E. Witten: *Two-dimensional gravity and intersection theory on mod-uli space*, in Surveys in Differential Geometry **1** (1991) 243–310

[Ye]   R. Ye: *Gromov's compactness theorem for pseudo holomorphic curves*, Trans. Amer. Math. Soc. **342** (1994) 671–694

Bernd Siebert,
Fakultät für Mathematik,
Ruhr-Universität Bochum,
D-44780 Bochum
Bernd.Siebert@ruhr-uni-bochum.de

A generic Torelli theorem
for the quintic threefold

Claire Voisin*

Contents

0.1   Introduction

This paper proves the generic Torelli theorem for the family of quintic hypersurfaces $X_5 \subset \mathbb{P}^4$:

Theorem 1 *The period map $\mathcal{P}_3$ defined on the quotient $U/\operatorname{PGL}(4)$, where $U \subset \mathbb{P}(H^0(\mathcal{O}_{\mathbb{P}^4}(5)))$ parametrizes smooth hypersurfaces, is of degree one onto its image.*

Recall that if B parametrizes a family $(X_b)_{b \in B}$ of compact Kähler varieties of dimension k, the period map

$$\mathcal{P}_k \colon B \to \mathcal{D}/\Gamma,$$

*Partially supported by the project "Algebraic Geometry in Europe" (AGE), Contract ERBCHRXCT 940557 sends $b \in B$ to the Hodge filtration $F^{\bullet} H^k(X_b)$ on $H^k(X_b, \mathbb{C}) \cong H^k(X_0, \mathbb{C})$. Here $\mathcal{D}$ is the period domain parametrizing all filtrations on $H^k(X_0, \mathbb{C})$ of given ranks, and Γ the group of automorphisms of $H^k(X_0, \mathbb{Z})$ preserving the intersection form, so that the isomorphism $H^k(X_b, \mathbb{Z}) \cong H^k(X_0, \mathbb{Z})$ is well defined up to Γ. It is known (Griffiths [13]) that $\mathcal{P}_k$ is holomorphic and satisfies Griffiths transversality.

Theorem 1 says that if $X \subset \mathbb{P}^4$ is a generic quintic threefold, any isomorphism of its polarized Hodge structure

$$\left(H^3(X, \mathbb{Z}), \langle \ , \ \rangle, F^{\bullet} H^3(X) \right) \cong \left(H^3(X', \mathbb{Z}), \langle \ , \ \rangle, F^{\bullet} H^3(X') \right)$$

with that of another quintic threefold X' is induced by a (projective) isomorphism $X \cong X'$.

We mention that Torelli statements for Calabi–Yau varieties may have an important role to play in mirror symmetry: at present, the mirror symmetry conjecture can be formulated in Hodge theoretic terms, as identifying the quantum variation of Hodge structure parametrized by $H^2(X^*, \mathbb{C})$, where X^* is a mirror Calabi–Yau threefold, with the variation of Hodge structure on $H^3(X, \mathbb{C})$ parametrized by the Kuranishi family of X. In this formulation it has been proved by Givental [8] when X^* is the quintic threefold and X its mirror (see [1]). It seems likely that Givental's method should also prove it when X is the quintic threefold and X^* its mirror. This last statement, together with our results would in fact prove a stronger conjecture, namely the fact that a complexified Kähler parameter on X^* determines a complex structure on X: indeed, what we show is that the generic quintic X is determined by its infinitesimal variation of Hodge structure, and it turns out that for Calabi-Yau threefolds, the infinitesimal variation of Hodge structure is equivalent to the isomorphism class of the Yukawa cubic (see below the discussion of the derivative of the period map) on the vector space $H^1(T_X)$. Moreover, the mirror symmetry conjecture identifies this with the quantum product on $H^2(X^*)$, which is determined by the complexified Kähler parameter on X^*.

The Torelli theorem for K3 surfaces was used in a similar way in Voisin [19] to construct the mirror map in this stronger sense for a particular class of Calabi–Yau threefolds. It seems that a Torelli statement could also play an important role in the new construction of mirror symmetry proposed by Strominger, Yau and Zaslow [17], and studied in Morrison [15] and Gross and Wilson [14].

0.2   The methods of Griffiths and Donagi

In the case of a Calabi–Yau threefold X (of which the quintic threefold is the most standard example), it is known that the local period map $\mathcal{P}_3$ from the Kuranishi family B of X to the local period domain is immersive. This follows from Griffiths' general description of the derivative of the period map at $0 \in B$, where $X = X_0$: by transversality, this derivative is described by a series of maps

$$\mu_p \colon T_{B,0} \to \operatorname{Hom}(H^{p,q}(X), H^{p-1,q+1}(X)) \quad \text{for } p + q = 3.$$

Griffiths proved that μ_p is the composite of the Kodaira–Spencer isomorphism $T_{B,0} \cong H^1(T_X)$ with the natural cup product map

$$H^1(T_X) \to \operatorname{Hom}(H^{p,q}(X), H^{p-1,q+1}(X)).$$

Now consider the period map $\mathcal{P}^{3,0}$ for holomorphic $(3,0)$ forms, which sends $t \in B$ to the line $H^{3,0} \subset H^3(X_0, \mathbb{C})$ based by the holomorphic $(3,0)$ form on X_t. It is obviously determined by $\mathcal{P}_3$, and by Griffiths transversality, its derivative is a map

$$\mathcal{P}_*^{3,0} \colon T_{B,0} \to \operatorname{Hom}\big(H^{3,0}(X), H^3(X,\mathbb{C})/H^{3,0}(X)\big),$$

with image contained in $\operatorname{Hom}(H^{3,0}(X), H^{2,1}(X))$, which can then be identified with μ_3. Now the cup product map

$$H^1(T_X) \to \operatorname{Hom}(H^{3,0}(X), H^{2,1}(X))$$

is clearly an isomorphism, since K_X is trivial, which implies that $\mathcal{P}_*^{3,0}$ is injective; therefore $\mathcal{P}_3$ is an immersion.

Apart from this general fact, there are very few results on the Torelli problem for Calabi–Yau threefolds. On the other hand, Donagi proved the generic Torelli theorem for many families of hypersurfaces in projective space, with a series of possible exceptions including Calabi–Yau hypersurfaces:

Theorem 2 (Donagi [4]) *The generic Torelli theorem holds for hypersurfaces in $\mathbb{P}^n$ of degree d, with the possible exception of the following cases:*

(a) *d divides $n + 1$;*

(b) *$d = 3$ and $n = 3$;*

(c) *$d = 4$ and $n \equiv 1 \mod 4$;*

(d) *$d = 6$ and $n \equiv 2 \mod 6$.*

(In fact, (d) has since been excluded, see Cox and Green [3].) In the remainder of this section, we sketch the ideas of Donagi's proof, and explain how his methods enable us to deduce Theorem 1 from the purely algebraic statement of Theorem 5.

Let X be a smooth hypersurface of degree d in $\mathbb{P}^n$, with equation $F = 0$. We write R_F for the *Jacobian ring* of F, that is, the quotient of the polynomial ring in $n+1$ variables $S = \bigoplus_{k \geq 0} S^k$ by the ideal J_F generated by the partial derivatives of F. We note for future use that $R_F = \sum_{d=0}^{\sigma} R_F^d$ is a graded complete intersection ring, with dth graded piece R_F^d, having element of top degree $\sigma(n+1)(d-2)$, and the multiplication $R_F^d \times R_F^{\sigma-d} \to R_F^{\sigma} = \mathbb{C}$ is a perfect pairing, defining a duality $(R_F^{\sigma-d})^* = R_F^d$ (see [2]).

Then if $n \geq 3$, and $d \geq 5$ if $n = 3$, the natural map

$$\{\text{infinitesimal deformations of } F\} \to \{\text{infinitesimal deformations of } X\}$$

defines an isomorphism

$$R_F^d \cong H^1(T_X).$$

The infinitesimal variation of Hodge structure of X, or the differential of the period map at $[X]$, is given as described above by a series of maps

$$\mu_p \colon H^1(T_X) \to \operatorname{Hom}(H_{\text{prim}}^{p,q}(X), H_{\text{prim}}^{p-1,q+1}(X)) \quad \text{for } p+q = n-1.$$

We have the following theorem:

Theorem 3 (Carlson and Griffiths [2]) *There are natural isomorphisms*

$$H_{\text{prim}}^{p,q}(X) \cong R_F^{(n-p)d-n-1} \quad \text{for } p+q = n-1,$$

which, up to universal coefficients, identify the μ_p with the multiplication maps

$$R_F^d \to \operatorname{Hom}(R_F^{(n-p)d-n-1}, R_F^{(n-p+1)d-n-1}). \tag{0.1}$$

It is known that, except if $d = 3$ and $n = 3$, the (polarized) period map for X is a local immersion. To prove that it is of degree 1 onto its image, it is enough to prove that if V and V' are simply connected open subsets of the quotient $U^0/\operatorname{PGL}(n+1)$ parametrizing smooth hypersurfaces without automorphisms up to projective equivalence, and $j \colon V \cong V'$ a given isomorphism such that $\mathcal{P}_{n-1} \circ j = \mathcal{P}_{n-1}$, where $\mathcal{P}_{n-1}$ is the local period map, then $V = V'$ (here we identify the local period domains for V and V' by the given isomorphism of polarized Hodge structures $H_{\text{prim}}^{n-1}(X_{t_0}) \cong H_{\text{prim}}^{n-1}(X_{j(t_0)})$ for some $t_0 \in V$). By translating the condition that $\mathcal{P}_{n-1*} \circ j_* = \mathcal{P}_{n-1*}$, it follows immediately that we have an induced isomorphism of infinitesimal variations of Hodge structure at t and $j(t)$.

Using Theorem 3, we see that this isomorphism of infinitesimal variations of Hodge structures gives a partial isomorphism between the Jacobian rings of F_t and $F_{j(t)}$:

$$R_{F_t}^d \longrightarrow \mathrm{Hom}(R_{F_t}^{(n-p)d-n-1}, R_{F_t}^{(n-p+1)d-n-1})$$
$$\downarrow \qquad\qquad\qquad \downarrow \qquad\qquad\qquad\qquad (0.2)$$
$$R_{F_{j(t)}}^d \longrightarrow \mathrm{Hom}(R_{F_{j(t)}}^{(n-p)d-n-1}, R_{F_{j(t)}}^{(n-p+1)d-n-1}),$$

where the second vertical arrow is an isomorphism induced by isomorphisms $R_{F_t}^{(n-p)d-n-1} \cong R_{F_{j(t)}}^{(n-p)d-n-1}$ for each p. In other words, knowing the Hodge structure and infinitesimal variation of Hodge structure of X_t determines *some graded pieces* of the multiplication map of the Jacobian ring R_{F_t}. The key ingredient in going from this to the proof of Theorem 2 is the following result, which allows us in many cases to recover the whole Jacobian ring from the above data:

Theorem 4 (Donagi–Green symmetrizer lemma, cf. [6]) *Let R be the quotient of S by a regular sequence of $n+1$ homogeneous elements of degree $d-1$; write $\sigma = (d-2)(n+1)$ for the top degree of R, and define*

$$T_{a,b} = \{\varphi \in \mathrm{Hom}(R^a, R^b) \mid w\varphi(v) = v\varphi(w) \text{ for all } v,w \in R^a\}.$$

Then $T_{a,b} = R^{b-a}$ if $a+b \le \sigma$ and $b+d-1 \le \sigma$.

This means that multiplication by elements of R^{b-a} is uniquely determined by the multiplication map $R^a \times R^b \to R^{a+b}$. In the range considered in Theorem 2, Donagi and Green [6] apply Theorem 4 to diagram (0.2) to recover isomorphisms between other pieces of the Jacobian rings of F_t and $F_{j(t)}$, and eventually to obtain a whole isomorphism between the Jacobian rings of F_t and $F_{j(t)}$ themselves; the degree 1 piece of this isomorphism then gives an isomorphism $\alpha\colon S^1 \cong S^1$ inducing $\alpha^{d-1}\colon S^{d-1} \cong S^{d-1}$ such that $\alpha^{d-1}(J_{F_t}^{d-1}) = J_{F_{j(t)}}^{d-1}$. It is then immediate to conclude that F_t and $F_{j(t)}$ are projectively equivalent. Thus Donagi's result is that, in the range of Theorem 2, the existence of the commutative diagrams (0.2) implies that $X_t \cong X_{j(t)}$ for generic t. See Donagi's survey [5] for more details.

The argument does not work for the series (a) of possible exceptions to Theorem 2, because when d divides $n+1$, the symmetrizer lemma applied to the multiplication maps (0.1) can never give new pieces of the Jacobian ring of degree not divisible by d.

0.3 Start of our proof

From now on we consider the case of the quintic threefold. This is a Calabi–Yau threefold, and its infinitesimal variation of Hodge structure is described by the maps

$$\mu_3 \colon H^1(T_X) \to \mathrm{Hom}(H^{3,0}(X), H^{2,1}(X)),$$
$$\mu_2 \colon H^1(T_X) \to \mathrm{Hom}(H^{2,1}(X), H^{1,2}(X)),$$
$$\mu_1 \colon H^1(T_X) \to \mathrm{Hom}(H^{1,2}(X), H^{0,3}(X)).$$

We proved above that the first map is an isomorphism, and the last map is the same isomorphism, in view of the duality isomorphisms

$$H^{0,3}(X) = H^{3,0}(X)^* \quad \text{and} \quad H^{1,2}(X) = H^{2,1}(X)^*.$$

Thus the only algebraic invariants of the infinitesimal variation of Hodge structure of X are contained in the map

$$\mu \colon H^1(T_X) \otimes H^1(T_X) \to H^1(T_X)^*,$$

well defined up to a coefficient depending on the choice of a generator of $H^{3,0}(X)$, which is obtained from the map

$$H^1(T_X) \otimes H^1(\Omega_X^2) \to H^2(\Omega_X)$$

given by μ_2, after using the isomorphism μ_3 to substitute $H^1(T_X)$ for $H^1(\Omega_X^2)$, and μ_1 to substitute $H^1(T_X)^*$ for $H^2(\Omega_X)$. The map μ is easily seen to come from a cubic form on $H^1(T_X)$, the *Yukawa coupling*. It follows from the above that this cubic determines the whole infinitesimal variation of Hodge structure. Theorem 3 identifies $\mu \colon S^2 H^1(T_X) \to H^1(T_X)^*$ with the multiplication map

$$\mu \colon S^2 R_F^5 \to R_F^{10}, \tag{0.3}$$

(for simplicity, we write xy for $\mu(x,y) \in R_F^{10}$ in what follows) followed by the duality isomorphism $R_F^{10} \cong (R_F^5)^*$.

To extend Donagi's method to the case of the quintic, it is enough to show that, at least for generic F, the multiplication map

$$\mu_{1,4} \colon S^1 \otimes R_F^4 \to R_F^5 \tag{0.4}$$

is uniquely determined by the map μ (0.3) and the isomorphism $R_F^{10} \cong (R_F^5)^*$. In other words, we need to prove that an isomorphism $R_F^5 \cong R_{F'}^5$ compatible with μ induces isomorphisms $S^1 \cong S^1$ and $R_F^4 \cong R_{F'}^4$ compatible with $\mu_{1,4}$. Indeed, if we can do this iterated application of the symmetrizer lemma allows us to reconstruct the whole Jacobian ring R_F, and hence F itself.

We view $\mu_{1,4}$ as an inclusion $R_F^4 \subset \mathrm{Hom}(S^1, R_F^5) \cong (R_F^5)^5$, where the last isomorphism involves choosing a basis of S^1, and write $W \subset (R_F^5)^5$ for the image of this inclusion. From now on, W is the main unknown in our proof;

our first attempt is to try to determine W from the following properties (a–g), which are stated in terms of the known map μ and isomorphism $R_F^{10} \cong (R_F^5)^*$.

First, some notation. The typical element of W is $w = (w_1, \ldots, w_5) = (x_1\varphi, \ldots, x_5\varphi)$, where $\varphi \in R_F^4$ and $x_1, \ldots, x_5$ are coordinates on $\mathbb{P}^4$. For $t = (t_1, \ldots, t_5) \in \mathbb{C}^5$ we write $w_t = \sum_i t_i w_i \in R_F^5$, which corresponds simply to the product $A\varphi$ of φ with a linear form $A \in S^1$. We think of $w \mapsto w_t$ as defining a *projection* $p_t \colon W \to R_F^5$ of $W \subset (R_F^5)^5$ to one factor R_F^5, and its image $p_t(W)$. In Section 2, we refer to this image as a *component* of W; in particular when t is one of the basis vectors $t = e_i$ of $\mathbb{C}^5$, we call $p_i(W) = p_{e_i}(W)$ the *ith component* of $W \subset (R_F^5)^5$. In a similar way, two elements $t, t' \in \mathbb{C}^5$ give rise to a *2-term projection* $W_{t,t'} = \{(w_t, w_{t'}) \mid w \in W\} \subset (R_F^5)^2$.

Properties of W

(a) $\dim W = \dim R_F^4 = 65 = \binom{8}{4} - 5$.

(b) *The symmetrizer property.* For $w, w' \in W$ we have

$$w_i w_j' = w_j w_i' \in R_F^{10} \quad \text{for } i, j = 1, 2, \ldots, 5.$$

More generally, $w_t w_{t'}' = w_{t'} w_t'$ for any $t, t' \in \mathbb{C}^5$. Indeed, w, w' correspond to elements φ, φ' of R_F^4, and t, t' to elements A, A' of S^1. Then in R_F^5, we have the equalities: $w_t = A\varphi$, $w_{t'} = A'\varphi$, $w_t' = A\varphi'$ and $w_{t'}' = A'\varphi'$, so that the result is clear.

(b') More precisely,

$$W = \{ z \in (R_F^5)^5 \mid z_i w_j = z_j w_i \text{ for all } i, j \text{ and all } w \in W \}.$$

This follows from the symmetrizer lemma (Theorem 4).

(c) $\dim R_F^5 \cdot W = 135$, where $R_F^5 \cdot W \subset (R_F^{10})^5$ is defined using the natural map $R_F^5 \otimes (R_F^5)^5 \to (R_F^{10})^5$. This holds because $R_F^5 \cdot W$ is isomorphic to R_F^9, which by Koszul has dimension $\binom{13}{4} - 5\binom{9}{4} + 10\binom{5}{4} = 135$.

The next property is less obvious. Assume that X is generic, and for generic $t, t' \in \mathbb{C}^5$, define the *partial symmetrizer* of the 2-term projection $W_{t,t'}$ by

$$W_{t,t'}' = \{ (z_1, z_2) \in (R_F^5)^2 \mid z_1 w_t = z_2 w_{t'} \in R_F^{10} \text{ for all } (w_t, w_{t'}) \in W_{t,t'} \};$$

then (b) says that $W_{t,t'} \subset W_{t,t'}'$. The following fact, while technical, is very helpful in the remainder of our calculations:

(d) $\operatorname{codim}(W_{t,t'} \subset W'_{t,t'}) = 2$ and $\operatorname{codim}(R_F^5 W_{t,t'} \subset R_F^5 W'_{t,t'}) = 2$; moreover, we can identify $R_F^5 W'_{t,t'}$ with the orthogonal complement of $W_{t,t'}$ in $(R_F^{10})^2$ and vice versa, using the pairing $\langle z, w \rangle = z_1 w_2 - z_2 w_1$ between $(R_F^5)^2$ and $(R_F^{10})^2$, so that $R_F^5 W'_{t,t'} / R_F^5 W_{t,t'}$ and $W'_{t,t'}/W_{t,t'}$ are dual to one another. Equivalently, for generic $\varphi \in R_F^5$, the skew pairing on $W'_{t,t'}/W_{t,t'}$ defined by $\langle z, w \rangle_\varphi = \varphi(z_1 w_2 - z_2 w_1) \in R_F^{15}$ is nondegenerate. This form vanishes when φ belongs to $p_t(W) + p_{t'}(W)$.

We sketch the proof of (d) later. The other properties that we use are the following:

(e) For generic $t \neq 0 \in \mathbb{C}^5$, the projection $p_t \colon W \to R_F^5$ given by $w \mapsto w_t$ is injective. Furthermore, the map $\mathbb{C}^5 \to \operatorname{Hom}(W, R_F^5)$ is injective.

Indeed, W is identified with R_F^4, and t with a generic element A of S^1; the map p_t is simply the multiplication by A from R_F^4 to R_F^5. Thus it is enough to exhibit one example where this multiplication map is injective: take X to be the Fermat hypersurface, and A the form $\sum_i X_i$. The second statement is simply Macaulay's theorem (see [20]).

(f) For any $z \in (R_F^5)^5$, the map $\nu_z \colon W \to R_F^{10}$ given by $w \mapsto \sum_i z_i w_i$ is zero or has rank ≥ 21.

Indeed, let x_i be the basis of S^1 giving the isomorphism $S^1 \cong \mathbb{C}^5$; using the isomorphism $W \cong R_F^4$, it is immediate to see that $\nu_z \colon R_F^4 \to R_F^{10}$ is identified with the multiplication map by $\sum_i x_i z_i \in R_F^6$. So the statement is that for generic X, and any nonzero element $P \in R_F^6$, the multiplication map $\mu_P \colon R_F^4 \to R_F^{10}$ has rank ≥ 21, which is not difficult to prove: in fact this bound, which we need for technical reasons, is not at all sharp. For example, for the Fermat quintic, the only polynomials not satisfying this bound are, up to a permutation of the coordinates, of the form $X_0^3 X_1^2 A$, for A a linear form. Then an infinitesimal computation shows that these polynomials P with rank $\mu_P \leq 20$ will not continue to exist for a generic deformation of the Fermat.

(g) For any four linearly independent elements $t_1, \ldots, t_4 \in \mathbb{C}^5$, we have $\sum_i p_{t_i}(W) = R_F^5$.

Indeed, four independent elements of S^1 generate the ideal of a point in $\mathbb{P}^4$, and since J_F^5 has no base point, they generate S^5 modulo J_F^5, or equivalently, they generate R_F^5.

Our plan in the next sections is to prove the following

Theorem 5 *Suppose that F is generic, and let $W \subset (R_F^5)^5$ be a subspace satisfying (a–g) above. Then there exists an isomorphism $S^1 \cong \mathbb{C}^5$ which identifies W with $R_F^4 \subset \operatorname{Hom}(S^1, R_F^5)$.*

This shows how to reconstruct $\mu_{1,4}$ from μ and the isomorphism $R_F^{10} \cong (R_F^5)^*$: the properties above are stated using only this data: the theorem shows the uniqueness of $W \subset (R_F^5)^5$ satisfying these properties, up to $\mathrm{Aut}\,\mathbb{C}^5$: choose such a W; put $R_F^4 = W$, $S^1 = \mathbb{C}^5$ and define $\mu_{1,4} \colon S^1 \otimes R_F^4 \to R_F^5$ as the natural map $\mathbb{C}^5 \otimes W \to R_F^5$. Then, as in Donagi's arguments mentioned above, the symmetrizer lemma allows us to recover the multiplication maps

$$\mu_{1,k} \colon S^1 \otimes R_F^k \to R_F^{k+1} \quad \text{for } k \leq 4,$$

giving the whole Jacobian ring of F, and therefore allows us to recover F itself. Thus Theorem 5 implies our main result Theorem 1.

Proof of (d) t, t' are identified with generic elements A, B of S^1; for any integer i, define the space $H_{A,B}^i$ as the middle cohomology of the complex

$$R_F^{i-1} \xrightarrow{(A,B)} R_F^i \times R_F^i \xrightarrow{B-A} R_F^{i+1}.$$

In degree $i = 5$, clearly $W_{t,t'}$ is identified with $\mathrm{Im}(A, B)$ and $W'_{t,t'}$ with $\mathrm{Ker}(B - A)$, so that $W'_{t,t'}/W_{t,t'}$ is naturally isomorphic to $H_{A,B}^5$. It is easy to prove that for generic A, B and X, $(A, B) \colon R_F^4 \to R_F^5 \times R_F^5$ is injective and $B - A \colon R_F^5 \times R_F^5 \to R_F^6$ is surjective. (This last fact means that the 5 partial derivatives of a generic quintic F, together with two generic linear forms, generate an ideal containing all sextic polynomials; this can be checked easily, for example using Macaulay.) Thus, since $\dim R_F^4 = 65$, $\dim R_F^5 = 101$, $\dim R_F^6 = 135$, it follows that $\dim W'_{t,t'}/W_{t,t'} = 2$.

Now it is immediate to see using the selfduality $R_F^k \cong (R_F^{15-k})^*$ of the ring R_F that $H_{A,B}^{10}$ is naturally dual to $H_{A,B}^5$, the duality being given by the pairing $\langle\ ,\ \rangle$ between $R_F^5 \times R_F^5$ and $R_F^{10} \times R_F^{10}$. On the other hand, $H_{A,B}^{10} \cong W_{t,t'}^{\perp}/R_F^5 W_{t,t'}$ by definition, and $R_F^5 W'_{t,t'} \subset W_{t,t'}^{\perp}$. For $\varphi \in R_F^5$, the multiplication map by $\varphi \colon H_{A,B}^5 \to H_{A,B}^{10}$ is skewsymmetric, and since $\dim H_{A,B}^5 = 2$, it is either zero or an isomorphism; in this last case, $\langle\ ,\ \rangle_\varphi$ is nondegenerate. It is easy to show that it is generically nonzero, hence we conclude that $R_F^5 W'_{t,t'} = W_{t,t'}^{\perp}$ and that $\langle\ ,\ \rangle_\varphi$ is generically nondegenerate. □

0.4   Overall plan of the proof

Although Theorem 5 is stated for generic F, for which W satisfies the above properties (a–g), we prove it via an appropriate specialization F_0, when (a–g) no longer hold for W_0. We now discuss the logic of this reduction. First, (a–g) are used in Section 1 in connection with the following definition:

Definition 1 Given a subspace $W \subset (R_F^5)^5$, the *conductor* of S^1 to W is the subspace $[W : S^1] \subset (R_F^4)^5$ defined by

$$[W : S^1] := \big\{ z \in (R_F^4)^5 \mid Az \in W \text{ for all } A \in S^1 \big\}.$$

Section 1, Proposition 1 gives the lower bound $\dim[W : S^1] \geq 19$, assuming that W satisfies (a–g). The next definition collects together for use in Section 2 the properties of W_0 that survive the specialization from F to F_0.

Definition 2 Let F_0 be the equation of a nonsingular quintic threefold. We say that a subspace $W_0 \subset (R_{F_0})^5$ has *Property P* if it satisfies (a), (b) of 0.3, and the following supplementary conditions (c'), (e) and (h):

(c') $\dim R_{F_0}^5 W_0 \leq 135$.

(e) The first projection $W_0 \to R_{F_0}^5$ and the map $\mathbb{C}^5 \to \mathrm{Hom}(W_0, R_{F_0}^5)$ are both injective.

(h) The conductor $Z_0 = [W_0 : S^1]$ has dimension ≥ 19 (see Section 1, Proposition 1).

We justify these properties. Let $U \subset \mathbb{P}(H^0(\mathcal{O}_{\mathbb{P}^4}(5)))$ be the open set parametrizing smooth hypersurfaces; over U, we can define the bundles $\mathcal{R}^5$ and $\mathcal{R}^{10}$, with fibres R_F^5 and R_F^{10} over $F \in U$. The multiplication in the ring R_F for any F gives a map

$$\mu \colon S^2 \mathcal{R}^5 \to \mathcal{R}^{10}.$$

In the Grassmannian of 65 dimensional subspaces of $(\mathcal{R}^5)^5$, the set of spaces W satisfying (b) and (c') is Zariski closed. It follows that if $W \subset (R_F^5)^5$ (defined over some finite cover of U) satisfies (a), (b), (c) for generic F, when F specializes to F_0, the corresponding limit W_0 of W satisfies (a), (b) and (c').

Now assume that the generic W satisfies the remaining properties. By specializing along a suitable curve, it can be seen that we can impose the following conditions on W_0. Since we may assume using the Aut $\mathbb{C}^5$ action and (e) that the first projection $\mathrm{pr}_1 \colon W \to R_F^5$ is injective for generic (W, F), we may also assume this is true for (W_0, F_0): for this, we consider the spaces W_t as subspaces of $R_{F_t}^5$ for generic t via the first projection; we then define W_0 as the limiting subspace of $R_{F_0}^5$. Moreover, since the map $\mathbb{C}^5 \to \mathrm{Hom}(W, R_F^5)$ is injective for generic W, we may also assume this to be true for W_0. In fact, for this it suffices to define the subspace $\mathbb{C}^5 \subset \mathrm{Hom}(W_0, R_{F_0}^5)$ as the limit of the subspaces $\mathbb{C}^5 \subset \mathrm{Hom}(W_t, R_{F_t}^5)$

When we specialize W to W_0, the dimension of the conductor $[W : S^1]$ can of course only jump up, so our specialized W_0 will satisfy (h).

An important role in the proof is played by the following definition.

Definition 3 The subspace $W_0 \subset (R_{F_0})^5$ is *degenerate* for X_4 if one component of $W_0 \subset (R_{F_0}^5)^5$ (in the sense explained in 0.3) is contained in the ideal (X_4), or in other words, the composite

$$\mathbb{C}^5 \to \mathrm{Hom}(W_0, R_{F_0}^5) \to \mathrm{Hom}(W_0, R_{F_0}^5/(X_4)) \qquad (0.5)$$

is not injective.

Section 2, Propositions 2–3 describe all the possible degenerate subspaces $W_0 \subset (R_{F_0}^5)^5$ satisfying property P. There turn out to be just two possibilities: *either* W_0 is the image of $R_{F_0}^4$ in $\mathrm{Hom}(S^1, R_{F_0}^5)$ for some isomorphism $S^1 \cong \mathbb{C}^5$, so that W_0 satisfies the conclusion of Theorem 5; *or* W_0 is a certain explicitly described degenerate limit of these under the action of $\mathrm{Aut}\,\mathbb{C}^5$. In Section 3, we use an infinitesimal argument to prove that any deformation of W_0 to a subspace W_t satisfying property P for generic t satisfies the conclusion of Theorem 5, whenever W_0 is degenerate or not.

The specialization F_0 we use is of the form

$$F_0(X_0, \ldots, X_4) = X_4^5 + G(X_0, \ldots, X_3), \qquad (0.6)$$

with G a generic polynomial of degree 5. In this case $R_{F_0}^5/(X_4)$ occuring in (0.5) is simply R_G^5, the degree 5 piece of the Jacobian ring of G. The final Section 4 proves that for F_0 of the form (0.6), any subspace $W_0 \subset (R_{F_0}^5)^5$ satisfying (a–g) above is degenerate, which combined with the results above concludes the proof of Theorem 5.

Remark To keep the paper to a reasonable length, we have chosen to skip the proofs of a certain number of technical lemmas giving lower bounds for the ranks of the multiplication map by a nonzero homogeneous element of the Jacobian ring of a generic hypersurface of degree 5 in $\mathbb{P}^3$.

1   Dimension of the conductor $Z = [W : S^1]$

Proposition 1 *Let F be a generic quintic, and $W \subset (R_F^5)^5$ a subspace satisfying (a–g) of 0.3. Then $\dim[W : S^1] \geq 19$ (see Definition 1).*

The rest of this section is devoted to the proof of Proposition 1. We first show the following result.

Lemma 1.1 *For generic $A, B \in S^1$, we have $[W : S^1] = [W : (A, B)]$.*

Proof If $z \in [W : (A, B)]$ then $z_i w_j - z_j w_i \in R_F^9$ is annihilated by A, B for any $w \in W$; but the surjectivity of $B - A \colon R_F^5 \otimes R_F^5 \to R_F^6$ used above implies by duality the injectivity of the map $(A, B) \colon R_F^9 \to R_F^{10} \times R_F^{10}$. Hence $z_i w_j - z_j w_i = 0 \in R_F^9$, which by (b') implies that $z \in [W : S^1]$. $\square$

We use the notation $H_{A,B}^k$ introduced in 0.3 in the proof of (d). We have shown that $H_{A,B}^5$ is two dimensional and that a generic $\varphi \in R_F^5$ gives a skew linear isomorphism $H_{A,B}^5 \to H_{A,B}^{10}$. Consider the map $B - A \colon W \times W \to S^1 W \subset (R_F^6)^5$. Its kernel $\mathrm{Ker}(B - A)$ admits a natural projection α to $(H_{A,B}^5)^5$. We have the next result:

Lemma 1.2 *The map* $\alpha \colon \mathrm{Ker}(B - A) \to (H_{A,B}^5)^5$ *is the zero map.*

Proof Suppose that $Bw - Aw' = 0$; then we set

$$\alpha((w, w')) = (\alpha_1, \ldots, \alpha_5) := ((w_1, w_1'), \ldots, (w_5, w_5')).$$

Now we claim that $\langle \alpha_i, \alpha_j \rangle_\varphi = 0$ for all i, j, where $\langle \ , \ \rangle_\varphi$ is the (generically nondegenerate) pairing introduced in (d): indeed $\langle \alpha_i, \alpha_j \rangle_\varphi = \varphi(w_i w_j' - w_j w_i') \in R_F^{15}$, which is zero by (b). It follows that the α_i generate a subspace of $H_{A,B}^5$ of dimension at most one; thus after a change of basis of $\mathbb{C}^5$, we may assume $\alpha_i = 0$ for $i \leq 4$. This means that for some $z_1, \ldots, z_4 \in R_F^4$, we have $w = (Az_1, \ldots, Az_4, w_5)$ and $w' = (Bz_1, \ldots, Bz_4, w_5')$. Now this implies that

$$\chi_i \alpha_5 = 0 \quad \text{in } R_F^5 \quad \text{for any } \chi = (\chi_1, \ldots, \chi_5) \in W \text{ and any } i \leq 4.$$

This is because by (b), $\chi_i w_5 = \chi_5 w_i = \chi_5 Az_i$ and $\chi_i w_5' = \chi_5 w_i' = \chi_5 Bz_i$. But by (g), we know that $\sum_{i \leq 4} \mathrm{pr}_i(W) = R_F^5$. So α_5 is annihilated by R_F^5, and hence is zero. $\square$

It follows from Lemmas 1.1 and 1.2 that

$$\dim S^1 W \geq 2 \dim W - \dim[W : S^1]. \tag{1.1}$$

Let $W_{12} = \mathrm{pr}_{12}(W)$ and $W_{12}' = W_{e_1,e_2}'$ for a generic choice of basis of $\mathbb{C}^5$. We note first that $\mathrm{pr}_{12} \colon S^1 W \to S^1 W_{12}$ is injective; indeed by (b), $\dim R_F^5 W = 135$, while $\dim R_F^5 W_{12} = \dim W_{12}^{\perp} - 2 = 135$ by (d). Thus $\mathrm{pr}_{12} \colon R_F^5 W \to R_F^5 W_{12}$ must be injective, and so is $\mathrm{pr}_{12} \colon S^1 W \to S^1 W_{12}$.

Now we prove the following fact.

Lemma 1.3 $\dim S^1 W_{12}' / S^1 W_{12} \geq 6$.

Proof It is enough to show that for x, y independent in W'_{12}/W_{12}, we have $\dim S^1 x/S^1 y \geq 3$, where $S^1 x/S^1 y$ is the image of $S^1 x$ in $(S^1 W'_{12}/S^1 W_{12})/S^1 y$. Let K be the kernel of the natural map $S^1 \to S^1 x/S^1 y$; for $\alpha \in K$ and $P \in R_F^4$, we have $\langle P\alpha x, y \rangle = 0$, so that $P\alpha x = 0$ in $R_F^5 W'_{12}/R_F^5 W_{12}$, by (d), since obviously also $\langle P\alpha x, x \rangle = 0$ in $R_F^5 W'_{12}/R_F^5 W_{12}$. This implies that $P\alpha$ annihilates W'_{12}/W_{12}, since multiplication by $P\alpha$ is a skew morphism. (We use the notation and results of (d).) Now assume $\dim K \geq 3$. We know that $\text{pr}_1(W) + \text{pr}_2(W) + R_F^4 K$ annihilates W'_{12}/W_{12}, so it must be a proper subspace of R_F^5. In particular $R_F^5 K$ must be a proper subspace of R_F^5.

Now we use the following fact which holds for generic X, as follows from an easy dimension count:

Fact 1.4 *The set of lines $\Delta \subset \mathbb{P}^4$ such that I_Δ does not generate R_F^5 is one dimensional, and for such a line, I_Δ generates a hyperplane of R_F^5.*

So under the assumption $\dim K \geq 3$, we find that there is a line Δ as above, for which W_1 is contained in $I_\Delta(5)$. Since the set of such lines is one dimensional, and the basis of $\mathbb{C}^5$ is generic, for a fixed such Δ, there is a hypersurface in $\mathbb{P}(\mathbb{C}^5)$ consisting of t such that $W_t \subset I_\Delta(5) \mod J_F^5$. This contradicts (g). $\square$

Now let $A, B \in S^1$ and $q_1, q_2 \in S^2$ be generic; we prove the following:

Lemma 1.5   (i) $[S^1 W_{12} : (A, B)] = W_{12}$;

   (ii) $[S^1 W'_{12} : (A, B)] = W'_{12}$;

   (iii) $[S^1 W_{12} : (q_1, q_2)] = [W_{12} : S^1]$;

   (iv) $[S^1 W'_{12} : (q_1, q_2)] = [W'_{12} : S^1]$;

   (v) $[W'_{12} : S^1] = [W_{12} : S^1] = \text{pr}_{12}([W : S^1])$.

Proof (i) and (ii): let $(z_1, z_2) \in [S^1 W'_{12} : (A, B)]$; then for $w \in W$ we have that $z_1 w_2 - z_2 w_1 \in R_F^{10}$ is annihilated by A and B. Since the map $(A, B)\colon R_F^{10} \to R_F^{11} \times R_F^{11}$ has kernel of dimension 6, it follows that the map $W \to R_F^{10}$ given by $w \mapsto z_1 w_2 - z_2 w_1$ has rank ≤ 6. By (f) it vanishes identically, hence $(z_1, z_2) \in [S^1 W'_{12} : S^1]$, which proves (ii). Now if $(z_1, z_2) \in [S^1 W_{12} : (A, B)]$ we know that $(z_1, z_2) \in W'_{12}$ and its image in W'_{12}/W_{12} is annihilated by A, B. Using (d) and the fact that A, B are generic, we conclude that $(z_1, z_2) \in W_{12}$. Indeed, if A annihilates a nonzero element of W'_{12}/W_{12}, $R_F^4 A$ annihilates W'_{12}/W_{12}, and clearly this is not true for generic A.

The proof of (iii) and (iv) are completely similar, so we only prove (v). If $z \in [W'_{12} : S^1]$, consider the map $S^1 \to W'_{12}/W_{12}$ given by $A \mapsto Az$: its kernel has dimension ≥ 3. But then for any $C \in S^1$, the annihilator of $Cz \in W'_{12}/W_{12}$ has dimension ≥ 3, which implies that $Cz = 0$, by the proof of Lemma 1.3. Hence $z \in [W_{12} : S^1]$. Finally let $z \in [W_{12} : S^1]$; for generic $A, B \in S^1$ we have $w = (Az_1, Az_2, w_3, w_4, w_5) \in [W_{12} : S^1]$ and $w'(Bz_1, Bz_2, w'_3, w'_4, w'_5) \in [W_{12} : S^1]$. So the first two components of $Bw - Aw' \in S^1W$ vanish. But we have already used the fact that $\mathrm{pr}_{12} \colon S^1W \to S^1W_{12}$ is injective; it follows that $Bw - Aw' = 0$ hence $w_i = Az_i$ and $w'_i = Bz_i$ for $i = 3, 4, 5$ for some $z_i \in R_F^4$, using Lemma 1.2. Then $(z_1, \ldots, z_5) \in [W : (A, B)]$, hence $(z_1, \ldots, z_5) \in [W : S^1]$ using (i) above. Therefore $z \in \mathrm{pr}_{12}([W : S^1])$. $\square$

Define $H^i_{q_1, q_2}$ as the middle cohomology group of the sequence

$$R_F^{i-2} \overset{(q_1, q_2)}{\longrightarrow} R_F^i \times R_F^i \overset{q_2 - q_1}{\longrightarrow} R_F^{i+2}.$$

There are natural maps

$$
\begin{aligned}
\alpha &\colon \mathrm{Ker}[B - A \colon S^1W_{12} \times S^1W_{12} \to S^2W_{12}] \to (H^6_{A,B})^2, \\
\alpha' &\colon \mathrm{Ker}[B - A \colon S^1W'_{12} \times S^1W'_{12} \to S^2W'_{12}] \to (H^6_{A,B})^2, \\
\beta &\colon \mathrm{Ker}[q_2 - q_1 \colon S^1W_{12} \times S^1W_{12} \to S^3W_{12}] \to (H^6_{q_1,q_2})^2, \\
\beta' &\colon \mathrm{Ker}[q_2 - q_1 \colon S^1W'_{12} \times S^1W'_{12} \to S^3W'_{12}] \to (H^6_{q_1,q_2})^2.
\end{aligned}
$$

From Lemma 1.5 we deduce the following inequalities:

Corollary 1.6

$$
\begin{aligned}
\dim S^2W_{12} &\geq 2\dim S^1W_{12} - \dim[S^1W_{12} : S^1] - \mathrm{rank}\,\alpha \\
&= 2\dim S^1W_{12} - 65 - \mathrm{rank}\,\alpha, \\
\dim S^2W'_{12} &\geq 2\dim S^1W'_{12} - \dim[S^1W'_{12} : S^1] - \mathrm{rank}\,\alpha' \\
&= 2\dim S^1W'_{12} - 67 - \mathrm{rank}\,\alpha',
\end{aligned}
$$

$$
\begin{aligned}
\dim S^3W_{12} &\geq 2\dim S^1W_{12} - \dim[S^1W_{12} : (q_1, q_2)] - \mathrm{rank}\,\beta \\
&= 2\dim S^1W_{12} - \dim[W_{12} : S^1] - \mathrm{rank}\,\beta, \\
\dim S^3W'_{12} &\geq 2\dim S^1W'_{12} - \dim[S^1W'_{12} : (q_1, q_2)] - \mathrm{rank}\,\beta' \\
&= 2\dim S^1W'_{12} - \dim[W_{12} : S^1] - \mathrm{rank}\,\beta'.
\end{aligned}
$$

Using Lemma 1.3 and inequality (1.1), we get then the following inequalities

Corollary 1.7

$$\dim S^2 W_{12} \geq 4 \dim W - 65 - 2 \dim[W : S^1] - \operatorname{rank}\alpha$$
$$= 195 - 2 \dim[W : S^1] - \operatorname{rank}\alpha,$$
$$\dim S^2 W'_{12} \geq 4 \dim W - 67 + 12 - 2 \dim[W : S^1] - \operatorname{rank}\alpha'$$
$$= 205 - 2 \dim[W : S^1] - \operatorname{rank}\alpha',$$
$$\dim S^3 W_{12} \geq 4 \dim W - 3 \dim[W : S^1] - \operatorname{rank}\beta$$
$$= 260 - 3 \dim[W : S^1],$$
$$\dim S^3 W'_{12} \geq 272 - 3 \dim[W : S^1] - \operatorname{rank}\beta'.$$

Note that on $H^6_{A,B}$, there is a natural skew pairing with values in R_F^{12}, defined by $\langle (z_1, z_2), (z'_1, z'_2) \rangle = z_1 z'_2 - z_2 z'_1$. One can show that for generic $\varphi \in R_F^3$ (and generic A, B, X) the pairing $\langle\ ,\ \rangle_\varphi = \varphi \langle\ ,\ \rangle$ with value in $R_F^{15} = \mathbb{C}$ is nondegenerate. Similarly we can define a nondegenerate skew pairing $\langle\ ,\ \rangle_\varphi$ on $H^6_{q_1,q_2}$. Hence there is an induced symmetric pairing on $H^6_{A,B} \times H^6_{A,B}$, defined by

$$\Big((\alpha_1, \alpha_2), (\alpha'_1, \alpha'_2)\Big)_\varphi = \langle \alpha_1, \alpha'_2 \rangle_\varphi + \langle \alpha'_1, \alpha_2 \rangle_\varphi,$$

and similarly for $H^6_{q_1,q_2}$. We have the following result.

Lemma 1.8 *Im α and Im α' are orthogonal under $\langle\ ,\ \rangle_\varphi$, as are Im β and Im β'.*

This is immediate using (b) and the definition of the pairing. □

Since we have rank $H^6_{A,B} = 14$, and rank $H^6_{q_1,q_2} = 50$, we deduce from Lemma 1.8 that

$$\operatorname{rank}\alpha + \operatorname{rank}\alpha' + \operatorname{rank}\beta + \operatorname{rank}\beta' \leq 128,$$

which together with Corollary 1.7 gives

Corollary 1.9

$$\dim S^2 W_{12} + \dim S^2 W'_{12} + \dim S^3 W_{12} + \dim S^3 W'_{12} \geq 804 - 10 \dim[W : S^1].$$

But note that by definition $S^2 W_{12} \subset R_F^7 \times R_F^7$ and $S^3 W_{12} \subset R_F^8 \times R_F^8$ are orthogonal with respect to the perfect pairing $\langle (\alpha_1, \alpha_2), (\beta_1, \beta_2) \rangle = \alpha_1 \beta_2 - \alpha_2 \beta_1 \in R_F^{15} \cong \mathbb{C}$. Similarly, $S^2 W'_{12} \subset R_F^7 \times R_F^7$ and $S^3 W'_{12} \subset R_F^8 \times R_F^8$ are orthogonal with respect to the same pairing. Since $\dim R_F^7 = \dim R_F^8 = 155$, we conclude that

$$\dim S^2 W_{12} + \dim S^2 W'_{12} + \dim S^3 W_{12} + \dim S^3 W'_{12} \leq 4 \dim R_F^7 = 620,$$

Using Corollary 1.9, we conclude that

$$620 \geq 804 - 10 \dim[W : S^1],$$

hence $\dim[W : S^1] \geq 19$ and Proposition 1 is proved. □

2   The degenerate case

In this section we assume that W_0 is degenerate in the sense of Definition 3 of 0.4. We first prove the following result.

Proposition 2 *If W_0 is degenerate and satisfies Property P of Definition 2, then one of the following cases holds:*

(i) *W_0 is the image of $R_{F_0}^4$ in $(R_{F_0}^5)^5$ for some isomorphism $S^1 \cong \mathbb{C}^5$.*

(ii) *Up to a change of basis of $\mathbb{C}^5$, W_0 contains the 5-tuples $(X_4 z, 0, \ldots, 0)$ for any $z \in R_{F_0}^4$ and for some $T \in R_{F_0}^4$ (well defined modulo X_4), and for some A_i for $i = 1, \ldots, 4$ independent modulo X_4 in S^1, W_0 contains the 5-tuples $(AT, X_4^3 A_i A)$ for any $A \in S^1$.*

Remark It is immediate to see that such a W_0 satisfies Property P. (b) holds because $X_4^4 = 0$ in R_{F_0}. In fact one can check that any W_0 as in the proposition can be realized as the limit of a family $W_t \subset (R_{F_t}^5)^5$ in Case (i).

By assumption, the first projection $\mathrm{pr}_1 \colon W_0 \to R_{F_0}^5$ is injective, thus of rank 65, and since rank $X_4 R_{F_0}^4 = 61$, the first component $\mathrm{pr}_1(W_0)$ of W_0 (see 0.3) is not contained in (X_4). We assume first that

there exists a component of W_0 contained in (X_4) but not in (X_4^2);

we may of course assume this is the second component $\mathrm{pr}_2(W_0)$ of W_0. We prove first the following proposition.

Proposition 3 *Under this supplementary assumption, W_0 is in Case (i) of Proposition 2.*

The proof of this proposition involves various technical steps, but the idea is the following: an element of W_0 is of the form $w = (w_1, X_4 z, w_3, w_4, w_5)$, with $z \in R_{F_0}^4$. Now we have the relations given by property (b)

$$X_4 (w_i z' - w_i' z) = 0 \quad \text{for } w, w' \in W_0, \tag{2.1}$$

which are part of the assumptions of the symmetrizer lemma (Theorem 4) for the correspondence $z \mapsto w_i$, and we would like to show that there are enough such relations to conclude that this correspondence is of the form $w_i = A_i z$ for some $A_i \in S^1$.

In doing this, we meet two problems: first, we need to show that enough elements z' are involved in the relations (2.1). Secondly, the factor X_4 in front of these relations makes them weakened symmetrizer relations; however, we show that they induce enough true symmetrizer relations in R_G.

This leads us to split the proof into two main steps. The first is Lemma 2.1, where we show that if w_i is divisible by X_4^k then z is also divisible by X_4^k. This shows that the graded pieces of W_0 for the filtration given by the order of divisibility by X_4 are also generated by elements of the form $X_4^k(w_1, X_4 z, w_3, w_4, w_5)$, with $z \in R_{F_0}^{4-k}$ and $w_i \in R_{F_0}^{5-k}$. This step allows us to translate relations (2.1) into symmetrizer relations in R_G.

The second step is Lemma 2.7, where we show that there exists $A_i \in S^1$ such that $w_i = A_i z$ mod X_4. This is done by showing that (2.1) gives enough symmetrized relations in R_G to apply the symmetrizer lemma. The conclusion of the proof of the Proposition is then comparatively easy.

We write W' for the 2-term projection $\mathrm{pr}_{12}(W_0) \subset (R_{F_0}^5)^2$. An element of W' is of the form $(w_1, x_4 z_2)$. Using (h), we prove the following:

Lemma 2.1 *If $w_1 \in X_4^i R_{F_0}^{5-i}$ then $z_2 \in X_4^i R_{F_0}^{4-i}$.*

The proof proceeds through Lemmas 2.2–2.6.

Proof of Lemma 2.1 Set $Z_0 = [W_0 : S^1]$. We know that $\dim Z_0 \geq 19$, and of course the first projection is injective on Z_0, since it is on W_0. Thus $\dim \mathrm{pr}_1(Z_0) = \dim Z' \geq 19$, where $Z' = \mathrm{pr}_{12}(Z_0)$. We define a filtration on W' and Z' using the order w.r.t. X_4 of the first component; the successive ranks of the associated graded objects will be denoted by $\alpha, \beta, \gamma, \delta$ for W', and by $\alpha', \beta', \gamma', \delta'$ for Z'. So a basis of W' can be listed as

$$\left. \begin{array}{ll} \alpha \text{ elements} & (w_1, X_4 z_2) \\ \beta \text{ elements} & (X_4 z_1, X_4 z_2') \\ \gamma \text{ elements} & (X_4^2 t_1, X_4 z_2'') \\ \delta \text{ elements} & (X_4^3 q_1, X_4 z_2''') \end{array} \right\}, \quad \text{where } w_1, z_1, t_1, q_1 \not\equiv 0 \text{ mod } X_4.$$

Similarly, a suitable basis of Z' can be listed as

$$\left. \begin{array}{ll} \alpha' \text{ elements} & (z_1', X_4 t_2) \\ \beta' \text{ elements} & (X_4 t_1', X_4 t_2') \\ \gamma' \text{ elements} & (X_4^2 q_1', X_4 t_2'') \\ \delta' \text{ elements} & (X_4^3 A_1, X_4 t_2''') \end{array} \right\}, \quad \text{where } z_1', t_1', q_1', A_1 \not\equiv 0 \text{ mod } X_4.$$

Of course, $\alpha + \beta + \gamma + \delta = 65$ and $\alpha' + \beta' + \gamma' + \delta' \geq 19$. We first assume $\alpha \geq 12$: (b) gives $w_1 t_2' = 0$ in R_G^8 for any $(w_1, X_4 z_2) \in W'$ and $(X_4 t_1', X_4 t_2') \in Z'$, since $X_4 w_1 t_2' - X_4^2 t_1' z_2 = 0$ in $R_{F_0}^9$. Similarly $w_1 t_2'' = w_1 t_2''' = 0$ in R_G^8 for $(X_4^2 q_1', X_4 t_2'')$, $(X_4^3 A_1, X_4 t_2''')$ elements of Z'. Next, we use the following result.

Lemma 2.2 *If G is generic, for any nonzero $t \in R_G^3$, the multiplication map $\mu_t : R_G^5 \to R_G^8$ by t has rank ≥ 29, that is, μ_t has kernel of dimension ≤ 11*

(since $\dim R_G^5 = 40$). Equivalently, by duality, the map $\mu_t \colon R_G^4 \to R_G^7$ has kernel of dimension ≤ 2. $\square$

It follows that if $\alpha \geq 12$, we have $t_2' = t_2'' = t_2''' = 0$ in R_G^3. We now write $t_2' = X_4 q_2', t_2'' = X_4 q_2'', t_2''' = X_4 q_2'''$ and show exactly as above that $q_2'' = q_2''' = 0$ in R_G^2, since $(X_4^2 q_1', X_4^2 q_2'')$ and $(X_4^3 A_1, X_4^2 q_2''') \in Z'$. Similarly we write $q_2''' = X_4 A_2$ and find that $A_2 \equiv 0 \bmod X_4$ since $(X_4^3 A_1, X_4^3 A_2) \in Z'$. In other words the property stated in the lemma is true for Z', and we can list a suitable basis of Z' as

$$
\left.
\begin{array}{ll}
\alpha' \text{ elements} & (z_1', X_4 t_2) \\
\beta' \text{ elements} & (X_4 t_1', X_4^2 q_2) \\
\gamma' \text{ elements} & (X_4^2 q_1', X_4^3 A_2) \\
\delta' \text{ elements} & (X_4^3 A_1, 0)
\end{array}
\right\}, \quad \text{where } z_1', t_1', q_1', A_1 \not\equiv 0 \bmod X_4.
$$

We now note that, again by (b), $z_2' t_1' = 0$ in R_G^7 for $(X_4 z_1, X_4 z_2') \in W'$ and $(X_4 t_1', X_4^2 q_2) \in Z'$, since $X_4^2 z_2' t_1' - X_4^3 q_2 z_1 = 0$ in $R_{F_0}^9$. Now we use the following result.

Lemma 2.3 *If G is generic, for any nonzero $z \in R_G^4$ the multiplication map $\mu_z \colon R_G^3 \to R_G^7$ by z has kernel of dimension at most one. It follows by duality that $\mu_z \colon R_G^5 \to R_G^9$ has kernel of dimension ≤ 21.* $\square$

So if $\beta' \geq 2$, we have $z_2' = 0$ in R_G^4, and similarly $z_2'' = z_2''' = 0$ in R_G^4. We write then $z_2'' = X_4 t_2''$, $z_2''' = X_4 t_2'''$ and using $\alpha \geq 12$ and Lemma 2.2, show that $t_2'' = t_2''' = 0$ in R_G^3 since $(X_4^2 t_1, X_4^2 t_2''), (X_4^3 q_1, X_4^2 t_2''') \in W'$. Then writing $t_2''' = X_4 q_2'''$, we find again that $q_2''' = 0$ in R_G^2, since $(X_4^3 q_1, X_4^3 q_2''') \in W'$. So if $\alpha \geq 12$ and $\beta' \geq 2$, Lemma 2.1 is true. It also holds if $\alpha \geq 12$ and $\gamma' \geq 1$, using the following lemma.

Lemma 2.4 *If G is generic, for any nonzero $z \in R_G^4$, the multiplication map $\mu_z \colon R_G^2 \to R_G^6$ by z is injective.* $\square$

So assume $\alpha \geq 12$ but $\beta' \leq 1$, $\gamma' = 0$. In this case we have $\alpha' \geq 14$, since $\delta' \leq 4$. On the other hand, $z_2' z_1' = 0$ in R_G^8 for $(X_4 z_1, X_4 z_2') \in W'$ and $(z_1', X_4 t_2) \in Z'$, by (b), since $X_4^2 z_1 t_2 - X_4 z_1' z_2' = 0$ in $R_{F_0}^9$. Now we use the following result.

Lemma 2.5 *If G is generic, for any nonzero $z \in R_G^4$ the multiplication map $\mu_z \colon R_G^4 \to R_G^8$ by z has kernel of dimension ≤ 11.* $\square$

It follows that $z_2' = 0$ in R_G^4. Similarly we find that $z_2'' = z_2''' = 0$ in R_G^4. Exactly as above we conclude that Lemma 2.1 is true in this case.

It remains to consider the case where $\alpha \leq 11$. Note that $\gamma \leq 20 = \dim R_G^3$ and $\delta \leq 10 = \dim R_G^2$, so $\alpha \leq 11$ implies $\beta \geq 24$. Now for $(X_4 z_1, X_4 z_2') \in W'$

and $(w_1, X_4 z_2) \in W'$, we have $w_1 z_2' = 0$ in R_G^9, by (b), since $X_4 w_1 z_2' - X_4^2 z_1 z_2 = 0$ in $R_{F_0}^{10}$, hence the space spanned by the z_2' in R_G^4 has dimension ≤ 19 by the following lemma.

Lemma 2.6 *If G is generic, for any nonzero $w \in R_G^5$ the multiplication map $\mu_w \colon R_G^4 \to R_G^9$ by w has kernel of dimension ≤ 19. Equivalently, by duality, the map $\mu_w \colon R_G^3 \to R_G^8$ has kernel of dimension ≤ 8.* $\square$

End of proof of Lemma 2.1   It follows that there exists $(X_4 z_1', X_4 z_2'') \in W'$ with $z_2'' \equiv 0 \bmod X_4$ and $z_1' \not\equiv 0 \bmod X_4$. Then for any $(X_4 z_1, X_4 z_2') \in W'$, we have $z_1' z_2' = 0$ in R_G^8, by (b), hence the space generated by the z_2' in R_G^4 has dimension ≤ 11 by Lemma 2.5. But then the space generated by the z_1' in R_G^4 for $(X_4 z_1', X_4 z_2'') \in W'$, with $z_2' \equiv 0 \bmod X_4$ has dimension ≥ 13. Still using Lemma 2.5 and (b), we conclude that $z_2' = 0$ in R_G^4 for any $(X_4 z_1, X_4 z_2') \in W'$. It is then easy to conclude that W' satisfies the conclusion of Lemma 2.1. $\square$

By Lemma 2.1, a suitable basis of W' can be listed as

$$
\left.
\begin{array}{ll}
\alpha \text{ elements} & (w_1, X_4 z_2) \\
\beta \text{ elements} & (X_4 z_1, X_4^2 t_2) \\
\gamma \text{ elements} & (X_4^2 t_1, X_4^3 q_2) \\
\delta \text{ elements} & (X_4^3 q_1, 0)
\end{array}
\right\}, \quad \text{where } w_1, z_1, t_1, q_1 \not\equiv 0 \bmod X_4.
$$

We want to show the following result.

Lemma 2.7 *There exists $A \in S^1$, with $A \not\equiv 0 \bmod X_4$ such that*

$$
w_1 \equiv A z_2, \quad z_1 \equiv A t_2 \quad \text{and} \quad t_1 \equiv A q_2 \quad \bmod X_4.
$$

The proof proceeds through Lemmas 2.8–2.11. We begin with the following statement.

Lemma 2.8   (i) $z_2 \equiv 0 \ \bmod X_4 \implies w_1 \equiv 0 \ \bmod X_4.$

(ii) $t_2 \equiv 0 \ \bmod X_4 \implies z_1 \equiv 0 \ \bmod X_4.$

(iii) $q_2 \equiv 0 \ \bmod X_4 \implies t_1 \equiv 0 \ \bmod X_4.$

Proof   We only give the proof of (i), the others being similar, or in fact even easier: so assume $z_2 \equiv 0 \bmod X_4$ but $w_1 \not\equiv 0 \bmod X_4$; then for any $(X_4 z_1, X_4^2 t_2) \in W'$, we have $w_1 t_2 = 0$ in R_G^8, using (b), since $X_4^2 z_1 z_2 - X_4^2 w_1 t_2 = 0$ in $R_{F_0}^{10}$, and $z_2 \equiv 0 \bmod X_4$. Similarly, for any $(X_4^2 t_1, X_4^3 q_2) \in W'$, we have $w_1 q_2 = 0$ in R_G^7. Now we use Lemma 2.6 and the following result.

Lemma 2.9 *If G is generic, for any nonzero $w \in R_G^5$ the multiplication map $\mu_w: R_G^2 \to R_G^7$ by w has kernel of dimension at most one.* $\square$

This implies that the space generated by the t_2 in R_G^3 (respectively, by the q_2 in R_G^2) has dimension ≤ 8 (respectively 1).

Suppose $\beta \geq 12$: then the second projection $\text{pr}_2^1: W_1'/W_2' \to R_G^3$ (which is well defined as a consequence of Lemma 2.1) has kernel of dimension ≥ 4, where W_i' is the filtration defined at the beginning of the proof of Lemma 2.1. Now let $(X_4 z_1, X_4^2 t_2)$ with $t_2 \equiv 0 \bmod X_4$, be in this kernel; then using (b), we see that z_1 annihilates the image of pr_2^1: indeed, for $(X_4 z_1', X_4^2 t_2') \in W_1'$, we have $X_4^3 z_1 t_2' - X_4^3 z_1' t_2 = 0$ in $R_{F_0}^{10}$, and $t_2 \equiv 0 \bmod X_4$. By Lemma 2.2, we conclude that $\text{pr}_2^1 = 0$. But then for any element $(X_4 z_1, X_4^2 t_2) \in W_1'$, and any $(w_1, X_4 z_2) \in W'$, we have $z_1 z_2 = 0$ in R_G^8 (by (b) and by $t_2 \equiv 0 \bmod X_4$). By Lemma 2.5 and $\beta \geq 12$ we conclude that $z_2 \equiv 0 \bmod X_4$, which contradicts our assumption that the second component of W' is not contained in X_4^2. Using the fact that the second projection $\text{pr}_2^2: W_2'/W_3' \to R_G^2$ has rank at most one, we conclude similarly that $\gamma \geq 3$ is absurd (we use Lemma 2.3). So we must have $\beta \leq 11$ and $\gamma \leq 2$; but $\delta \leq 10 = \dim R_G^2$, hence $\alpha \geq 42 > \text{rank}\, R_G^5$, which is absurd. $\square$

If we use the notation $\text{pr}_2^0: W'/W_1' \to R_G^4$, $\text{pr}_2^1: W_1'/W_2' \to R_G^3$ and $\text{pr}_2^2: W_2'/W_3' \to R_G^2$ for the second projections (modulo X_4), Lemma 2.8 gives the following

Corollary 2.10 $\text{rank}\, \text{pr}_2^0 + \text{rank}\, \text{pr}_2^1 + \text{rank}\, \text{pr}_2^2 \geq 55$.

This follows from $\alpha + \beta + \gamma + \delta = 65$ and $\delta \leq 10$. $\square$

Since $\text{rank}\, \text{pr}_2^1 + \text{rank}\, \text{pr}_2^2 \leq 30 = \text{rank}\, R_G^3 + \text{rank}\, R_G^2$, we conclude that $\text{rank}\, \text{pr}_2^0 \geq 25$. This allows us to show the following:

Lemma 2.11 *For $(X_4^2 t_1, X_4^3 q_2), (X_4^2 t_1', X_4^3 q_2') \in W_2'$, we have*

$$t_1 q_2' - t_1' q_2 = 0 \quad \text{in } R_G^5.$$

Proof Since by (b), $w_1 q_2 - t_1 z_2 = 0$ in R_G^7 and $w_1 q_2' - t_1' z_2 = 0$ in R_G^7 for any $(w_1, X_4 z_2) \in W'$, we have $z_2(t_1 q_2' - t_1' q_2) = 0$ in R_G^9 for any $z_2 \in \text{Im}\, \text{pr}_2^0$. From $\text{rank}\, \text{pr}_2^0 \geq 25$, one concludes that $t_1 q_2' - t_1' q_2 = 0$ in R_G^5 using Lemma 2.6. $\square$

Proof of Lemma 2.7 Now assume that $\text{rank}\, \text{pr}_2^1 \geq 18$ or $\text{rank}\, \text{pr}_2^2 \geq 8$. Then $\text{Im}\, \text{pr}_2^1$ generates R_G^5 (respectively, $\text{Im}\, \text{pr}_2^2$ generates R_G^5). Applying the symmetrizer lemma (Theorem 4) to G and to $R_G^2(W_1'/W_2') \subset R_G^6 \times R_G^5$ (respectively, $R_G^3(W_2'/W_3') \subset R_G^6 \times R_G^5$), which surject on the second factor and by Lemma 2.11 satisfy (b), that is, the assumptions of the symmetrizer lemma, we find that there exists $A \in S^1/(X_4)$ such that $z_1 \equiv A t_2 \bmod X_4$ for all $(X_4z_1, X_4^2t_2) \in W_1'/W_2'$ (respectively, there exists $A \in S^1/(X_4)$ such that $t_1 \equiv Aq_2$ mod X_4 for all $(X_4^2t_1, X_4^3q_2) \in W_2'/W_3'$). In either case, we conclude that $w_1 \equiv Az_2$ mod X_4 for $(w_1, X_4z_2) \in W'$, since by (b) $w_1 - Az_2 \in R_G^5$ is annihilated by a 18 dimensional subspace of R_G^5 (respectively, by an 8 dimensional subspace of R_G^2). Since z_2 varies in a 25 dimensional subspace of R_G^4, one concludes easily that $z_1 \equiv At_2$ and $t_1 \equiv Aq_2$ mod X_4.

Finally, if rank $\mathrm{pr}_2^2 \leq 7$ and rank $\mathrm{pr}_2^1 \leq 17$, we have rank $\mathrm{pr}_2^0 \geq 31$ by Corollary 2.10, that is, pr_2^0 is surjective. But then we can apply the symmetrizer lemma to $W'/W_1' \subset R_G^5 \times R_G^4$ and we can conclude the proof as above. $\square$

Proof of Proposition 3 It is easy to conclude from Lemma 2.7 that Proposition 3 holds if the composite map $\mathbb{C}^5 \to \mathrm{Hom}(W_0, R_G^5)$ has rank 4; indeed, applying Lemma 2.7 to a generic component of W' in place of the first, we conclude that after a change of basis of $\mathbb{C}^5$, there exists elements $A_1, \ldots, A_4$ of S^1 such that elements of W_0 have the form $w = (w_1, \ldots, w_4, X_4z_5)$, with $w_i \equiv X_4^l A_i z_5'$ mod X_4^{l+1} when $z_5 = X_4^l z_5'$, except maybe if $l \geq 3$, in which case $w_i \in (X_4^3)$. In fact, this last restriction can easily be removed, because if $X_4^3(w_1', \ldots, w_4', 0) \in W_0$, by (b), $z_5(A_i w_j' - A_j w_i') \equiv 0$ mod X_4 for at least one nonzero $z_5 \in R_G^4$. By Lemma 2.3, this implies that the space generated by the $A_i w_j' - A_j w_i'$ in R_G^3 has dimension at most one. But clearly the A_i have to be independent modulo X_4, otherwise $\mathbb{C}^5 \to \mathrm{Hom}(W_0, R_G^5)$ will not have rank 4; it is then immediate to conclude that in fact $A_i w_j' - A_j w_i'$ all vanish. Hence by the symmetrizer lemma, we find that $w_i' \equiv A A_i$ mod X_4 for some $A \in S^1$.

Now it follows from this that W_0 can be deformed to $W_0' = R_{F_0}^4 \subset \mathrm{Hom}(S^1, R_{F_0}^5)$, by a deformation preserving (b): for this we use the one parameter group of automorphisms of the ring R_F given by multiplication of X_4 by $\lambda \in \mathbb{C}^*$ and note that the limit of $\lambda(W_0)$ as λ tends to 0 is equal to the W_0' corresponding to the basis of S^1 given by $A_1', \ldots, A_i', X_4$, where A_i' is the unique element of S^1 such that $A_i' \equiv A_i$ mod X_4, and not involving the coordinate X_4. We conclude then that $W_0 = W_0'$ for some choice of basis of S^1 using the following easy rigidity lemma:

Lemma 2.12 *Choose a basis of S^1; the only (small) deformations of $W_0' = R_{F_0}^4 \subset \mathrm{Hom}(S^1, R_{F_0}^5) = (R_{F_0}^5)^5$ preserving (b) come from a change of basis of S^1.* $\square$

So to conclude the proof of Proposition 3, we have to exclude the existence of another component of W_0 contained in (X_4). We may assume such a component is the third one; now let W' be the projection $\mathrm{pr}_{123}(W_0) \subset (R_{F_0}^5)^3$. As above, we put a filtration W_i' on W', where W_i' is the set of elements whose first component lies in (X_4^i), and denote by $\alpha, \beta, \gamma, \delta$ the ranks of the successive quotients; using Lemmas 2.1, 2.7, we find that a suitable basis of W' can be listed as follows:

$$
\left.
\begin{array}{ll}
\alpha \text{ elements} & (w_1, X_4 z_2, X_4 z_3) \\
\beta \text{ elements} & (X_4 z_1, X_4^2 t_2, X_4^2 t_3) \\
\gamma \text{ elements} & (X_4^2 t_1, X_4^3 q_2, X_4^3 q_3) \\
\delta \text{ elements} & (X_4^3 q_1, 0, 0)
\end{array}
\right\} ,
\tag{2.2}
$$

and there exists $A_2, A_3 \in S^1$ such that $w_1 \equiv A_2 z_2 \equiv A_3 z_3$, $z_1 \equiv A_2 t_2 \equiv A_3 t_3$ and $t_1 \equiv A_2 q_2 \equiv A_3 q_3$ mod X_4. It is easy to show that this implies that $A_2 \equiv A_3$ mod X_4, using the fact that $\alpha + \beta + \gamma \geq 55$. It follows that some combination of the second and the third component lies in (X_4^2). Of course we can assume the third component lies in (X_4^2). We use now Lemma 2.14 below to conclude that the third component is in fact contained in (X_4^3). We then list a suitable basis of W' as

$$
\left.
\begin{array}{ll}
\alpha \text{ elements} & (w_1, X_4 z_2, X_4^3 q_3) \\
\beta \text{ elements} & (X_4 z_1, X_4^2 t_2, X_4^3 q_3') \\
\gamma \text{ elements} & (X_4^2 t_1, X_4^3 q_2, X_4^3 q_3'') \\
\delta \text{ elements} & (X_4^3 q_1, 0, X_4^3 q_3''')
\end{array}
\right\} ,
\tag{2.3}
$$

with $w_1 \equiv A z_2$ mod X_4 etc. In particular, we find that $\alpha \geq 25$, since $\beta \leq 20$, $\gamma \leq 10$ and $\delta \leq 10$. Now clearly by (b), $w_1 q_3' = w_1 q_3'' = w_1 q_3''' = 0$ in R_G^7, in the notation of (2.3). We now use the following lemma:

Lemma 2.13 *If G is generic, for any nonzero $q \in R_G^2$ the multiplication map $\mu_q \colon R_G^5 \to R_G^7$ by q has kernel of dimension ≤ 9.* $\square$

It follows that $q_3' = q_3'' = q_3''' = 0$ in R_G^2. Now it follows that the third projection $W'/W_1' \to R_G^2$ is well defined and has kernel of dimension ≥ 15. If $(w_1', X_4 z_2', 0)$ belongs to this kernel, by (b), $w_1' q_3 = 0$ in R_G^7 for any q_3, and using Lemma 2.13 again, this implies that $q_3 = 0$ in R_G^2. But then the third component of W' vanishes identically, which contradicts the fact that the map $\mathbb{C}^5 \to \mathrm{Hom}(W_0, R_{F_0}^5)$ is injective. Hence Proposition 3 is proved. $\square$

Proof of Proposition 2 By Proposition 3, to prove Proposition 2, it remains only to study the case where any component of W_0 contained in (X_4) is contained in (X_4^2). We split the proof in several lemmas. We first show (Lemma 2.14) that any component contained in (X_4^2) is contained in (X_4^3). We then show (Lemma 2.15) that if some component is contained in (X_4^3), then at most one component is not contained in X_4. We then conclude that W_0 has the form (ii) of Proposition 2. We begin with the following result.

Lemma 2.14 *Any component of W_0 contained in (X_4^2) is contained in (X_4^3).*

Proof We may of course assume it is the second component. We again set
$W' = \mathrm{pr}_{12}(W_0) \subset (R^5_{F_0})^2$ and introduce the filtration W'_i used above (by the
order of vanishing in X_4 of the first component), with successive quotients of
ranks $\alpha, \beta, \gamma, \delta$. Then by assumption, a suitable basis of W' can be listed as

$$
\left.
\begin{array}{ll}
\alpha \text{ elements} & (w_1, X_4^2 t_2) \\
\beta \text{ elements} & (X_4 z_1, X_4^2 t_2') \\
\gamma \text{ elements} & (X_4^2 t_1, X_4^2 t_2'') \\
\delta \text{ elements} & (X_4^3 q_1, X_4^2 t_2''')
\end{array}
\right\}.
\tag{2.4}
$$

Assume $\alpha > 11$; using (b), in the notation of (2.4), we must have $w_1 t_2' = w_1 t_2'' = w_1 t_2''' = 0$ in R^8_G, and by Lemma 2.2, it follows that $t_2' = t_2'' = t_2''' = 0$
in R^3_G. Similarly, writing $t_2'' = X_4 q_2''$, $t_2''' = X_4 q_2'''$, we find that $q_2'' = q_2''' = 0$ in
R^2_G. So a suitable basis of W' can in fact be represented as

$$
\left.
\begin{array}{ll}
\alpha \text{ elements} & (w_1, X_4^2 t_2) \\
\beta \text{ elements} & (X_4 z_1, X_4^3 q_2) \\
\gamma \text{ elements} & (X_4^2 t_1, 0) \\
\delta \text{ elements} & (X_4^3 q_1, 0)
\end{array}
\right\}.
$$

In this notation, if $q_2 \equiv 0 \bmod X_4$, then $z_1 t_2 = 0$ in R^7_G, since $X_4^3 w_1 q_2 - X_4^3 z_1 t_2 = 0$ in $R^{10}_{F_0}$. If there is at least one nonzero t_2, it follows by Lemma 2.2
that the space generated by such z_1 in R^4_G has dimension ≤ 2, which implies
$\beta \leq 12$. One shows similarly that $\alpha \leq 20$, and since $\gamma \leq 20 = \mathrm{rank}\, R^3_G$ and
$\delta \leq 10 = \mathrm{rank}\, R^2_G$, we contradict $\alpha + \beta + \gamma + \delta = 65$.

Next consider the case where $\alpha \leq 11$; then $\beta \geq 24$. Using the notation of
(2.4), $w_1 t_2' = 0$ in R^8_G, which by Lemma 2.6 implies that the space generated
by the t_2' modulo X_4 has dimension ≤ 8. It follows that the space generated
by the t_2' modulo X_4^2 has dimension ≤ 18, hence that there is a 6 dimensional
space of z_1 modulo X_4 such that $(X_4 z_1, 0) \in W'$. But these z_1 annihilate
$t_2 \in R^3_G$ for $(w_1, X_4^2 t_2) \in W'$. Hence $t_2 \equiv 0 \bmod X_4$ by Lemma 2.2 and we
conclude again that the second component is contained in (X_4^3). $\square$

We now show the following:

Lemma 2.15 *Assume there is one component of W_0 contained in (X_4^3).
Then the map $\mathbb{C}^5 \to \mathrm{Hom}(W_0, R^5_G)$ has rank 1.*

Proof It has rank at least 1, since the first component, being of rank 65,
cannot be contained in (X_4). Now assume it has rank ≥ 2. Up to a change
of basis of $\mathbb{C}^5$, we may assume that the two first components are independent
modulo (X_4), and the third is contained in (X_4^3). Now we use a proposition
proved in Section 4 (cf. Proposition 6).

Proposition 4 *Let $W \subset (R_{F_0}^5)^2$ be a 65 dimensional subset satisfying (b). Then, if the map $\mathbb{C}^2 \to \mathrm{Hom}(W, R_G^5)$ is injective, and the first projection $W \to R_{F_0}^5$ is injective, we have*

$$w_1 \in (X_4^i) \iff w_2 \in (X_4^i) \quad \text{for } (w_1, w_2) \in W.$$

Let W' be the projection of W_0 on the first three factors. According to Proposition 4, a basis of W' can be represented as follows:

$$
\left.
\begin{array}{ll}
\alpha \text{ elements} & (w_1, w_2, X_4^3 q_3) \\
\beta \text{ elements} & (X_4 z_1, X_4 z_2, X_4^3 q_3') \\
\gamma \text{ elements} & (X_4^2 t_1, X_4^2 t_2, X_4^3 q_3'') \\
\delta \text{ elements} & (X_4^3 q_1, X_4^3 q_2, X_4^3 q_3''')
\end{array}
\right\}
\tag{2.5}
$$

with w_1, w_2 independent modulo X_4, z_1, z_2 independent modulo X_4, etc. Now it is easy to see that α must be strictly larger than 10. Furthermore, in the notation of (2.5), $w_1 q_3' = w_1 q_3'' = w_1 q_3''' = 0$ in R_G^7, and this implies that $q_3' = q_3'' = q_3''' = 0$ in R_G^2, by Lemma 2.13. Now, since $\alpha > 10 = \mathrm{rank}\, R_G^2$, there exists $(w_1', w_2', X_4^3 q_3') \in W'$, with $q_3' \equiv 0 \bmod X_4$ and $w_1' \not\equiv 0 \bmod X_4$. Then for any $(w_1, w_2, X_4^3 q_3) \in W'$, by (b), $w_1' q_3 = 0$ in R_G^7, and this implies that the space generated by the q_3 in R_G^2 has dimension at most one, by Lemma 2.9. But then there is set of $(w_1', w_2', 0) \in W'/W_1'$ of dimension ≥ 10 and by Lemma 2.13 the condition $w_1' q_3 = 0$ in R_G^7 implies that $q_3 = 0$. So the third component should vanish identically, which is absurd, since the map $\mathbb{C}^5 \to \mathrm{Hom}(W_0, R_{F_0}^5)$ is injective. $\square$

It now follows from Lemmas 2.15, 2.14 and Proposition 3 that if there is one component of W_0 contained in (X_4^3), there is in fact a 4 dimensional set of such components so that after a change of basis of $\mathbb{C}^5$, we may filter W_0 by the order in X_4 of the first component and represent elements of W_0 as follows

$$
\left.
\begin{array}{ll}
\alpha \text{ elements} & (w_1, X_4^3 q_2, \dots, X_4^3 q_5) \\
\beta \text{ elements} & (X_4 z_1, X_4^3 q_2', \dots, X_4^3 q_5') \\
\gamma \text{ elements} & (X_4^2 t_1, X_4^3 q_2'', \dots, X_4^3 q_5'') \\
\delta \text{ elements} & (X_4^3 q_1, X_4^3 q_2''', \dots, X_4^3 q_5''')
\end{array}
\right\}
\tag{2.6}
$$

End of proof of Proposition 2 To conclude the proof of Proposition 2, we want to show now that $q_i' = q_i'' = q_i''' \equiv 0 \bmod X_4$ and that w_1, q_i are in Case (ii) of Proposition 2. First of all we have the result:

Lemma 2.16 $\alpha \leq 10$.

Proof If $\alpha > 10$ the second projection $W_0/(W_0)_1 \to R_G^2$ is not injective. So there exists $(w_1', 0, X_4^3 q_i) \in W_0$ with $w_1' \neq 0$ in R_G^5. Then in the notation of (2.6), $w_1' q_2 = 0$ in R_G^7, and by Lemma 2.9 this implies that the space generated by the q_2 in R_G^2 has dimension at most one. Hence the space generated by the w_1' as above in R_G^5 has dimension ≥ 10, and since it annihilates q_2, we conclude that $q_2 = 0$ in R_G^2 by Lemma 2.13. Also since $w_1 q_i' = w_1 q_i'' = w_1 q_i''' = 0$ in R_G^7, and $\alpha \geq 11$, we find that $q_i' = q_i'' = q_i''' = 0$ in R_G^2, so that the second component of W_0 vanishes, in contradiction with our assumptions. $\square$

Now note that in the notation of (2.6), $w_1 q_i' = w_1 q_i'' = w_1 q_i''' = 0$ in R_G^7, which by Lemma 2.9 implies that the space generated by the q_i', q_i'', q_i''' in R_G^2 has dimension at most one, and by Lemma 2.13, it is zero if $\alpha = 10$.

From Lemma 2.16, it follows that $\beta + \gamma + \delta \geq 55$, with equality possible only if the q_i', q_i'', q_i''' vanish. This implies that for any i, j there is at least a 54 dimensional subspace of $R_{F_0}^4$ consisting of elements z such that $(X_4 z, X_4^3 q_1, \dots, X_4^3 q_4) \in W_0$, with $q_i \equiv q_j \equiv 0 \mod X_4$. Then it is immediate to conclude

Corollary 2.17 *The subspace of $R_{F_0}^5 W_0$ made of elements of the form*

$$(X_4 \varphi, X_4^3 \chi_1, \dots, X_4^3 \chi_4) \quad with \quad \chi_i = \chi_j = 0 \quad in \ R_G^7$$

maps surjectively by the first projection onto $X_4 R_{F_0}^9$.

Note that $X_4 R_{F_0}^9$ has dimension 91. We want now to use (c') which says that $\dim R_{F_0}^5 W_0 \leq 135$: using Corollary 2.17, it implies that the projection onto the product of the $(i+1)$st and $(j+1)$st factors of $R_{F_0}^5 W_0$ has dimension ≤ 44. Now we have the following.

Lemma 2.18 *Let $(q_1, q_2), (q_1', q_2') \in R_G^2 \times R_G^2$; assume that $\dim R_G^5(q_1, q_2) + R_G^5(q_1', q_2') \leq 44$. Then $q_1 q_2' - q_2 q_1' = 0$ in R_G^4.*

Proof We may assume that q_1', q_2' are not proportional. Let $Z \subset R_G^5 \times R_G^5$ be the kernel of the map

$$(q_1, q_2) + (q_1', q_2'): R_G^5 \times R_G^5 \to R_G^7 \times R_G^7.$$

The first projection of Z in R_G^5 has dimension ≥ 36, using Lemma 2.9, and $\dim R_G^5 = 40$. On the other hand, it is easy to see that it annihilates $q_1 q_2' - q_2 q_1'$. Hence $q_1 q_2' - q_2 q_1' = 0$ in R_G^4 by Lemma 2.3. $\square$

Combining Corollary 2.17 and Lemma 2.18, we find

Corollary 2.19 *Let $(w_1, X_4^3 q_1, \dots, X_4^3 q_5), (w_1', X_4^3 q_1', \dots, X_4^3 q_5') \in W_0$; then $q_i q_j' - q_i' q_j = 0$ in R_G^4.*

It follows from the corollary that if $q_i \equiv 0 \bmod X_4$, but $(q_1, \ldots, q_4) \not\equiv 0$ mod X_4, any other q_i' vanishes, because for generic G, the product of two nonzero elements of R_G^2 does not vanish in R_G^4. Since no component of W_0 vanishes, we find that if $(q_1, \ldots, q_4) \not\equiv 0 \bmod X_4$, the q_i are independent modulo X_4. It is then easy to conclude from Corollary 2.19, that if $\mathrm{rank}(\mathrm{pr}_{2345}(W_0)) \geq 2$ there exists $(A_1, \ldots, A_4) \in (S^1)^4$, well defined modulo X_4, such that any $X_4^3(q_1, \ldots, q_4) \in \mathrm{pr}_{2345}(W_0)$ is of the form $A X_4^3(A_1, \ldots, A_4)$ for some $A \in S^1$.

Now the rank of $\mathrm{pr}_{2345}(W_0)$ cannot be equal to 1; for otherwise there is

$$(w_1', 0, \ldots, 0) \in W_0 \quad \text{with } w_1' \neq 0 \text{ in } R_G^5;$$

then for $(w_1, X_4^3 q_i) \in W_0$, we have $w_1' q_i = 0$ in R_G^7, which by Lemma 2.9 implies that all the q_i are proportional. But then the four last components of W_0 are not independent, since they have one dimensional projection in $(R_{F_0}^5)^4$, generated by $(X_4^3 q_i)$ and the q_i are not independent in $R_{F_0}^5$.

Now assume $(w_1', 0, \ldots, 0) \in W_0$ and $(w_1, X_4^3 A A_i) \in W_0$, with $A \not\equiv 0$ mod X_4: then $w_1' A A_i = 0$ in R_G^7, and using the fact that the A_i are independent modulo X_4, we conclude by Macaulay's theorem that $A w_1' = 0$ in R_G^6 hence that $w_1' \equiv 0 \bmod X_4$, using the following lemma.

Lemma 2.20 *If G is generic, for any $w \neq 0$ in R_G^5 and $A \neq 0$ in $S^1/(X_4)$, we have $Aw \neq 0$ in R_G^6.* $\square$

Since $\mathrm{rank}\, \mathrm{pr}_1(W_0) = 65$ and $\mathrm{rank}\, X_4 R_G^4 = 61$, we find by the above that the projection pr_{2345} has rank ≥ 4, hence that its image is made of the $(X_4^3 A A_i)$ for arbitrary $A \in S^1$. Finally, since $w_1 \not\equiv 0 \bmod X_4$ implies that $A \not\equiv 0 \bmod X_4$, we can consider w_1 modulo X_4 as a function of A, and (b) implies that $w_1(A) A' - w_1(A') A \equiv 0 \bmod X_4$, since it is annihilated by all the A_i. The symmetrizer lemma implies then that there exists $T \in R_G^4$, such that $w_1(A) = TA$ in R_G^5.

Finally, $A T q_i' = A T q_i'' = A T q_i''' = 0$ in R_G^7 for any $A \in S^1$, and this implies that $T q_i' = T q_i'' = T q_i''' = 0$ in R_G^6 by Macaulay's theorem. But then $q_i' = q_i'' = q_i''' = 0$ in R_G^2, by Lemma 2.4. Hence W_0 is in Case (ii) of Proposition 2, and this proves Proposition 2. $\square$

3 Proof of Proposition 2 $\Longrightarrow$ Theorem 5

Proposition 5 *Assume W_0 is in Case (i) or (ii) of Proposition 2. Then for a generic deformation F of F_0, any deformation $W \subset (R_F^5)^5$ of $W_0 \subset (R_{F_0}^5)^5$ as a 65 dimensional subspace satisfying (b) and (c) must be identified with the subspace $R_F^4 \subset \mathrm{Hom}(S^1, R_F^5)$, for some isomorphism $S^1 \cong \mathbb{C}^5$.*

Proof If W_0 is in Case (i) of Proposition 2, the result is easy. Indeed, we know Lemma 2.12 that W_0 is infinitesimally rigid modulo $\operatorname{Aut} \mathbb{C}^5$ as a subspace of $(R_{F_0}^5)^5$ satisfying (b). Thus consider the variety $\mathcal{G} \to U$ parametrizing 65 dimensional subspaces of $(\mathcal{R}^5)^5$ satisfying (b) and (c') with respect to $\mu \colon S^2 \mathcal{R}^5 \to \mathcal{R}^{10}$. Its infinitesimal dimension at (W_0, F_0) is $\leq 24 + \dim U$. But this is the dimension of the variety $\mathcal{G}' \subset \mathcal{G}$ parametrizing $W \cong R_F^4 \subset \operatorname{Hom}(S^1, R_F^5)$. Hence we must have $\mathcal{G}' = \mathcal{G}$ in a neighbourhood of (W_0, F_0).

Thus we only need to study Case (ii). We note first that any $W_0 \subset (R_{F_0}^5)^5$ in Case (ii) of Proposition 2 is in the Zariski closure of $\mathcal{G}'$: it is enough to consider a family $W_t \cong R_{F_t}^4 \subset (R_{F_t}^5)^5$ for a suitable one parameter family F_t, and for a family of isomorphisms $S^1 \cong_t \mathbb{C}^5$ given by a basis $A_0(t), \dots, A_4(t)$ of S^1 such that $A_0(0) = X_4$, and to take the limit W_0 of W_t such that $\operatorname{pr}_1 \colon W_0 \to R_{F_0}^5$ is injective. (This supposes of course making elementary transformations over 0 on the bundle $\mathcal{W}$ with fibre W_t.) Then one finds that the limiting W_0 has the expected form, with $T = \frac{\partial}{\partial X_4}\left(\frac{dF_t}{dt}\big|_{t=0}\right)$.

Furthermore it is easy to see that the map $\mathcal{G}' \to U$ has relative dimension 24, and is proper over the open set of U where the following property is satisfied:

> The multiplication map $\mu_X \colon R_F^4 \to R_F^5$ by X is injective for any nonzero $X \in S^1$.

So, to conclude that $\mathcal{G} \subset \overline{\mathcal{G}'}$ near (W_0, F_0) (the statement of Proposition 5), it is enough to prove the following.

Lemma 3.1 *The dimension of the Zariski tangent space $T_{\mathcal{G}, (W_0, F_0)}$ to $\mathcal{G}$ at (W_0, F_0) equals $\dim U + 24$.*

Proof Choosing a lifting of $R_{F_0}^5$ to S^5 gives a trivialization of the bundle $\mathcal{R}^5$ near F_0. An infinitesimal deformation of (W_0, F_0) is then described by maps $h, h_i \colon S^1 \to R_{F_0}^5$ and $k, k_i \colon R_{F_0}^4 \to R_{F_0}^5$, together with a deformation $F_0 + \varepsilon H$ of F_0. The corresponding deformation $W_\varepsilon \subset (R_{F_\varepsilon}^5)^5 \cong (R_{F_0}^5)^5$ is generated by the vectors

$$(TA + \varepsilon h(A), A_i X_4^3 A + \varepsilon h_i(A)) \quad \text{for } A \in S^1$$
$$\text{and} \quad (X_4 z + \varepsilon k(z), \varepsilon k_i(z)) \quad \text{for } z \in R_{F_0}^4.$$

Of course the infinitesimal deformation is trivial if $H = 0$ and $(h(A), h_i(A))$, $(k(z), k_i(z)) \in W_0$. Differentiating the equations given by (b), we get the following equations

(i) $X_4 z k_i(z') = X_4 z' k_i(z)$.

(ii) $ATk_i(z) = X_4zh_i(A) + A_iX_4^3Ak(z) - \frac{1}{5}\frac{\partial H}{\partial X_4}A_iAz.$

(iii) $A_iX_4^3Ak_j(z) = A_jX_4^3Ak_i(z).$

(iv) $A_iX_4^3Ah_j(A') + A_jX_4^3A'h_j(A) = A_iX_4^3A'h_j(A) + A_jX_4^3Ah_j(A').$

(v) $h(A)A_iX_4^3A' + ATh_i(A') = h(A')A_iX_4^3A + A'Th_i(A).$

These equations have to hold in $R_{F_0}^{10}$ for any $A, A' \in S^1$ and for any $z, z' \in R_{F_0}^4$. Applying the symmetrizer lemma to (i), we conclude that $k_i(z) = Y_iz + X_4^3k_i'(z)$ for some $Y_i \in S^1$. The condition (ii) modulo X_4 then gives

$$ATY_iz = -\frac{1}{5}\frac{\partial H}{\partial X_4}A_iAz \quad \text{in } R_G^{10},$$

which implies that $TY_i = -\frac{1}{5}\frac{\partial H}{\partial X_4}A_i$ in R_G^5. Using the fact that the map $\mu_T \colon R_G^2 \to R_G^6$ is injective (Lemma 2.4), we conclude that $A_iY_j = A_jY_i$ in R_G^2, and hence that $Y_i = \alpha A_i + \alpha_iX_4$, by the symmetrizer lemma. Furthermore $-\frac{1}{5}\frac{\partial H}{\partial X_4} = \alpha T$ in R_G^4: we write $-\frac{1}{5}\frac{\partial H}{\partial X_4} = \alpha T + X_4H'$.
Condition (ii) modulo X_4^3 then gives

$$AT\alpha_iX_4z = X_4zh_i(A) + X_4H'Az \quad \text{in } R_{F_0}^{10}/(X_4^3),$$

i.e., $h_i(A) = \alpha_iTA - H'A$ in $R_{F_0}^5/(X_4^2)$. Hence $h_i(A) = \alpha_iTA - H'A + X_4^2h_i'(A)$. Then Condition (ii) gives

$$ATk_i'(z) = zh_i'(A) + A_iAk(z) \quad \text{in } R_G^7 \tag{3.1}$$

It follows that $z(Ah_i'(A') - A'h_i'(A)) = 0$ in R_G^8 for any $z \in R_G^4$. Hence by the symmetrizer lemma there exists $q_i \in R_G^2$ such that $h_i'(A) \equiv q_iA \bmod X_4$. But then (3.1) gives $z(A_jq_i - A_iq_j) = 0$ in R_G^7, whenever $k_i'(z) = k_j'(z) = 0$ in R_G^2. By Lemma 2.2, it follows that $(A_jq_i - A_iq_j) = 0$ in R_G^3, and hence by the symmetrizer lemma that $q_i \equiv A_iB \bmod X_4$. Hence $h_i'(A) \equiv A_iBA \bmod X_4$, and (3.1) then gives

$$Tk_i'(z) = zBA_i + A_ik(z) \quad \text{in } R_G^6 \tag{3.2}$$

Then $A_ik_j'(z) - A_jk_i'(z) \in R_G^3$ is annihilated by T. But $\mu_T \colon R_G^3 \to R_G^7$ has kernel of dimension at most one. It is easy to conclude from this that $A_ik_j'(z) - A_jk_i'(z)$ must in fact vanish identically in R_G^3, so that by the symmetrizer lemma, $k_i'(z) \equiv A_ik''(z) \bmod X_4$, for some function $k'' \colon R_{F_0}^4 \to S^1$. Finally (3.2) gives $Tk''(z) \equiv zB + k(z) \bmod X_4$. Summarizing, we have found

$$k(z) \equiv Tk''(z) - zB \bmod X_4,$$
$$k_i(z) = \alpha A_iz + X_4^3A_ik''(z) + \alpha_iX_4z,$$
$$h_i(A) = \alpha_iTA - H'A + X_4^2A_iAB + X_4^3h_i'''(A),$$

$$\text{and} \quad -\frac{1}{5}\frac{\partial H}{\partial X_4} = \alpha T + X_4H' \quad \text{in } R_{F_0}^4.$$

Now $(Tk''(z), X_4^3 A_i k''(z)) \in W_0$, and $(X_4\varphi, 0, \ldots 0) \in W_0$. So the deformation induced by $(k(z), k_i(z)) \in (R_{F_0}^5)^5$ depends only on $(k(z) \mod X_4, k_i(z))$, and not on $k''(z)$. It follows that $(k(z), k_i(z)) \in (R_{F_0}^5)^5/W_0$ depends only on B modulo X_4, α, α_i. Hence it varies in a 9 dimensional vector space. Let us now consider the deformations of (W_0, F_0) for which the corresponding infinitesimal deformation of F_0 is trivial, and $\alpha, \alpha_i, B \equiv 0 \mod X_4$. We now use (c') to describe them: notice that for W_0 in Case (ii) we have in fact $\dim R_{F_0}^5 W_0 = 135$. Hence the infinitesimal version of condition (c') gives the following

Suppose that $\varphi, \psi \in R_{F_0}^5$ and $(h(A), h_i(A)), (h(A'), h_i(A'))$ are infinitesimal deformations of $(TA, A_i AX_4^3), (TA', A_i A' X_4^3) \in W_0$. Assume that

$$\varphi(TA, A_i AX_4^3) + \psi(TA', A_i A' X_4^3) = 0 \quad in \; (R_{F_0}^{10})^5.$$

Then $\varphi(h(A), h_i(A)) + \psi(h(A'), h_i(A')) \in R_{F_0}^5 W_0$.

In particular, for any $z \in R_{F_0}^4$, we have

$$zA'(TA, X_i X_4^3 A) + (-zA)(TA', X_i X_4^3 A') = 0.$$

Hence we must have $zA'(h(A), h_i(A)) - zA(h(A'), h_i(A')) \in R_{F_0}^5 W_0$.
Now, since $\alpha = 0, B = 0, H' = 0$, we have $h_i(A) = X_4^3 h_i''(A)$, hence

$$zA'(h(A), X_4^3 h_i''(A)) - zA(h(A'), X_4^3 h_i''(A')) \in R_{F_0}^5 W_0.$$

Clearly this is equivalent to: there exists $\varphi \in R_G^6$ depending on z, A, A', such that

$$z(A'h(A) - Ah(A')) = T\varphi, \quad z(A'h_i''(A) - Ah_i''(A')) = A_i\varphi \quad in \; R_G^7. \quad (3.3)$$

From the second equation, we deduce that

$$z(A_j(A'h_i''(A) - Ah_i''(A')) - A_i(A'h_j''(A) - Ah_j''(A'))) = 0 \quad in \; R_G^8,$$

for any $z \in R_G^4$, hence by the symmetrizer lemma, that

$$(A'h_i''(A) - Ah_i''(A')) = A_i h''(A, A'), \quad (3.4)$$

for some bilinear function $h'' \colon S^1 \times S^1 \to R_G^2$. This function then satisfies the following condition

$$A_1 h''(A_2, A_3) - A_2 h''(A_1, A_3) + A_3 h''(A_1, A_2) = 0$$

in R_G^3 for any $A_i \in S^1$ from which it is easy to conclude that

$$h''(A, A') = A'h'''(A) - Ah'''(A'),$$

for some function $h''' : S^1 \to S^1/(X_4)$. But then by (3.4) we get

$$A'\big(h_i''(A) - A_ih'''(A)\big) - A\big(h_i''(A') - A_ih'''(A')\big) = 0 \quad \text{in } R_G^3,$$

and applying the symmetrizer lemma once more, we conclude that $h_i''(A) \equiv A_ih'''(A) + B_iA \bmod X_4$ for some $B_i \in S^1$. The second equality in (3.3) then gives $\varphi(z, A, A') = z(A'h'''(A) - Ah'''(A'))$ and the first $A'h(A) - Ah(A') = T(A'h'''(A) - Ah'''(A'))$ in R_G^6, that is:

$$A'(h(A) - Th'''(A)) - A(h(A') - Th'''(A')) = 0 \quad \text{in } R_G^6,$$

which by the symmetrizer lemma gives $h(A) - Th'''(A) \equiv \chi A \bmod X_4$ for some $\chi \in R_G^4$. In conclusion, we get

$$h_i(A) = X_4^3(A_ih'''(A) + B_iA), \quad h(A) = \chi A + Th'''(A).$$

Working modulo W_0, this is also:

$$h_i(A) = X_4^3 B_i A, \quad h(A) = \chi A.$$

Thus these deformations depend on the 16 parameters $B_i \in S^1$ and the 31 parameters $\chi \in R_G^4$, with the relation: $B_i = \eta A_i$, $\chi = \eta T$.

Since the allowed deformations for F_0 vary in the space of codimension 31 defined by the condition $\frac{\partial H}{\partial X_4} = 0$ in R_G^4, when α, α_i, B vanish, we have found that the tangent space to $\mathcal{G}$ at (W_0, F_0) has dimension $31 + 15 + \dim U - 31 + 9 = \dim U + 24$, and Lemma 3.1 is proved. □

4   Proof of the degeneracy property

To complete the proof of Theorem 5, it remains to prove that if F_0 is of the form (0.6), its W_0 is degenerate in the sense of Definition 3 (see 0.4). We start by proving the following result.

Proposition 6 *Let $V \subset (R_{F_0}^5)^2$ be a 65 dimensional subspace, and assume that*

1. *the first projection $\mathrm{pr}_1 : V \to R_{F_0}^5$ is injective;*

2. *$w_1w_2' - w_1'w_2 = 0$ in $R_{F_0}^{10}$ for all $w, w' \in V$;*

3. *the map $\mathbb{C}^2 \to \mathrm{Hom}(V, R_G^5)$ has rank 2.*

Then for $w = (w_1, w_2) \in V$

$$w_1 \in X_4^i R_{F_0}^{5-i} \iff w_2 \in X_4^i R_{F_0}^{5-i}$$

In particular, assume that $W_0 \subset (R_{F_0}^5)^5$ satisfies (b), (e) and is non-degenerate; then

$$X_4^i(z_1, \ldots, z_5) \in W_0 \implies$$

$z_1, \ldots, z_5$ *are either linearly independent or all 0 in R_G^{5-i}.*

Proof We filter V by the order w.r.t. X_4, so that $V_i = V \cap (X_4^i R_{F_0}^{5-i})^2$, and denote by $\alpha, \beta, \gamma, \delta$ the ranks of the successive quotients. We have to show that if $w = X_4^i(z_1, z_2) \in V_i \setminus V_{i+1}$, then z_1, z_2 are independent in R_G^{5-i}. We only prove the case $i = 0$, the others being similar. So assume $w_2 \equiv 0$ mod X_4, but $w_1 \not\equiv 0$ mod X_4.

First of all, $\delta \leq 2$; indeed if $(X_4^3 q_1, X_4^3 q_2) \in V$ then $w_1 q_2 = 0$ in R_G^7, since $X_4^3 w_1 q_2 - X_4^3 w_2 q_1 = 0$ in $R_{F_0}^{10}$ and $w_2 \equiv 0$ mod X_4, which by Lemma 2.9 implies that the image of the second projection $V_3 \to R_G^2$ has dimension ≤ 1. But if $X_4^3(q_1, 0) \in V$ is in the kernel of this map, $q_1 w_2' = 0$ in R_G^7 for any $(w_1', w_2') \in V$. Since there is at least one nonzero w_2' by the nondegeneracy assumption, we conclude that this map has kernel of dimension at most one, so that $\delta \leq 2$.

Suppose now $\gamma \geq 9$; for $X_4^2(t_1, t_2) \in V_2$ we have $w_1 t_2 = 0$ in R_G^8, and by Lemma 2.6, this implies that the image of the second projection $V_2 \to R_G^3$ has dimension ≤ 8. Hence there exists $X_4^2(t_1', t_2') \in V_2$ with $t_2' = 0$ in R_G^3 and $t_1' \neq 0$ in R_G^3. Now for $(X_4 z_1, X_4 z_2) \in V_1$ we have $t_1' z_2 = 0$ in R_G^7, since $X_4^3(t_1' z_2 - t_2' z_1) = 0$ in $R_{F_0}^{10}$ and $t_2' \equiv 0$ mod X_4, which by Lemma 2.2 implies that the image of the second projection $V_1 \to R_G^4$ has dimension ≤ 2. If $\beta \geq 5$, there exists $(X_4 z_1', X_4 z_2'), (X_4 z_1'', X_4 z_2''), (X_4 z_1''', X_4 z_2''') \in V_1$, with $z_2' \equiv z_2'' \equiv z_2''' \equiv 0$ mod X_4, and z_1', z_1'', z_1''' independent in R_G^4. Then for any $X_4^2(t_1, t_2) \in V_2$ we have $z_1' t_2 = z_1'' t_2 = z_1''' t_2 = 0$ in R_G^7, which by Lemma 2.2 implies that the second projection $V_2 \to R_G^3$ vanishes. But then from $\gamma \geq 9$, we conclude that there is a 9 dimensional subspace of R_G^3 made of elements t_1 such that $X_4^2(t_1, t_2) \in V_2$, with $t_2 \equiv 0$ mod X_4. But we have for any such t_1 and any $(w_1, w_2) \in V$, $t_1 w_2 = 0$ in R_G^8, which by Lemma 2.6 implies that $w_2 = 0$ in R_G^5, and this contradicts the nondegeneracy assumption. So we conclude that $\gamma \geq 9 \implies \beta \leq 4$. Also, it is easy to show using Lemma 2.6 that $\gamma \leq 16$. Hence $\gamma \geq 9 \implies \beta + \gamma \leq 20$, and since $\delta \leq 2$ we find $\alpha \geq 65 - 22 = 43 > \operatorname{rank} R_G^5$. But this is absurd because the image of V/V_1 in $R_G^5 \times R_G^5$ is totally isotropic for any of the pairings $\langle w, w' \rangle_\varphi = \varphi(w_1 w_2' - w_2 w_1') \in R_G^{12} \cong \mathbb{C}$, where $\varphi \in R_G^2$, and for generic G and φ, this pairing is nondegenerate.

Thus we must have $\gamma \leq 8$ and $\alpha + \beta \geq 55$. Now we claim that $\beta \leq 19$. Indeed we have $w_1 z_2 = 0$ in R_G^9 for $(X_4 z_1, X_4 z_2) \in V_1$, and by Lemma 2.6 this implies that the second projection $V_1 \to R_G^4$ has rank ≤ 19. So if $\beta > 19$, there is an element $(X_4 z_1', X_4 z_2') \in V_1$, with $z_2' \equiv 0 \bmod X_4$ but $z_1' \not\equiv 0 \bmod X_4$. But this implies easily that $\beta \leq 22$ and $\gamma = 0$, using Lemmas 2.5 and 2.2. Then $\alpha \geq 41$ and this is absurd, since V/V_1 is a totally isotropic subspace of $R_G^5 \times R_G^5$ for any of the generically nondegenerate pairings

$$((u,v),(u',v')) \mapsto q(uv' - u'v) \in R_G^{12} \cong \mathbb{C}, \quad \text{for } q \in R_G^2.$$

Thus we must have $\beta \leq 19$, hence $\alpha \geq 65 - 19 - 8 - 2 = 36$. This implies in fact that $\delta \leq 1$: indeed let $K \subset R_G^5$ be the image of the second projection map $\mathrm{pr}_2 \colon V \to R_G^5$, and let $L \subset R_G^5$ be the image by the first projection $\mathrm{pr}_1 \colon V \to R_G^5$ of $\mathrm{Ker}\,\mathrm{pr}_2$. Then by assumption, both K and L are nonzero and $KL = 0$ in R_G^{10}. Furthermore $\alpha = \dim K + \dim L$. We know that the second projection $\mathrm{pr}_2^3 \colon V_3 \to R_G^2$ has rank at most one, and is annihilated by the space L introduced above. Its kernel is annihilated by the space K so has dimension at most one by Lemma 2.9. If both the image and the kernel of pr_2^3 are nonzero, we find that $\dim K \leq 9$, and $\dim L \leq 9$ by Lemma 2.13. Hence $\alpha \leq 18$, a contradiction.

From $\delta \leq 1$, we conclude that $\alpha \geq 37$. But in the above notation this contradicts $\alpha = \dim K + \dim L$ with K, L nontrivial, and the following result.

Lemma 4.1 *Let G be generic and K, L two nontrivial subspaces of R_G^5 such that $KL = 0$ in R_G^{10}; then $\dim K + \dim L \leq 36$.* $\square$

Thus Proposition 6 is proved. $\square$

So what we have to show now is the following:

Proposition 7 *There does not exist a 65 dimensional subspace $W \subset (R_{F_0}^5)^5$, satisfying the following conditions:*

(i) $\mathrm{pr}_{1|W} \colon W \to R_{F_0}^5$ *is injective.*

(ii) *Filter W by $W_i = W \cap (X_4^i R_{F_0}^{5-i})^5$; then for $w = (X_4^i z_i) \in W_i \setminus W_{i+1}$, the z_i are independent in R_G^{5-i}.*

(iii) $w_i w_j' = w_j w_i'$ *in $R_{F_0}^{10}$ for any i, j and any $w, w' \in W$.*

The first step is the following

Lemma 4.2 *If such W exists, then*

$$t_i t_j' - t_j t_i' = 0 \text{ in } R_G^6 \quad \text{for all } t = X_4^2(t_1, \ldots, t_5), \ t' = X_4^2(t_1', \ldots, t_5') \in W_2.$$

Proof We prove this by contradiction: assuming Lemma 4.2 is not true, we have first the following properties:

(a) $\dim(S^1W_2/W_3 + W_1/W_2) \geq \dim W_1/W_2 + 6$.

(b) $\dim(S^2W_2/W_3 + W/W_1) \geq \dim W/W_1 + 8$.

Indeed, let $X, Y \in S^1/(X_4)$ such that $Xt + Yt' = (z_1, \ldots, z_5) \in W_1/W_2$; then $z_it_j - z_jt_i = 0$ in R_G^7, hence $Y(t_it_j' - t_jt_i') = 0$ in R_G^7. Similarly $X(t_it_j' - t_jt_i') = 0$ in R_G^7. But the annihilator in $S^1/(X_4)$ of any nonzero element of R_G^6 has dimension at most one, since G is generic, hence it follows that the set of such pairs (X, Y) has dimension ≤ 2, so that $\dim S^1t + S^1t' \mod W_1/W_2 \geq 6$. Property (b) is proved similarly using the fact that the annihilator in $S^2/(X_4)$ of any nonzero element of R_G^6 has dimension ≤ 6.

Now we show the following:

For any nonzero $(z_1, \ldots, z_5) \in S^1W_2/W_3 + W_1/W_2 \subset (R_G^4)^5$, the z_i are independent in R_G^4.

Indeed, assume by contradiction that $z_1 \neq 0$ and $z_5 = 0$; then for any elements $(w_1, \ldots, w_5)$ in W/W_1 and $X_4(z_1', \ldots, z_5')$ in W_1/W_2, by (iii), we have $z_1w_5 = 0$ in R_G^9 and $z_1z_5' = 0$ in R_G^8; using Lemmas 2.3 and 2.5 and (ii), this implies $\alpha \leq 21$ and $\beta \leq 12$. But then $\gamma + \delta \geq 32$ and this is absurd because by (ii), $\gamma \leq \dim R_G^3 - 4 = 16$ and $\delta \leq \dim R_G^2 - 4 = 6$.

This last fact implies that $\dim(S^1W_2/W_3 + W_1/W_2) \leq \dim R_G^4 - 4 = 27$, hence by inequality (a), we conclude that $\beta \leq 21$. Next, we show:

For any nonzero $(w_1, \ldots, w_5) \in S^2W_2/W_3 + W/W_1 \subset (R_G^5)^5$, the w_i are independent in R_G^5, unless $\alpha \leq 31$ and $\beta \leq 19$.

This follows in fact from Lemmas 2.9 and 2.6 and (ii); indeed by (iii), if $(w_1', \ldots, w_5') \in S^2W_2/W_3 + W/W_1$ with $w_5' = 0$ in R_G^5 and $w_1' \neq 0$ in R_G^5, we have $w_1'w_5 = 0$ in R_G^{10} and $w_1'z_5 = 0$ in R_G^9 for any $(w_1, \ldots, w_5) \in W/W_1$ and $(z_1, \ldots, z_5) \in W_1/W_2$.

In the first case $\dim S^2W_2/W_3 + W/W_1 \leq \dim R_G^5 - 4 = 36$ hence by inequality (b) above, $\dim W/W_1 \leq 28$. Since $\beta \leq 21$, in either case $\alpha + \beta \leq 50$, which implies $\gamma + \delta \geq 15$. Now we show:

$\delta \leq 1$ and $\dim[W_2/W_3 : (X, Y)] \leq 1$ for generic $X, Y \in S^1/(X_4)$.

This is proved as follows: let $(q_1, \ldots, q_5), (q_1', \ldots, q_5')$ be elements of W_3 or $[W_2/W_3 : (X, Y)]$. It is immediate to see that

$$q_iw_j - q_jw_i = q_i'w_j - q_j'w_i = 0 \text{ in } R_G^7 \quad \text{for } (w_1, \ldots, w_5) \in W/W_1,$$

for all i, j. (In the first case, this follows from (iii); in the second, (iii) implies that these elements of R_G^7 are annihilated by X and Y.) This implies that $q_i q_j' - q_j q_i' \in R_G^4$ is annihilated by w_i, w_j. Thus if it is nonzero we have $\alpha \leq 21$ by Lemma 2.3, and then $\alpha + \beta \leq 42$. So $\gamma + \delta \geq 23$, which contradicts the inequality $\gamma + \delta \leq 22$ proved above. Now we conclude by the following easy result.

Lemma 4.3 *There do not exist independent elements* $(q_1, \ldots, q_5), (q_1', \ldots, q_5')$ *of* $(R_G^2)^5$ *satisfying: the* q_i *are independent in* R_G^2 *and* $q_i q_j' = q_j q_i'$ *in* R_G^4 *for all* i, j.

So we have now proved that $\gamma \geq 14$ and $\dim[W_2/W_3 : (X, Y)] \leq 1$, and this implies that $\dim XW_2/W_3 + YW_2/W_3 \geq 27$. But $S^1 W_2/W_3$ cannot be equal to $XW_2/W_3 + YW_2/W_3$ for generic X, Y because this would imply that the composite map: $S^1 W_2/W_3 \xrightarrow{\text{pr}_1} R_G^4 \to R_G^4/(X, Y)$ is not surjective, while the assumption $\gamma \geq 14$ implies that the composite map $W_2/W_3 \xrightarrow{\text{pr}_1} R_G^3 \to R_G^3/(X, Y)$ is surjective for generic X, Y (cf. Green [11]). It follows that $\dim S^1 W_2/W_3 \geq 28$, which is absurd because we have already proved that $\dim S^1 W_2/W_3 \leq 27$. Hence Lemma 4.2 is proved. $\square$

Next we prove the following result.

Lemma 4.4 *Suppose that* G *is generic, and that* $Z \subset (R_G^3)^5$ *satisfies the following:*

(i) *For any nonzero* $(t_1, \ldots, t_5) \in Z$, *the* t_i *are independent in* R_G^3.

(ii) *For* $(t_1, \ldots, t_5), (t_1', \ldots, t_5') \in Z$, *we have* $t_i t_j' - t_j t_i' = 0$ *in* R_G^6.

Then $\dim Z \leq 7$.

Proof In this proof S^k denotes the degree k component of the polynomial ring $S/(X_4)$ in four variables. Let X be generic in S^1. One shows that for generic $P \in (R_G^6/(X))^*$, the pairing $\langle \ , \ \rangle_P$ on $R_G^3/(X)$ defined by $\langle \alpha, \beta \rangle_P = P(\alpha\beta)$ is nondegenerate. Now assume that Z is as above, but $\dim Z \geq 8$. Since $\dim R_G^3/(X) = \dim S^3/(X) = 10$, there exists $(t_1, \ldots, t_5) \in Z$ such that the t_i are not independent in $R_G^3/(X)$. After a change of basis of $\mathbb{C}^5$ we may assume that $t_1 \equiv 0 \mod X$. Let r be the dimension of the kernel of the composite

$$f: Z \xrightarrow{\text{pr}_1} R_G^3 \to R_G^3/(X).$$

Then for any $t \in \text{Ker } f$ and $t' \in Z$ we have $t_i t_1' = t_1 t_i' = 0$ in $R_G^6/(X)$, hence $\langle t_i, t_1' \rangle_P = 0$, which implies that the space generated by the t_i modulo

X with $t \in \operatorname{Ker} f$ has dimension $\leq 10 - 8 + r = 2 + r$. Since $\dim \operatorname{Ker} f = r$ it follows that there exists $t \in \operatorname{Ker} f$ such that $t_2, \dots, t_5$ are linearly dependent in $R_G^3/(X)$. After a change of basis of $\mathbb{C}^5$, we may assume that $t_1 \equiv t_2 \equiv 0$ mod X, that is, $t = (Xq_1, Xq_2, t_3, t_4, t_5)$ for some $q_2, q_3 \in S^2$.

Let us show that $q_1, q_2 \in S^2$ cannot have a common factor: indeed, if $q_1 = AY$ and $q_2 = BY$ for some $A, B, Y \in S^1$, then $At_2' = Bt_1'$ in R_G^4 for any $t' \in Z$, since this is annihilated by $0 \neq XY \in S^2$. But the map $(B - A): R_G^3 \times R_G^3 \to R_G^4$ has kernel of dimension ≤ 11, and should contain $\operatorname{pr}_{12}(Z)$. Since it contains the 10 dimensional subspace $(A, B)R_G^2$, this would imply that Z contains a subspace Z' of dimension ≥ 7 such that $\operatorname{pr}_{12}(Z') = \{(A, B)q$ for $q \in Q \subset R_G^2\}$. For $t \in Z'$, we can view t_3, t_4, t_5 as functions of $q \in Q$, and $qt_i(q') = q't_i(q) \in R_G^5$. Since $\operatorname{rank} Q \geq 7$, $R_G^3 Q \subset R_G^5$ is at least a hyperplane, and it is not difficult to adapt the symmetrizer lemma to conclude that there must exist $A_i \in S^1$ for $i = 3, 4, 5$ such that $t_i(q) = A_i q \in R_G^3$. But then A, B, A_i are not independent in S^1, and this contradicts (i). So q_1, q_2 cannot have a common factor. Now we use the following result.

Lemma 4.5 *Suppose that G is generic, and that $q_1, q_2 \in R_G^2 = S^2$ have no common factor. Then the map $(q_2 - q_1): R_G^3 \times R_G^3 \to R_G^5$ has kernel of dimension ≤ 10.* $\square$

Since $t = (Xq_1, Xq_2, t_3, t_4, t_5) \in Z$, by (ii) and the fact that multiplication $\mu_X: R_G^5 \to R_G^6$ by X is injective, we now have $q_1 t_2' - q_2 t_1' = 0$ in R_G^5 for any $t' \in Z$. So $\operatorname{pr}_{12}(Z)$, of dimension ≥ 8, is contained in $\operatorname{Ker}(q_2 - q_1)$, which has dimension ≤ 10 and contains the 4 dimensional subspace $S^1 \cdot (q_1, q_2)$. It follows that there exists $Y \in S^1$ not proportional to X such that $Y(q_1, q_2) \in \operatorname{pr}_{12}(Z)$. Let $t' = (Yq_1, Yq_2, t_3', t_4', t_5') \in Z$. Then the first two components of $Yt - Xt'$ vanish, and it follows easily using (ii) that $Yt - Xt' = 0$ in $(R_G^4)^5$. But the space $H^3_{X,Y}$, defined as the middle cohomology group of the sequence

$$0 \to R_G^2 \xrightarrow{(X,Y)} R_G^3 \times R_G^3 \xrightarrow{Y-X} R_G^4$$

has dimension at most one; hence it follows that after a change of basis of $\mathbb{C}^5$, we have $t = (Xq_1, \dots, Xq_4, t_5)$ and $t' = (Yq_1, \dots, Yq_4, t_5')$ for some $q_3, q_4 \in R_G^2$. Thus we have proved that for generic X there is $t \in Z$ such that at least four components of t vanish modulo (X). Hence for generic X and T in S^1, after a change of basis of $\mathbb{C}^5$ there exists $t = (Xq_1, Xq_2, Xq_3, t_4, t_5) \in Z$ and $t' = (Tq_1', Tq_2', Tq_3', t_4', t_5') \in Z$. Then $q_i q_j' = q_i' q_j$ in R_G^4 for $i, j \leq 3$, and since we know that q_i, q_j do not have a common factor, this implies easily that (q_1, q_2, q_3) and (q_1', q_2', q_3') are proportional in $(R_G^2)^3$. Hence we may assume they are equal. But then $Xt' - Tt$ must vanish in $(R_G^4)^5$, and since X and T are generic the space $H^3_{X,T}$ is zero. It follows that in fact

$t = X(q_1, \ldots, q_5)$ and $t' = T(q_1, \ldots, q_5)$ for some $(q_1, \ldots, q_5) \in (R_G^2)^5$. Then for any $z = (z_1, \ldots, z_5) \in Z$, we have $q_i z_j - q_j z_i = 0$ in R_G^5 for any $i, j \leq 5$.

Now let Z' be a subspace of Z supplementary to $X(q_1, \ldots, q_5)$, so that $\dim Z' \geq 7$. Since $\dim R_G^3/(X) = 10$, there is a $t' \in Z'$ such that the t_i' are not independent in $R_G^3/(X)$. We may assume $t_1' \equiv 0 \bmod X$, and using the property $q_1 t_i' = q_i t_1'$ in R_G^5, we conclude that $q_1 t_i' = 0$ in $R_G^5/(X)$. Now we use the result:

Lemma 4.6 *Suppose that G and $X \in S^1$ are generic. Then for any nonzero $q \in R_G^2/(X)$ the multiplication map $\mu_q \colon R_G^3/(X) \to R_G^5/(X)$ has kernel of dimension ≤ 3.* $\square$

So if $q_1 \not\equiv 0$ modulo (X), we see that the t_i' generate a subspace of $R_G^3/(X)$ of dimension ≤ 3, so that at least two components of t' are divisible by X. We then conclude as before that $t' = X(q_1', \ldots, q_5')$; then $(q_1, \ldots, q_5)$ and $(q_1', \ldots, q_5') \in (R_G^2)^5$ are not proportional with independent components, and satisfy $q_i q_j' - q_i' q_j = 0$ in R_G^4, which contradicts Lemma 4.3. Hence we have $q_1 \equiv 0 \bmod X$.

Thus we have shown that for generic X, there exists nonzero $q = XA \in \langle q_1, \ldots, q_5 \rangle \subset R_G^2 = S^2$. We then conclude that $[\langle q_1, \ldots, q_5 \rangle : S^1] \neq \{0\}$. So after a change of basis of $\mathbb{C}^5$, we have $(q_1, \ldots, q_5) = (AA_1, \ldots, AA_4, q_5)$ for some $A \in S^1$ and independent $A_i \in S^1$. Then for $z \in Z$ we have $A_i z_j = A_j z_i$ in R_G^4 for $i, j \leq 4$, which by the symmetrizer lemma implies that there exists $q \in R_G^2$ such that $(z_1, \ldots, z_4) = q(A_1, \ldots, A_4)$. Since $\dim Z \geq 8$, q runs through a subspace Q of S^2 of dimension ≥ 8. Again we view z_5 as a function of $q \in Q$, satisfying $z_5(q')q = z_5(q)q'$ for any $q, q' \in Q$. Since Q generate R_G^5 by [11], we can apply the symmetrizer lemma to conclude that $z_5(q) = A_5 q$ for some $A_5 \in S^1$. But then the A_i for $i = 1, \ldots, 5$ are not independent in S^1, which contradicts (i). This concludes the proof of Lemma 4.4. $\square$

Proof of Proposition 7 By Lemmas 4.2 and 4.4 we have $\gamma \leq 7$. We also have $\delta \leq 1$. Indeed if $\delta \geq 2$, there exist independent elements

$$(q_1, \ldots, q_5), (q_1', \ldots, q_5') \in W_3.$$

Then by Lemma 4.3, there exist i, j such that $q_i q_j' - q_j q_i' \neq 0$ in R_G^4. But it annihilates $w_i, w_j \in R_G^5$ for any $(w_1, \ldots, w_5) \in W/W_1$. Hence by property (ii) and Lemma 2.3, we have $\alpha \leq 20$. From $\gamma \leq 7, \delta \leq 6$ we conclude that $\beta \geq 65 - 20 - 6 - 7 = 32$, which is absurd, since $\beta \leq 31 = \dim R_G^4$.

Hence we have $\alpha + \beta \geq 57$. This implies easily that for any nonzero $(w_1, \ldots, w_5) \in S^1(W_1/W_2)$, the w_i are independent in R_G^5, so that

$$\dim S^1(W_1/W_2) \leq \dim R_G^5 - 4 = 36.$$

Now if $\beta \geq 22$, one can adapt the proof of Lemma 1.2 to show that for generic $X, Y \in S^1/(X_4)$ the kernel of the map

$$(X + Y)\colon W_1/W_2 \times W_1/W_2 \to S^1(W_1/W_2)$$

is equal to $(-Y, X)[W_1/W_2 : (X, Y)]$. Since

$$\dim W_1/W_2 \geq 22, \quad \dim S^1(W_1/W_2) \leq 36,$$

this implies $\dim[W_1/W_2 : (X, Y)] \geq 8$. On the other hand, $[W_1/W_2 : (X, Y)] \subset (R_G^3)^5$ satisfies the assumptions of Lemma 4.4, because if

$$(t_1, \ldots, t_5), (t_1', \ldots, t_5') \in [W_1/W_2] : (X, Y),$$

then $t_i t_j' - t_j t_i'$ is annihilated by X^2, Y^2, XY, hence must be zero. Therefore $\dim[W_1/W_2 : (X, Y)] \geq 8$ is impossible, and we must have $\beta \leq 21$, hence $\alpha \geq 36$. We claim this is impossible: if $\alpha \geq 36$, the image of $\mathrm{pr}_1\colon W/W_1 \to R_G^5$ has codimension ≤ 4, hence $\dim S^1 \mathrm{pr}_1(W/W_1) \geq \dim R_G^6 - 3$ by [11]. It follows that there exists a nonzero $(z_1, \ldots, z_5) \in S^1(W/W_1)$ for which the z_i are not independent in R_G^6. We may assume that $z_5 = 0$, and then we have $z_i w_5 = z_5 w_i = 0$ in R_G^{11} for any $w \in W$. So the z_i for $i \leq 4$ are orthogonal to $S^1 \mathrm{pr}_5(W/W_1)$ in R_G^6 under the perfect pairing $R_G^6 \times R_G^6 \to R_G^{12} \cong \mathbb{C}$. Since $S^1 \mathrm{pr}_5(W/W_1)$ has codimension ≤ 3 in R_G^6, we conclude that the z_i for $i \leq 4$ are not independent in R_G^6. So we may assume that $z_4 = 0$. But then the z_i for $i \leq 3$ are orthogonal to $S^1(\mathrm{pr}_4(W/W_1) + \mathrm{pr}_4(W/W_1))$ which has codimension ≤ 2 in R_G^6, since by property (ii) and $\alpha \geq 36$, $\mathrm{pr}_4(W/W_1) + \mathrm{pr}_4(W/W_1)$ has codimension ≤ 3 in R_G^5. So the z_i for $i \leq 3$ are not independent in R_G^6. Continuing in this way, we show finally that $(z_1, \ldots, z_5) = 0$, which is a contradiction. $\square$

References

[1] P. Candelas, X.C. de la Ossa, P.S. Green, L. Parkes: A pair of Calabi-Yau manifolds as an exactly soluble superconformal field theory, Nucl. Phys. **B359** (1991), 21–74

[2] J. Carlson, P. Griffiths: Infinitesimal variations of Hodge structures and the global Torelli problem, in *Géométrie algébrique*, Angers 1979, A. Beauville Ed., Sijthoff and Noordhoff (1980), 51–76

[3] D. Cox, M. Green: Polynomial structures and generic Torelli for projective hypersurfaces, Comp. Math. **73** (1990), 121–124

[4] R. Donagi: Generic Torelli theorem for projective hypersurfaces, Comp. Math. **50** (1983), 325–353

[5] R. Donagi: Generic Torelli and variational Schottky, in Topics in transcendental algebraic geometry, Ann. of Math. Studies **106**, Princeton 1984, 239–258

[6] R. Donagi, M. Green: A new proof of the symmetrizer lemma and a stronger weak Torelli theorem, J. Diff. Geom. **20** (1984), 459–461

[7] R. Friedman: On threefolds with trivial canonical bundle, in *Complex geometry and Lie theory* (Sundance UT, 1989), Proc. Symp. Pure Math. **53**, 103–134

[8] A. B. Givental: Equivariant Gromov–Witten invariants, Internat. Math. Res. Notices **13** (1996), 613–663

[9] M. Green: Koszul cohomology and the geometry of projective varieties, J. Diff. Geom. **20** (1984), 279–289

[10] M. Green: The period map for hypersurface sections of high degree of an arbitrary variety, Comp. Math. **55** (1984), 135–156

[11] M. Green: Restrictions of linear series to hyperplanes, and some results of Macaulay and Golzmann, in *Algebraic curves and projective geometry*, Trento 1988, E. Ballico and C. Ciliberto Eds., Lecture Notes in Mathematics **1389**, 76–86

[12] P. Griffiths: On the periods of certain rational integrals. I and II, Ann. of Math. **90** (1969), 460–541

[13] P. Griffiths: Periods of integrals on algebraic manifolds. I and II, Amer. J. Math. **90** (1968), 568–626 and 805–865

[14] M. Gross and P. M. H. Wilson: Mirror symmetry via 3 tori for a class of Calabi–Yau threefolds, Math. Ann. **309** (1997), 505–531.

[15] D. Morrison: The geometry underlying mirror symmetry, this volume, pp. 283–310

[16] C. Peters, J. Steenbrink: Infinitesimal variations of Hodge structure and the generic Torelli problem for projective hypersurfaces, in *Classification of algebraic and analytic manifolds*, K. Ueno Ed., Progress in Math. **39** Birkhäuser (1983), pp 399–463.

[17] A. Strominger, S. T. Yau and E. Zaslow: Mirror symmetry is T-duality, Nucl. Phys. B **479** (1996), 243–259.

[18] C. Voisin: Variations of Hodge structure of Calabi–Yau threefolds, "Lezioni Lagrange", Roma 1996, Publications of the Scuola Normale Superiore di Pisa.

[19] C. Voisin: Miroirs et involutions sur les surfaces K3, in *Journées de géométrie algébrique d'Orsay, juillet 1992*, A. Beauville, O. Debarre and Y. Laszlo Eds., Astérisque **218** (1993),

[20] IVHS: Infinitesimal variations of Hodge structure, Comp. Math. **50** (1983):

I. J. Carlson, M. Green, P. Griffiths, J. Harris: Infinitesimal variations of Hodge structure, 109–205

II. P. Griffiths, J. Harris: Infinitesimal invariants of Hodge classes, 207–265

III. P. Griffiths: Determinantal varieties and the infinitesimal invariant of normal functions, 267–324

Claire Voisin,
Institut de Mathématiques (UMR 7586),
Case 247, 4, place Jussieu, 75252 Paris Cedex 05
e-mail: voisin@math.jussieu.fr

Flops, Type III contractions and Gromov–Witten invariants on Calabi–Yau threefolds

P.M.H. Wilson

1 Introduction

In this paper, we investigate Gromov–Witten invariants associated to exceptional classes for primitive birational contractions on a Calabi–Yau threefold X. As already remarked in [18], these invariants are locally defined, in that they can be calculated from knowledge of an open neighbourhood of the exceptional locus of the contraction; intuitively, they are the numbers of rational curves in such a neighbourhood. In §2, we make this explicit in the case of Type I contractions, where the exceptional locus is by definition a finite set of rational curves. Associated to the contraction, we have a flop; we deduce furthermore in Proposition 2.1 that the changes to the basic invariants (the cubic form on $H^2(X, \mathbb{Z})$ given by cup product, and the linear form given by cup product with the second Chern class c_2) under the flop are explicitly determined by the Gromov–Witten invariants associated to the exceptional classes.

The main results of this paper concern the Gromov–Witten invariants associated to classes of curves contracted under a Type III primitive contraction. Recall [17] that a primitive contraction $\varphi \colon X \to \overline{X}$ is of Type III if it contracts down an irreducible divisor E to a curve of singularities C. For X a smooth Calabi–Yau threefold, such contractions were studied in [18]; in particular, it was shown there that the curve C is smooth and that E is a conic bundle over C. We denote by $2\eta \in H_2(X, \mathbb{Z})/\text{Tors}$ the numerical class of a fibre of E over C. In the case when E is a $\mathbb{P}^1$-bundle over C, this may in fact be a primitive class, and so the notation is at slight variance with that adopted in §2, where η is assumed to be the primitive class. In the case when the class of a fibre is not primitive (for instance, when E is not a $\mathbb{P}^1$-bundle over C), the primitive class contracted by φ will be η. We denote the Gromov–Witten numbers associated to η and 2η by n_1 and n_2, with the convention that $n_1 = 0$ if 2η is the primitive class. The above conventions

465

have been adopted so as to achieve consistency of notation for all Type III contractions.

If the genus g of the curve C is strictly positive, under a general holomorphic deformation of the complex structure on X, the divisor E disappears leaving only finitely many of its fibres, and (except in the case of elliptic quasiruled surfaces, where all the Gromov–Witten invariants vanish) we have a Type I contraction. The results of §2 may then be applied to deduce the Gromov–Witten invariants associated to the classes $m\eta$ for $m > 0$. These are all determined by the Gromov–Witten numbers n_1 and n_2, and explicit formulas for n_1 and n_2 are given in Proposition 3.3; in particular $n_2 = 2g - 2$.

The formulas for n_1 and n_2 remain valid also for $g = 0$, although the slick proof given in Proposition 3.3 for the case $g > 0$ no longer works. The formula for n_1 is proved for all values of $g(C)$ by local deformation arguments in Theorem 3.5. Verifying that $n_2 = -2$ in the case when $g(C) = 0$ is rather more difficult, and involves the technical machinery of moduli spaces of stable pseudoholomorphic maps and the virtual neighbourhood method, as used in [2, 9] in order to construct Gromov–Witten invariants for general symplectic manifolds. In particular, we shall need a cobordism result from [13], which we show in Theorem 4.1 applies directly in the case where no singular fibre of E is a double line. The general case may be reduced to this one by making a suitable almost complex small deformation of complex structure. In §5, we give an application of our calculations. In [18], it was shown that if X_1, X_2 are Calabi–Yau threefolds which are symplectic deformations of each other (and general in their complex moduli), then their Kähler cones are the same. Now we can deduce (Corollary 5.1) that corresponding codimension one faces of these cones have the same contraction type.

The author thanks Yongbin Ruan for the benefit of conversations concerning material in §4 and his preprint [13].

2   Flops and Gromov–Witten invariants

If X is a smooth Calabi–Yau threefold with Kähler cone $\mathcal{K}$, then the nef cone $\overline{\mathcal{K}}$ is locally rational polyhedral away from the cubic cone

$$W^* = \{ D \in H^2(X, \mathbb{R}) \; ; \; D^3 = 0 \};$$

moreover, the codimension one faces of $\overline{\mathcal{K}}$ (not contained in W^*) correspond to primitive birational contractions $\varphi \colon X \to \overline{X}$ of one of three different types [17].

In the numbering of [17], Type I contractions are those where only a finite number of curves (in fact $\mathbb{P}^1$s) are contracted. The singular threefold $\overline{X}$ then has a finite number of cDV (compound Du Val) singularities. Whenever one has such a small contraction on X, there is a flop of X to a different birational model X', also admitting a birational contraction to $\overline{X}$; moreover, identifying $H^2(X', \mathbb{R})$ with $H^2(X, \mathbb{R})$, the nef cone of X' intersects the nef cone of X along the codimension one face which defines the contraction to $\overline{X}$ [6, 7]. It is well known [7] that X' is smooth, projective and has the same Hodge numbers as X, but that the finer invariants, such as the cubic form on $H^2(X, \mathbb{Z})$ given by cup product, and the linear form on $H^2(X, \mathbb{Z})$ given by cup product with $c_2(X) = p_1(X)$, will in general change. Recall that, when X is simply connected, these two forms along with $H^3(X, \mathbb{Z})$ determine the diffeomorphism class of X up to finitely many possibilities [14], and that if furthermore $H_2(X, \mathbb{Z})$ is torsion free, this information determines the diffeomorphism class precisely [16].

When the contraction $\varphi \colon X \to \overline{X}$, corresponding to such a *flopping face* of $\overline{\mathcal{K}}$, contracts only isolated $\mathbb{P}^1$s with normal bundle $(-1, -1)$ (that is, $\overline{X}$ has only simple nodes as singularities), then it is a standard calculation to see how the above cubic and linear forms (namely the cup product $\mu \colon H^2(X, \mathbb{Z}) \to \mathbb{Z}$, and the form $c_2 \colon H^2(X, \mathbb{Z}) \to \mathbb{Z}$) change on passing to X' under the flop. Since any flop is an isomorphism in codimension one, we have natural identifications

$$H^2(X', \mathbb{R}) \cong \operatorname{Pic}_{\mathbb{R}}(X') \cong \operatorname{Pic}_{\mathbb{R}}(X) \cong H^2(X, \mathbb{R}).$$

If we are in the case where the exceptional curves $C_1, \ldots, C_N$ are isolated $\mathbb{P}^1$s with normal bundle $(-1, -1)$, and if we denote by D' the divisor on X' corresponding to D on X, then

$$(D')^3 = D^3 - \sum (D \cdot C_i)^3 \quad \text{and} \quad c_2(X') \cdot D' = c_2(X) \cdot D + 2 \sum D \cdot C_i \,.$$

This is an easy verification – see for instance [1].

Proposition 2.1 *Suppose that X is a smooth Calabi–Yau threefold, and $\varphi \colon X \to \overline{X}$ is any Type I contraction, with X' denoting the flopped Calabi–Yau threefold. The cubic and linear forms $(D')^3$ and $D' \cdot c_2(X')$ on X' are then explicitly determined by the cubic and linear forms D^3 and $D \cdot c_2(X)$ on X, and the 3-point Gromov–Witten invariants Φ_A on X, for $A \in H_2(X, \mathbb{Z})$ ranging over classes which vanish on the flopping face.*

Remark 2.2 This is essentially the statement from physics that the A-model 3-point correlation function on $\mathcal{K}(X)$ may be analytically continued to give the A-model 3-point correlation function on $\mathcal{K}(X')$.

Proof We use the ideas from [18]; in particular, we know that on a suitable open neighbourhood of the exceptional locus of φ, there exists a small holomorphic deformation of the complex structure for which the exceptional locus splits up into disjoint $(-1, -1)$-curves ([18], Proposition 1.1).

Let $A \in H_2(X, \mathbb{Z})$ be a class with $\varphi_* A = 0$. The argument from [18], Section 1 then shows how the Gromov–Witten invariants $\Phi_A(D, D, D)$ can be calculated from local information. Having fixed a Kähler form ω on X, a small deformation of the holomorphic structure on a neighbourhood of the exceptional locus may be patched together in a C^∞ way with the original complex structure to yield an almost complex structure tamed by ω, and the Gromov–Witten invariants can then be calculated in this almost complex structure. The Gromov Compactness Theorem is used in this argument to justify the fact that all of the pseudoholomorphic rational curves representing the class A have images which are $(-1, -1)$-curves in the deformed local holomorphic structure.

Here we also implicitly use the Aspinwall–Morrison formula for the contribution to Gromov–Witten invariants from multiple covers of infinitesimally rigid $\mathbb{P}^1$s, now proved mathematically by Voisin [15]. So if $n(B)$ denotes the number of $(-1, -1)$-curves representing a class given B, then

$$\Phi_A(D, D, D) = (D \cdot A)^3 \sum_{kB=A} n(B)/k^3,$$

where the sum is taken over all integers $k > 0$ and classes $B \in H_2(X, \mathbb{Z})$ such that $kB = A$. So if $H_2(X, \mathbb{Z})$ is torsion free and A is the primitive class vanishing on the flopping face, this says that

$$\Phi_{mA}(D, D, D) = (D \cdot A)^3 \sum_{d|m} n(dA)d^3.$$

Recall that the Gromov–Witten invariants used here are the ones (denoted $\tilde{\Phi}$ in [12]) which count marked parametrized curves satisfying a perturbed pseudoholomorphicity condition. Knowledge of the numbers $n(A)$ for the classes A with $\varphi_* A = 0$ determines the Gromov–Witten invariants Φ_A for classes A with $\varphi_* A = 0$, and vice-versa.

If we can now show that the local contributions to $(D')^3$ and $D' \cdot c_2(X')$ are well-defined and invariant under the holomorphic deformations of complex structure we have made locally, then the obvious formulas for them will hold. Let $\eta \in H_2(X, \mathbb{Z})/\text{Tors}$ be the primitive class with $\varphi_* \eta = 0$ and n_d denote the total number of $(-1, -1)$-curves on the deformation which have numerical class $d\eta$; the n_d are therefore nonnegative integers (cf. [10], Remark 7.3.6). Then

$$(D')^3 = D^3 - (D \cdot \eta)^3 \sum_{d>0} n_d d^3, \tag{2.1.1}$$

$$D' \cdot c_2(X') = D \cdot c_2(X) + 2(D \cdot \eta) \sum_{d>0} n_d d. \tag{2.1.2}$$

To justify the premise in the first sentence of the paragraph, the basic result needed is that of local conservation of number, as stated in [3], Theorem 10.2.

For calculating the change in D^3 for instance, let X now denote the neighbourhood of the exceptional locus of φ and $\pi \colon \mathcal{X} \to B$ the small deformation under which the exceptional locus splits up into $(-1, -1)$-curves. So we have a regular embedding (of codimension six)

$$
\begin{array}{ccc}
\mathcal{X} & \hookrightarrow & \mathcal{X} \times \mathcal{X} \times \mathcal{X} = \mathcal{Y} \\
\downarrow & & \downarrow \\
B & = & B
\end{array}
$$

In order to calculate the triple products $D_1' \cdot D_2' \cdot D_3'$ from $D_1 \cdot D_2 \cdot D_3$ and the numbers n_d, we may assume *wlog* that the D_i are very ample, and so in particular we get effective divisors $\mathcal{D}_1$, $\mathcal{D}_2$ and $\mathcal{D}_3$ on $\mathcal{X}/B$. Applying [8], Theorem 11.10, we can flop in the family $\mathcal{X} \to B$, hence obtaining a deformation $\mathcal{X}' \to B$ of the flopped neighbourhood X'. We wish to calculate the local contribution to $D_1' \cdot D_2' \cdot D_3'$; with the notation as in [3], Theorem 10.2, we have a fibre square

$$
\begin{array}{ccc}
\mathcal{W} & \longrightarrow & \mathcal{D}_1' \times \mathcal{D}_2' \times \mathcal{D}_3' \\
\downarrow & & \downarrow \\
\mathcal{X}' & \longrightarrow & \mathcal{X}' \times \mathcal{X}' \times \mathcal{X}'
\end{array}
$$

with $\mathrm{Supp}(\mathcal{W}) = \bigcap \mathrm{Supp}(\mathcal{D}_i')$. Furthermore, we may assume that the divisors $\mathcal{D}_i$ were chosen so that $\mathcal{D}_1 \cap \mathcal{D}_2 \cap \mathcal{D}_3$ has no points in $\mathcal{X}$, and so in particular $\mathcal{W}$ is proper over B. Letting $D_i'(t)$ denote the restriction of $\mathcal{D}_i'$ to the fibre X_t', we therefore have a well-defined local contribution to $D_1'(t) \cdot D_2'(t) \cdot D_3'(t)$ (concentrated on the flopping locus of X_t'), which is moreover independent of $t \in B$. Thus by making the local calculation as in (7.4) of [1], we deduce that

$$
D_1 \cdot D_2 \cdot D_3 - D_1' \cdot D_2' \cdot D_3' = (D_1 \cdot \eta)(D_2 \cdot \eta)(D_3 \cdot \eta) \sum_{d>0} n_d d^3
$$

as required.

The proof for $c_2 \cdot D$ is similar. Here we consider the graph $\widetilde{X} \subset X \times X'$ of the flop, with $\pi_1 \colon \widetilde{X} \to X$ and $\pi_2 \colon \widetilde{X} \to X'$ denoting the two projections, and $E \subset \widetilde{X}$ the exceptional divisor for both π_1 and π_2. Then $\pi_2^*(T_{X'})|_{\widetilde{X} \backslash E} = \pi_1^*(T_X)|_{\widetilde{X} \backslash E}$, and so in particular $c_2(\pi_2^* T_{X'}) - c_2(\pi_1^* T_X)$ is represented by a 1-cycle Z on E. Suppose *wlog* that D is very ample, and that D' denotes the corresponding divisor on X'. Set $\pi_1^* D = \widetilde{D}$ and $\pi_2^* D' = \widetilde{D} + F$, with F supported on E. Then $c_2(X') \cdot D' = c_2(\pi_2^* T_{X'}) \cdot (\widetilde{D} + F)$. Hence

$$
c_2(X') \cdot D' - c_2(X) \cdot D = c_2(\pi_2^* T_{X'}) \cdot F + Z \cdot \widetilde{D} = c_2((\pi_2^* T_{X'})|_F) + (Z \cdot \widetilde{D})_E
$$

where the right-hand side is purely local. Note the slight abuse of notation here that F denotes also the fixed *scheme* for the linear system $|\pi_2^* D'|$.

Now taking X to be a local neighbourhood of the flopping locus, and taking a small deformation $\mathcal{X} \to B$ as before, we obtain families $\mathcal{X}'$, $\widetilde{\mathcal{X}}$, $\mathcal{D}$, $\mathcal{E}$, $\mathcal{F}$ and $\mathcal{Z}$ over B (corresponding to X', $\widetilde{X}$, D, E, F and Z). For ease of notation, we shall use π_1 and π_2 also for the morphisms of families $\widetilde{\mathcal{X}} \to \mathcal{X}$, respectively $\widetilde{\mathcal{X}} \to \mathcal{X}'$. Applying [3], Theorem 10.2 to the family of vector bundles $(\pi_2^* T_{\mathcal{X}'/B})|_{\mathcal{F}}$ on the scheme $\mathcal{F}$ over B yields that $c_2((\pi_2^* T_{\mathcal{X}'/B})|_{F_t})$ is independent of $t \in B$. Noting that $\widetilde{\mathcal{D}} \hookrightarrow \widetilde{\mathcal{X}}$ is a regular embedding, we apply the same theorem to the fibre square

$$
\begin{array}{ccc}
\widetilde{\mathcal{D}} \times_{\widetilde{\mathcal{X}}} \mathcal{E} & \longrightarrow & \mathcal{E} \\
\downarrow & & \downarrow \\
\widetilde{\mathcal{D}} & \longrightarrow & \widetilde{\mathcal{X}}
\end{array}
$$

and the cycle $\mathcal{Z}$ on $\mathcal{E}$. This yields that $(Z_t \cdot \widetilde{D}_t)_{E_t}$ on E_t is independent of $t \in B$, where by definition

$$
Z_t = c_2(\pi_2^* T_{\mathcal{X}'/B})|_{X_t} - c_2(\pi_1^* T_{\mathcal{X}/B})|_{X_t}.
$$

Thus the local contribution to $D'(t) \cdot c_2(X_t')$ is well-defined and independent of t, and so we need only make the local calculation for generic t (where the exceptional locus of the flop consists of disjoint $(-1, -1)$-curves). This calculation may be found in [1], (7.4).

Speculation 2.3 There are reasons for believing that only the numbers n_1 and n_2 are nonzero, and hence that the Gromov–Witten invariants associated to classes $m\eta$ for $m > 2$ all arise from multiple covers. If this speculation is true, then the changes under flopping to the cubic form and the linear form would be determined by these two integers, and conversely.

3   Type III contractions and Gromov–Witten invariants

The main results of this paper concern the Gromov–Witten invariants associated to classes of curves contracted under a Type III primitive contraction. Recall [17] that a primitive contraction $\varphi \colon X \to \overline{X}$ is of Type III if it contracts down an irreducible divisor E to a curve of singularities C. For X a smooth Calabi–Yau threefold, such contractions were studied in [18]; in particular, it was shown there that the curve C is smooth and that E is a conic bundle over C. We denote by $2\eta \in H_2(X, \mathbb{Z})/\operatorname{Tors}$ the numerical class of a fibre of E over C. As explained in the Introduction, we denote by n_1 and n_2 the Gromov–Witten numbers associated to the classes η and 2η, where $n_1 = 0$ if E is a $\mathbb{P}^1$-bundle over C. If the generic fibre of E over C is reducible (consisting of two lines, each with class η), then, except in two cases, it follows from the arguments of [18], §4 that, by making a global holomorphic deformation of the complex structure, we may reduce down to the case where the generic fibre of E over C is irreducible. The two exceptional cases are:

(a) $g(C) = 1$ and E has no double fibres.

(b) $g(C) = 0$ and E has two double fibres.

However, Case (a) is an *elliptic quasi-ruled* surface in the terminology of [18], and hence disappears completely under a generic global holomorphic deformation. In particular, we know that all the Gromov–Witten invariants Φ_A are zero, for $A \in H_2(X, \mathbb{Z})$ having numerical class $m\eta$ for any $m > 0$.

In Case (b), E is a nonnormal generalized del Pezzo surface $\overline{\mathbb{F}}_{3;2}$ of degree 7 (see [18]). As argued there however, we may make a holomorphic deformation in a neighbourhood of E so that E deforms to a *smooth* del Pezzo surface of degree 7, and where the class η is then represented by either of two 'lines' on the del Pezzo surface (which are $(-1, -1)$-curves on the threefold); hence $n_1 = 2$. In fact, the smooth del Pezzo surface is fibred over $\mathbb{P}^1$ with one singular (line pair) fibre. The arguments we give below may be applied locally (more precisely with the global almost complex structure obtained by suitably patching the local small holomorphic deformation on an open neighbourhood of E with the original complex structure), and the Gromov–Witten invariants may be calculated as if the original contraction φ had contracted such a smooth del Pezzo surface of degree 7. In particular, $n_1 = 2$ comes from the two components of the singular fibre (Theorem 3.5), and $n_2 = -2$ is proved in §4 (see also Remark 3.4).

Let us therefore assume that the generic fibre of E over C is irreducible, and so in particular $E \to C$ is obtained from a $\mathbb{P}^1$-bundle over C by means of blowups and blowdowns. Moreover E itself is a conic bundle over C, and so its singular fibres are either line pairs or double lines.

Lemma 3.1 *In the above notation, E has only singularities on the singular fibres of the map $E \to C$. When the singular fibre is a line pair, we have an A_n singularity at the point where the two components meet (we include here the possibility $n = 0$ when the point is a smooth point of E). When the singular fibre is a double line, we have a D_n singularity on the fibre (here we need to include the case $n = 2$, where we in fact have two A_1 singularities, and $n = 3$, where we have an A_3 singularity).*

Proof The proof is obvious, once the correct statement has been found. The statement of this result in [17] omits (for fibre a double line) the cases D_n for $n > 2$.

Lemma 3.2 *Suppose that $E \to C$ as above has a_r fibres which are line pairs with an A_r singularity and b_s fibres which are double lines with a D_s singularity (for $r \geq 0$ and $s \geq 2$), then*

$$K_E^2 = 8(1 - g) - \sum_{r \geq 0} a_r(r + 1) - \sum_{s \geq 2} b_s s,$$

where g denotes the genus of C.

This enables us to give a slick calculation of the Gromov–Witten invariants when the base curve has genus $g > 0$. In this case, it was shown in [17] that for a generic deformation of X, only finitely many fibres from E deform, and hence the Type III contraction deforms to a Type I contraction. Thus Gromov–Witten numbers n_1 and n_2 may be defined as in Section 1, and are nonnegative integers.

Proposition 3.3 *When $g > 0$, we have*

$$n_1 = 2 \sum_{r \geq 0} a_r(r + 1) + 2 \sum_{s \geq 2} b_s s \quad and \quad n_2 = 2g - 2.$$

Proof We take a generic 1-parameter deformation of X, for which the Type III contraction deforms to a Type I contraction. We therefore have a diagram

$$
\begin{array}{ccc}
\mathcal{X} & \longrightarrow & \overline{\mathcal{X}} \\
\downarrow & & \downarrow \\
\Delta & = & \Delta
\end{array}
$$

where $\Delta \subset \mathbb{C}$ denotes a small disc. Since the singular locus of $\overline{\mathcal{X}}$ consists only of curves of cDV singularities, we may again apply [8], Theorem 11.10 to deduce the existence of a (smooth) flopped fourfold $\mathcal{X}' \to \overline{\mathcal{X}}$. The induced family $\mathcal{X}' \to \Delta$ is given generically by flopping the fibres, and at $t = 0$ it is easily checked that $X_0' \cong X_0$; this operation is often called an *elementary transformation* on the family. Identifying the groups $H^2(X_t, \mathbb{Z}) \cong H^2(X_t', \mathbb{Z})$ as before, this has the effect (at $t = 0$) of sending E to $-E$ (cf. the discussion in [5], §3.3). So if E' denotes the class in $H^2(X_t', \mathbb{Z})$ corresponding to the class E in $H^2(X_t, \mathbb{Z})$, we have $(E')^3 = -E^3$. For $t \neq 0$, we just have a flop, and so

$(E')^3$ can be calculated from equation (2.1.1), namely $(E')^3 = E^3 + n_1 + 8n_2$. Therefore, using Lemma 3.2

$$n_1 + 8n_2 = -2E^3 = 16(g-1) + 2\sum_{r\geq 0} a_r(r+1) + 2\sum_{s\geq 2} b_s s.$$

Similarly, we have $c_2(X') \cdot E' = -c_2(X) \cdot E$, and so from equation (2.1.2) it follows that $2n_1 + 4n_2 = 2c_2 \cdot E$. An easy calculation of the right-hand side then provides the second equation

$$2n_1 + 4n_2 = 8(g-1) + 4\sum_{r\geq 0} a_r(r+1) + 4\sum_{s\geq 2} b_s s.$$

Solving for n_1 and n_2 from these two equations gives the desired result.

Remark 3.4 This result remains true even when $g = 0$, although the slick proof given above is no longer valid. The formula for n_1 is checked in Theorem 3.5 by local deformation arguments (for which the genus g is irrelevant), showing that the contribution to n_1 from a line pair fibre with A_r singularity is $2(r+1)$, and from a double line fibre with D_s singularity is $2s$. Let $A \in H_2(X, \mathbb{Z})$ denote the class of a fibre of $E \to C$. Observe that any pseudo-holomorphic curve representing the numerical class η will be a component of a singular fibre of $E \to C$. Moreover, the components l of a singular fibre represent the same class in $H_2(X, \mathbb{Z})$, and so in particular twice this class is A. Thus the Aspinwall–Morrison formula (as proved in [15]) yields the contribution to the Gromov–Witten invariants $\Phi_A(D, D, D)$ from double covers, purely in terms of n_1 and $D \cdot A$. The difference may be regarded as the contribution to $\Phi_A(D, D, D)$ from simple maps, and taking this to be $n_2(D \cdot A)^3$ determines the number n_2 (in §4, we shall see how n_2 may be determined directly from the moduli space of simple stable holomorphic maps). If $g > 0$, the above argument shows that this is in agreement with our previous definition, and yields moreover the equality $n_2 = 2g - 2$. The fact that $n_2 = -2$ when $g = 0$ requires a rather more subtle argument involving technical machinery – see Theorem 4.1. I remark that the value $n_2 = -2$ is needed in physics, and that there is also a physics argument justifying it (see [4], §5.2 and [5], §3.3) – essentially, it comes down to a statement about the A-model 3-point correlation functions. In §4 below, we give a rigorous mathematical proof of the assertion.

Theorem 3.5 *The formula for n_1 in Proposition 3.3 is valid irrespective of the value of the genus $g = g(C)$.*

Proof By making a holomorphic deformation of the complex structure on an open neighbourhood U in X of the singular fibre Z of $E \to C$, we may calculate the contribution to n_1 from that singular fibre – see [18], (4.1). The deformation of complex structure is obtained as in [18] by considering the one dimensional family of Du Val singularities in $\overline{X}$, and deforming this family locally in a suitable neighbourhood $\overline{U}$ of the dissident point. Our assumption is that the family $\overline{U} \to \Delta$ has just an A_1 singularity on $\overline{U}_t$ for $t \neq 0$, and we may assume also that $\overline{U} \to \Delta$ is a good representative (in the sense explained in [18]). The open neighbourhood U is then the blowup of $\overline{U}$ in the smooth curve of Du Val singularities ([18], p. 569). The contribution to n_1 may be calculated locally, and will not change when we make small holomorphic deformations of the complex structure on U, which in turn corresponds to making small deformations to the family $\overline{U} \to \Delta$.

First we consider the case where the singular fibre Z is a line pair – from this, it will follow that the dissident singularity on $\overline{U}$ is a cA_n singularity with $n > 1$, and that $\overline{U}$ has a local analytic equation of the form

$$x^2 + y^2 + z^{n+1} + tg(x, y, z, t) = 0$$

in $\mathbb{C}^3 \times \Delta$ (here t is a local coordinate on Δ, and $x = y = z = 0$ the curve C of singularities). For $t \neq 0$, we have an A_1 surface singularity, which implies that g must contain a term of the form $t^r z^2$ for some $r \geq 0$. By an appropriate analytic change of coordinates, we may then assume that $\overline{U}$ has a local analytic equation of the form

$$x^2 + y^2 + z^{n+1} + t^{r+1} z^2 + th(x, y, z, t) = 0,$$

where h consists of terms which are at least cubic in x, y, z. By making a small deformation of the family $\overline{U} \to \Delta$, we may reduce to the case $n = 2$, that is, $\overline{U}$ having local equation $x^2 + y^2 + z^3 + t^{r+1} z^2 + th = 0$. At this stage, we could in fact also drop the term th (an easy check using the versal deformation family of an A_2 singularity), but this will not be needed.

We now make a further small deformation to get $\overline{U}_\varepsilon \subset \mathbb{C}^3 \times \Delta$ given by a polynomial

$$x^2 + y^2 + z^3 + t^{r+1} z^2 + \varepsilon z^2 + th = x^2 + y^2 + z^2(z + t^{r+1} + \varepsilon) + th.$$

This then has $r + 1$ values of t for which the singularity is an A_2 singularity – for other values of t, it is an A_1 singularity. If we blow up the singular locus of $\overline{U}_\varepsilon$, we therefore obtain a smooth exceptional divisor for which $r + 1$ of the fibres over Δ are line pairs. By the argument of [18], (4.1), this splitting of the singular fibre into $r + 1$ line pair singular fibres of the simplest type can be achieved by a local holomorphic deformation on a suitable open neighbourhood of the fibre in the original threefold X.

It is however clear that a line pair coming from a dissident cA_2 singularity of the above type contributes precisely two to the Gromov–Witten number n_1 – one for each line in the fibre. In terms of equations, we have a local equation for $\overline{X}$ of the form $x^2 + y^2 + z^3 + wz^2 = 0$; deforming this to say $x^2 + y^2 + z^3 + wz^2 + \varepsilon w = 0$, we get two simple nodes, and hence two disjoint $(-1, -1)$-curves on the resolution.

The argument of [18], (4.1) shows that the Gromov–Witten number n_1 may be calculated purely from these local contributions, and so the total contribution to n_1 from the line pair singular fibre of E with A_r singularity is indeed $2(r + 1)$, as claimed.

For the case of the singular fibre Z of E being a double line, the dissident singularity must be cE_6, cE_7, cE_8, or cD_n for $n \geq 4$. Thus $\overline{U}$ has a local analytic equation of the form $f(x, y, z) + tg(x, y, z, t)$ in $\mathbb{C}^3 \times \Delta$ for f a polynomial of the appropriate type (t a local coordinate on Δ, and $x = y = z = 0$ the curve of singularities). To simplify matters, we may deform f to a polynomial defining a D_4 singularity, and hence make a small deformation of the family to one in which the dissident singularity is of type cD_4. We then have a local analytic equation of the form

$$x^2 + y^2 z + z^3 + tg(x, y, z, t) = 0.$$

For $t \neq 0$, we have an A_1 singularity, and so the terms of g must be at least quadratic in x, y, z. Moreover, by changing the x-coordinate, we may take the equation to be of the form

$$x^2 + y^2 z + z^3 + t^a y^2 + t^b yz + t^c z^2 + th(x, y, z, t) = 0,$$

with a, b, c positive, and where the terms of h are at least cubic in x, y, z. The fact that the blowup U of $\overline{U}$ in C is smooth is easily checked to imply that $a = 1$. Since

$$ty^2 + 2t^b yz = t(y + t^{b-1} z)^2 - t^{2b-1} z^2,$$

we have an obvious change of y-coordinate which brings the equation into the form

$$x^2 + y^2 z + z^3 + ty^2 + t^r z^2 + th_1(x, y, z, t) = 0,$$

where $r = \min\{c, 2b - 1\}$ and h_1 has the same property as h.

When we blow up $\overline{U}$ along the curve $x = y = z = 0$, we obtain an exceptional locus E with a double fibre over $t = 0$, on which we have a D_{r+1} singularity (including the case $r = 1$ of two A_1 singularities, and $r = 2$ of an A_3 singularity). Moreover, this was also true of our original family, since the small deformation of f we made did not affect the local equation of the exceptional locus.

Moreover, by adding a term $\varepsilon_1 y^2 + \varepsilon_2 z^2$, we may deform our previous equation to one of the form

$$x^2 + y^2(z + t + \varepsilon_1) + z^2(z + t^r + \varepsilon_2) + th_1(x, y, z, t) = 0.$$

When $t + \varepsilon_1 = 0$, we have an A_3 singularity, and when $t^r + \varepsilon_2 = 0$, an A_2 singularity. Moreover, when we blow up the singular locus of this deformed family, the resulting exceptional divisor is smooth and has line pair fibres for these $r + 1$ values of t. Thus, as seen above, the contribution to n_1 from the original singular fibre (a double line with a D_{r+1} singularity) is $2(r + 1)$ as claimed.

4   Calculation of n_2 for Type III contractions

Let $\varphi \colon X \to \overline{X}$ be a Type III contraction on a Calabi–Yau threefold X, which contracts a divisor E to a (smooth) curve C of genus g. When $g > 0$, it was proved in Proposition 3.3 that the Gromov–Witten number n_2 (defined for arbitrary genus via Remark 3.4) is $2g - 2$. The purpose of this Section is to extend this result to include the case $g = 0$ (C is isomorphic to $\mathbb{P}^1$), and to prove $n_2 = 2g - 2$ in general.

Arguing as in [18], it is clear that the desired result is a local one, depending only on a neighbourhood of the exceptional divisor E. As remarked in §3, we may then always reduce down to the case that the generic fibre of $E \to C$ is irreducible. If all the fibres of $E \to C$ are smooth (so E is a $\mathbb{P}^1$-bundle over C), the fact that $n_2 = 2g - 2$ was proved in Proposition 5.7 of [11], using a cobordism argument. This latter result was extended by Ruan in [13], Proposition 2.10, using the theory of moduli spaces of stable maps and the virtual neighbourhood technique (cf. [2, 9]). If the singular fibres of $E \to C$ are line pairs, Ruan's result applies directly. We prove below that the linearized Cauchy–Riemann operator has constant corank for the stable (unmarked) rational curves given by the fibres of E over C, and hence by Ruan's result that there is an obstruction bundle $\mathcal{H}$ on C, with n_2 determined by the Euler class of $\mathcal{H}$. By Dolbeault cohomology, there is a natural identification of $\mathcal{H}$ with the cotangent bundle T_C^* on C, and hence the formula for n_2 follows. We note however that for Ruan's result to hold, we do not need an integrable almost complex structure on X. Provided we have a natural identification between the cokernel of the linearized Cauchy–Riemann operator and the cotangent space at the corresponding point of C, we can still deduce that $n_2 = 2g - 2$. In the general case of a Type III contraction which has double fibres, we show below that we can make a small local deformation of the almost complex structure on X so that E deforms to a family of pseudoholomorphic rational curves over C with at worst line pair singular fibres, and for which Ruan's method applies.

Theorem 4.1 *For any Type III contraction* $\varphi\colon X \to \overline{X}$, *the Gromov–Witten number* $n_2 = 2g - 2$.

Proof We saw above that we may assume that the generic fibre of $E \to C$ is irreducible. Furthermore, we initially assume also that the singular fibres are all line pairs, and later reduce the general case to this one.

We let J denote the almost complex structure on X, which we know is integrable (at least in a neighbourhood of E), and tamed by a symplectic form ω. Let $A \in H_2(X, \mathbb{Z})$ be the class of a fibre of $E \to C$. Adopting the notation from [13], we consider the moduli space $\overline{\mathcal{M}}_A(X, J) = \overline{\mathcal{M}}_A(X, 0, 0, J)$ of stable unmarked rational holomorphic maps, a compactification of the space of (rigidified) pseudoholomorphic maps $\mathbb{CP}^1 \to X$, representing the class A. The theory of stable maps, as explained in Section 3 of [13], goes through for unmarked stable maps, by taking each component of the domain as a bubble component, and adding marked points (in addition to the double points) as in [13] in order to stabilize the components (thus taking a local slice of the automorphism group).

In the case that all the singular fibres of $E \to C$ are line pairs, $\overline{\mathcal{M}}_A(X, J)$ has two components, one corresponding to simple maps and the other to double covers. It is now a simple application of Gromov compactness to see that these two components are disjoint, since a sequence of double cover maps cannot converge to a simple map. A similar argument will show that for all almost complex structures J_t in some neighbourhood of $J = J_0$, the moduli space $\overline{\mathcal{M}}_A(X, J_t)$ will consist of two disjoint components, one corresponding to the simple maps and the other to the double covers.

Since any stable unmarked rational holomorphic map must be an embedding, it is clear that the component $\overline{\mathcal{M}}'_A(X, J)$ corresponding to the simple maps can be identified naturally with the smooth base curve C, and that for all almost complex structures in some neighbourhood of $J = J_0$, the moduli space $\overline{\mathcal{M}}'_A(X, J_t)$ of simple unmarked stable holomorphic maps is compact. The Gromov–Witten invariant n_2 that we seek can then be defined via Ruan's virtual neighbourhood invariant μ_S, and may be evaluated on (X, J) by using [13], Proposition 2.10.

Let us now go into more details of this. We consider C^∞ stable maps $f \in \overline{B}_A(X) = \overline{B}_A(X, 0, 0)$ in the sense of [13], Definition 3.1, where Ruan shows later in the same Section that the naturally stratified space $\overline{B}_A(X)$ satisfies a property which he calls *virtual neighbourhood technique admissible* or *VNA*, and as he says, for the purposes of the virtual neighbourhood construction, behaves as if it were a Banach V-manifold. Since any simple marked holomorphic stable map f in $\overline{\mathcal{M}}'_A(X, J)$ is forced to be an embedding, we may restrict our attention to C^∞ stable maps whose domain Σ comprises at most two $\mathbb{P}^1$s. We stratify $\overline{B}_A(X)$ according to the combinatorial type D of the domain Σ. Thus any $f \in \overline{\mathcal{M}}'_A(X, J)$ belongs to one of two strata of $\overline{B}_A(X)$.

In general, for k-pointed C^∞ stable maps of genus g, Ruan shows that for any given combinatorial type D, the substratum $B_D(X, g, k)$ is a Hausdorff Fréchet V-manifold ([13], Proposition 3.6). As mentioned above, he needs to add extra marked points in order to stabilize the nonstable components of the domain Σ, thus taking a local slice of the action of the automorphism group on the unstable marked components of Σ. Moreover, the tangent space $T_f B_D(X, g, k)$ is identified with $\Omega^0(f^*T_X)$, as defined in his equation [13], (3.29). The tangent space $T_f \overline{B}_A(X, g, k)$ can then be defined as $T_f B_D(X, g, k) \times \mathbb{C}_f$, where $\mathbb{C}_f$ is the space of gluing parameters (see [13], equation before (3.67)).

In our case, however, things are a bit simpler. Given $f \in \overline{\mathcal{M}}'_A(X, J)$ with domain Σ consisting of two $\mathbb{P}^1$s, the tangent space $T_f \overline{B}_A(X)$ is of the form $\Omega^0(f^*T_X) \times \mathbb{C}$, and we have a neighbourhood $\widetilde{U}_f$ of f in $\overline{B}_A(X)$ defined by [13], (3.43), consisting of stable maps $\overline{f}^{v,w}$ parametrized locally by

$$\{ w \in \Omega^0(f^*T_X) \; ; \; \|w\|_{C^1} < \varepsilon' \}$$

(corresponding to deformations within the stratum $B_D(X)$), and by $v \in \mathbb{C}_f^\varepsilon$ (an ε-ball in $\mathbb{C}_f = \mathbb{C}$ giving the gluing parameter at the double point). This then corresponds to the above decomposition of $T_f \overline{B}_A(X)$ into two factors. On the first factor, the linearization $D_f \overline{\partial}_J$ of the Cauchy–Riemann operator restricts to

$$\overline{\partial}_{J,f} \colon \Omega^0(f^*T_X) \to \Omega^{0,1}(f^*T_X)$$

in the notation of [13]. The index of this operator may be calculated using Riemann–Roch on each component of Σ (cf. the proof of Lemma 3.16 in [13], suitably modified to take account of the extra marked points), and is seen to be -2.

Let us now consider the stable maps $f^v = f^{v,0}$ for $v \in \mathbb{C}_f^\varepsilon \setminus \{0\}$. These are stable maps $\mathbb{CP}^1 \to X$ which differ from f only in small discs around the double point, and in this sense are approximately holomorphic. Set $v = re^{i\theta}$; then the gluing to get $f^v \colon \Sigma^v \to X$ is only performed in discs around the double point of radius $2r^2/\rho$ in the two components (ρ a suitable constant). It can then be checked for any $2 < p < 4$ that $\|\overline{\partial}_J(f^v)\|_{L^p_1} \le Cr^{4/p}$ (see [13] Lemma 3.23, and [10] Lemma A.4.3), from which it follows that the linearization

$$L_A = D_f \overline{\partial}_J$$

of the Cauchy–Riemann operator should be taken as zero on the factor $\mathbb{C}_f$ in $T_f \overline{B}_A(X)$. Thus we deduce that the index of L_A is zero, and that coker L_A is same as the cokernel of $\overline{\partial}_{J,f} \colon \Omega^0(f^*T_X) \to \Omega^{0,1}(f^*T_X)$, which by Dolbeault cohomology may be identified as

$$H^1(f^*T_X) = H^1(Z, T_{X|_Z}),$$

where Z is the fibre of $E \to C$ (over a point $x \in C$) corresponding to the image of f.

We note that these are exactly the same results as are obtained in the smooth case, when Σ consists of a single $\mathbb{P}^1$. Here, we need to add three marked points to stabilize Σ, and Riemann–Roch then gives immediately that the index of L_A is zero.

Observe that Z is a complete intersection in X, and so for our purposes is as good as a smooth curve. Via the obvious exact sequence, $H^1(T_{X|_Z})$ may be naturally identified with $H^1(N_{Z/X})$, which in turn may be naturally identified with $H^0(N_{Z/X})^*$ (since $K_Z = \bigwedge^2 N_{Z/X}$, we have a perfect pairing $H^0(N_{Z/X}) \times H^1(N_{Z/X}) \to H^1(K_Z) \cong \mathbb{C}$). Observing that $H^0(N_{Z/X}) = H^0(\mathcal{O}_Z \oplus \mathcal{O}_Z(E)) \cong \mathbb{C}$, we know that coker L_A has complex dimension one and is naturally identified with $T^*_{C,x}$, the dual of the tangent space at x to the Hilbert scheme component C. This we have seen is true for all $f \in \overline{\mathcal{M}}'_A(X, J)$.

We now apply [13], Proposition 2.10, (2) to our set-up, where $C = \overline{\mathcal{M}}'_A(X, J) = \mathcal{M}_S = S^{-1}(0)$ for S the Cauchy–Riemann section of $\overline{\mathcal{F}}_A(X)$ (as constructed in [13], §3) over a suitable neighbourhood of $\mathcal{M}_S$ in $\overline{B}_A(X)$. The above calculations verify that the conditions of Proposition 2.10, (2) are satisfied, with $\operatorname{ind}(L_A) = 0$, $\dim(\operatorname{coker} L_A) = 2$ and $\dim(\mathcal{M}_S) = 2$. Moreover, we deduce that the obstruction bundle $\mathcal{H}$ on $\mathcal{M}_S$ is just the cotangent bundle T^*_C on C.

The Gromov–Witten number n_2 may then be defined to be $\mu_S(1)$. It follows from the basic Theorem 4.2 from [13] that this is independent of any choice of tamed almost complex structure and is a symplectic deformation invariant. Thus by considering a small deformation of the almost complex structure and using [13], Proposition 2.10, (1), it is the invariant n_2 that we seek. Applying Ruan's crucial Proposition 2.10, (2), the invariant can be expressed as

$$\mu_S(1) = \int_{\mathcal{M}'_A(X,J)} e(T^*_C),$$

from which it follows that $n_2 = 2g - 2$ as claimed.

The general case (where $E \to C$ also has double fibres) can now be reduced to the case considered above. Suppose we have a point $Q \in C$ for which the corresponding fibre is a double line. We choose an open disc $\Delta \subset C$ with centre Q, and a neighbourhood U of Z in X, with U fibred over Δ, the fibre U_0 over Q containing the fibre Z. Letting $\overline{U} \to \Delta$ denote the image of U under φ, a family of surface Du Val singularities, we make a small deformation $\overline{U} \to \Delta'$ of $\overline{U}$, as in the proof of Theorem 3.5 of this paper, and in this way obtain a holomorphic deformation $\mathcal{U} \to \Delta'$ of U under which $E_0 = E_{|\Delta}$ deforms to a family of surfaces E_t ($t \in \Delta'$), all fibred over Δ, and with at worst line pair singular fibres for $t \neq 0$. Considering $\overline{U} \to \Delta \times \Delta'$ as a two parameter deformation of the surface singularity $\overline{U}_0$, we may take a good representative and apply Ehresmann's fibration theorem (with boundary) to the corresponding resolution $\mathcal{U} \to \Delta \times \Delta'$ (cf. [18], proof of Lemma 4.1). In this way, we may assume that $\mathcal{U} \to \Delta \times \Delta'$ is differentiably trivial over the base. In particular, the family $\mathcal{U} \to \Delta'$ is also differentiably trivial, and hence determines a holomorphic deformation of the complex structure on a fixed neighbourhood U of Z, where $U \to \Delta$ is also differentiably trivial.

We perform this procedure for each singular fibre $Z_1, \ldots, Z_N$ of $E \to C$, obtaining, for each i, an open neighbourhood U_i of Z_i fibred over $\Delta_i \subset C$, and a holomorphic complex structure J_i on U_i with the properties explained above (of course, if Z_i is a line pair, we may take J_i to be the original complex structure J). Let $\frac{1}{2}\overline{\Delta}_i$ denote the closed subdisc of Δ_i with half the radius, $C^* = C \backslash \bigcup_{i=1}^{N} \frac{1}{2}\overline{\Delta}_i$, and $E^* = E_{|C^*} \to C^*$ the corresponding open subset of E. We then take a tubular neighbourhood $U^* \to C^*$ of $E^* \to C^*$, equipped with the original complex structure J. By taking deformations to be sufficiently small and shrinking radii of tubular neighbourhoods if necessary, all these different complex structures may be patched together in a C^∞ way (tamed by the symplectic form) over the overlaps in C. In this manner, we obtain an open neighbourhood W of E in X, and a tamed almost complex structure J' on W, which is a small deformation of the original complex structure J and which satisfies the following properties:

(a) Each singular fibre Z_i of $E \to C$ has an open neighbourhood $U_i \subset W$ fibred over $\Delta_i \subset C$ with J' inducing an integrable complex structure on each fibre (thus $U_i \to \Delta_i$ is a C^∞ family of holomorphic surface neighbourhoods).

(b) The almost complex structure J' is integrable in a smaller neighbourhood $U_i' \subset U_i$ of each singular fibre, with the corresponding family $U_i' \to \Delta_i'$ being holomorphic.

(c) On the complement of $\bigcup U_i$ in W, the almost complex structure J' coincides with the original complex structure J.

(d) E deforms to a C^∞ family of pseudoholomorphic rational curves $E' \to C$ in (W, J'), with generic fibre $\mathbb{CP}^1$ and the only singular fibres being line pairs. Moreover, we may assume that any such singular fibre is contained in one of the above open sets U_i'.

Of course, we may now patch J' on W with the original complex structure J on X to get a global tamed almost complex structure on X, which we shall also denote by J'. Provided we have taken our deformations sufficiently small, the standard argument via Gromov compactness ensures that any pseudoholomorphic stable map representing the class A has image contained in a fibre of $E' \to C$.

The theory of [13] applies equally well to almost complex structures, and hence to our almost complex manifold X' with complex structure J'. Clearly, all the calculations remain unchanged for stable maps whose image (a fibre of $E' \to C$) has a neighbourhood on which J' is integrable, and in particular this includes all the singular fibres. Suppose therefore that $f \colon \mathbb{CP}^1 \to X'$ is a pseudoholomorphic rational curve whose image Z is contained in an overlap $U_i \setminus U_i'$ (where J' may be nonintegrable). The linearized Cauchy–Riemann operator L_A still has index zero, since by the argument of [10], p. 24, the calculation via Riemann–Roch continues to give the correct value. We therefore need to show that coker L_A is still identified naturally as $T^*_{C,x}$, and hence that the obstruction bundle is $\mathcal{H} = T^*_C$ as before.

Setting $U = U_i$ and $\Delta = \Delta_i$, we know that $U \to \Delta$ is locally (around the image Z of f) a C^∞ family of holomorphic surface neighbourhoods. Moreover, the linearized Cauchy–Riemann operator $L_A = D_f \colon C^\infty(f^*T_U) \to \Omega^{0,1}(f^*T_U)$ fits into the following commutative diagram (with exact rows)

$$
\begin{array}{ccccccccc}
0 & \to & C^\infty(f^*T_{U/\Delta}) & \to & C^\infty(f^*T_U) & \to & C^\infty(g^*T_\Delta) & \to & 0 \\
& & \downarrow \bar{\partial}_f & & \downarrow D_f & & \downarrow & & \\
0 & \to & \Omega^{0,1}(f^*T_{U/\Delta}) & \to & \Omega^{0,1}(f^*T_U) & \to & \Omega^{0,1}(g^*T_\Delta) & \to & 0
\end{array}
$$

where g is the constant map on $\mathbb{CP}^1$ with image the point $x \in \Delta$, and where the fibre of E' over x is Z. Let us denote by U_x the corresponding holomorphic surface neighbourhood, the fibre of U over x. The cokernel of

$$
\bar{\partial}_f \colon C^\infty(f^*T_{U/\Delta}) \to \Omega^{0,1}(f^*T_{U/\Delta})
$$

is then naturally identified via Dolbeault cohomology with $H^1(T_{U_x}|_Z) \cong H^1(N_{Z/U_x})$. This latter space is in turn naturally identified with $H^1(N_f) \cong H^0(N_f)^* \cong T^*_{C,x}$.

I claim now that J' may be found as above for which coker L_A has the correct dimension (namely real dimension two) for all fibres of $E' \to C$. Since

L_A has index zero and ker L_A has dimension at least two, we need to show that the dimension of coker L_A is not more than two. This follows by a Gromov compactness argument. Suppose that, however close we take J' to J, the dimension is too big for some fibre of $E' \to C$. We can then find sequences of almost complex structures J'_ν (with the properties (a)–(d) described above) converging to $J = J_0$, and pseudoholomorphic rational curves $f_\nu \colon \mathbb{CP}^1 \to (X, J'_\nu)$ at which coker L_A has real dimension > 2. By construction, the image of such a map is not contained in any U'_i (since J'_ν would then be integrable on some neighbourhood of the image, and then we know that coker L_A has the correct dimension). Thus the image of f_ν has nontrivial intersection with the compact set $X \setminus \bigcup U'_i$. By Gromov compactness, the f_ν may be assumed to converge to a pseudoholomorphic rational curve on (X, J) whose image is not contained in any U'_i. This is therefore just an embedding $f \colon \mathbb{CP}^1 \to (X, J)$ of some smooth fibre of $E \to C$, at which we know that coker L_A has real dimension precisely two; this then is a contradiction. A similar argument, via Gromov compactness, then yields the fact that J' may be found as above such that the linear map $\mathrm{coker}(\overline{\partial}_f) \to \mathrm{coker}(D_f)$ is an isomorphism for all smooth fibres of $E' \to C$, since this is true for all the smooth fibres of $E \to C$ on (X, J).

For such a J', we deduce that coker L_A is naturally identified with $T^*_{C,x}$ for all fibres, and hence the obstruction bundle identified as T^*_C. The previous argument may then be applied directly with the almost complex structure J', showing that the symplectic invariant n_2 is $2g - 2$ in general. The proof of Theorem 4.1 is now complete.

5   An application to symplectic deformations of Calabi–Yaus

If X is a Calabi–Yau threefold which is general in moduli, we know that any codimension one face of its nef cone $\overline{\mathcal{K}}(X)$ (not contained in the cubic cone W^*) corresponds to a primitive birational contraction $\varphi \colon X \to \overline{X}$ of Type I, II or III$_0$, where Type III$_0$ denotes a Type III contraction for which the genus of the curve C of singularities on $\overline{X}$ is zero.

In [18], we studied Calabi–Yau threefolds which are symplectic deformations of each other. One of the results proved there (Theorem 2) said that if X_1 and X_2 are Calabi–Yau threefolds, general in their complex moduli, which are symplectic deformations of each other, then their Kähler cones are the same. The proof of this essentially came down to showing that certain Gromov–Witten invariants associated to exceptional classes were nonzero. Using the much more precise information obtained in this paper, we are able to make a stronger statement.

Corollary 5.1 *With the notation as above, any codimension one face (not contained in W^*) of $\overline{\mathcal{K}}(X_1) = \overline{\mathcal{K}}(X_2)$ has the same contraction type (Type I, II or III_0) on X_1 as on X_2.*

Proof The fact that Type II faces correspond is easy, since for D in the interior of such a face, the quadratic form $q(L) = D \cdot L^2$ is degenerate, which is not the case for D in the interior of a Type I or Type III_0 face. Stating it another way, if we consider the Hessian form associated to the topological cubic form μ, then h is a form of degree $\rho = b_2$ which has a linear factor corresponding to each Type II face. Thus the condition that a face is of Type II is topologically determined.

The result will therefore follow if we can show that a face of the nef cone which is Type I for one of the Calabi–Yau threefolds is not of Type III_0 for the other. However, for a Type I face, we saw in §2 that n_d is always nonnegative; for a Type III_0 face, we proved in Theorem 4.1 that $n_2 = -2$. Since Gromov–Witten invariants are invariant under symplectic deformations, the result is proved.

Remark 5.2 It is still an open question whether there exist examples of Calabi–Yau threefolds X_1 and X_2 which are symplectic deformations of each other but not in the same algebraic family.

References

[1] R. Friedman, *On threefolds with trivial canonical bundle*. In *Complex geometry and Lie theory (Sundance, UT 1989)*, pp. 103–154. Proc. Symp. Pure Math. **53**, AMS Providence R.I. 1991.

[2] K. Fukaya and K. Ono, *Arnold conjecture and Gromov–Witten invariants*, Warwick preprint 29/1996, Kyoto Univ. preprint (1996).

[3] W. Fulton, *Intersection Theory*. Springer-Verlag, Berlin Heidelberg New York Tokyo, 1984.

[4] O.J. Ganor, D.R. Morrison and N. Seiberg, *Branes, Calabi–Yau spaces, and toroidal compactifications of the $N = 1$ six-dimensional E_8 theory*, Nuclear Phys. B **487** (1996) 93–140.

[5] S. Katz, D.R. Morrison and M.R. Plesser, *Enhanced gauge symmetry in type II string theory*, Nuclear Phys. B **477** (1996) 105–140.

[6] Y. Kawamata, *Crepant blowing-up of 3-dimensional canonical singularities and its application to degenerations of surfaces*, Ann. of Math. **127** (1988) 93–163.

[7] J. Kollár, *Flips, flops, minimal models, etc.*, Surveys in differential geometry (Cambridge, MA, 1990), Suppl. J. Diff. Geom. **1** (1991) 113–199.

[8] J. Kollár and S. Mori, *Classification of three-dimensional flips*, Jour. Amer. Math. Soc. **5** (1992) 533–703.

[9] J. Li and G. Tian, *Virtual moduli cycles and Gromov–Witten invariants of general symplectic manifolds*, Duke eprint, alg-geom/9608032.

[10] D. McDuff and D. Salamon, *J-holomorphic curves and quantum cohomology*, Univ. Lecture Series, Vol. 6, Amer. Math. Soc. 1994.

[11] Y. Ruan, *Symplectic topology and extremal rays*, Geom. Funct. Anal. **3** (1993) 395–430.

[12] Y. Ruan, *Topological sigma model and Donaldson type invariants in Gromov theory*, Duke J. Math. **83** (1996) 461–500.

[13] Y. Ruan, *Virtual neighborhoods and pseudo-holomorphic curves*, Duke eprint, alg-geom/9611021.

[14] D. Sullivan, *Infinitesimal computations in topology*, Publ. Math. Inst. Hautes Etud. Sci. **47** (1977) 269–331.

[15] C. Voisin, *A mathematical proof of a formula of Aspinwall and Morrison*, Comp. Math. **104** (1996) 135–151.

[16] C.T.C. Wall, *Classification problems in differential topology V. On certain 6-manifolds*, Invent. math. **1** (1966) 355–374.

[17] P.M.H. Wilson, *The Kähler cone on Calabi–Yau threefolds*, Invent. math. **107** (1992) 561–583. *Erratum*, Invent. math. **114** (1993) 231–233.

[18] P.M.H. Wilson, *Symplectic deformations of Calabi–Yau threefolds*, J. Diff. Geom. **45** (1997) 611–637.

Mathematics Subject Classification (1991):
14J10, 14J15, 14J30, 32J17, 32J27, 53C15, 53C23, 57R15, 58F05

P.M.H. Wilson,
Department of Pure Mathematics, University of Cambridge,
16 Mill Lane, Cambridge CB2 1SB, UK
e-mail: pmhw@dpmms.cam.ac.uk

Printed in the United States
By Bookmasters